Ocean Acidification

EDITED BY

Jean-Pierre Gattuso

Senior Research Scientist at Centre National de la Recherche Scientifique (CNRS) and Université Pierre et Marie Curie – Paris 6, France

Lina Hansson

Project Manager EPOCA (European Project on Ocean Acidification), Centre National de la Recherche Scientifique (CNRS), France

OXFORD
UNIVERSITY PRESS

OXFORD

UNIVERSITY PRESS

Great Clarendon Street, Oxford OX2 6DP

Oxford University Press is a department of the University of Oxford.
It furthers the University's objective of excellence in research, scholarship,
and education by publishing worldwide in

Oxford New York

Auckland Cape Town Dar es Salaam Hong Kong Karachi
Kuala Lumpur Madrid Melbourne Mexico City Nairobi
New Delhi Shanghai Taipei Toronto

With offices in

Argentina Austria Brazil Chile Czech Republic France Greece
Guatemala Hungary Italy Japan Poland Portugal Singapore
South Korea Switzerland Thailand Turkey Ukraine Vietnam

Oxford is a registered trade mark of Oxford University Press
in the UK and in certain other countries

Published in the United States
by Oxford University Press Inc., New York

British Library Cataloguing in Publication Data

Data available

Library of Congress Cataloging in Publication Data

Data available

Typeset by SPI Publisher Services, Pondicherry, India
Printed in Spain
on acid-free paper by
Grafos SA

ISBN 978-0-19-959108-4 (Hbk.)
      978-0-19-959109-1 (Pbk.)

3 5 7 9 10 8 6 4 2

Foreword

Much of the discussion regarding the impacts of the ongoing build-up of fossil-fuel CO_2 in our atmosphere has been focused on issues such as the loss of coastal property as the result of rising sea level, the further desiccation of the earth's dry lands, and the intrusion of destructive insects into our forests. However, there is another potential threat to the well-being of our planet that has received less attention, namely the gradual acidification of the ocean. The fate of much of the CO_2 we produce will be to enter the ocean. In a sense, we are fortunate that ocean water is endowed with the capacity to absorb far more CO_2 per litre than were it salt free. Instead of just dissolving, as it would in distilled water, the CO_2 reacts with carbonate and borate ions dissolved in the sea, thereby increasing the concentration of bicarbonate ions. That's the good part, for already roughly 30% of the CO_2 we have produced by burning coal, oil, and natural gas has been taken up by the sea. Over the next few hundred years, this fraction will increase, thus reducing its climatic impacts.

But there is a flip side to this. As the excess CO_2 content of seawater increases, its hydrogen ion concentration undergoes a parallel increase. Consequently its acidity rises. This increase will make life more difficult for those marine organisms that are housed in solid $CaCO_3$. The clams, oysters, and mussels we dine on live in shells made of $CaCO_3$; the coccolithophores and foraminifera which are at the base of the marine food chain on which fish depend are also protected by $CaCO_3$. The coral reefs that form one of the major habitats for what we refer to as our planet's biodiversity are built of $CaCO_3$.

To form their $CaCO_3$ housing, marine organisms combine calcium ions with carbonate ions. Although like the O_2 content of our atmosphere, the calcium content of our ocean is so vast that there is no chance that our activities will significantly change it this is not the case for carbonate ion. The fossil-fuel CO_2 already taken up by surface-ocean water has used up about 30% of the available carbonate ions and, as time goes on, this fraction will increase.

Fortunately, the surface waters of the ocean are highly supersaturated with respect to the mineral-phase calcite formed by coccolithophores, foraminifera, and oysters. While the extent of supersaturation with respect to the mineral aragonite formed by clams, mussels, and coral is somewhat smaller, there still is a healthy margin before the build-up of CO_2 becomes large enough to force these organisms to build their houses in water undersaturated with respect to aragonite.

So, one might ask, what's the problem? Unless we raise the CO_2 level in the atmosphere to a level which would create huge problems for us and our fellow inhabitants of the continents, most of the surface ocean will remain supersaturated with respect to both aragonite and calcite. The answer is that it's not merely an either/or situation where organisms happily produce their $CaCO_3$ houses as long as the water is supersaturated, and go over the brink of incapability when the water becomes undersaturated. Instead, experiments show that as the concentration of carbonate ions goes down, so does the rate at which at least some types of marine organisms are able to precipitate $CaCO_3$. So, the challenge is to establish, for each type of organism, how they respond to lowered carbonate concentration, and then to determine how it affects their ability to hold on to what we refer to as their ecological niche.

Concern focuses on the world's coral reefs. It is clear that they are under stress created by overfishing, excess nitrate availability, flooding with turbid river water, etc. The impacts of this extra stress become clear, particularly during hot summers. The dinoflagellates which inhabit the coral polyps vanish causing the reef to lose its colour. Some reefs recover from this so-called bleaching. Others do not. The steady reduction in carbonate ions associated with ocean acidification is likely to aggravates the situation, adding yet another challenge to already threatened corals.

What can we do? Of course, ideally we would strive to slow down and eventually bring to a halt the rise in CO_2. But, as the opponents of action currently hold the upper hand, during the next couple of decades the rate of increase of CO_2 is likely to accelerate rather than slow. This being the case, it behoves us to increase our efforts to understand the impacts of ocean acidification. Not only will this put us in a better position to educate the public regarding the dangers it holds, but it will also allow us to evaluate whether there are actions we could take to counter the coming impacts. This book brings together much of our knowledge regarding ocean acidification and also points out the gaps that could be filled by further research.

Wallace S. Broecker
Newberry Professor of Earth and
Environmental Sciences
Lamont-Doherty Earth Observatory of
Columbia University

Preface

In less than fifteen years ocean acidification has emerged as a key research priority for marine science and has recently begun to gain visibility in political agendas. Even though the history of anthropogenic ocean acidification as a research topic is brief, ocean acidification was triggered, like global warming, by the industrial revolution more than 200 years ago. Since 1800, the oceans have absorbed about one-third of the carbon dioxide produced by man. Without this moderating capacity of the oceans, atmospheric levels of carbon dioxide would have been considerably higher and, consequently, the effect on Earth's climate more pronounced. As ocean acidification continues at a rate that is probably unprecedented in the history of Earth, our understanding of its possible impacts on marine life slowly becomes clearer. The picture that appears is complex, but indicates negative consequences on some marine organisms and ecosystems.

The number of papers on ocean acidification increased considerably after 2004. With a total number of more than 800 papers, the time is ripe to produce the first authoritative book on ocean acidification. Structured around fifteen chapters, this book takes the reader through the history and recent research on ocean acidification, from the chemical background to the biological and biogeochemical consequences. It is organized in four parts, reflecting the pluri-disciplinary character of the subject: chemistry (Chapters 1–3), biology (Chapters 4–10),

biogeochemistry (Chapters 11 and 12), and the policy implications, and mitigation scenarios (Chapters 13 and 14). Finally, Chapter 15 provides a summary of the knowns and the unknowns, as well as perspectives.

We are very grateful to the authors for their dedication and for producing landmark contributions. All chapters were reviewed by at least one of the authors of another chapter as well as by ourselves. We are also indebted to Joanie A. Kleypas and Stephen V. Smith who reviewed Chapter 1, to Helen Findlay, who reviewed Chapter 7, and to the publication team at Oxford University Press for its support. Last but not least, we are very grateful for the support of the European Commission through the European Project on Ocean Acidification (EPOCA) during the preparation of this book. All EPOCA participants have contributed to providing a very lively and exciting working environment to both of us.

This is a contribution to two core projects of the International Geosphere-Biosphere Programme (IGBP), the Integrated Marine Biogeochemistry and Ecosystem Research (IMBER) project, and the Surface Ocean Lower Atmosphere Study (SOLAS), that have jointly developed the SOLAS-IMBER Ocean Acidification (SIOA) working group.

Last but not least, we thank our families for their support and patience.

Jean-Pierre Gattuso and Lina Hansson
Villefranche-sur-mer,
10th June 2011

Contents

List of abbreviations

AAnP	aerobic anoxygenic phototrophs
ACD	aragonite compensation depth
AE	Cl^-/HCO_3^- exchanger
AR4 [IPCC]	Fourth Assessment Report
ARISA	automated ribosomal intergenic spacer analysis
ASFA	Aquatic Science and Fisheries Abstracts
ASH	aragonite saturation horizon
BATS	Bermuda Atlantic Time-Series Station
BIOACID	Biological Impacts of Ocean Acidification
CA	carbonic anhydrase
CBD	[UN] Convention on Biological Diversity
CCA	crustose coralline algae
CCD	calcite compensation depth
CCMs	carbon-concentrating mechanisms
CCN	cloud condensation nuclei
CCS	CO_2 capture and storage
cDOM	chromophoric DOM
CDR	carbon dioxide removal
CFC	chlorofluorocarbon
CMIP	Coupled Model Intercomparison Project
CSH	calcite saturation horizon
DGGE	denaturing gradient gel electrophoresis
DMS	dimethyl sulphide
DMSP	dimethyl sulphoniopropionate (the precursor molecule to DMS)
DOC	dissolved organic carbon
DOM	dissolved organic matter
EPA	US Environmental Protection Agency
EPOCA	European Project on Ocean Acidification
ESF	European Science Foundation
ESTs	expressed sequence tags
ESTOC	European Time Series in the Canary Islands
FACE	free-air CO_2 enrichment experiments
FISH	fluorescence *in situ* hybridization
FOCE	free ocean CO_2 enrichment
FORAM	Federal Ocean Acidification Research and Monitoring
GDP	gross domestic product
GHGs	greenhouse gases

GLODAP	global, three-dimensional gridded data product
HA	H^+-ATPase
HOT	Hawaii Ocean Time-Series
IAP	Interacademy Panel
IAMs	integrated assessment models
IMBER	Integrated Marine Biogeochemistry and Ecosystem Research
IOC	Intergovernmental Oceanographic Commission
IPCC	Intergovernmental Panel on Climate Change
IPSL	Institut Pierre Simon Laplace
JGOFS	Joint Global Ocean Flux Study
LGM	Last Glacial Maximum
LOICZ	Land–Ocean Interactions in the Coastal Zone
LOSCAR	long-term ocean-atmosphere-sediment carbon cycle reservoir model
MBL	marine boundary layer
MCCIP	Marine Climate Change Impacts Partnership
MPT	mid-Pleistocene transition
MSA	methane sulphonic acid
MSIA	methane sulphinic acid
NADW	North Atlantic Deep Water
NBC	Na^+/HCO_3^- cotransporter
NCAR	National Centre for Atmospheric Research
NEAC	north-eastern Arctic cod
NHE	Na^+/H^+ exchangers
NOAA	National Oceanic and Atmospheric Administration
NSC	North Sea cod
NSF	National Science Foundation
OCB	ocean carbon and biogeochemistry
OCLTT	oxygen- and capacity-limited thermal tolerance
OCMIP	Ocean Carbon Cycle Model Intercomparison Project
OMZs	oxygen minimum zones
OTUs	operational taxonomic units
PAGES	Past Global Changes
PAL	present atmospheric level
PeECE	Pelagic Ecosystem Enrichment Experiment
PETM	Palaeocene–Eocene Thermal Maximum
PFTs	plankton functional types
PIC	particulate inorganic carbon
PIUB	University of Bern Physics Institute
POC	particulate organic carbon
POM	particulate organic matter
ppmv	parts per million by volume
REDD	Reducing Emissions from Deforestation and Degradation
RF	radiative forcing
RI	respiration index
RubisCO	ribulose-1,5-bisphosphate carboxylase oxygenase
RUG	Reference User Group
SIOA	SOLAS-IMBER Working Group on Ocean Acidification
SOCM	shallow-water ocean carbonate model

SOLAS	Surface Ocean–Lower Atmosphere Study
SRES	Special Report on Emissions Scenarios
SRM	solar radiation management
TAR	Third Assessment Report [of the IPCC]
TEP	transparent exopolymeric particles
TRFLP	terminal fragment length polymorphism
UKOA	UK Ocean Acidification research programme
UNEP	United Nations Environment Programme
UNFCCC	UN Framework Convention on Climate Change
USGS	US Geological Survey
VOCs	volatile organic compounds
WMO	World Meteorological Organization
WOCE	World Ocean Circulation Experiment

List of contributors

Andreas Andersson: Scripps Institution of Oceanography, University of California San Diego, Hubbs Hall Room 2150, 9500 Gilman Dr., La Jolla, CA 92093-0202, USA
aandersson@ucsd.edu

James P. Barry: Monterey Bay Aquarium Research Institute, 7700 Sandholdt Road, Moss Landing, CA 95039 USA
barry@mbari.org

Jelle Bijma: Marine Biogeosciences, Alfred Wegener Institute for Polar and Marine Research, Am Handelshafen 12, D-27570 Bremerhaven, Germany
jelle.bijma@awi.de

Kelvin Boot: Plymouth Marine Laboratory, Prospect Place, The Hoe, Plymouth PL1 3DH, UK
kelota@pml.ac.uk

Laurent Bopp: Laboratoire des Sciences du Climat et de l'Environnement, LSCE/IPSL, UMR CEA/CNRS/UVSQ, L'Orme des Merisiers, Bât. 712, 91191 Gif-sur-Yvette Cedex, France
laurent.bopp@lsce.ipsl.fr

Woodward W. Fischer: Division of Geological and Planetary Sciences, California Institute of Technology, Pasadena CA 91125, USA
wfischer@caltech.edu

Thomas L. Frölicher: Climate and Environmental Physics, Physics Institute, University of Bern, Sidlerstrasse 5, CH-3012 Bern, Switzerland
Present address: Program in Atmospheric and Oceanic Sciences, Princeton University, 300 Forrestal Road, Princeton, NJ, USA
froelicher@climate.unibe.ch

Reidun Gangstø: Federal Office of Meteorology and Climatology MeteoSwiss, Kraehbuehlstrasse 58, PO Box 514, CH-8044 Zurich, Switzerland
reidun.gangsto@meteoswiss.ch

Jean-Pierre Gattuso: (1) INSU-CNRS, Laboratoire d'Océanographie, F-06234, Villefranche-sur-Mer Cedex, France; (2) UPMC Univ. Paris 06, Observatoire Océanologique, F-06234, Villefranche-sur-Mer Cedex, France
gattuso@obs-vlfr.fr

Marion Gehlen: Laboratoire des Sciences du Climat et de l'Environnement, LSCE/IPSL, UMR CEA/CNRS/UVSQ, L'Orme des Merisiers, Bât. 712, 91191 Gif-sur-Yvette Cedex, France
marion.gehlen@lsce.ipsl.fr

Nicolas Gruber: Environmental Physics, Institute of Biogeochemistry and Pollutant Dynamics, ETH Zürich, Universitätstrasse 16, 8092 Zurich, Switzerland
nicolas.gruber@env.ethz.ch

Magda Gutowska: Physiologie, Kiel University, Hermann-Rodewald-Strasse 5, 24118 Kiel, Germany
m.gutowska@physiologie.uni-kiel.de

Jason M. Hall-Spencer: Marine Institute, University of Plymouth, Plymouth PL4 8AA, UK
jason.hall-spencer@plymouth.ac.uk

Lina Hansson: (1) INSU-CNRS, Laboratoire d'Océanographie, F-06234, Villefranche-sur-Mer Cedex, France; (2) UPMC Univ. Paris 06, Observatoire Océanologique, F-06234, Villefranche-sur-Mer Cedex, France
hansson@obs-vlfr.fr

Frances Hopkins: Plymouth Marine Laboratory, Prospect Place, The Hoe, Plymouth PL1 3DH, UK
fhop@pml.ac.uk

Atsushi Ishimatsu: Institute for East China Sea Research, Nagasaki University, Tairamachi, Nagasaki, 851-2213, Japan
a-ishima@nagasaki-u.ac.jp

Fortunat Joos: (1) Climate and Environmental Physics, Physics Institute, University of Bern,

Sidlerstrasse 5, CH-3012 Bern, Switzerland; (2) Oeschger Centre for Climate Change Research, University of Bern, Zähringerstrasse 25, CH-3012 Bern, Switzerland
joos@climate.unibe.ch

Vassilis Kitidis: Plymouth Marine Laboratory, Prospect Place, The Hoe, Plymouth PL1 3DH, UK
vak@pml.ac.uk

Andrew H. Knoll: Department of Organismic and Evolutionary Biology, Harvard University, Cambridge, MA 02138, USA
aknoll@oeb.harvard.edu

Peter Liss: School of Environmental Sciences, University of East Anglia, Norwich NR4 7TJ, UK
p.liss@uea.ac.uk

Magnus Lucassen: Integrative Ecophysiology, Alfred Wegener Institute for Polar and Marine Research, Am Handelshafen 12, 27570 Bremerhaven, Germany
magnus.lucassen@awi.de

Fred T. Mackenzie: Department of Oceanography, School of Ocean and Earth Science and Technology, University of Hawaii, Honolulu, HI 968 22, USA
fredm@soest.hawaii.edu

Xavier Mari: (1) IRD, UMR 5119 ECOSYM, Université Montpellier II, cc 093, Place Bataillon, 34095 Montpellier, France
Present address: Institute of Biotechnology, Environmental Biotechnology Laboratory, 18 Hoang Quoc Viet Street, Cau Giay, Hanoi, Vietnam
xavier.mari@ird.fr

Frank Melzner: Leibniz Institute of Marine Science (IFM-GEOMAR), Biological Oceanography, Hohenbergstrasse 2, 24105 Kiel, Germany
fmelzner@ifm-geomar.de

Philip Nightingale: Plymouth Marine Laboratory, Prospect Place, The Hoe, Plymouth PL1 3DH, UK
pdn@pml.ac.uk

James Orr: Laboratoire des Sciences du Climat et de l'Environnement, LSCE-IPSL, CEA-CNRS-UVSQ, CEA Saclay, l'Orme, Bat. 712, F-91191 Gif-sur-Yvette, France
james.orr@lsce.ipsl.fr

Andreas Oschlies: Leibniz Institute of Marine Science (IFM-GEOMAR), Marine Biogeochemistry, Düsternbrooker Weg 20, 24105 Kiel, Germany
aoschlies@ifm-geomar.de

Gian-Kasper Plattner: Climate and Environmental Physics, Physics Institute, University of Bern, Sidlerstrasse 5, CH-3012 Bern, Switzerland
plattner@climate.unibe.ch

Hans-Otto Pörtner: Integrative Ecophysiology, Alfred Wegener Institute for Polar and Marine Research, Am Handelshafen 12, 27570 Bremerhaven, Germany
hans.poertner@awi.de

Andy Ridgwell: School of Geographical Sciences, University of Bristol, University Road, Bristol BS8 1SS, UK
andy@seao2.org

Ulf Riebesell: Leibniz Institute of Marine Sciences (IFM-GEOMAR), Düsternbrooker Weg 20, 24105 Kiel, Germany
uriebesell@ifm-geomar.de

Brad Seibel: Biological Sciences, University of Rhode Island, Kingston, RI 02881, USA
seibel@uri.edu

John I. Spicer: Marine Biology and Ecology Research Centre, School of Marine Science and Engineering, University of Plymouth, Plymouth PL4 8AA, UK
j.i.spicer@plymouth.ac.uk

Marco Steinacher: (1) Climate and Environmental Physics, Physics Institute, University of Bern, Sidlerstrasse 5, CH-3012 Bern, Switzerland; (2) Oeschger Centre for Climate Change Research, University of Bern, Zähringerstrasse 25, CH-3012 Bern, Switzerland
steinacher@climate.unibe.ch

Philippe D. Tortell: Department of Earth and Ocean Sciences, University of British Columbia, 2146 Health Sciences Mall, Vancouver, B.C., Canada V6T 1Z3
pdtortell@gmail.com

Carol Turley: Plymouth Marine Laboratory, Prospect Place, The Hoe, Plymouth PL1 3DH, UK
ct@pml.ac.uk

Markus G. Weinbauer: (1) INSU-CNRS, Laboratoire d'Océanographie, F-06234, Villefranche-sur-Mer Cedex, France; (2) UPMC Univ Paris 06,

Observatoire Océanologique, F-06234, Villefranche-sur-Mer Cedex, France
wein@obs-vlfr.fr

Stephen Widdicombe: Plymouth Marine Laboratory, Prospect Place, The Hoe, Plymouth PL1 3DH, UK
swi@pml.ac.uk

Richard E. Zeebe: School of Ocean and Earth, Science and Technology, Department of Oceanography, University of Hawaii at Manoa, 1000 Pope Road, MSB 504, Honolulu, HI 96822, USA
zeebe@soest.hawaii.edu

CHAPTER 1

Ocean acidification: background and history

Jean-Pierre Gattuso and Lina Hansson

1.1 Introduction

The ocean and the atmosphere exchange massive amounts of carbon dioxide (CO_2). The pre-industrial influx from the ocean to the atmosphere was 70.6 Gt C yr^{-1}, while the flux in the opposite direction was 70 Gt C yr^{-1} (IPCC 2007). Since the Industrial Revolution an anthropogenic flux has been superimposed on the natural flux.

The concentration of CO_2 in the atmosphere, which remained in the range of 172–300 parts per million by volume (ppmv) over the past 800 000 years (Lüthi *et al.* 2008), has increased during the industrial era to reach 387 ppmv in 2009. The rate of increase was about 1.0% yr^{-1} in the 1990s and reached 3.4% yr^{-1} between 2000 and 2008 (Le Quéré *et al.* 2009). Future levels of atmospheric CO_2 mostly depend on socio-economic parameters, and may reach 1071 ppmv in the year 2100 (Plattner *et al.* 2001), corresponding to a fourfold increase since 1750. As pointed out over 50 years ago, 'human beings are now carrying out a large scale geophysical experiment of a kind that could not have happened in the past nor be reproduced in the future' (Revelle and Suess 1957).

Anthropogenic CO_2 has three fates. In the years 2000 to 2008, about 29% was absorbed by the terrestrial biosphere and 26% by the ocean, while the remaining 45% remained in the atmosphere (Le Quéré *et al.* 2009). The accumulation of CO_2 in the atmosphere increases the natural greenhouse effect and generates climate changes (IPCC 2007). It is estimated that the surface waters of the oceans have taken up 118 Pg C, or about 25% of the carbon generated by human activities since 1800 (Sabine *et al.* 2004). By taking CO_2 away from the atmosphere, the oceanic and terrestrial sinks mitigate climatic changes. Should their efficiency decrease, more CO_2 would remain in the atmosphere, generating larger climate perturbations.

This book has four main groups of chapters. The first group reviews the past (Zeebe and Ridgwell, Chapter 2), recent, and future (Orr, Chapter 3) changes in seawater carbonate chemistry. The second group of chapters reviews the biological and ecological consequences of ocean acidification, looking at the response of calcifiers to past ocean acidification events (Knoll and Fischer, Chapter 4), the response of heterotrophic microorganisms (Weinbauer *et al.*, Chapter 5) as well as pelagic (Riebesell and Tortell, Chapter 6), benthic (Andersson *et al.*, Chapter 7), nektonic (Pörtner *et al.*, Chapter 8), and sedimentary (Widdicombe *et al.*, Chapter 9) organisms and ecosystems, and the effects on biodiversity and ecosystem function (Barry *et al.*, Chapter 10). The chapters of the third group summarize the biogeochemical consequences of ocean acidification, both on the production of climate-relevant trace gases (Hopkins *et al.*, Chapter 11) and overall biogeochemical cycles (Gehlen *et al.*, Chapter 12). Chapters of the fourth group address the societal consequences of ocean acidification (Turley and Boot, Chapter 13) and the impact of climate change mitigation on future projections (Joos *et al.*, Chapter 14). The final chapter summarizes what is known and what is unknown, and provides recommendations for future research (Gattuso *et al.*, Chapter 15).

The aim of the present chapter is to introduce ocean acidification in a broad context, starting with the chemical background (Section 1.2 and Box 1.1) and its effects on biological and biogeochemical processes (Section 1.3). The history of ocean acidification research and a bibliometric analysis follow (Section 1.4). Finally, the potential societal and policy implications are outlined (Section 1.5).

1.2 What is ocean acidification?

Ocean acidification refers to a reduction in the pH of the ocean over an extended period, typically decades or longer, caused primarily by the uptake of CO_2 from the atmosphere, but it can be caused by other chemical additions or subtractions from the ocean. This book focuses on anthropogenic ocean acidification, which refers to the component of pH reduction caused by human activity.

Once dissolved in seawater, CO_2 is a weak acid which generates a number of changes in seawater chemistry, primarily carbonate chemistry (see Box 1.1). It increases the concentration of bicarbonate ions ([HCO_3^-]) and dissolved inorganic carbon (C_T),

and lowers the pH, the concentration of carbonate ions ([CO_3^{2-}]), and the saturation state of the three major carbonate minerals present in shells and skeletons:

$$CO_2 + H_2O + CO_3^{2-} \rightarrow 2HCO_3^{2-}. \tag{1.1}$$

The concentration of protons ([H^+]), which is proportional to the ratio [HCO_3^-]/[CO_3^{2-}] (see Eq. B1.3 in Box 1.1), increases and pH decreases. Mean surface-ocean pH expressed on the total hydrogen ion scale (pH_T) has decreased from approximately 8.2 to 8.1 between pre-industrial time and the 1990s, and may reach 7.8 in 2100 (Gattuso and Lavigne 2009). The expression 'ocean acidification' refers to the decrease in pH, but does not imply that the pH of surface-ocean waters will become acidic (below 7.0) any time soon. The whole process could equally be referred to as 'carbonation' as it increases the concentration of dissolved inorganic carbon.

Box 1.1 Chemistry of the seawater–carbonate system

Richard Zeebe and Jean-Pierre Gattuso

The chemistry of inorganic carbon in seawater is relatively complex and is only briefly described here. Comprehensive information can be found, for example, in Zeebe and Wolf-Gladrow (2001) and Millero (2006).

Dissolved inorganic carbon

Dissolved inorganic carbon is mostly present in three inorganic forms in seawater: free aqueous carbon dioxide (CO_2(aq)), bicarbonate (HCO_3^-), and carbonate (CO_3^{2-}) ions. A minor form is true carbonic acid (H_2CO_3) with a concentration of less than 0.3% of [CO_2(aq)]. The sum of [CO_2(aq)] and [H_2CO_3] is denoted as [CO_2]. The majority of dissolved inorganic carbon in the ocean is in the form of HCO_3^- (>85%; Fig. B1.1). Gaseous carbon dioxide (CO_2(g)), and [CO_2] are related by Henry's law in thermodynamic equilibrium:

$$CO_2(g) = CO_2; K_0 \tag{B1.1}$$

where K_0 is the solubility coefficient of CO_2 in seawater. The concentration of dissolved CO_2 and the fugacity of gaseous CO_2, fCO_2, then obey the equation [CO_2] = $K_0 \times fCO_2$. The fugacity is practically equal to the partial pressure, pCO_2 (within ~1%). The dissolved carbonate species are related by:

$$CO_2 + H_2O = HCO_3^- + H^+; K_1^* \tag{B1.2}$$

$$HCO_3^- = CO_3^{2-} + H^+; K_2^*. \tag{B1.3}$$

The pK^*s (= $-\log_{10}(K^*)$) of the stoichiometric dissociation constants of carbonic acid in seawater are pK_1^* = 5.94 and pK_2^* = 9.13 at temperature T = 15°C, salinity S = 35, and surface pressure P = 1 atm on the total pH scale (Lueker et al. 2000). At a typical surface-seawater pH_T of 8.2, the speciation between [CO_2], [HCO_3^-], and [CO_3^{2-}] is hence 0.5%, 89%, and 10.5%, showing that most of the dissolved CO_2 is in the form of HCO_3^- and not CO_2. The sum of the dissolved carbonate species is denoted as total dissolved inorganic carbon (DIC $\equiv \Sigma CO_2 \equiv TCO_2 \equiv C_T$):

$$C_T = [CO_2] + [HCO_3^-] + [CO_3^{2-}] \tag{B1.4}$$

Total alkalinity

Total alkalinity (A_T) is related to the charge balance in seawater:

$$A_T = [HCO_3^-] + 2[CO_3^{2-}] + [B(OH)_4^-] + [OH^-] - [H^+] + \text{minor compounds}. \tag{B1.5}$$

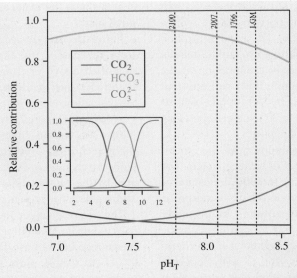

Figure B1.1   Bjerrum plot showing the relative contributions of carbon dioxide (CO_2), bicarbonate (HCO_3^-), and carbonate (CO_3^{2-}) to the dissolved inorganic carbon as a function of pH_T at 15°C and a salinity of 35. K_1 and K_2 were calculated according to Lueker *et al.* (2000). The dashed vertical lines indicate the average open-ocean surface pH_T during the Last Glacial Maximum (LGM), 1766, 2007, and 2100 (see Table 1.1). The figure is drawn with the Bjerrum function 'seacarb' (Lavigne and Gattuso, 2010).

pH

pH is a measure of ocean acidity ($pH = -\log_{10}[H^+]$) which can be reported using different scales: National Bureau of Standards (pH_{NBS}), seawater (pH_{SWS}), free (pH_F), and total (pH_T) scales. The pH values on the NBS and SWS scales are, respectively, about 0.15 units higher and about 0.01 units lower than on the total scale. This makes data compilation and analysis difficult because some conversions from one scale to another are not straightforward. The total scale is recommended (Dickson 2010) and is the scale used in this book whenever possible.

Calcium carbonate

Three major biogenic calcium carbonate ($CaCO_3$) minerals occur in seawater: aragonite, calcite, and magnesian calcite (Mg-calcite). Aragonite is about 1.5 times more soluble than calcite. Mg-calcite is a variety of calcite with magnesium ions substituted for calcium ions. Its solubility is lower than that of calcite at low (< 4%) mole fractions of

magnesium whereas it is higher at high (>12%) mole fractions (Dickson 2010). The dissolution equilibrium is:

$$CaCO_3 = Ca^{2+}(aq) + CO_3^{2-}(aq) \qquad (B1.6)$$

with the equilibrium constant defined as the solubility product for calcite or aragonite:

$$K_{sp}^* = [Ca^{2+}]_{sat}[CO_3^{2-}]_{sat}. \qquad (B1.7)$$

The $CaCO_3$ saturation state is defined as the ratio between the *observed* ion product and the *expected* ion product when the solution is in equilibrium with a particular calcium carbonate mineral:

$$\Omega = [Ca^{2+}][CO_3^{2-}]/K_{sp}^*. \qquad (B1.8)$$

Seawater is in equilibrium with that mineral when $\Omega = 1$, supersaturated when $\Omega > 1$ (which promotes inorganic precipitation), and is undersaturated when $\Omega < 1$ (which promotes inorganic dissolution).

The increase in surface-ocean dissolved CO_2 is proportional to that in the atmosphere (upon equilibration after about 1 year) but the increase in C_T is not. This is a result of the buffer capacity of seawater. The relative change of dissolved CO_2 to the

relative change of C_T in seawater in equilibrium with atmospheric CO_2 is described by the so-called Revelle factor according to which a doubling in atmospheric CO_2 only leads to an increase in C_T of about 10% (Zeebe and Wolf-Gladrow 2001; see also

Chapter 3). Egleston *et al.* (2010) have shown that the absorption of CO_2 from the atmosphere will be less efficient and that $[CO_2]$, $[H^+]$, and saturation states will be more sensitive to environmental changes. For example, careful examination of the buffer factor Π_D (Frankignoulle 1994) indicates that the pH changes generated by CO_2 uptake will be magnified as C_T increases (a 30% increase between 1766 and 2100; see also Chapter 3).

The deposition of compounds from the atmosphere can also alter the pH of surface waters. For example, 'acid rain' resulting from the deposition of sulphur and nitrogen compounds led to widespread acidification of surface freshwaters in the early 1970s (e.g. Wright 2003) and to significant biotic responses (Hendrey 1984). Alterations of the chemistry of the surface ocean due to anthropogenic nitrogen and sulphur deposition represent only a small fraction of the acidification resulting from the uptake of anthropogenic CO_2 (Doney *et al.* 2007). However, the impacts can be much more substantial in some areas of the coastal ocean, of the order of 10–50% or more of the anthropogenic CO_2-driven

changes near the major source regions and in marginal seas.

Future decrease of seawater pH can result either from the passive uptake of atmospheric CO_2 as described above or from purposeful dumping of liquid CO_2 into the deep ocean (for CO_2 disposal). Several studies have investigated the impact of the very low pH (below 7) that such a dumping would induce on various functions and organisms (e.g. Bibby *et al.* 2007; Widdicombe and Needham 2007). Only the changes in the carbonate chemistry generated by the passive uptake of atmospheric CO_2 are addressed in the present chapter.

The recent changes in the carbonate chemistry inferred from modelling studies can actually be measured. In the past decades, pH decreased by 0.0017 to 0.0024 units per year, depending on location and whether the whole year or only winter data are considered (Bates and Peters 2007; Santana-Casiano *et al.* 2007; Dore *et al.* 2009; Olafsson *et al.* 2009; see Chapter 3). Another important outcome of ocean acidification for calcifying organisms and the global calcium carbonate cycle

Table 1.1 Average changes in the carbonate chemistry of surface seawater from 1766 to 2100 (Gattuso and Lavigne 2009). Total alkalinity, CO_2 partial pressure (pCO_2), salinity, and temperature were fixed and used to derive all other parameters using the seacarb software (Lavigne and Gattuso 2010) and the dissociation constant of carbonic acid of Lueker *et al.* (2000). It is assumed that the ocean and atmosphere are in equilibrium with respect to CO_2. Values of temperature, salinity, total alkalinity, and total phosphate in 1766, 2007, and 2100 are from Plattner *et al.* (2001) prescribing historical CO_2 records and non-CO_2 radiative forcing from 1766 to 1990 and using the A2 IPCC SRES emissions scenario (Nakićenović *et al.* 2000) thereafter. pCO_2 in 1766 and 2100 are from Plattner *et al.* (2001), while values for 2007 are from Keeling *et al.* (2008). The concentration of total silicate was calculated using the gridded data reported by Garcia *et al.* (2006) between 0 and 10 m and weighing the averages using the surface area of each grid cell. pH is expressed on the total scale.

Parameter	Unit	LGM	1766	2007	2100
Temperature	°C	17.2	18.3	18.9	21.4
Salinity		36	34.9	34.9	34.7
Total phosphate	10^{-6} mol kg^{-1}	0.66	0.66	0.63	0.55
Total silicate	10^{-6} mol kg^{-1}	7.35	7.35	7.35	7.35
Total alkalinity	10^{-6} mol kg^{-1}	2399	2326	2325	2310
CO_2 partial pressure (seawater)	µatm	180	267	384	793
$[CO_2]$	10^{-6} mol kg^{-1}	6.26	9.05	12.8	24.7
$[HCO_3^-]$	10^{-6} mol kg^{-1}	1660	1754	1865	2020
$[CO_3^{2-}]$	10^{-6} mol kg^{-1}	299	231	186	118
Dissolved inorganic carbon	10^{-6} mol kg^{-1}	1966	1994	2064	2162
pH_T		8.33	8.2	8.07	7.79
$[H^+]$	10^{-9} mol kg^{-1}	4.589	6.379	8.600	16.13
Calcite saturation	—		7.1	5.5	4.5	2.8
Aragonite saturation	—		4.6	3.6	2.9	1.8

LGM, Last Glacial Maximum.

is the decrease in the saturation state of calcium carbonate of about 20% between 1766 and 2007 and a projected further potential decline of about 40% by 2100 (Table 1.1). Consequently, the $CaCO_3$ saturation horizon (the depth at which the saturation is 1) was, depending on the $CaCO_3$ mineral considered, 80 to 200 m shallower in 1994 than in pre-industrial times (Feely *et al.* 2004). Recent changes are most dramatic in surface waters of high-latitude areas where seasonal undersaturation of aragonite has been reported. Model projections indicate that undersaturation of aragonite will become a widespread feature of the Southern (Orr *et al.* 2005) and Arctic (Steinacher *et al.* 2009) oceans within the next decades (see Chapter 3).

1.3 The biological and biogeochemical processes that are potentially affected

Changes in the carbonate chemistry of seawater can have a wide range of effects, some of which may be mediated through disturbances in the acid–base status of affected organisms. The extracellular pH of body fluids in animals and the intracellular pH of various organs or unicellular organisms are usually tightly regulated, but the capacity of regulatory mechanisms can be overwhelmed. pH plays a key role in many physiological processes such as ion transport, enzyme activity, and protein function. Many intracellular enzymes are pH-sensitive and display a pH optimum around the physiological range (Madshus 1988). For example, the activity of phosphofructokinase, a key enzyme in the glycolytic pathway, exhibits a 10- to 20-fold reduction when pH decreases by as little as 0.1 units below the physiological pH optimum (Trivedi and Danforth 1966). Other direct effects of ocean acidification could occur when one or several reactant(s) in a physiological process, such as calcification and photosynthesis, is a carbon species. Calcification is often described by Eq. 1.2a, which may give the wrong impression that it could be stimulated by ocean acidification as the concentration of bicarbonate rises. However, the ultimate reaction at the site where $CaCO_3$ is precipitated is controlled by the concentration of carbonate ions (Eq. 1.2b), and hence the $CaCO_3$ saturation state

$$Ca^{2+} + 2HCO_3^- \rightarrow CaCO_3 + CO_2 + H_2O \quad (1.2a)$$

$$Ca^{2+} + CO_3^- \rightarrow CaCO_3. \quad (1.2b)$$

Both parameters decrease with increasing ocean acidification, which can trigger a decline in the calcification rate. In contrast, the reverse reaction of dissolution of calcium carbonate is favoured by the decrease in the $CaCO_3$ saturation state generated by ocean acidification.

Photosynthesis is another process which can be directly affected by changes in the carbonate system. The ultimate source of inorganic carbon is carbon dioxide according to Eqs 1.3a and 1.3b (depending on the source of nitrogen):

$$106CO_2 + 16NH_4^+ + HPO_4^{2-} + 106H_2O$$
$$= C_{106}H_{263}O_{110}N_{16}P + 106O_2 + 14H^+ \quad (1.3a)$$

$$106CO_2 + 16NO_3^- + HPO_4^{2-} + 122H_2O$$
$$+ 18H^+ = C_{106}H_{263}O_{110}N_{16}P + 138O_2. \quad (1.3b)$$

However, CO_2 is often in very limited supply in seawater and several organisms have developed carbon-concentrating mechanisms (CCMs; Reinfelder 2011). CCMs use bicarbonate, which is available in large amounts in seawater, to concentrate inorganic carbon intracellularly. The increase in CO_2 and bicarbonate with increasing ocean acidification could therefore have significant effects on photosynthesis. It can be stimulated by elevated CO_2 in species lacking a CCM or when the CCM is not operating optimally. It is also possible that even bicarbonate users favour the use of CO_2 over HCO_3^- because CCMs are energetically costly to operate (see Chapter 6).

Ocean acidification could also have indirect effects on many biological processes. For example, photosynthesis requires nutrients other than CO_2 (Eqs 1.3a and 1.3b), such as nitrogen, phosphorus, or iron (not shown in the equations). Lower ocean pH has an impact on the thermodynamics and kinetics of metals and some nutrients in seawater, resulting in changes in their speciation, behaviour, and fate (e.g. Millero *et al.* 2009, and Chapter 6). These changes could affect the availability and toxicity of metals to marine organisms. For example, chemical equilibria (Millero *et al.* 2009) suggest an increased bioavailability of iron, an important micronutrient for primary production, at lower seawater pH. This is supported by some (Breitbarth *et al.* 2010) but not all (Shi *et al.* 2010) experimental data. Another example of a possible indirect effect is the tight link that may exist between processes such as photosynthesis and calcification. If photosynthesis stimulates calcification, a stimulation of photosynthesis (which removes CO_2 and so tends to increase CO_3^{2-}) by ocean acidification could benefit calcification by mitigating the limited supply of carbonate ions. Conversely, if calcification stimulates photosynthesis, a decrease in calcification generated by ocean acidification could have a negative effect on photosynthesis. Such cascading, indirect effects could occur at both the organism and community levels.

1.4 A short history of ocean acidification research

1.4.1 The early days

The distribution of pH in the oceans, its changes with depth, tide, and other physical and biological processes, and the impact of the changes on organisms were studied early in the 20th century. Some studies even pre-date the definition of pH (technically, p[H] as the initial definition was based on concentration rather than activity as used today) by Sørensen (1909). For example, Moore *et al.* (1906) investigated the effect of 'alkalies [*sic.*] and acids' on growth and cell division in the fertilized eggs of a sea urchin.

A few key early studies can be mentioned, some of which were remarkably innovative even by today's standards. McClendon (1917) showed that the oxygen consumption of certain marine invertebrates varies with pH. He subsequently reported that the pH range compatible with the life of seaweed is rather broad, may be different for different species, that little was known about the effect on different life stages, and that there may be interaction with other environmental factors (McClendon 1918). He also found that corals from deep waters are smaller, more fragile, and deposit less $CaCO_3$ than those of shallow waters, and proposed that this was related to the decline of pH with depth (McClendon 1918).

Gail (1919) reported results from a study which looked at the combined effect of changes in pH and temperature on survival, reproduction, and growth of a macroalga (*Fucus*; Fig. 1.1). Observations were made over a few weeks, which enabled an assessment of acclimation. Germination was more successful at pH values above 7.4 and below 8.6 at all temperatures considered. The maximum germination occurred in seawater at pH values between 8.0 and 8.2 and decreased on either side of these. The growth of spores as well as larger plants of *Fucus* was inhibited when the pH value was below 7.2.

Powers (1920) suggested that the abundance of fauna at certain localities in the Puget Sound region and the average size of species such as barnacles collected at different localities were dependent on pH. Two years later, he proposed that the 'ability of marine fishes to absorb oxygen at low tension from the sea water is more or less dependent upon the hydrogen ion concentration of the water' (Powers 1922). Bouxin (1926a) found that the larval development of a sea urchin was unaffected by changes in pH within the range of 7.3 to 8.1, was slower below this threshold and came to a halt at a pH of 6.4. Skeletal growth was slower below 7.3, and dissolution occurred below 6.4 (Bouxin 1926a, 1926b).

Prytherch (1929) showed that pH in Milford Harbor (CT, USA) was lowest at low tide and highest close to high tide (7.2 vs 8.2). He noted that the failure of oysters to spawn at low tide was correlated to relatively high temperature (above 20°C) and low pH. Spawning occurred near the times of high tide at a similar temperature but much higher pH value (7.8 or above). He concluded that the most important factors controlling oyster spawning are 'the temperature of the water, the range of tide, and the hydrogen-ion concentration'. Rubey (1951) summarized a large part of the earlier studies and concluded:

If only one-one hundredth of all this buried carbon dioxide [fossil carbon] were suddenly added to today's atmosphere and ocean (that is, if the amount in the present atmosphere and ocean were suddenly increased... seven-fold, from 1.3×10^{20} to 9.1×10^{20} g), it would have profound effects on the chemistry of sea

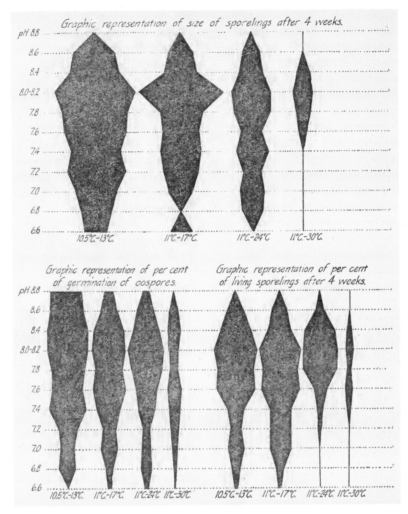

Figure 1.1   Size, percentage germination, and percentage survival of spores of the macroalga *Fucus* sp. as a function of pH and temperature at the end of a 4-week experiment (Gail 1919).

water and on the organisms living in the sea. The first effect would be to change the average pH of sea water from about 8.2 to 5.9. . . . This acid water would be much less than saturated with calcium carbonate, and thus further changes would follow. Eventually, when equilibrium was re-established, the partial pressure of carbon dioxide in the atmosphere would be about 110 [sic.] times its present value, and the pH of the sea water would end up at an average of about 7.0.

The effects of these changes on living organisms would be drastic. If the supposed increase

of carbon dioxide happened suddenly, it would probably mean wholesale extinction of many of the marine species of today. If, however, the increase were gradual, so that organisms could adapt themselves by generations of evolutionary changes, the effects would be much less disastrous—but perhaps no less clearly recorded in the physiological adaptations of the surviving forms. From the paleontologic record it appears improbable that any change so drastic as an abrupt sevenfold increase of carbon dioxide has happened, at least since the beginning of the Cambrian.

...For many forms of life, the concentration of bicarbonate and hydrogen ions and the carbon-dioxide tension are among the most critical factors in their chemical environment. A number of higher marine animals (the herring, for example) are extremely sensitive to small changes in the pH of their environment. A large proportion of the eggs of some marine animals remains unfertilized if the acidity of sea water departs more than about 0.5 pH from normal. Lower organisms are commonly less sensitive; but many species of mollusks, sea urchins, Medusa, diatoms, bacteria, algae, and others seem unable to tolerate a range of more than about 1 unit of pH. Recent workers attribute a larger part of these observed biologic effects to the carbon-dioxide tension or to the concentration of bicarbonate ions than to hydrogen-ion concentration directly.

This was the first time that the early literature had been explored, even though briefly. Obviously, the results collected in the early days are not of great utility for assessing the effects of future ocean acidification because of the relatively uncertain measurements of the carbonate system and the large pH changes that were investigated. Some obvious errors were also made. For example, it is now known that the decline in coral calcification with increasing depth is mostly controlled by decreasing irradiance rather than pH (Gattuso *et al.* 1999). It is nevertheless very interesting to note that many research questions which are given top priority today (e.g. multifactorial perturbation experiments, acclimation, and adaptation) had already been considered and sometimes investigated decades ago.

1.4.2 Modern research

The goal of this subsection is to provide a brief overview of how recent research on ocean acidification developed and to identify seminal publications. Readers are invited to consult the subsequent chapters for in-depth reviews. Modern research, arbitrarily set from the 1950s, began with the landmark paper of Revelle and Suess (1957) which showed that the uptake of anthropogenic CO_2 decreases the oceans' buffering capacity. In other words, other

parameters remaining the same, the oceans capture less atmospheric CO_2 as ocean acidification proceeds. In their study of calcium carbonate precipitation on the Bahama Banks, Broecker and Takahashi (1966) were the first to report that the concentration of CO_3^{2-}, which decreases at elevated $p$$CO_2$ (Table 1.1), could control community calcification. This finding was subsequently confirmed in coral reefs (e.g. Smith and Pesret 1974) and reef mesocosms (e.g. Langdon *et al.* 2000; Leclercq *et al.* 2000).

Brewer (1978) derived a method for estimating the original atmospheric equilibration signal of water masses using total alkalinity and the concentration of dissolved inorganic carbon and correcting for the perturbations of the CO_2 system due to respiration, carbonate dissolution, and nitrate addition. This method provided the first direct evidence that the increase in atmospheric CO_2 leads to a corresponding increase in the $p$$CO_2$ of Antarctic intermediate water and revealed a propagation of the atmospheric CO_2 signal northwards. Several other approaches were subsequently designed (see Álvarez *et al.* 2009).

Recent changes in the carbonate chemistry of seawater have been documented using time-series data, first from the western North Atlantic (Bates 2001) and subsequently from the eastern North Atlantic (Santana-Casiano *et al.* 2007), northern Pacific (Dore *et al.* 2009), and the Icelandic Sea (Olafsson *et al.* 2009). The instrumental record begun only recently, but palaeo-oceanographic tracers have been used to extend estimates back in time. The first key papers are those of Spivack *et al.* (1993) who used the boron isotopic composition, Barker and Elderfield (2002) who used foraminiferal size-normalized shell weight, and Yu *et al.* (2007) who proposed the use of the boron:calcium ratio to assess changes in carbonate chemistry. Modelling has also been used to reconstruct past changes in the carbonate system using simple box models (Zeebe and Westbroek 2003) or more sophisticated process-driven ecosystem models (e.g. Andersson *et al.* 2003).

Manipulative experiments have been performed to assess the response of organisms to changes in seawater carbonate chemistry. Many of the first experiments carried out used $p$$CO_2$ levels that were unrealistically high in the context of anthropogenic ocean acidification and provided little information on parameters of the carbonate system other than pH. However, important results were found (and subsequently forgotten). For example, Swift and Taylor (1966) reported that the growth of a coccolithophore followed a bell-shaped response as a function of pH. Such a bell-shaped response was subsequently found to describe the effect of ocean acidification on the calcification rate of a coccolithophore (Langer *et al.* 2006). The first purposeful ocean acidification experiment was performed by Agegian (1985) who showed a negative impact of ocean acidification on the growth of coralline algae. A relatively large number of manipulative experiments have been carried out on zooxanthellate scleractinian corals. The first one, published in 1998, related the decline of calcification to the decrease of the saturation state of calcium carbonate manipulated through changes in the calcium concentration (Gattuso *et al.* 1998). Subsequent studies altered the saturation state through changes in the concentration of carbonate which better mimic ocean acidification. Riebesell *et al.* (2000) were the first to report on the effect of ocean acidification on photosynthesis and calcification in two phytoplanktonic species (coccolithophores). Nitrogen fixation is also a key process which has been shown to respond to ocean acidification (Levitan *et al.* 2007; see also Chapter 6).

A relatively large number of perturbation experiments have been reported in recent years, but results have not always been consistent. For example, some calcifiers do not seem to be affected by ocean acidification (e.g. Langer *et al.* 2006; Iglesias-Rodriguez *et al.* 2008; Ries *et al.* 2009; Rodolfo-Metalpa *et al.* 2010). Several hypotheses have been proposed to explain these discrepancies: species differences which could be related to different calcification mechanisms, methodological differences, and misinterpretation of the data (see Chapters 6 and 7).

In addition to reconstructing past changes in the carbonate system, modelling tools are critical for projecting future changes in the carbonate system and the biogeochemical impacts of ocean acidification. Caldeira and Wickett (2003) published a seminal paper in which they compared the timing and magnitude of past and future changes in ocean pH and popularized the term 'ocean acidification', which had originally been introduced by Broecker and Clark (2001). Ocean carbon cycle models were used to estimate changes in the carbonate system in the recent past and the 21st century (e.g. Orr *et al.* 2005), identify the geographical areas which are most at risk (Steinacher *et al.* 2009), and project future changes in calcification (Kleypas *et al.* 1999) and feedbacks on the global carbon cycle (e.g. Heinze 2004).

1.4.3   Bibliometric analysis

The recent attention paid to ocean acidification by the scientific community, policymakers, and the media, as well as increased funding and the launch of several national and international research projects, have spurred a steep increase in research efforts, starting from around 1990 and accelerating during the last few years. For the purpose of this chapter, a bibliographic database including scientific articles, books, and book chapters on ocean acidification was compiled in order to illustrate trends in the publication effort. Dissertations, reports, abstracts of presentations at meetings, and popular articles were excluded. The complete list of references is available in Appendix 1.1 at http://ukcatalogue.oup.com/product/9780199591091.do.

A total of 846 articles (from 1906 to 2010) were included in this study. A series of keywords describing the content of each article, for example the type of organism or process studied, was used to categorize articles and extract statistical information from the database. The primary categories used to classify papers are shown in Fig. 1.2A. The main ones are: 'biological response', 'biogeochemistry', 'paleo' (papers using a palaeo-oceanographic approach; mainly reconstruction of past carbonate chemistry), 'modelling', 'chemistry' (including, for example, time-series and cruise data), and 'review' (articles that discuss and synthesize results from other papers). Obviously, cataloguing an article into one of these specific categories can sometimes be difficult and articles frequently required the allocation of several of the above keywords to fully describe their content.

The 'biological response' category was further divided into subcategories of taxonomic groups, and processes and parameters studied (Fig. 1.2B to D). Keywords indicating the type of study (laboratory or field, mesocosms, and gene expression/genetic diversity) were added to articles when appropriate (mostly for 'biological response' and 'chemistry'

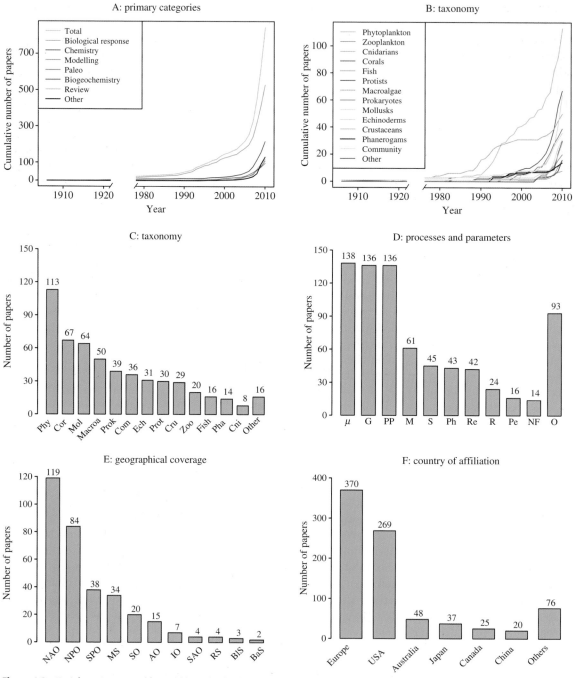

Figure 1.2   Key information extracted from a bibliographic database on ocean acidification. A, Numbers of articles published on ocean acidification (cumulative), separated into six areas of interest. The black line represents articles discussing mitigation, policy, socio-economy, fisheries, methods, education as well as articles that could not be categorized into any of the study areas identified. 'Biological response' articles were further subdivided into categories of taxonomic group (B and C) and processes and parameters (D; see key below). E, Geographical distribution of study sites (see key below). F, Country of affiliation of first author. Key to D: μ, growth; G, calcification; PP, primary production; M, morphology/morphometry; S, survival; Ph, physiology; Re, reproduction; R, respiration; Pe, performance; NF, nitrogen fixation; O, other. Key to E: NAO, North Atlantic Ocean; NPO, North Pacific Ocean; SPO, South Pacific Ocean; MS, Mediterranean Sea; SO, Southern Ocean; AO, Arctic Ocean; IO, Indian Ocean; SAO, South Atlantic Ocean; RS, Red Sea; BlS, Black Sea; BaS, Baltic Sea.

papers). Likewise, the 'modelling' papers were allocated keywords describing the nature of the model study (global, regional, or community). Keywords describing the geographical coverage of studies were added to 'biological response' and 'chemistry' papers, but only when the location was clearly stated (Fig. 1.2E). These keywords were not used for studies using long-term laboratory-cultured organisms or when organisms were collected remotely from the laboratory where the study took place. Finally, the articles were tagged with the country of affiliation of its first author (Fig. 1.2F).

1.4.3.1   Total number of papers on ocean acidification

To the authors' knowledge, the first paper directly addressing an ocean acidification-related question was published in 1906 (Moore *et al.* 1906). With the exception of this paper and a handful of other early

articles (see Section 1.4.1), the field started to receive increased attention only at the end of the 1980s, and most ocean acidification papers have only been published in the last few years (Fig. 1.2A). From 1989 to 2003, the rate of publication remained relatively stable at an average of 9 articles per year. From 2004 and onwards, the field experienced a sharp increase in the number of publications from 22 articles in 2004 to 213 articles in 2010. This corresponds to a more than ninefold increase in the number of papers published per year between 2004 and 2010 and a 43-fold increase from 1989 to 2010. It is worth noting that 79% of all papers were published after 2003.

In March 2011, the time at which the database was analysed, 105 articles were already published or in press, indicating that the total number of articles would increase even further in 2011. It must be pointed out that the total number of scientific

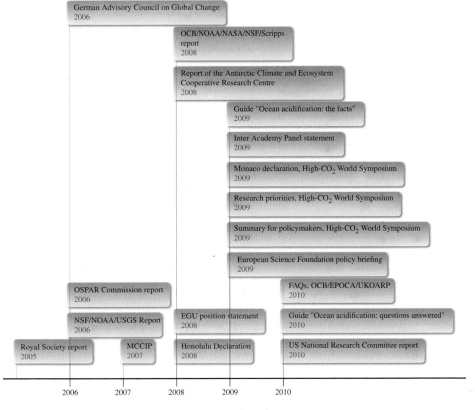

Figure 1.3 Recent key reports and conferences on ocean acidification (full references are available in Appendix 1.2 at http://ukcatalogue.oup.com/product/9780199591091.do).

articles also increased from 2004 to 2010. For example, the Aquatic Science and Fisheries Abstracts (ASFA) database, published by Cambridge Scientific Abstracts, comprises 53 357 papers published in 2004 and 88 962 papers published in 2010 (as of 1st June 2011). Therefore, the increase in the publication rate in the field of ocean acidification was considerably higher than in marine science in general (482% vs. 67% increase). This wave of research effort devoted to ocean acidification in the last few years can probably be explained by the increased attention from society, largely thanks to several important reports and the outcome of international conferences (Section 1.5.2, Fig. 1.3), resulting in increased funding opportunities. In addition to peer-reviewed scientific publications, recommendations from such reports and conferences have probably contributed to the launch of several large-scale research projects (see Section 1.5.3.1).

1.4.3.2 Type of study

The biological response to ocean acidification has dominated the publication efforts (63% of the total number of articles in the database; Fig. 1.2A). Articles dealing with palaeo-oceanography, modelling, biogeochemistry, and chemistry, as well as review papers, represented 13, 11, 10, 25, and 14%, respectively, of the total number of publications. The fact that several studies were allocated two or more keywords (see above) explains that the sum of the percentages exceeds 100%.

The trends of articles on the biological response to ocean acidification follow the overall tendency with a fairly stable number of publications per year from 1989 to 2003 (on average eight articles per year) followed by a steep increase from 2004 to 2010 (18 and 112 papers, respectively, a more than sixfold increase; Fig. 1.2A). For the other categories (palaeo-oceanography, modelling, chemistry, biogeochemistry, and review), the number of papers increased slowly between 1988 and 2004 (2 and 14 articles, respectively) and more drastically from 2005 to 2010 (28 and 164 papers, respectively; corresponding to an almost sixfold increase). The relatively solid knowledge that already exists on the chemical aspects of ocean acidification compared with the poor understanding of the consequences for marine life might explain the large

amount of work devoted to the biological response to ocean acidification during recent years.

1.4.3.3 Taxonomic groups

Studies have historically been focused on macroalgae (Fig. 1.2B), with the first study published in 1919 and a steep yearly increase between 1989 and 1994 (an average of 3 new articles per year), but interest in this group has somewhat decreased in recent years and articles now more frequently study phytoplankton and corals. Overall, phytoplankton is the most studied group, with a total of 113 articles, against 67 papers on corals, 64 on molluscs and 50 on macroalgae. The obvious early choices of model organism were the calcifiers and photosynthesizers, because both calcification and photosynthesis could be directly affected by the increased CO_2 in seawater (Section 1.3). Studies of other organisms that might show more subtle effects of ocean acidification (e.g. on performance, reproduction, etc.) were limited until very recent years (data not shown). This is also reflected by the imbalance in the number of studies devoted to primary production, growth, and calcification compared with other processes (a total of 136, 138, and 136 papers, respectively, compared with e.g. 43 on physiology, 42 on reproduction, and 14 on nitrogen fixation; Fig. 1.2D). The large number of articles on phytoplankton probably reflects the numerous studies carried out on *Emiliana huxleyi* and other coccolithophores, the major pelagic calcifiers, because of their possible sensitivity to changes in ocean carbonate chemistry and their significance in the biogeochemical cycles of carbon, calcium carbonate, and sulphur.

1.4.3.4 Study region

When looking at the geographical distribution of study sites, the North Atlantic, North Pacific, South Pacific, and the Mediterranean (38, 27, 12, and 11%, respectively, of the articles indicating a geographical location) have been investigated to a larger extent than the Indian, South Atlantic, Arctic, and Southern Oceans, and the Baltic, Red, and Black Seas (18%; Fig. 1.2E). The number of articles that studied polar regions (Arctic and Southern Oceans; 5 and 7%, respectively) is relatively low considering that these regions have been identified as the most sensitive to ocean acidification (Chapter 3; Orr *et al.* 2005; Steinacher *et al.* 2009).

1.4.3.5   Modelling papers
The modelling studies have largely focused on global models: 62 out of 96 papers used a global model, against 25 papers using regional models. Only six studies modelled a community response to ocean acidification. Regional models are important for estimating the socio-economic consequences and costs of ocean acidification, since these are largely region-specific (Cooley *et al.* 2009).

1.4.3.6   Methods used
The majority of the studies on the biological response and chemistry, reporting new results, were conducted in the laboratory (68% of articles compared with about 33% field studies). Forty-six papers used a mesocosm approach (including 32 in the field and 14 in the laboratory). The keyword 'mesocosm' was only attributed when the word mesocosm was explicitly mentioned in the article. This imbalance illustrates the need to conduct additional mesocosm experiments in order to complement laboratory studies. While laboratory experiments are appropriate for some types of studies, mesocosm and other larger-scale field studies provide another dimension to the interpretation of results, thanks to the inclusion of trophic interactions and an environment closer to a natural one. Few papers (23 or fewer than 6%) used molecular techniques to study changes in gene expression and genetic diversity in response to ocean acidification. However, the first paper was published only in 2006; these methods are likely to become more important in the near future (see Chapter 5).

1.4.3.7   Affiliation of first author
European countries contributed 44% and the USA 32% of the publications on ocean acidification from 1906 to 2010 (based on affiliation of the first author; Fig. 1.2F). Within Europe, 34% of the articles were produced by UK scientists and 25% by German scientists. Funding to launch major research programmes such as EPOCA (the European Project on OCean Acidification), even though recent, could partly explain the greater contribution from European countries. With the Federal Ocean Acidification

Research and Monitoring (FOARAM) Act and the US projects that will result from it, one might expect the US contribution to increase in subsequent years.

To conclude, research efforts seem unbalanced in many aspects, including taxonomic groups, processes studied, and geographical study sites. For example, high-latitude areas are under-investigated although they have been identified as the regions potentially most sensitive to ocean acidification. Other trends include a disproportionately large research effort on phytoplankton and on the processes of photosynthesis and calcification.

1.5   Risks and policy implications

Ocean acidification, often referred to as 'the other CO_2 problem', has stood in the shadow of global warming for many years and is only recently beginning to gain increased consideration from policy-makers, politicians, the media, and the general public. The coverage of ocean acidification in the Fourth Assessment Report of the Intergovernmental Panel on Climate Change (IPCC), published in 2007, was limited to just a few out of several thousands of pages. The body of literature published in this research domain does not permit an understanding as robust as the one produced by the multitude of studies that have been carried out on the effects of climate change. However, it has allowed the detection of a second, and possibly as serious, facet of the consequences of anthropogenic CO_2 emissions. While the number of ocean acidification studies increases and a better understanding of its consequences becomes available, scientists continue to work for the consideration of this phenomenon in the post-Kyoto negotiations and in the Fifth Assessment Report of the IPCC. The rest of this section briefly introduces the socio-economic risks, the dissemination approaches, and the policy implications associated with ocean acidification. A full coverage of these issues is available in Chapter 13.

1.5.1   Risks to society and the economy

The economic, cultural, societal, and nutritional dependency of humans on the oceans, from the fishing and tourism industries to recreational activities and as a vital protein source, puts us in a vulnerable situation. Disrupting the oceanic system could have potentially catastrophic consequences for humankind. Risks to economic and societal welfare include disturbances to seafood resources and the deterioration of coral reefs, both of which could lead to losses to the fishing and tourism industry, decreased biodiversity, and threatened food security for populations relying to a large extent on seafood as a protein source (Cooley *et al.* 2009; Kleypas and Yates 2009). At a US Senate hearing in April 2010, representatives of the tourism, fishing, and diving industries testified to their concern about the impact of ocean acidification on their activities. Two general examples of socio-economic risks associated with ocean acidification are briefly introduced below. The reader is referred to Chapter 13 for a thorough analysis.

1.5.1.1 *Commercial seafood species and fisheries*

The potential threat of ocean acidification to the seafood industry is twofold and can be expressed in terms of: (1) direct impacts on commercially valuable species, and (2) indirect effects via perturbations to the marine food web. The most studied direct effect on seafood is associated with the process of calcification. It has been shown that many calcareous organisms will have problems forming their shells or skeletons in an ocean rich in CO_2. For example, laboratory experiments on calcifying species of shellfish with a commercial value, such as the blue mussel, *Mytilus edulis*, and the Pacific oyster, *Crassostrea gigas*, have shown a decrease in calcification of 25 and 10%, respectively, for conditions of ocean chemistry projected for the end of the 21st century (Gazeau *et al.* 2007). However, the extrapolation of laboratory results to the natural environment is difficult, and some species are unaffected (Miller *et al.* 2009). The effects of ocean acidification on molluscs therefore remain unknown, despite the fact that the aquaculture industry represents a billion dollar industry worldwide and food security for millions of people. In addition to consequences for the ability of species to calcify their shell or skeleton, ocean acidification may have direct (effects on metabolism or reproduction) and indirect (prey or habitat loss) negative impacts on economically important finfish (Cooley and Doney 2009a). The first-sale value of the global fisheries production

was estimated at US\$91.2 billion in 2006 (FAO 2009), indicating the potentially large economic consequences of decreased seafood harvests.

Fisheries and seafood industries represent both commercial and recreational interests, that contribute an economic value via travel and purchases of permits and equipment. Furthermore, numerous jobs depend on fishery industries, from fishermen to affiliated industries such as processing, transportation, preparation, and sales (Cooley and Doney 2009a).

Disturbances to trophic dynamics may pose an indirect threat to fisheries. For example, shelled pteropods play a key role in some marine food webs and their decline or disappearance might cause perturbations in trophic levels important to humans. The consequences of changing food web structure are potentially dire. Even if the direct effect of an unbalanced food web is difficult to estimate, it is clear that the future success of top predators will depend on their capacity to alter their food sources and whether alternative prey will be available (i.e. species that are resistant to or unaffected by ocean acidification and have the nutritional qualities required to function as a substitute; Cooley and Doney 2009a).

1.5.1.2 *Coral reefs*

Coral reefs are among the ecosystems most sensitive to the combined effects of elevated pCO_2 and temperature (Hoegh-Guldberg *et al.* 2007; Cooley and Doney 2009a; see also Chapter 15). Several studies have shown that coral reef ecosystems might shift from a state of calcium carbonate construction to erosion under the pressure of ocean acidification (e.g. Silverman *et al.* 2009). Their damage or loss could have direct and indirect detrimental consequences, including impacts on commercially or ecologically important fishes that feed and reproduce in reefs, on the particularly rich biodiversity inhabiting the reefs, and on the tourism industry and protein supply for millions of people, most of whom live in the poorest areas of the world (Donner and Potere 2007; Kleypas and Yates 2009). Coral reefs also protect other important biodiversity-rich ecosystems such as mangroves and seagrass beds, and play a key role in protecting shorelines from erosion and inundation. Cesar *et al.* (2003) estimated the potential net benefit from coral reefs (taking into account fisheries, coastal protection, tourism, and biodiversity value) at US\$30 billion yr^{-1}.

1.5.2   Outreach to policymakers and the general public

Once the risks are identified, the transfer of scientific knowledge from researchers to policymakers and other end-users is a critical but challenging step. Policy briefings and reports as well as 'reference user groups' are two examples of efficient methods for sharing new scientific findings.

Since the Royal Society report was published in 2005, a milestone in the broader visibility of ocean acidification, several reports and policy briefs have been produced, including the European Science Foundation (ESF) policy briefing recommending European actions and the Monaco Declaration, signed by 155 scientists from 26 nations and supported by HSH Prince Albert II of Monaco, that urges political leaders to act (Fig. 1.3). Among the most recent briefs is 'Ocean acidification – the facts', launched during the Fifteenth Session of the Conference of the Parties (COP 15) in Copenhagen in December 2009 and its follow-up: 'Ocean acidification – Questions Answered' launched in Monaco in November 2010. Such documents have contributed significantly to the increased attention paid to ocean acidification recently and have made important recommendations for policy design (see below).

An innovative approach is used by several ocean acidification projects. It involves a 'reference user group' (RUG) to disseminate scientific results outside the circle of the scientific community to other possible end-users such as policy advisers, decision-makers, industrial leaders, schools, media, and the general public. The creation of a RUG, with membership from industry, governmental and non-governmental organizations, and foundations, has turned out to be a successful means to shorten pathways of information exchange between researchers and end-users (Turley 1999; Chapter 13).

1.5.3   Policy implications

The risks introduced by ocean acidification to environmental, financial, and social structures have led to suggestions that they be included in various types of policy arena.

1.5.3.1   Research policies

Several reports and policy briefs have highlighted the high level of uncertainty concerning the impacts of ocean acidification and insisted on the need for increased research efforts and international cooperation (e.g. The Royal Society 2005; Anonymous 2009; Bijma *et al.* 2009). New international research programmes have been launched and recommendations on allocation of funding have been put forward. Reports also highlight the importance of promoting exchange between scientists and economists in order to assess the socio-economic consequences of ocean acidification, and between scientists and policymakers in order to ensure that the most up-to-date information underlies developing policies. Finally, actions to coordinate national and international research efforts to share technical facilities and information, and to develop training activities and workshops, have been suggested.

These recommendations have led to a number of research, synthesis, organizational, coordinative, and educational efforts worldwide. Several of the major international marine science programmes such as IMBER (Integrated Marine Biogeochemistry and Ecosystem Research), SOLAS (Surface Ocean-Lower Atmosphere Study), LOICZ (Land–Ocean Interactions in the Coastal Zone), and PAGES (Past Global Changes) have developed working groups, programmes, websites, or educational resources devoted to ocean acidification. Current major research projects on ocean acidification include EPOCA launched in May 2008 as the first large-scale research project on ocean acidification, the German project BIOACID (Biological Impacts of Ocean Acidification), and a UK-wide programme in Europe, as well as the OCB (Ocean Carbon and Biogeochemistry) Office and the FOARAM Act, in the USA where a national programme is due to be launched in 2011. This list is far from exhaustive and many other smaller projects exist or will be launched in the coming years. The need for international coordination to support new research programmes have been recognized, for example via the SOLAS-IMBER Working Group on Ocean Acidification (SIOA), launched in 2009 with the tasks of coordinating international research efforts on ocean acidification and undertaking synthesis activities at the international level. The organization of international conferences, such as the 'Ocean in a High-CO_2 World' symposia, have also contributed to international cooperation and visibility in policy networks, in particular via the products resulting from the meetings, such as the Monaco Declaration and the 'Research priorities for ocean acidification' revisited and updated during the meetings.

1.5.3.2 *Mitigation and CO₂ emissions policies*

The changes in ocean carbonate chemistry generated by ocean acidification are occurring at such a rapid rate that there is little time to produce the level and amount of knowledge necessary to provide an unambiguous input to policymakers. Policy action is therefore required before a comprehensive understanding is available, and application of the precautionary principle is advocated by many. Strategies to reduce anthropogenic CO_2 emissions are the most efficient and perhaps the only current option to halt ocean acidification, and the implementation of large-scale agreements to decrease CO_2 emissions is often the major recommendation made in policy briefs (e.g. Anonymous 2009). The failure to reach binding agreement at the international level illustrates the difficulties of implementing this recommendation. However, recommendations on emission policies led to the recognition of CO_2 as a pollutant by the US Environmental Protection Agency (EPA) in April 2009.

Potential alternative methods exist to reduce the impacts of the increasing CO_2 emissions, including carbon capture and storage (IPCC 2005) and various geo-engineering techniques (The Royal Society 2009; see Chapter 14). The Sleipner project launched in 1996 is an example of the former approach. CO_2 from the natural gas produced at the Statoil-operated gas field 'Sleipner Vest' in the North Sea is injected into a saline aquifer, located above the Sleipner gas field (IPCC 2005). An example of a geo-engineering technique has been provided by Crutzen (2006), who described the injection of sulphur particles into the stratosphere in order to reflect incoming solar energy and thereby decrease the temperature on earth. However, geo-engineering approaches were proposed to mitigate climate change, not ocean acidification. In fact, such methods would enable continued and increasing CO_2 emissions, and would increase rather than decrease ocean acidification.

1.5.3.3 *Socio-economic and adaptation policies*

Reports also highlight the need to evaluate the costs to society and work on ways to adapt society

to consequences that are likely to occur as a result of ocean acidification (Bijma *et al.* 2009). The potential cost to society is not clear, but there is a serious risk that the implementation of actions will become increasingly difficult and expensive as decisions are delayed. It will be critical to put a figure on costs for action versus inaction and to evaluate the consequences of political decisions (Anonymous 2009). The inclusion of ocean acidification as a risk factor in fishery management plans, to anticipate a potential threat and act in time to avoid it, is an example of a concrete recommendation for sectoral economic policies (Cooley and Doney 2009b).

Environmental protection policies also have a part to play. Loss of biodiversity is not only an ecological concern but is also associated with an economic value, even if it is difficult to estimate (but see Cesar *et al.* 2003). In order to preserve biodiversity, detection and protection of areas that have a naturally higher ability to tackle ocean acidification and thus have the potential to be used as refuges in the future has been suggested (Ocean Acidification Reference User Group 2009). It has also been recommended to devote resources and time to find means to improve the ability of the ocean to resist ocean acidification (i.e. to create increased ecosystem resilience) and support conservation efforts (The Royal Society 2005; Ocean Acidification Reference User Group 2009).

1.6 Conclusions

Ocean acidification is a new field of research which has implications for a very large number of scientific and socio-economic subdisciplines (Fig. 1.4). Despite the impressive development of this field in the past few years, both in terms of research initiatives and publications, only a few consequences are known with a high level of confidence (e.g. chemistry), whereas many (e.g. the biological and biogeochemical responses, and the impact on socio-economics) are known with only a moderate level of confidence (see Chapter 15). The goal of this book is to review all aspects of ocean acidification research, summarize the current understanding, identify gaps, and provide recommendations for future research and international coordination.

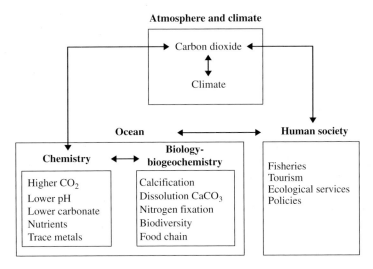

Figure 1.4 Highly simplified view of the relationships between the atmosphere, ocean, and human activities in the context of ocean acidification. For the sake of clarity, just a few examples are shown. The relationships between the atmosphere and ocean apply in both past and modern times.

1.7 Acknowledgements

Anne-Marin Nisumaa is gratefully acknowledged for her assistance to draw figure 1.2. We thank Kelvin Boot, Peter Liss and Stephen Smith for comments on an early draft of this chapter. This work is a contribution to the 'European Project on Ocean Acidification' (EPOCA) which received funding from the European Community's Seventh Framework Programme (FP7/2007–2013) under grant agreement n°211384.

References

Agegian, C.R. (1985). *The biogeochemical ecology of Porolithon gardineri (Foslie)*, 178 pp. PhD Thesis, University of Hawaii, Honolulu, HI.

Álvarez, M., Monaco, C.L., Tanhua, T. *et al.* (2009). Estimating the storage of anthropogenic carbon in the subtropical Indian Ocean: a comparison of five different approaches. *Biogeosciences*, **6**, 681–703.

Andersson, A., Mackenzie, F.T., and Ver, L.M. (2003). Solution of shallow-water carbonates: an insignificant buffer against rising atmospheric CO_2. *Geology*, **31**, 513–16.

Anonymous (2009). *Monaco Declaration*. Prince Albert II of Monaco Foundation, Monaco.

Barker, S. and Elderfield, H. (2002). Foraminiferal calcification response to glacial–interglacial changes in atmospheric CO_2. *Science*, **297**, 833–6.

Bates, N.R. (2001). Interannual variability of oceanic CO_2 and biogeochemical properties in the Western North Atlantic subtropical gyre. *Deep-Sea Research (Part II, Topical Studies in Oceanography)*, **48**, 1507–28.

Bates, N.R. and Peters, A.J. (2007). The contribution of atmospheric acid deposition to ocean acidification in the subtropical North Atlantic Ocean. *Marine Chemistry*, **107**, 547–58.

Bibby, R., Cleall-Harding, P., Rundle, S., Widdicombe, S., and Spicer, J. (2007). Ocean acidification disrupts induced defences in the intertidal gastropod *Littorina littorea*. *Biology Letters*, **3**, 699–701.

Bijma, J., Barange, M., Brander, L. *et al.* (2009). *Impacts of ocean acidification*, 12 pp. European Science Foundation, Strasbourg, France.

Bouxin, H. (1926a). Action des acides sur le squelette des larves de l'oursin *Paracentrotus lividus*. Influence du pH. *Comptes Rendus des Séances de la Société de Biologie*, **94**, 453–5.

Bouxin, H. (1926b). Action des acides sur les larves de l'oursin *Paracentrotus lividus*. Étude morphologique. *Comptes Rendus des Séances de la Société de Biologie*, **94**, 451–3.

Breitbarth, E., Bellerby, R.J., Neill, C.C. *et al.* (2010). Ocean acidification affects iron speciation during a coastal seawater mesocosm experiment. *Biogeosciences*, **7**, 1065–73.

Brewer, P.G. (1978a). Direct observation of the oceanic CO_2 increase. *Geophysical Research Letters*, **5**, 997–1000.

Broecker, W. and Clark, E. (2001). A dramatic Atlantic dissolution event at the onset of the last glaciation.

Geochemistry Geophysics Geosystems, **2**, 1065, doi: 10.1029/2001GC000185.

Broecker, W.S. and Takahashi, T. (1966). Calcium carbonate precipitation on the Bahama Banks. *Journal of Geophysical Research*, **71**, 1575–602.

Caldeira, K. and Wickett, M.E. (2003). Anthropogenic carbon and ocean pH. *Nature*, **425**, 365.

Cesar, H.S.J., Burke, L.M., and Pet-Soede, L. (2003). *The economics of worldwide coral reef degradation*, 23 pp. Cesar Environmental Economics Consulting (CEEC), Arnhem, The Netherlands.

Cooley, S.R. and Doney, S.C. (2009a). Ocean acidification's impact on fisheries and societies: a U.S. perspective. *Current, the Journal of Marine Education*, **25**, 15–19.

Cooley, S.R. and Doney, S.C. (2009b). Anticipating ocean acidification's economic consequences for commercial fisheries. *Environmental Research Letters*, **4**, 1–8.

Cooley, S.R., Kite-Powell, H.L. and Doney, S.C. (2009). Ocean acidification's potential to alter global marine ecosystem services. *Oceanography*, **22**, 172–81.

Crutzen, P.J. (2006). Albedo enhancement by stratospheric sulfur injections: a contribution to resolve a policy dilemma? *Climatic Change*, **77**, 211–19.

Dickson, A. (2010). The carbon dioxide system in seawater: equilibrium chemistry and measurements. In: U. Riebesell, V. Fabry, L. Hansson, and J.-P. Gattuso (eds) *Guide for best practices in ocean acidification research and data reporting*, pp. 17–40. Office for Official Publications of the European Communities, Luxembourg.

Doney, S.C., Mahowald, N., Lima, I. *et al.* (2007). Impact of anthropogenic atmospheric nitrogen and sulfur deposition on ocean acidification and the inorganic carbon system. *Proceedings of the National Academy of Science USA*, **104**, 14580–5.

Donner, S.D. and Potere, D. (2007). The inequity of the global threat to coral reefs. *Bioscience*, **57**, 214–15.

Dore, J.E., Lukas, R., Sadler, D.W., Church, M.J., and Karl, D.M. (2009). Physical and biogeochemical modulation of ocean acidification in the central North Pacific. *Proceedings of the National Academy of Science USA*, **106**, 12235–40.

Egleston, E.S., Sabine, C.L., and Morel, F.M.M. (2010). Revelle revisited: buffer factors that quantify the response of ocean chemistry to changes in DIC and alkalinity. *Global Biogeochemical Cycles*, **24**, GB1002, doi:10.1029/2008GB003407.

FAO (2009). *The state of fisheries and aquaculture 2008*, 196 pp. Food and Agriculture Organization of the United Nations, Rome, Italy.

Feely, R.A., Sabine, C.L., Lee, K. *et al.* (2004). Impact of anthropogenic CO_2 on the $CaCO_3$ system in the oceans. *Science*, **305**, 362–6.

Frankignoulle, M. (1994). A complete set of buffer factors for acid/base CO_2 system in seawater. *Journal of Marine Systems*, **5**, 111–18.

Gail, F.W. (1919). Hydrogen ion concentration and other factors affecting the distribution of *Fucus*. *Publication Puget Sound Biological Station*, **2**, 287–306.

Garcia, H.E., Locarnini, R.A., Boyer, T.P., and Antonov, J.I. (2006). World Ocean Atlas 2005. Volume 4: Nutrients (phosphate, nitrate, silicate). In: S. Levitus (ed.), *NOAA atlas NESDIS 64*, pp. 1–396. US Government Printing Office, Washington, DC.

Gattuso, J.-P. and Lavigne, H. (2009). Technical note: approaches and software tools to investigate the impact of ocean acidification. *Biogeosciences*, **6**, 2121–33.

Gattuso, J.-P., Allemand, D., and Frankignoulle, M. (1999). Photosynthesis and calcification at cellular, organismal and community levels in coral reefs: a review on interactions and control by carbonate chemistry. *American Zoologist*, **39**, 160–83.

Gattuso, J.-P., Frankignoulle, M., Bourge, I., Romaine, S., and Buddemeier, R.W. (1998). Effect of calcium carbonate saturation of seawater on coral calcification. *Global and Planetary Change*, **18**, 37–46.

Gazeau, F., Quiblier, C., Jansen, J.M., Gattuso, J.-P., Middelburg, J.J., and Heip, C.H.R. (2007). Impact of elevated CO_2 on shellfish calcification. *Geophysical Research Letters*, **34**, L07603, doi:10.1029/2006GL028554.

Heinze, C. (2004). Simulating oceanic $CaCO_3$ export production in the greenhouse. *Geophysical Research Letters*, **31**, L16308, doi:10.1029/2004GL020613.

Hendrey, G.R. (1984). *Early biotic responses to advancing lake acidification*, 173 pp. Butterworth, Boston.

Hoegh-Guldberg, O., Mumby, P.J., Hooten, A.J. *et al.* (2007). Coral reefs under rapid climate change and ocean acidification. *Science*, **318**, 1737–42.

Iglesias-Rodriguez, M.D., Halloran, P.R., Rickaby, R.E.M. *et al.* (2008). Phytoplankton calcification in a high-CO_2 world. *Science*, **320**, 336–40.

IPCC (2007). Climate Change 2007: The Physical Science Basis. Contribution of Working Group I to the Fourth Assessment Report of the Intergovernmental Panel on Climate Change (eds Solomon, S., D. Qin, M. Manning, Z. Chen, M. Marquis, K.B. Averyt, M. Tignor and H.L. Miller). Cambridge University Press, Cambridge.

IPCC (2005). IPCC Special Report on Carbon Dioxide Capture and Storage. Prepared by Working Group III of the Intergovernmental Panel on Climate Change (eds Metz, B., O. Davidson, H. C. de Coninck, M. Loos, and L. A. Meyer). Cambridge University Press, Cambridge.

Keeling, R.F., Piper, S.C., Bollenbacher, A.F., and Walker, J.S. (2008). *Atmospheric CO_2 records from sites in the SIO* *air sampling network*, Carbon Dioxide Information Analysis Center, Oak Ridge National Laboratory, US Department of Energy, Oak Ridge, TN, USA.

Kleypas, J.A. and Yates, K.K. (2009). Coral reefs and ocean acidification. *Oceanography*, **22**, 108–17.

Kleypas, J.A., Buddemeier, R.W., Archer, D., Gattuso, J.-P., Langdon, C., and Opdyke, B.N. (1999). Geochemical consequences of increased atmospheric CO_2 on coral reefs. *Science*, **284**, 118–20.

Langdon, C., Takahashi, T., Marubini, F. *et al.* (2000). Effect of calcium carbonate saturation state on the rate of calcification of an experimental coral reef. *Global Biogeochemical Cycles*, **14**, 639–54.

Langer, G., Geisen, M., Baumann, K.H. *et al.* (2006). Species-specific responses of calcifying algae to changing seawater carbonate chemistry. *Geochemistry, Geophysics, Geosystems*, **7**, Q09006, doi:10.1029/2005GC001227.

Lavigne, H. and Gattuso, J.-P. (2010). *seacarb: seawater carbonate chemistry with R*. R package version 2.3.3. http://cran-project.org/package=seacarb.

Leclercq, N., Gattuso, J.-P., and Jaubert, J. (2000). CO_2 partial pressure controls the calcification rate of a coral community. *Global Change Biology*, **6**, 329–34.

Le Quéré, C., Raupach, M.R., Canadell, J.G. *et al.* (2009). Trends in the sources and sinks of carbon dioxide. *Nature Geoscience*, **2**, 831–6.

Levitan, O., Rosenberg, G., Setlik, I. *et al.* (2007). Elevated CO_2 enhances nitrogen fixation and growth in the marine cyanobacterium *Trichodesmium*. *Global Change Biology*, **13**, 531–8.

Lueker, T.J., Dickson, A.G., and Keeling, C.D. (2000). Ocean pCO_2 calculated from dissolved inorganic carbon, alkalinity, and equations for K_1 and K_2: validation based on laboratory measurements of CO_2 in gas and seawater at equilibrium. *Marine Chemistry*, **70**, 105–19.

Lüthi, D., Le Floch, M., Bereiter, B. *et al.* (2008). High-resolution carbon dioxide concentration record 650,000–800,000 years before present. *Nature*, **453**, 379–82.

McClendon, J.F. (1917). *Physical chemistry of vital phenomena*, 240 pp. Princeton University Press, Princeton, NJ.

McClendon, J.F. (1918). On changes in the sea and their relation to organisms. *Carnegie Institution of Washington Publication*, **252**, 213–58.

Madshus, I.H. (1988). Regulation of intracellular pH in eukaryotic cells. *Biochemistry Journal*, **250**, 1–8.

Miller, A.W., Reynolds, A.C., Sobrino, C., and Riedel, G.F. (2009). Shellfish face uncertain future in high CO_2 world: influence of acidification on oyster larvae calcification and growth in estuaries. *PLoS ONE*, **4**, e5661.

Millero, F.J. (2006). *Chemical oceanography*. CRC/Taylor and Francis, Boca Raton, FL.

Millero, F.J., Woosley, R., DiTrolio, B., and Waters, J. (2009). Effect of ocean acidification on the speciation of metals in seawater. *Oceanography*, **22**, 72–85.

Moore, B., Roaf, H.E., and Whitley, E. (1906). On the effects of alkalies and acids, and of alkaline and acid salts, upon growth and cell division in the fertilized eggs of *Echinus esculentus*. A study in relationship to the causation of malignant disease. *Proceedings of the Royal Society of London. Series B: Biological Sciences*, **77**, 102–36.

Nakićenović, N., Alcamo, J., Davis, G. *et al.* (2000). *Special report on emissions scenarios: a special report of Working Group III of the Intergovernmental Panel on Climate Change*, 599 pp. Cambridge University Press, Cambridge.

Ocean Acidification Reference User Group (2009). *Ocean acidification: the facts. A special introductory guide for policy advisers and decision makers* (ed. D.d'A. Laffoley and J.M. Baxter), 12 pp. European Project on Ocean Acidification (EPOCA).

Olafsson, J., Olafsdottir, S.R., Benoit-Cattin, A., Danielsen, M., Arnarson, T.S., and Takahashi, T. (2009). Rate of Iceland Sea acidification from time series measurements. *Biogeosciences*, **6**, 2661–8.

Orr, J.C., Fabry, V.J., Aumont, O. *et al.* (2005). Anthropogenic ocean acidification over the twenty-first century and its impact on calcifying organisms. *Nature*, **437**, 681–6.

Plattner, G.K., Joos, F., Stocker, T.F., and Marchal, O. (2001). Feedback mechanisms and sensitivities of ocean carbon uptake under global warming. *Tellus B*, **53**, 564–92.

Powers, E.B. (1920). The variation of the condition of seawater, especially the hydrogen ion concentration, and its relation to marine organisms. *Washington State University Puget Sound Biological Station Publication*, **2**, 369–85.

Powers, E.B. (1922). The physiology of the respiration of fishes in relation to the hydrogen ion concentration of the medium. *Journal of General Physiology*, **4**, 305–17.

Prytherch, H.F. (1929). Investigation of the physical conditions controlling spawning of oysters and the occurrence, distribution, and setting of oyster larvae in Milford Harbor, Connecticut. *US Bureau of Fisheries, Bulletin*, **44**, 429–503.

Reinfelder, J.R. (2011). Carbon concentrating mechanisms in eukaryotic marine phytoplankton. *Annual Review of Marine Science*, **3**, 291–315.

Revelle, R. and Suess, H.E. (1957). Carbon dioxide exchange between atmosphere and ocean and the question of an increase of atmospheric CO_2 during the past decades. *Tellus*, **9**, 18–27.

Riebesell, U., Zondervan, I., Rost, B., Tortell, P.D., and Morel, F.M.M. (2000). Reduced calcification of marine plankton in response to increased atmospheric CO_2. *Nature*, **407**, 364–7.

Ries, J.B., Cohen, A.L., and McCorkle, D.C. (2009). Marine calcifiers exhibit mixed responses to CO_2-induced ocean acidification. *Geology*, **37**, 1131–4.

Rodolfo-Metalpa, R., Martin, S., Ferrier-Pagès, C., and Gattuso, J.-P. (2010). Response of the temperate coral *Cladocora caespitosa* to mid- and long-term exposure to pCO_2 and temperature levels projected for the 2100 AD. *Biogeosciences*, **7**, 289–300.

Rubey, W.W. (1951). Geologic history of sea water: an attempt to state the problem. *Bulletin of the Geological Society of America*, **62**, 1111–48.

Sabine, C.L., Feely, R.A., Gruber, N. *et al.* (2004). The oceanic sink for anthropogenic CO_2. *Science*, **305**, 367–71.

Santana-Casiano, J.M., González-Dávila, M., Rueda, M.J., Llinás, O., and González-Dávila, E.F. (2007). The interannual variability of oceanic CO_2 parameters in the northeast Atlantic subtropical gyre at the ESTOC site. *Global Biogeochemical Cycles*, **21**, GB1015, doi:10.1029/2006GB002788.

Shi, D., Xu, Y., Hopkinson, B.M., and Morel, F.M. (2010). Effect of ocean acidification on iron availability to marine phytoplankton. *Science*, **327**, 676–9.

Silverman, J., Lazar, B., Cao, L., Caldeira, K., and Erez, J. (2009). Coral reefs may start dissolving when atmospheric CO_2 doubles. *Geophysical Research Letters*, **36**, L05606, doi:10.1029/2008GL036282.

Smith, S.V. and Pesret, F. (1974). Processes of carbon dioxide flux in the Fanning Island lagoon. *Pacific Science*, **28**, 225–45.

Sørensen, S.P.L. (1909). Études enzymatiques. II. Sur la mesure et l'importance de la concentration des ions hydrogène dans les réactions enzymatiques. *Comptes Rendus des Travaux du Laboratoire Carlsberg*, **8**, 1–168.

Spivack, A.J., You, C.-F., and Smith, H.J. (1993). Foraminiferal boron isotope ratios as a proxy for surface ocean pH over the past 21 Myr. *Nature*, **363**, 149–51.

Steinacher, M., Joos, F., Frölicher, T.L., Plattner, G.-K., and Doney, S.C. (2009). Imminent ocean acidification projected with the NCAR global coupled carbon cycle-climate model. *Biogeosciences*, **6**, 515–33.

Swift, E. and Taylor, W.R. (1966). The effect of pH on the division rate of the coccolithophorid *Cricosphaera elongata*. *Journal of Phycology*, **2**, 121–5.

The Royal Society (2005). *Ocean acidification due to increasing atmospheric carbon dioxide*, 60 pp. The Royal Society, London.

The Royal Society (2009). *Geoengineering the climate: science, governance and uncertainty*, 82 pp. The Royal Society, London.

Trivedi, B. and Danforth, W.H. (1966). Effect of pH on the kinetics of frog muscle phosphofructokinase. *Journal of Biological Chemistry*, **241**, 4110–12.

Turley, C.M. (1999). Taking research into policy making. *Proceedings of the Third European Marine Science and Technology Conference*, **6**, pp. 13–19, EU Directorate General Science, Research and Development, Brussels.

Widdicombe, S. and Needham, H.R. (2007). Impact of CO_2-induced seawater acidification on the burrowing activity of *Nereis virens* and sediment nutrient flux. *Marine Ecology Progress Series*, **341**, 111–22.

Wright, R.F. (2003). Predicting recovery of acidified freshwaters in Europe and Canada. *Hydrology and Earth System Sciences*, **7**, 429–30.

Yu, J.M., Elderfield, H., and Hönisch, B. (2007). B/Ca in planktonic foraminifera as a proxy for surface seawater pH. *Paleoceanography*, **22**, PA2202, doi:10.1029/2006PA001347.

Zeebe, R.E. and Westbroek, P. (2003). A simple model for the $CaCO_3$ saturation state of the ocean: the 'Strangelove', the 'Neritan', and the 'Cretan' Ocean. *Geochemistry Geophysics Geosystems*, **4**, 1104, doi:10.1029/2003GC0000538.

Zeebe, R.E. and Wolf-Gladrow, D.A. (2001). *CO_2 in seawater: equilibrium, kinetics, isotopes*, 346 pp. Elsevier, Amsterdam.

CHAPTER 2

Past changes in ocean carbonate chemistry

Richard E. Zeebe and Andy Ridgwell

2.1 Introduction

Over the period from 1750 to 2000, the oceans have absorbed about one-third of the carbon dioxide (CO_2) emitted by humans. As the CO_2 dissolves in seawater, the oceans become more acidic and between 1750 and 2000, anthropogenic CO_2 emissions have led to a decrease of surface-ocean total pH (pH_T) by ~0.1 units from ~8.2 to ~8.1 (see Chapters 1 and 3). Surface-ocean pH_T has probably not been below ~8.1 during the past 2 million years (Hönisch *et al.* 2009). If CO_2 emissions continue unabated, surface-ocean pH_T could decline by about 0.7 units by 2300 (Zeebe *et al.* 2008). With increasing CO_2 and decreasing pH, carbonate ion (CO_3^{2-}) concentrations decrease and those of bicarbonate (HCO_3^-) rise. With declining CO_3^{2-} concentration ($[CO_3^{2-}]$), the stability of the calcium carbonate ($CaCO_3$) mineral structure, used extensively by marine organisms to build shells and skeletons, is reduced. Other geochemical consequences include changes in trace metal speciation (Millero *et al.* 2009) and even sound absorption (Hester *et al.* 2008; Ilyina *et al.* 2010).

Do marine organisms and ecosystems really 'care' about these chemical changes? We know from a large number of laboratory, shipboard, and mesocosm experiments, that many marine organisms react in some way to changes in their geochemical environment like those that might occur by the end of this century (see Chapters 6 and 7). Generally (but not always), calcifying organisms produce less $CaCO_3$, while some may put on more biomass. Extrapolating such experiments would lead us to expect potentially significant changes in ecosystem structure and nutrient cycling. But can one really extrapolate an instantaneous environmental change to one occurring on a timescale of a century? What capability, if any, do organisms have to adapt to future ocean acidification which is occurring on a slower timescale than can be replicated in the laboratory? Simultaneous changes in ocean temperature and nutrient supply as well as in organisms' predation environment may create further stresses or work to ameliorate the effect of changes in ocean chemistry. Either way, the impact in the future ocean may further diverge from projections based on simple manipulation experiments.

We know that the chemistry of the ocean has varied in the past (see below). We also know something about the past composition of marine ecosystems and sometimes we can even infer changes in individuals' physiological state, such as growth rate. In this regard, the geological record may provide clues about what the future will hold for changes in ocean chemistry and their effects on marine life. When studying the geological record, however, the critical task is to identify an appropriate analogue for the future. Among other things, this requires a basic understanding of ocean chemistry controls during long-term steady states versus transient events, because carbonate chemistry parameters do not have to vary with the same relationship if either the rate of change or the initial chemistry is very different. Future versus past comparisons conducted without sufficient appreciation about how the carbon cycle and ocean chemistry is regulated on geological timescales may ultimately lead to invalid conclusions. In case of aberrations, knowledge of the magnitude and timescale of the acidification event is necessary, otherwise geological periods or events may be studied that are unsuitable for comparison with future scenarios.

Another reason for studying past changes in ocean chemistry is that it allows us to evaluate the current anthropogenic perturbation in the context of earth's history. For instance, we can ask questions such as: what was the amplitude of natural variations in ocean chemistry immediately prior to industrialization (say during the Holocene, the past 12 000 years)? Are there past events that are comparable in magnitude and timescale to the present ocean acidification caused by humans, or is it unprecedented in earth's history? How did the carbon cycle, climate, and ocean chemistry respond to massive carbon input in the past—and on what timescale was the carbon removed from the ocean–atmosphere system by natural sequestration?

In this chapter, basic controls on ocean carbonate chemistry on different timescales will be reviewed. Changes in ocean chemistry during various geological eras will be investigated and past ocean acidification events will be examined. *Note, however, that this chapter's focus is on past changes in ocean chemistry, not solely on past ocean acidification events.* Evidence of biotic responses to past changes in ocean chemistry will be discussed only sporadically here (for more information see Chapter 4). For consistency, 'Myr' is used in this book to denote both geological dates and durations, although 'Ma' is recommended to denote geological dates.

2.2   Seawater carbonate chemistry

The basics of seawater carbonate chemistry have been outlined in Chapter 1 and will not be repeated here. Additional information can be found elsewhere (see e.g. Zeebe and Wolf-Gladrow 2001). In this section, we will focus on a few fundamentals and subtleties of seawater carbonate chemistry that will aid the discussion to follow.

2.2.1   CaCO$_3$ saturation state of seawater, Ω

The CaCO$_3$ saturation state of seawater is expressed by Ω:

$$\Omega = \frac{[Ca^{2+}]_{sw} \times [CO_3^{2-}]_{sw}}{K_{sp}^*} \qquad (2.1)$$

where $[Ca^{2+}]_{sw}$ and $[CO_3^{2-}]_{sw}$ are the concentrations of Ca^{2+} and CO_3^{2-} in seawater and K_{sp}^* is the solubility product of calcite or aragonite, the two major forms of CaCO$_3$, at the *in situ* conditions of temperature, salinity, and pressure. Values of $\Omega > 1$ signify supersaturation and $\Omega < 1$ signifies undersaturation. Because K_{sp}^* increases with pressure (the temperature effect is small) there is a transition of the saturation state from $\Omega > 1$ (calcite rich) to $\Omega < 1$ (calcite depleted) sediments at depth.

2.2.2   Two parameters are required to determine the carbonate chemistry

At thermodynamic equilibrium, the carbonate system can be described by six fundamental parameters: dissolved inorganic carbon (C_T), total alkalinity (A_T), [CO$_2$], [HCO$_3^-$], [CO$_3^{2-}$], and [H$^+$] (see Box 1.1). The concentration of OH$^-$ and pCO$_2$ can be readily calculated using the dissociation constant of water and Henry's law. Given the first and second dissociation constants of carbonic acid and the definitions of C_T and A_T, we have four equations with six unknowns. Thus, if the values of two parameters are known, we are left with four equations and four unknowns and the system can be solved. The fundamental rule follows that *two carbonate system parameters are required to determine the carbonate chemistry,* one parameter is insufficient.

Ignorance of this rule has led to misinformation in the literature. For instance, future atmospheric CO$_2$ concentrations have been compared with pCO$_2$ levels during the Cretaceous (~145 to ~65 Myr), which may have been as high as, say 2000 ppmv. While at some point in the future, atmospheric CO$_2$ levels might approach values similar to those during the Cretaceous, this by no means implies similar surface-ocean chemistry. A surface ocean with $C_T =$ 2.4 mmol kg^{-1} in equilibrium with an atmosphere at pCO$_2 = 2000$ ppmv would have a calcite saturation state Ω_c of 1.1 (temperature $T = 15$°C, salinity $S = 35$). However, at a higher C_T of 4.9 mmol kg^{-1}, the calcite saturation Ω_c would be 4.5 (at the same T and S). The latter example illustrates possible Cretaceous seawater conditions and shows that such an ocean would have had surface waters with a favourable carbonate mineral saturation state, despite high pCO_2. For simplicity, the above figures are based on modern calcium concentrations; for variable calcium see Tyrrell and Zeebe (2004). The bottom line is that comparisons of seawater chemistry between the Cretaceous, for instance, and the near future cannot be based on one carbonate system parameter alone (see Section 2.4.5).

2.2.3 Effect of CaCO$_3$ production and dissolution on carbonate chemistry

$CaCO_3$ precipitation decreases C_T and A_T in a ratio of 1:2, and, counterintuitively, increases $[CO_2]$ although the inorganic carbon concentration has decreased. Dissolution has the reverse effect. For a qualitative understanding, consider the reaction

$$Ca^{2+} + 2HCO_3^- \rightarrow CaCO_3 + CO_2 + H_2O \quad (2.2)$$

which indicates that CO_2 is liberated during $CaCO_3$ precipitation. Quantitatively, however, the conclusion that $[CO_2]$ in solution is increasing by one mole per mole of $CaCO_3$ precipitated is incorrect because of buffering. The correct analysis takes into account the decrease of C_T and A_T in a ratio 1:2 and the buffer capacity of seawater. That is, the medium becomes more acidic because the decrease in total alkalinity outweighs that of total inorganic carbon and hence $[CO_2]$ increases. For instance, at surface-seawater conditions ($C_T = 2000$ µmol kg^{-1}, pH$_T$ = 8.2, T = 15°C, S = 35), $[CO_2]$ increases by only ~0.03 µmol per µmol $CaCO_3$ precipitated (for more details, see Zeebe and Wolf-Gladrow 2001).

As a result, production of $CaCO_3$ in the surface ocean and its transport to depth tends to *increase* atmospheric CO_2. This process represents one component of the ocean's biological carbon pump and has been dubbed the 'CaCO$_3$ counter pump' because of its reverse effect relative to the organic carbon pump, which tends to reduce atmospheric CO_2. One ironic consequence of this is that if marine calcifiers were to disappear, it would constitute a small negative feedback on rising atmospheric CO_2 levels in the short term. It is also important that the ocean carbonate pump on the timescale discussed above

merely leads to shifts in the vertical distribution of ocean C_T and A_T, rather than changing their inventories. This process can be important in changing surface-ocean chemistry and reducing atmospheric CO_2 on timescales shorter than ~10 000 years (see Section 2.3.2). On a million year timescale, on the other hand, the burial of $CaCO_3$ in marine sediments represents one major pathway for removing carbon from the ocean–atmosphere system (see Section 2.3.7).

2.2.4 Temperature and salinity

The abiotic environment plays a significant role in setting the carbonate chemistry state, particularly at the surface. For instance, CO_2 is less soluble at higher temperatures, leading to out-gassing to the atmosphere and hence locally reduced C_T. Conversely, CO_2 uptake takes place preferentially in colder waters and C_T is higher. Hence warm regions tend to have higher $[CO_3^{2-}]$ and be more saturated with respect to carbonate minerals than colder regions. As surface-ocean temperatures have varied in the past, both globally as well as regionally (latitudinally), so has carbonate chemistry. Also related to changes in climate is the importance of salinity, as adding (or subtracting) freshwater will reduce (increase) the concentration of C_T and A_T in a 1:1 ratio (they are conservative quantities). For instance, the larger ice volume at the time of the Last Glacial Maximum, equivalent to the removal of around 3% of the water from the ocean (and storage primarily in the great ice sheets of the Northern Hemisphere), would have acted to increase $[CO_2]$ and hence atmospheric pCO_2, and just at a time when ice core records of pCO_2 show it was at a record low. A multitude of other factors must then come into play to counter the salinity effect and further drive pCO_2 down to glacial concentrations (see Kohfeld and Ridgwell 2009).

2.3 Controls on ocean carbonate chemistry

Under most natural conditions, the ocean inventories of C_T and A_T determine the whole-ocean carbonate chemistry. Changes in the C_T and A_T inventories over time therefore constitute the major control on

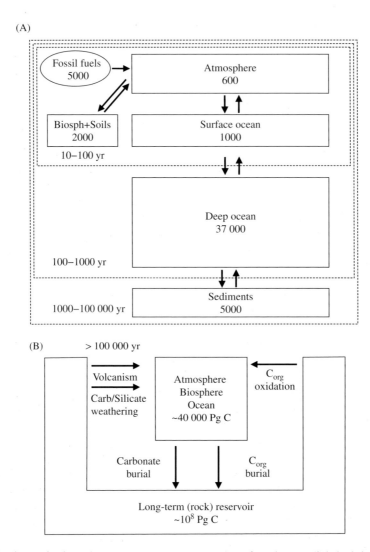

Figure 2.1 (A) The surface (exogenic) carbon cycle. Approximate reservoir sizes are in units of Pg C (1 Pg = 10^{15} g). The dashed boxes demarcate reservoirs involved in carbon exchange on the respective timescales. (B) The long-term carbon cycle.

the evolution of the carbonate system in seawater (for carbon cycling, see Fig. 2.1). The characterization of the dominant carbon and total alkalinity fluxes to and from the oceans on different timescales is hence fundamental to understanding controls on ocean CO_2 chemistry (e.g. Sundquist 1986).

2.3.1 Decadal to centennial timescale

On timescales shorter than about 100 years, the natural reservoirs that exchange carbon at the earth's surface include the atmosphere (pre-anthropogenic inventory ~600 Pg C; 1 Pg = 10^{15} g), the biosphere (~500 Pg C), soils (~1500 Pg C), and the surface ocean (~1000 Pg C). Combined, these reservoirs hold less than ~4000 Pg C (see Fig. 2.1A). Fossil fuel reserves, on the other hand, have been estimated at ~5000 Pg C (excluding hydrates). It is thus immediately clear that the release of several thousands of Pg C over a few hundred years will overwhelm the capacity of these surface reservoirs to absorb carbon.

2.3.2   Millennial timescale

On timescales of the order of ~1000 years, the deep ocean reservoir becomes an important component of the surface carbon cycle (modern whole ocean inventory ~38 000 Pg C, Fig. 2.1A). In fact, most of the anthropogenic carbon will eventually be absorbed by the ocean and neutralized by reaction with carbonate sediments (see below).

Once released to the atmosphere, it takes about a year for CO_2 to mix throughout the atmosphere. The very upper boundary of the ocean is nearly instantaneously in equilibrium with the overlying atmosphere; however, on average it takes somewhat less than a year for the ocean mixed-layer to equilibrate with the overlying atmosphere. Once in the mixed-layer, this CO_2 is available for transport to the deep ocean. Movement of cold seawater away from the surface in regions of deep-water formation such as the North Atlantic, as well as subduction associated with frontal systems (and the formation of, for example, Antarctic Intermediate Water) carries the excess C_T to the deep ocean. Replacement of these surface waters from lower latitudes (and warmer, lower C_T environments) or nutrient-depleted (and hence C_T depleted) upwelled waters, allows the cycle of C_T uptake and transport to be repeated. By this means, an anomaly in pH (and other carbonate chemistry parameters) is gradually propagated into the ocean interior. Once emissions have ceased and the ocean has had time to fully mix on a ~1000 yr timescale, a new equilibrium is established between ocean and atmosphere, with the partitioning of CO_2 in a roughly 1:3 ratio between atmosphere and ocean (Archer 2005). The greater the total release, the larger the exhaustion of oceanic buffering, and hence the greater final airborne fraction. It is thought that climate change will both warm the ocean surface and increase net precipitation and ice melting at high latitudes, with the result that vertical stratification in the ocean will increase at both low and high latitudes. This is expected to slow the propagation of the C_T and carbonate chemistry anomaly into the ocean interior. Furthermore, a warmer overall ocean will result in a higher airborne CO_2 fraction because of the effect of temperature on the solubility of CO_2.

After ~1000 years, with no additional processes operating, the ocean would be left with a reduced pH (i.e. increased hydrogen ion activity) and the atmosphere with higher pCO_2. The carbon has been distributed (or partitioned) between atmosphere and ocean. The subsequent steps of fossil fuel neutralization include dissolution of carbonate sediment in the deep sea and terrestrial weathering of carbonate and silicate minerals. Near-complete removal of fossil fuel carbon from the atmosphere will take tens to hundreds of thousands of years (e.g. Archer 2005; Uchikawa and Zeebe 2008; Zachos *et al.* 2008).

Under natural steady-state conditions, the oceanic inventories of C_T and A_T can be considered essentially constant on a timescale of ~1000 years. Exceptions to this are rapid carbon inputs from otherwise long-term storage reservoirs such as combustion of fossil fuel by humans. Further examples include catastrophes from possible impact events over carbonate platforms, or other abrupt carbon releases from geological reservoirs [e.g. during the Palaeocene–Eocene Thermal Maximum (PETM), see below]. In the case of rapid CO_2 addition to the ocean–atmosphere system, dissolution of carbonate sediment may occur on timescales shorter than their usual response time of >1000 yr (see Section 2.3.3).

2.3.3   Millennial to 100 000 yr timescale

On timescales of 1000 to 100 000 years, fluxes between reactive carbonate sediments (~5000 Pg C) and the oceans' inventories of C_T and A_T have to be considered as well. Oceanic inventories may vary, for instance, during glacial–interglacial cycles (see so-called calcite compensation below). The magnitude of these changes is, however, limited and so are the associated changes in ocean chemistry and atmospheric CO_2. The fate of anthropogenic CO_2 emissions on this timescale involves reaction of fossil fuel carbon with deep-sea carbonate sediments and terrestrial carbonates, which will ultimately facilitate removal of carbon from the ocean–atmosphere system (so-called fossil fuel neutralization).

Another mechanism that affects ocean carbonate chemistry on the millennial to 100 000 yr timescale is associated with changes in ocean carbon pumps (see below). This can lead to shifts in the vertical distribution of ocean C_T and A_T, while not affecting their inventories. This process is believed to be important for understanding changes in surface-ocean chemistry and atmospheric CO_2 on glacial–interglacial timescales. While surface-ocean changes during the glacial–interglacial cycles were ~80 µatm in pCO_2 and ~0.2 units in pH, deep-ocean carbonate chemistry changes were probably much smaller (see Section 2.4.2 and Zeebe and Marchitto 2010).

2.3.4   Tectonic (>100 000 yr) timescale

A large amount of carbon is locked up in the earth's crust as carbonate carbon (~70 × 10^6 Pg C) and as elemental carbon in shales and coals (~20 × 10^6 Pg C). On timescales >100 000 yr, this reservoir is active, and imbalances in the fluxes to and from this pool can lead to large changes in C_T, A_T, and atmospheric CO_2 (Fig. 2.1B). The balance between CO_2 consumption by subduction of marine sediments, net organic carbon burial, weathering, subsequent carbonate burial, and volcanic degassing of CO_2 is the dominant process controlling carbon fluxes on this timescale (e.g. Walker *et al.* 1981; Berner *et al.* 1983; Caldeira 1992; Zeebe and Caldeira 2008).

Figure 2.1 illustrates the fundamental difference between short-term carbon cycling on, for example, a 10 to 100 yr timescale (Fig. 2.1A) and long-term carbon cycling (Fig. 2.1B). The two distinct cycles involve vastly different reservoir sizes and different sets of controls on atmospheric CO_2 and ocean chemistry. It follows that carbon cycling and ocean chemistry conditions during long-term steady states (e.g. over millions of years) cannot simply be compared with rapid, transient events (e.g. over the next few centuries). Section 2.4.5 provides more details.

2.3.5   Calcite compensation

Calcite compensation maintains the balance between $CaCO_3$ weathering fluxes into the ocean and $CaCO_3$ burial fluxes in marine sediments on a timescale of 5000 to 10 000 years (e.g. Broecker and Peng 1989; Zeebe and Westbroek 2003). At steady state, the riverine flux of Ca^{2+} and CO_3^{2-} ions from weathering must be balanced by burial of $CaCO_3$ in the sea, otherwise $[Ca^{2+}]$ and $[CO_3^{2-}]$ would rise or fall. The feedback that maintains this balance works as follows. Assume there is an excess weathering

influx of Ca^{2+} and CO_3^{2-} over burial of $CaCO_3$. Then, the concentrations of Ca^{2+} and CO_3^{2-} in seawater increase which leads to an increase in the $CaCO_3$ saturation state. This in turn leads to a deepening of the saturation horizon and to an increased burial of $CaCO_3$ until the burial again balances the influx. The new balance is restored at higher CO_3^{2-} than before.

2.3.6   Biological pump

In nature, C_T and A_T are manipulated both by biotic and by abiotic factors. For instance, the removal of dissolved inorganic carbon from the ocean by phytoplankton at the ocean surface and the creation of carbon-based cellular organic components (carbohydrates, fats, proteins, etc.) drives down the CO_2 concentration. If the cells were subsequently degraded in the surface layer and compounds broken down back into their basic components, the C_T concentration would be restored and atmospheric pCO_2 unaffected. The net effect is zero if aerobic respiration balances photosynthetic production, because respiration and photosynthesis run the same equation in opposite directions (see Chapter 1). It does not quite happen like this in the real ocean, as on average about 10% of primary production (and C_T) removal escapes the surface layer and settles gravitationally into the ocean interior before being broken down ('remineralized'). Ultimately, ocean circulation works to bringing the excess C_T back to the surface, but a gradient is created with higher C_T at depth. Hence, the faster the rate of export of particulate organic carbon (POC) from the surface ocean, or the deeper it can sink without being degraded, the stronger the C_T gradient, the lower the surface-ocean CO_2 concentrations, and hence the lower the atmospheric pCO_2. This process is known loosely as the 'biological pump'.

Past changes in the strength of the biological pump have thus modulated the C_T concentration at the surface, and by inference the acidity (pH). For instance, it has been hypothesized that during the last glacial period, the strength of the biological pump was greater, meaning lower atmospheric pCO_2 and higher pH. Reconstructions of changes in ocean-surface pH based on the boron isotopic composition of marine carbonates (boron speciation in seawater being pH-sensitive) suggest a glacial surface ocean 0.1 to 0.2 pH units higher than during interglacial periods (e.g. Sanyal *et al.* 1995; Hönisch and Hemming 2005; Foster 2008). Of course, changes in the strength of the biological pump only repartition C_T, primarily vertically, meaning that deep-ocean pH should be slightly lower during glacial periods, all other things being equal (see Section 2.4.2).

2.3.7 Carbonate and silicate mineral weathering

Weathering of carbonate minerals on continents may be represented by

$$CaCO_3 + CO_2 + H_2O \rightarrow Ca^{2+} + 2HCO_3^- \qquad (2.3)$$

whereas the reverse reaction (Eq. 2.2) represents the precipitation and subsequent burial of carbonates in marine sediments. As described in Section 2.3.5, carbonate weathering (input to the ocean) and burial (output) are balanced via calcite compensation on a relatively short timescale (~10 000 yr). Note that for each mole of CO_2 taken up during $CaCO_3$ weathering, one mole of CO_2 is also released during $CaCO_3$ precipitation. The net carbon balance for the combined ocean–atmosphere system on timescales over which carbonate weathering is balanced by carbonate burial is therefore zero. For this reason, carbonate weathering and burial are often ignored in models of the long-term carbon cycle over millions of years (see, however, Ridgwell *et al.* 2003).

On the other hand, silicate mineral weathering and subsequent burial as calcium carbonate in marine sediments may be represented by

$$CaSiO_3 + CO_2 + H_2O \rightarrow CaCO_3 + SiO_2 + H_2O \quad (2.4)$$

which shows that on a net basis one mole of carbon in the form of CO_2 is removed from the atmosphere and buried as $CaCO_3$ in sediments. This cycle is balanced by input from volcanic degassing and net organic oxidation on a timescale of 10^5 to 10^6 years (see above).

When the cycle is out of balance, for instance during enhanced mineral weathering in response to elevated atmospheric CO_2, silicate weathering and

subsequent carbonate burial removes carbon from the atmosphere–ocean system. Thus, the silicate weathering cycle is ultimately responsible for sequestering carbon in the long term until a balance between sources and sinks is restored. In the case of large anthropogenic fossil fuel emissions (e.g. a total of 5000 Pg C), it will take hundreds of thousands of years for atmospheric CO_2 to return to climatically relevant levels of, say, 400 ppmv in the future. Note that the exact timing is difficult to forecast, mostly because of uncertainties in weathering parameterizations (e.g. Uchikawa and Zeebe 2008).

2.4 Long-term changes during earth's history (quasi-steady states)

In this section, we discuss long-term changes of ocean chemistry during earth's history. As mentioned above, this chapter's focus is on past changes in ocean chemistry, not solely on past ocean acidification events (to be discussed in Section 2.5). Over several thousands of years, the carbonate mineral saturation state of the oceans is controlled by the balance of carbonate mineral weathering on continents (input to the ocean) and carbonate burial (output) in ocean sediments (see Section 2.3.5 and, e.g., Broecker and Peng 1989; Zeebe and Westbroek 2003; Ridgwell and Schmidt 2010). This balance helps to establish fairly constant atmospheric CO_2 concentrations and ocean carbonate chemistry conditions on timescales >10 000 years. However, the entire system may not be in steady state with, for instance, long-term processes such as silicate weathering fluxes (hence the term 'quasi'-steady state).

2.4.1 The Holocene

It is instructive to consider ocean chemistry changes during the Holocene (the ~12 000 yr period prior to industrialization), not because of large variations and/or acidification events but because of the remarkable stability of the carbon cycle. It illustrates the stark contrast to the current anthropogenic disruption, representing a large and rapid carbon perturbation relative to the natural balance of the Holocene (see Fig. 2.2A). Ice core records reveal that Holocene atmospheric CO_2 varied at most between ~260 and ~280 ppmv, with $p CO_2$ gradually rising toward the present (Monnin *et al.* 2004). Ocean chemistry was also quite stable during the same interval. This has been indicated by deep-sea carbonate ion proxy records, although slightly larger changes than those expected to accompany the 20 ppmv rise in atmospheric CO_2 cannot be excluded (Broecker and Clark 2007).

Overall, the data suggest that the Holocene carbon cycle was in or close to steady-state conditions with generally minor imbalances in carbon sources and sinks, some of which were in response to the recovery from the last deglaciation (Elsig *et al.* 2009). Using the Holocene atmospheric CO_2 record, we have hindcast changes in ocean carbonate chemistry using the carbon cycle model LOSCAR

(Long-term Ocean-atmosphere-Sediment CArbon cycle Reservoir model; Zeebe *et al.* 2008, 2009). Our results indicate that Holocene ocean carbonate chemistry was nearly constant. For instance, we estimate that the calcite saturation state has varied by less than ~10%, and pH_T by less than ~0.04 units in the surface ocean over the past 10 000 years (Fig. 2.2B and C). In contrast, since the year 1750 anthropogenic CO_2 emissions have led to a decrease of surface-ocean pH_T by ~0.1 units within less than 300 years. If CO_2 emissions continue unabated, surface-ocean calcite saturation state will drop to about one-third of its pre-industrial value by 2300, while pH_T will decline by about 0.7 units (Fig. 2.2B and C).

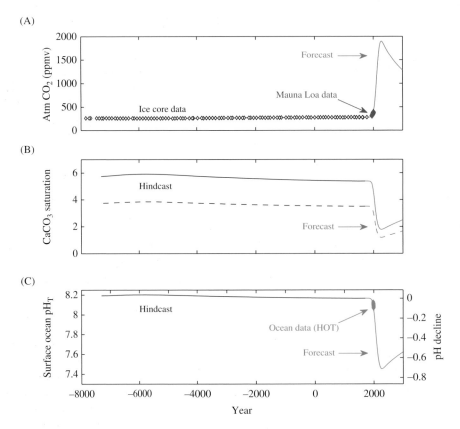

Figure 2.2   Holocene versus Anthropocene. (A) Open diamonds, Holocene atmospheric CO_2 concentrations from ice cores (Monnin *et al.* 2004); blue diamonds, CO_2 measured at Mauna Loa Observatory, Hawaii (Tans 2009); red line, forecast based on carbon input of 5000 Pg C over ~500 yr using the carbon cycle model LOSCAR (Zeebe *et al.* 2008; 2009). (B) Modelled (hindcast and forecast) $CaCO_3$ saturation state of surface seawater (see text). Solid and dashed lines: calcite and aragonite saturation. (C) Surface-ocean pH. Lines, model results; green diamonds, Hawaii ocean time-series (HOT) pH data at 25°C (Dore *et al.* 2009).

2.4.2 Glacial–interglacial changes

Over the past 800 000 years or so, atmospheric CO_2 has varied periodically between ~200 ppmv and ~280 ppmv (Siegenthaler *et al.* 2005; Lüthi *et al.* 2008). These glacial–interglacial cycles were accompanied by periodic changes in surface-ocean CO_2 chemistry, while deep-sea pH and carbonate ion concentration are believed to have been relatively stable (Zeebe and Marchitto 2010). Compared with interglacials, glacial surface-ocean conditions were characterized by lower temperatures, higher pH, and higher carbonate ion concentration (e.g. Sanyal *et al.* 1995; Hönisch and Hemming 2005; Foster 2008). For example, an interglacial surface-seawater sample at $T = 15°C$, $S = 35$, $C_T = 2000$ µmol kg^{-1}, and $A_T = 2284$ µmol kg^{-1}, has a pCO_2 of 280 µatm, $pH_T = 8.17$, and $[CO_3^{2-}] = 198$ µmol kg^{-1}. Corresponding glacial conditions may have been $T = 12°C$, $S = 36$, $C_T = 2006$ µmol kg^{-1}, and $A_T = 2353$ µmol kg^{-1}, which yields a pCO_2 of 200 µatm, $pH_T = 8.30$ and $[CO_3^{2-}] = 238$ µmol kg^{-1}. This indicates a difference in the glacial–interglacial saturation state of about 20%. Note that while this scenario assumes 3% higher glacial A_T, various other scenarios are possible, which would also modify the calculated pH change (e.g. Archer *et al.* 2000). Nevertheless, it illustrates the order of magnitude of glacial–interglacial changes in surface-ocean carbonate chemistry.

When considering the time evolution of the system over glacial–interglacial cycles, it is clear that surface-ocean pH and saturation state decline during the course of a deglaciation. One might thus think of a deglaciation as an 'acidification event', albeit a very slow and moderate one. In terms of rate and magnitude, it is important to realize that a deglaciation is not a past analogue for the current anthropogenic perturbation. For example, the rate of surface-ocean pH change during the most recent deglaciation may be estimated as 0.1 to 0.2 units per 10 000 years, or 0.001 to 0.002 units per century on average. In contrast, under business-as-usual CO_2 emissions, humans may cause a surface-ocean pH change of 0.7 units per 500 years, or 0.14 units per century on average. Thus, changes in surface-ocean chemistry during the Anthropocene are expected to be about three to seven times larger and 70 times faster than during a deglaciation.

The glacial–interglacial changes in surface-ocean CO_2 chemistry have been invoked to explain changes in shell weights of surface-dwelling foraminifera. For example, calcite shells of different planktonic foraminiferal species recovered from deep-sea sediment cores in the North Atlantic and Indian Ocean show higher shell weights during the last glacial period compared with the Holocene (e.g. Barker and Elderfield 2002; de Moel *et al.* 2009). These authors suggest that lower pCO_2 and elevated surface $[CO_3^{2-}]$ caused higher initial shell weights during the last glacial stage. On the other hand, shell weights of planktonic foraminifera have been used as an indicator of carbonate sediment dissolution and thus as a proxy for $[CO_3^{2-}]$ in the *deep sea*, rather than the *surface* (e.g. Broecker and Clark 2003). Clearly, the issue is complicated due to various factors, including possible effects of growth temperature, $[CO_3^{2-}]$, nutrients, and other environmental parameters on initial shell weight, as well as dissolution in sediments and/or the water column (e.g. Bijma *et al.* 2002). Interrelations between coccolithophore species, coccolith weight/chemistry, primary production, and the carbon cycle appear to be even more complex (for discussions see, e.g., Zondervan *et al.* 2001; Beaufort *et al.* 2007; Rickaby *et al.* 2007).

2.4.3 Pleistocene and Pliocene (~5 Myr to ~12 000 yr ago)

Ice-core records of atmospheric CO_2 are limited by the oldest samples available in Antarctic ice cores, which go back at most about 1 Myr. Beyond that, estimates of palaeo-pCO_2 levels and ocean chemistry have to rely on other proxies. Based on stable boron isotopes in foraminifera, glacial pCO_2 levels before the mid-Pleistocene Transition (MPT; ~1 Myr) were estimated to have been about 30 µatm higher than after the transition. Estimates of pre-MPT interglacial values appear similar to those obtained from ice cores during the late Pleistocene (Hönisch *et al.* 2009). Note that stable boron isotopes are actually a proxy for seawater pH and that one other CO_2 system parameter is required to reconstruct atmospheric CO_2. Regardless, the boron isotope data indicate that surface-ocean pH_{SWS} over the past 2 Myr has varied periodically between ~8.1 and ~8.3 (Hönisch *et al.* 2009). So far, no major excursions or ocean acidification events have been identified during the Pleistocene. Briefly, the available Pleistocene data indicate periodic variations in ocean carbonate chemistry during the past 2 Myr. These variations are part of the natural glacial–interglacial climate cycle and are restricted within remarkably stable lower and upper limits (between ~180 µatm and ~300 µatm for pCO_2 and between ~8.1 and ~8.3 for surface-ocean pH_T).

A few estimates of surface-ocean pH_T and atmospheric CO_2 are available for the Pliocene epoch (Bartoli *et al.* pers. comm.; Pagani *et al.* 2010). Stable boron isotopes indicate variations in surface-ocean pH_T between ~8.0 and ~8.3 and a gradual pCO_2 decline from 4.5 Myr to 2 Myr with extreme values ranging between ~200 and ~400 µatm (Bartoli *et al.* pers. comm.). Over the same time interval, alkenone data suggest a similar pCO_2 decline with extreme pCO_2 values ranging between ~200 and ~525 µatm (Pagani *et al.* 2010). Alkenone-based pCO_2 estimates derive from records of the carbon isotope fractionation that occurred during marine photosynthetic carbon fixation in the past. Several lines of evidence suggest that the carbon isotope fractionation depends on CO_2 levels (e.g. Pagani *et al.* 2010). Note that these reconstructions have large uncertainties. Nevertheless, taking the results at face value, one may estimate the maximum change in the surface-ocean saturation state of calcite (Ω_c) over the past 4 Myr. The cold periods may be characterized by pH_T = 8.3 and pCO_2 = 200 µatm, which yields Ω_c = 6.1 (T = 15°C, S = 35). The warm Pliocene periods (~4°C warmer than pre-industrial) may be characterized by pH_T = 8.0 and pCO_2 = 525 µatm, which yields Ω_c = 4.6 (T = 19°C, S = 35). By and large, the combined evidence for the Pliocene and Pleistocene suggests that over the past 4 Myr, ocean carbonate chemistry has experienced relatively slow changes on timescales >10 000 yr, with atmospheric CO_2 varying roughly between 200 ppmv and 500 ppmv.

2.4.4 The Cenozoic and beyond

A common approach to reconstructing ocean chemistry over the Cenozoic (the past ~65 Myr) is based on estimates of past atmospheric CO_2 concentrations and the carbonate mineral saturation state of the ocean (e.g. Sundquist 1986; Broecker and Sanyal 1998; Zeebe 2001). Deep-sea sediment cores reveal that the long-term steady-state position of the calcite compensation depth (CCD) over the past 100 to 150 Myr did not vary dramatically; it rather gradually deepened slightly toward the present (for a summary see Tyrrell and Zeebe 2004). This suggests a more or less constant carbonate mineral saturation state of the ocean over the Cenozoic, except for the Eocene–Oligocene transition (~34 Myr) when the CCD rapidly deepened permanently by several hundred metres. A recent study indicates a more dynamic CCD on shorter timescales, for instance during the Eocene in the Equatorial Pacific (Pälike *et al.* pers. comm.). Nevertheless, on long timescales, the carbonate chemistry of the ocean over the Cenozoic may be reconstructed based on saturation state estimates and palaeo-pCO_2 reconstructions (e.g. Tyrrell and Zeebe 2004; Ridgwell 2005; Goodwin *et al.* 2009; Stuecker and Zeebe 2010).

The details of the reconstructions can vary substantially, mostly depending on the different palaeo-pCO_2 estimates. However, several trends appear to be robust. Atmospheric CO_2 concentrations were higher during the early Cenozoic and have declined from a few thousand ppmv to 200 to 300 ppmv during the late Pleistocene (Fig. 2.3). While the surface-ocean saturation state was nearly constant over this period of time, surface-ocean pH_T was lower during the early Cenozoic (perhaps ~7.6) and has gradually increased to its modern value of about 8.2 (see Fig. 2.3 and Tyrrell and Zeebe 2004; Ridgwell and Zeebe 2005). Note that these are long-term trends which do not resolve possible large short-term variations during ocean acidification events such as the PETM (see below). Note also that low surface-ocean pH in the past during multimillion year periods of, for instance, the Palaeocene or Cretaceous, are no analogues for the centuries to come because of different seawater carbonate mineral saturation states (Section 2.4.5).

One could ask whether or not the long-term trends in ocean carbonate chemistry throughout the Cenozoic had an effect on the evolution of marine calcifying organisms. Note, however, that if species evolution was sensitive to carbonate saturation state, little effect is to be expected because saturation state appears to have been nearly constant over the Cenozoic. Regarding coccolithophores, a trend

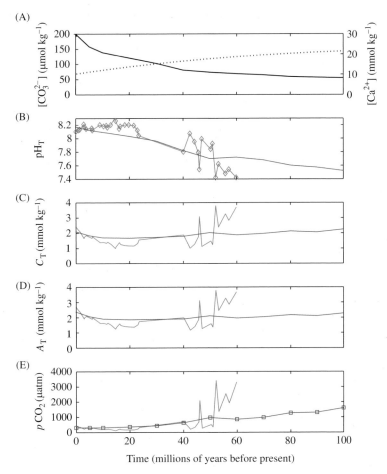

Figure 2.3 Carbonate chemistry reconstruction of surface seawater over the Cenozoic based on two different parameter combinations: (1) $[CO_3^{2-}]$ and pCO_2 (blue lines) and (2) $[CO_3^{2-}]$ and pH_T (green lines); see Tyrrell and Zeebe 2004. (A) Surface-ocean concentrations of CO_3^{2-} (solid line, derived from saturation state indicators) and Ca^{2+} (dotted line, from fluid inclusions). (B) pH_T based on combination 1 (blue line) and from stable boron isotopes (green diamonds, Pearson and Palmer 2000). (C) Total dissolved inorganic carbon. (D) Total alkalinity. (E) pCO_2 from GEOCARB-III (blue squares, Berner and Kothavala 2001) and based on combination 2 (green line).

toward smaller cell sizes in the Oligocene (~34 to ~23 Myr) relative to the Eocene (~55 to ~34 Myr) has been suggested based on deep-sea sediment records (Henderiks and Pagani 2008). These authors speculated that the size trend in the haptophyte algae may reflect a response to increased CO_2 limitation associated with the decline in atmospheric CO_2 across the Eocene–Oligocene transition. If so, this would represent a CO_2-related effect on photosynthesis and cell growth rather than on calcification. Regarding planktonic foraminifera, a trend toward larger test sizes in low-latitude species has been reported, particularly since the end of the Miocene (Schmidt *et al.* 2004). The

authors suggested that the size increase was a response to intensified surface-water stratification at low latitudes. Changes in Cenozoic carbonate chemistry appear unlikely to have caused the increase in shell size in foraminifera, emphasizing the importance of other functional correlations with shell size. For example, surface-water changes in CO_2 at the end of the Miocene (~23 to ~5 Myr) seem rather small compared with those during the Palaeogene. Thus, if dissolved CO_2 or pH would have been important factors in determining shell size in planktonic foraminifera, one would expect large changes during the Palaeogene (~65 to ~23 Myr), but this is not the case.

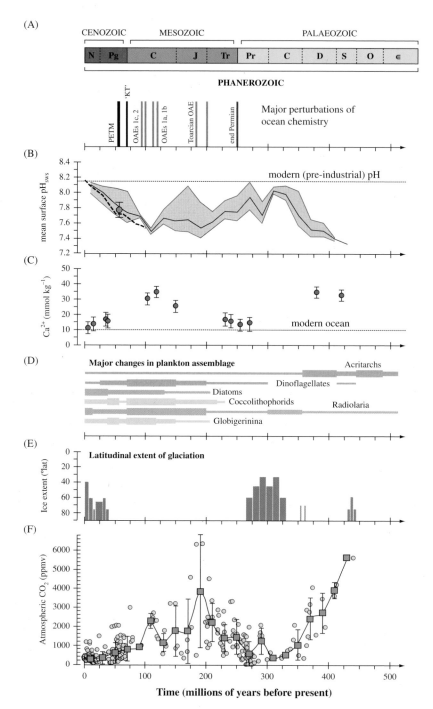

Figure 2.4 The geological context for past changes in ocean carbonate chemistry (modified after Kump *et al.* 2009). (A) Major global carbon cycle events during the past ~500 Myr. (B) Evolution of ocean surface pH$_{sws}$ (Tyrrell and Zeebe 2004; Ridgwell 2005; Ridgwell and Zeebe 2005, Ridgwell and Schmidt 2010). Black line: response of the global carbonate cycle to the mean palaeo-pCO$_2$ reconstruction; grey-filled envelope: response to the uncertainty (one standard deviation) in palaeo-pCO$_2$; orange-filled circle: estimates for the late Palaeocene. (C) Reconstructed Ca^{2+} concentrations (Lowenstein *et al.* 2001). (D) Major changes in plankton assemblages (Martin 1995). Calcifying taxa are highlighted in yellow with non-calcifying taxa shown in grey and blue. The rise during the early- to mid-Mesozoic of the importance of *Globigerinina* is shown as broadly representative of the timing of changes of planktic foraminiferal taxa in general, although the evolution of the first foraminifera taxon occurred somewhat earlier in the mid-Paleozoic (Martin 1995). (E) Latitudinal extent of glaciation (Crowley and Burke 1998). (F) Phanerozoic evolution of atmospheric pCO$_2$ reconstructed from proxy records (yellow-filled circles) by Royer *et al.* (2004). Palaeo-pCO$_2$ data have been binned into 20 Myr intervals, with the mean indicated by green squares and one standard deviation indicated by error bars. The geological timescale is delineated at the top.

Finally, one could ask whether long-term changes in ocean carbonate chemistry could have influenced, for instance, the distribution of coral reefs throughout the Cenozoic. Based on geological evidence in the continental United States, Opdyke and Wilkinson (1993) suggested a roughly 10° latitudinal reduction in areal extent of reefal/oolitic carbonate accumulation between the Cretaceous and the Holocene, with a gradual decrease toward the present. These authors mostly focused on carbonate mineral saturation state and sea-surface temperature as environmental parameters that control the latitudinal extent of reefs. In contrast, based on an extensive dataset of palaeolatitudinal distribution of reef sites, Kiessling (2001) concluded that neither the width of the tropical reef zone nor the total latitudinal range of reefs is correlated with palaeotemperature estimates. He inferred a fairly wide reef zone during the Cretaceous and early Palaeogene and an exceptionally wide tropical reef zone in the late Palaeocene and Eocene, relative to the modern latitudinal boundaries. The bottom line is that these studies do not seem to indicate any obvious relationships between the distribution of reefs and changes in, for example, seawater CO_2 or pH over the Cenozoic (see Fig. 2.3). As mentioned above, no significant relationships are in fact to be expected because the long-term saturation state of the ocean appears to have been nearly constant throughout the Cenozoic. In contrast, rapid short-term ocean acidification events such as the Palaeocene–Eocene Thermal Maximum (~55 Myr) have been identified as the cause for ancient reef crises (e.g. Kiessling and Simpson 2011).

Long-term changes in ocean carbonate chemistry over the past 500 Myr are summarized in Fig. 2.4 (see Ridgwell 2005; Ridgwell and Zeebe 2005).

2.4.5 Comparisons between the Cretaceous and the near future are invalid

As mentioned above, comparisons between the Cretaceous and the near future are frequently made to suggest that marine calcification will not be impaired in a future high-CO_2 world. The evidence cited for this is usually based on the occurrence of massive carbonate deposits during the Cretaceous such as the White Cliffs of Dover—carbonate formations that consist of coccolithophore calcite. Given the basics of carbon cycling and controls on seawater carbonate chemistry as reviewed above, it is obvious that such comparisons are invalid (see also Zeebe and Westbroek 2003; Ridgwell and Schmidt 2010). This applies not only to the Cretaceous, but in general to long-term, high-CO_2 steady states in the past. Briefly, because two CO_2 system parameters are required to set the carbonate chemistry, similar CO_2 concentrations do not imply similar carbonate chemistry conditions because of differences in, for example, saturation state (Section 2.2.2). The anthropogenic perturbation represents a transient event with massive carbon release over a few hundred years. In contrast, the Cretaceous, for instance, represents a long-term steady-state interval over millions of years. As a result, the timescales involved (centuries versus millions of years), reservoir sizes (a few thousand Pg C versus 10^8 Pg C), and controls on carbonate chemistry are fundamentally different (Section 2.3).

Ocean carbonate saturation state is generally well regulated by the requirement that on 'long' (>10 000 yr) timescales, sources (weathering) and sinks (shallow- and deep-water $CaCO_3$ burial) must balance (Ridgwell and Schmidt 2010). In contrast, as pH reflects the balance between dissolved CO_2 and carbonate ion concentration, it is governed primarily by pCO_2 (controlling CO_2 for given temperature) and Ca^{2+}/Mg^{2+} (controlling CO_3^{2-} for given Ω) rather than weathering. It follows, for instance, that there was no late Mesozoic carbonate crisis because Ω was probably high and decoupled from pH. Only events involving geologically 'rapid' (<10 000 yr) CO_2 release will overwhelm the ability of the ocean and sediments to regulate Ω, producing a coupled decline in both pH and saturation state, and hence providing an ocean acidification analogue relevant to the future.

2.5 Ocean acidification events in earth's history

We will use the term 'ocean acidification events' to describe episodes in earth's history that involve geologically 'rapid' changes of ocean carbonate chemistry on timescales shorter than ~10 000 years. In the following, we will limit our discussion to a few episodes that appear most relevant to us in relation to the ongoing acidification event caused by humans. Our list is not comprehensive, and there may be other events in earth's history that deserve more attention in the context of ocean acidification (see, e.g., Kump *et al.* 2009 and Chapter 4).

2.5.1 Aptian Oceanic Anoxic Event

One possible example of ocean acidification is the Aptian Oceanic Anoxic Event (OAE1a; ~120 Myr), an interval characterized by the widespread deposition of organic-rich sediments. It has been suggested that a marine calcification crisis occurred during OAE1a (e.g. Erba and Tremolada 2004). However, rather than being of a transient nature (e.g. showing a decay pattern after an initial perturbation), the event was long-lasting, with a total duration of ~1 Myr. The timescale of its onset has been estimated at ~20 to 44 kyr (Li *et al.* 2008; Mehay *et al.* 2009), and was most likely slower than the onset of the PETM, for example. Also, the substantial decline in nanoconid abundance (calcareous nanoplankton, proposed as an indicator of the calcification crisis) had already started ~1 Myr prior to the onset of the event (Erba and Tremolada 2004; Mehay *et al.* 2009). As pointed out in Section 2.4.5, the ocean carbonate saturation state is generally well buffered on timescales >10 000 yr, which makes it improbable that effects on calcification would have lasted over millions of years. This view is supported by the fact that other heavily calcified taxa peaked in abundance precisely during the interval of minimum nanoconid abundance (Erba and Tremolada 2004). Some species such as the coccolithophore *Watznaueria barnesiae* show little change in abundance during the onset of the event (Mehay *et al.* 2009).

2.5.2 End-Permian and K/T boundary

An excellent summary of the end-Permian mass extinction (~252 Myr) is provided by Knoll and Fischer in Chapter 4 and shall not be repeated here (see also Knoll 2003). The Cretaceous–Tertiary boundary (~65 Myr) represents another possible ocean acidification event, probably involving sulphur compounds which acidified the surface ocean. Undoubtedly, other dramatic perturbations

accompanied this event that affected marine life (see Caldeira 2007 for more details).

2.5.3 Palaeocene–Eocene Thermal Maximum

The PETM (~55 Myr) appears to be the closest analogue for the future that has so far been identified in the geological record. The onset of the PETM was marked by a global increase in surface temperatures by 5 to 9°C within a few thousand years (e.g. Kennett and Stott 1991; Thomas and Shackleton 1996; Zachos *et al.* 2003; Sluijs *et al.* 2006). At nearly the same time, a substantial carbon release occurred, as evidenced by a large drop in the $^{13}C/^{12}C$ ratio of surficial carbon reservoirs. The carbon release led to ocean acidification and widespread dissolution of deep-sea carbonates (e.g. Zachos *et al.* 2005; Zeebe *et al.* 2009; Ridgwell and Schmidt 2010). Different sources for the carbon input have been suggested, which has led to speculations concerning the mechanism. Some, such as volcanic intrusion, imply that the carbon drove the warming. Others, such as the destabilization of oceanic methane hydrates, imply that the carbon release is a feedback that can exacerbate warming (Dickens *et al.* 1995; Dickens 2000; Pagani *et al.* 2006). Note that regarding impacts on ocean acidification, it is of minor importance whether the carbon source was in the form of CO_2 or methane, as methane would have been oxidized rapidly to CO_2 in the water column and/or the atmosphere. Remarkably, even the lower estimates for the carbon release during the onset of the PETM (~1 Pg C yr^{-1}) and over the past 50 years from anthropogenic sources appear to be of similar order of magnitude.

The PETM exhibits several characteristics that are essential for a meaningful comparison with the anthropogenic perturbation: (1) it was a transient event with a rapid onset (not a long-term steady state); (2) it was associated with a large and rapid carbon input. Also, in contrast to aberrations that occurred in the more distant past, the PETM is relatively well studied because a number of well-preserved terrestrial and marine palaeorecords for this time interval are available (on the marine side accessible through ocean drilling). The PETM may therefore serve as a case study for ocean acidification caused by CO_2 released by human activities. However, it is important to keep in mind that the climatic and carbon cycle boundary conditions before the PETM were different from today's—including a different continental configuration, absence of continental ice, and a different base climate. Moreover, ocean carbonate chemistry prior to the event was different from modern conditions and the sensitivity to carbon perturbation was probably reduced (Goodwin *et al.* 2009; Stuecker and Zeebe 2010). These aspects limit the suitability of the PETM as a perfect future analogue. Nevertheless, the PETM provides invaluable information on the response of the carbon cycle, climate, and ocean carbonate chemistry to massive carbon input in the past. It also allows us to estimate the timescale over which carbon was removed from the ocean–atmosphere system by natural sequestration.

Different carbon input scenarios have been proposed for the PETM (e.g. Dickens *et al.* 1995; Panchuk *et al.* 2008; Zeebe *et al.* 2009). For example, the scenario proposed by Zeebe *et al.* (2009) requires an initial carbon pulse of about 3000 Pg C over ~6000 yr in order to be consistent with the timing and magnitude of stable carbon isotope records and deep-sea dissolution patterns. We have compared this PETM scenario

with a business-as-usual scenario of fossil fuel emissions of 5000 Pg C over ~500 yr (Fig. 2.5). Our results show that if the proposed PETM scenario roughly resembles the actual conditions during the onset of the event, then the effects on ocean chemistry, including surface-ocean saturation state, were less severe during the PETM than expected for the future (Zeebe and Zachos, pers. comm. 2007; Ridgwell and Schmidt 2010). As pointed out in Section 2.3, not only the magnitude but also the timescale of the carbon input is critical for its effect on ocean carbonate chemistry. The timescale of the anthropogenic carbon input is so short that the natural capacity of the surface reservoirs to absorb carbon is overwhelmed (Fig. 2.1A). As a result of a 5000 Pg C input over ~500 yr, the surface-ocean saturation state of calcite (Ω_c) would drop from about 5.4 to less than 2 within a few hundred years. In contrast, the PETM scenario suggests a corresponding decline of Ω_c from 5.5 to only about 4 within a few thousand years. We emphasize, however, that the PETM scenario may be subject to revision, depending on the outcome of future studies that will help to better constrain the timescale of the carbon input.

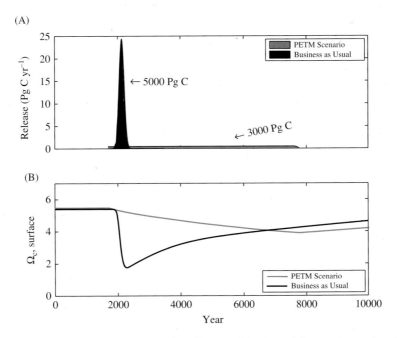

Figure 2.5   Palaeocene–Eocene Thermal Maximum (PETM) versus the Anthropocene. (A) Carbon emission scenarios as projected for the future (5000 Pg C over ~500 yr; Zeebe *et al.* 2008) and the PETM (3000 Pg C over 6 kyr; Zeebe *et al.* 2009). The onset of the PETM has been aligned with the onset of industrialization. (B) Changes in surface-ocean saturation state of calcite simulated with the LOSCAR model in response to the carbon input shown in (A).

Interestingly, our suggestion that the PETM carbon input had a moderate impact on surface-ocean saturation state is consistent with the results of Gibbs *et al.* (2006), who studied the origination and extinction of nanoplankton during the PETM. They concluded that the perturbation of the surface-water saturation state across the PETM was not detrimental to the survival of most calcareous nanoplankton taxa. In contrast, the Palaeocene–Eocene boundary marks a major extinction event of benthic foraminifera, affecting 30 to 50% of species globally (e.g. Thomas 2007). It is not clear, however, whether the benthic extinction was caused by changes in oxygenation, bottom-water temperatures, carbonate undersaturation as a result of the carbon input, and/or other factors (for discussion see Ridgwell and Schmidt 2010). In summary, the direct effects of ocean acidification on surface calcifiers during the PETM may have been limited because of a relatively 'slow' carbon input rate (slow on human timescales, rapid on geological timescales). Possible acidification effects on benthic organisms are as yet difficult to quantify because of competing effects from other environmental changes (see also Chapter 4).

As mentioned above, among the ocean acidification events hitherto identified in earth's history, the PETM may be the closest analogue for the future. Yet the evidence suggests that the rate of carbon input from human activities may exceed the rate of carbon input during the PETM. Thus, at present it seems that the ocean acidification event humans may cause over the next few centuries is unprecedented in the geological past for which sufficiently well-preserved palaeorecords are available.

2.6 Conclusions

Fossil fuel burning and the resulting input of CO_2 to the atmosphere has been referred to in the literature as a global 'geophysical experiment'. By definition, an experiment is 'An operation carried out under controlled conditions in order to discover an unknown effect or law, to test or establish a hypothesis, or to illustrate a known law'. In fact, none of this applies to fossil fuel burning. Conducting a

successful scientific experiment—in the sense of the above definition—requires purposeful, and in most cases clever and careful planning, design, and execution of the experiment. Moreover, the operator is usually able to terminate the experiment at any time if so desired (except in the case of an ill-designed experiment). The fact that humans are emitting CO_2 is a consequence of a fossil fuel-based economy. It seems that the only resemblance to an experiment is that the outcome is as yet unknown.

In this regard, the geological record can provide valuable information about the response of the earth system to massive and rapid carbon input, which should ultimately lead to improved future predictions. In particular, studies of past changes in ocean chemistry teach us a lesson about the effects that ocean acidification may have on marine life in the future. In addition, they provide the necessary background for assessing the current anthropogenic perturbation in the context of earth's history. Our assessment shows that when studying the past, a good understanding of the relevant timescales involved is of utmost importance. For instance, short-term carbon cycling on a timescale of 10 to 100 yr and long-term carbon cycling on a timescale of millions of years involve two distinct cycles with vastly different reservoir sizes and different sets of controls on atmospheric CO_2 and ocean chemistry. Thus, the pertinent timescales of palaeo-pCO_2 and palaeochemistry records require thorough examination to qualify as appropriate future analogues.

Our survey of long-term changes of ocean carbonate chemistry during earth's history (quasi-steady states) revealed that natural variations are generally slow and small on timescales relevant to the near future (see Fig. 2.6 for a summary). Because the ocean saturation state is usually well regulated and decoupled from pH over tens of thousands of years, past events that involve geologically 'rapid' changes of ocean carbonate chemistry are of particular interest. Among the ocean acidification events discussed here, the PETM may be the closest analogue for the future. However, the anthropogenic rate of carbon input appears to be greater than during any of the ocean acidification events identified so far, including the PETM.

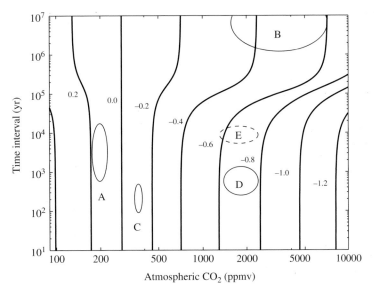

Figure 2.6   Estimated maximum change in surface-ocean pH (labelled contour lines) as a function of final atmospheric CO_2 pressure, and the transition time over which this CO_2 pressure is linearly approached from 280 ppmv (modified after Caldeira and Wickett 2003). A, Glacial–interglacial CO_2 changes; B, slow changes over the past 300 Myr; C, historical changes in ocean surface waters; D, unabated fossil-fuel burning over the next few centuries. E, The range of the timescale of carbon input and pCO_2 estimates during the Palaeocene–Eocene Thermal Maximum (PETM) (which was probably approached from a CO_2 level significantly higher than 280 ppmv). Note that surface-ocean pH changes were probably much smaller during the PETM than suggested in the figure because of lower sensitivity to carbon perturbations and substantially higher initial pCO_2 (e.g. Zeebe *et al.* 2009; Stuecker and Zeebe 2010).

2.7   Acknowledgements

We thank Jelle Bijma, Ken Caldeira, Jean-Pierre Gattuso, Lina Hansson, and Andy Knoll for comments on an earlier version of the manuscript. This work was supported by NSF grants OCE09–02869 and OCE09–27089 to REZ. It is a contribution to the 'European Project on Ocean Acidification' (EPOCA) which received funding from the European Community's Seventh Framework Programme (FP7/2007–2013) under grant agreement no. 211384.

References

Archer, D. (2005). Fate of fossil fuel CO_2 in geologic time. *Journal of Geophysical Research*, **110**, C09S05, doi:10.1029/2004JC002625.

Archer, D., Winguth, A., Lea, D.W., and Mahowald, N. (2000). What caused the glacial/interglacial atmospheric pCO_2 cycles? *Reviews of Geophysics*, **38**, 159–89.

Barker, S. and Elderfield, H. (2002). Foraminiferal calcification response to glacial–interglacial changes in atmospheric CO_2. *Science*, **297**, 833–6.

Beaufort, L., Probert, I., and Buchet, N. (2007). Effects of acidification and primary production on coccolith weight: implications for carbonate transfer from the surface to the deep ocean. *Geochemistry, Geophysics, Geosystems*, **8**, Q08011, doi:10.1029/2006GC001493.

Berner, R.A. and Kothavala, Z. (2001). GEOCARB III: a revised model of atmospheric CO_2 over Phanerozoic time. *American Journal of Science*, **304**, 397–437.

Berner, R.A., Lasaga, A.C., and Garrels, R.M. (1983). The carbonate-silicate geochemical cycle and its effect on atmospheric carbon dioxide over the past 100 million years. *American Journal of Science*, **283**, 641–83.

Bijma, J., Hönisch, B., and Zeebe, R.E. (2002). Impact of the ocean carbonate chemistry on living foraminiferal shell weight: Comment on 'Carbonate ion concentration in glacial-age deep waters of the Caribbean Sea' by W. S. Broecker and E. Clark. *Geochemistry, Geophysics, Geosystems*, **3**, 1064, doi:10.1029/2002GC000388.

Broecker, W.S. and Clark, E. (2003). Glacial-age deep sea carbonate ion concentrations. *Geochemistry, Geophysics, Geosystems*, **4**, 1047, doi:10.1029/2003GC000506.

Broecker, W. S. and Clark, E. (2007). Is the magnitude of the carbonate ion decrease in the abyssal ocean over the last 8 kyr consistent with the 20 ppm rise in atmospheric

CO$_2$ content? *Paleoceanography*, **22**, PA1202, doi:10.1029/2006PA001311.

Broecker, W.S. and Peng, T.-H. (1989). The cause of the glacial to interglacial atmospheric CO$_2$ change: a polar alkalinity hypothesis. *Global Biogeochemical Cycles*, **3**, 215–39.

Broecker, W. S. and Sanyal, A. (1998). Does atmospheric CO$_2$ police the rate of chemical weathering? *Global Biogeochemical Cycles*, **12**, 403–8.

Caldeira, K. (1992). Enhanced Cenozoic chemical weathering and the subduction of pelagic carbonate. *Nature*, **357**, 578–81.

Caldeira, K. (2007). What corals are dying to tell us about CO$_2$ and ocean acidification. *Oceanography*, **20**, 188–95.

Caldeira, K. and Wickett, M.E. (2003) Anthropogenic carbon and ocean pH. *Nature*, **425**, 365.

Crowley, T.J. and Burke, K.C. (1998). *Tectonic boundary conditions for climate reconstructions*, 285 pp. Oxford University Press, Oxford.

Dickens, G.R. (2000). Methane oxidation during the late Palaeocene Thermal Maximum. *Bulletin de la Societé Géologique de France*, **171**, 37–49.

Dickens, G.R., O'Neil, J.R., Rea, D.K., and Owen, R.M. (1995). Dissociation of oceanic methane hydrate as a cause of the carbon isotope excursion at the end of the Paleocene. *Paleoceanography*, **10**, 965–71.

Dore J.E., Lukas R., Sadler D.W., Church M.J., and Karl D.M. (2009). Physical and biogeochemical modulation of ocean acidification in the central North Pacific. *Proceedings of the National Academy of Sciences*, **106**, 12235–40.

Elsig, J., Schmitt, J., Leuenberger, D. *et al.* (2009). Stable isotope constraints on Holocene carbon cycle changes from an Antarctic ice core. *Nature*, **461**, 507–10.

Erba, E. and Tremolada, F. (2004). Nannofossil carbonate fluxes during the Early Cretaceous: phytoplankton response to nutrification episodes, atmospheric CO$_2$, and anoxia. *Paleoceanography*, **19**, PA1008, doi:10.1029/2003PA000884.

Foster, G.L. (2008). Seawater pH, pCO$_2$ and [CO$_3^{2-}$] variations in the Caribbean Sea over the last 130 kyr: a boron isotope and B/Ca study of planktic foraminifera. *Earth and Planetary Science Letters*, **271**, 254–66.

Gibbs, S.J., Bown, P.R., Sessa, J.A., Bralower, T., and Wilson, P. (2006). Nannoplankton extinction and origination across the Paleocene–Eocene Thermal Maximum. *Science*, **314**, 1770–3.

Goodwin, P., Williams, R.G., Ridgwell, A., and Follows, M.J. (2009). Climate sensitivity to the carbon cycle modulated by past and future changes in ocean chemistry. *Nature Geoscience*, **2**, 145–50.

Henderiks, J. and Pagani, M. (2008). Coccolithophore cell size and the Paleogene decline in atmospheric CO$_2$. *Earth and Planetary Science Letters*, **269**, 575–83.

Hester, K.C., Peltzer, E.T., Kirkwood, W.J., and Brewer, P.G. (2008). Unanticipated consequences of ocean acidification: a noisier ocean at lower pH. *Geophysical Research Letters*, **35**, L19601, doi:10.1029/2008GL034913.

Hönisch, B. and Hemming, N.G. (2005). Surface ocean pH response to variations in pCO$_2$ through two full glacial cycles. *Earth and Planetary Science Letters*, **236**, 305–14.

Hönisch, B., Hemming, N.G., Archer, D., Siddall, M., and McManus, J.F. (2009). Atmospheric carbon dioxide concentration across the mid-Pleistocene transition. *Science*, **324**, 1551–4.

Ilyina, T., Zeebe, R.E., and Brewer, P.G. (2010). Future ocean increasingly transparent to low-frequency sound owing to carbon dioxide emissions. *Nature Geoscience*, **3**, 18–22.

Kennett, J.P. and Stott, L.D. (1991). Abrupt deep-sea warming, palaeoceanographic changes and benthic extinctions at the end of the Palaeocene. *Nature*, **353**, 225–9.

Kiessling, W. (2001). Paleoclimatic significance of Phanerozoic reefs. *Geology*, **29**, 751–4.

Kiessling, W. and Simpson, C. (2011). On the potential for ocean acidification to be a general cause of ancient reef crises. *Global Change Biology*, **17**, 56–67.

Knoll, A.H. (2003). Biomineralization and evolutionary history. *Reviews in Mineralogy and Geochemistry*, **54**, 329–56.

Kohfeld, K.E., and Ridgwell, A. (2009). Glacial–interglacial variability in atmospheric CO$_2$. In: C. Le Quere and E.S. Saltzman (eds), *Surface ocean–lower atmosphere processes*, pp. 251–86. Geophysical Monograph 187. American Geophysical Union, Washington, DC.

Kump, L.R., Bralower, T.J., and Ridgwell, A. (2009). Ocean acidification in deep time. *Oceanography*, **22**, 94–107.

Li, Y.-X., Bralower, J.T., Montañez, I.P. *et al.* (2008). Toward an orbital chronology for the early Aptian Oceanic Anoxic Event (OAE1a, ~120 Ma). *Earth and Planetary Science Letters*, **271**, 88–100.

Lowenstein, T.K., Timofeeff, M.N., Brennan, S.T., and Hardie, L. A. (2001). Oscillations in Phanerozoic seawater chemistry: evidence from fluid inclusions. *Science*, **294**, 1086–8.

Lüthi, D., Le Floch, M., Bereiter, B. *et al.* (2008). High-resolution carbon dioxide concentration record 650,000-800,000 years before present. *Nature*, **453**, 379–82.

Martin, R.E. (1995). Cyclic and secular variation in microfossil biomineralization—clues to the biogeochemical evolution of Phanerozoic oceans. *Global and Planetary Change*, **11**, 1–23.

Mehay, S., Keller, C.E., Bernasconi, S.M. *et al.* (2009). A volcanic CO$_2$ pulse triggered the Cretaceous Oceanic Anoxic Event 1a and a biocalcification crisis. *Geology*, **37**, 819–22.

Millero, F.J., Woosley, R., DiTrolio, B., and Waters, J. (2009). Effect of ocean acidification on the speciation of metals in seawater. *Oceanography*, **22**, 72–85.

de Moel, H., Ganssen, G.M., Peeters, F.J.C., Jung, S.J.A., Brummer, G.J.A., Kroon, D., and Zeebe, R.E. (2009). Planktic foraminiferal shell thinning in the Arabian Sea due to anthropogenic ocean acidification? *Biogeosciences*, **6**, 1917–25.

Monnin E., Steig, E.J., Siegenthaler, U. *et al.* (2004). *EPICA Dome C ice core high-resolution Holocene and transition CO_2 data*. IGBP PAGES/World Data Center for Paleoclimatology, no. 2004-055. NOAA/NGDC, Boulder, CO.

Opdyke, B.N. and Wilkinson, B.H. (1993). Carbonate mineral saturation state and cratonic limestone accumulation. *American Journal of Science*, **293**, 217–34.

Pagani, M., Calderia, K., Archer, D., and Zachos, J.C. (2006). An ancient carbon mystery. *Science*, **314**, 1556–7.

Pagani, M., Liu, Z., LaRiviere, J., and Ravelo, A.C. (2010). High Earth-system climate sensitivity determined from Pliocene carbon dioxide concentrations. *Nature Geoscience*, **3**, 27–30.

Panchuk, K., Ridgwell, A., and Kump, L.R. (2008). Sedimentary response to Paleocene-Eocene Thermal Maximum carbon release: a model–data comparison. *Geology*, **36**, 315–18.

Pearson, P.N., and Palmer, M.R. (2000). Atmospheric carbon dioxide concentrations over the past 60 million years. *Nature*, **406**, 695–9.

Rickaby, R.E.M., Bard, E., Sonzogni, C. *et al.* (2007). Coccolith chemistry reveals secular variations in the global ocean carbon cycle? *Earth and Planetary Science Letters*, **253**, 83–95.

Ridgwell, A. (2005). A mid Mesozoic revolution in the regulation of ocean chemistry. *Marine Geology*, **217**, 339–57.

Ridgwell, A. and Schmidt, D. (2010). Past constraints on the vulnerability of marine calcifiers to massive carbon dioxide release. *Nature Geoscience*, **3**, 196–200.

Ridgwell, A. and Zeebe, R.E. (2005). The role of the global carbonate cycle in the regulation and evolution of the Earth system. *Earth and Planetary Science Letters*, **234**, 299–315.

Ridgwell, A.J., Kennedy, M.J., and Caldeira, K. (2003). Carbonate deposition, climate stability, and Neoproterozoic ice ages. *Science*, **302**, 859–62.

Royer, D.L., Berner, R.A., Montanez, I.P., Tabor, N.J., and Beerling, D.J. (2004). CO_2 as a primary driver of Phanerozoic climate change. *GSA Today*, **14**, 4–10.

Sanyal, A., Hemming, N.G., Hanson, G.N., and Broecker, W.S. (1995). Evidence for a higher *p*H in the glacial ocean from boron isotopes in foraminifera. *Nature*, **373**, 234–6.

Schmidt, D.N., Thierstein, H.R., Bollmann, J., and Schiebel, R. (2004). Abiotic forcing of plankton evolution in the Cenozoic. *Science*, **303**, 207–10.

Siegenthaler, U., Stocker, T.F., Monnin, E. *et al.* (2005). Stable carbon cycle-climate relationship during the Late Pleistocene. *Science*, **310**, 1313–17.

Sluijs, A., Schouten, S., Pagani, M. *et al.* (2006). Subtropical Arctic Ocean temperatures during the Palaeocene/Eocene thermal maximum. *Nature*, **441**, 610–13.

Stuecker, M.F. and Zeebe, R.E. (2010). Ocean chemistry and atmospheric CO_2 sensitivity to carbon perturbations throughout the Cenozoic. *Geophysical Research Letters*, **37**, L03609, doi:10.1029/2009GL041436.

Sundquist, E.T. (1986). Geologic analogs: their value and limitations in carbon dioxide research. In: J.R. Trabalka and D.E. Reichle (eds), *The changing carbon cycle: a global analysis*, pp. 371–402. Springer-Verlag, New York.

Tans, P. (2009). *Trends in atmospheric carbon dioxide*. NOAA/ESRL. Avaialable at: http://www.esrl.noaa.gov/gmd/ccgg/trends/

Thomas, E. (2007). Cenozoic mass extinctions in the deep sea: what disturbs the largest habitat on Earth? *Geological Society of America Special Papers*, **424**, 1–23.

Thomas, E. and Shackleton, N.J. (1996). The Paleocene–Eocene benthic foraminiferal extinction and stable isotope anomalies. *Geological Society of London Special Publication*, **101**, 401–41.

Tyrrell, T. and Zeebe, R.E. (2004). History of carbonate ion concentration over the last 100 million years. *Geochimica et Cosmochimica Acta*, **68**, 3521–30.

Uchikawa, J. and Zeebe, R.E. (2008). Influence of terrestrial weathering on ocean acidification and the next glacial inception. *Geophysical Research Letters*, **35**, L23608, doi:10.1029/2008GL035963.

Walker, J.C.G., Hays, P.B., and Kasting, J.F. (1981). Negative feedback mechanism for the long-term stabilization of earth's surface temperature. *Journal of Geophysical Research*, **86**, 9776–82.

Zachos, J.C., Wara, M.W., Bohaty, S. *et al.* (2003). A transient rise in tropical sea surface temperature during the Paleocene-Eocene Thermal Maximum. *Science*, **302**, 1551–4.

Zachos, J.C., Röhl, U., Schellenberg, S.T. *et al.* (2005). Rapid acidification of the ocean during the Paleocene-Eocene Thermal Maximum. *Science*, **308**, 1611–15.

Zachos, J.C., Dickens, G.R., and Zeebe, R.E. (2008). An early Cenozoic perspective on greenhouse warming and carbon-cycle dynamics. *Nature*, **451**, 279–83.

Zeebe, R.E. (2001). Seawater *p*H and isotopic paleotemperatures of Cretaceous oceans. *Palaeogeography, Palaeoclimatology, Palaeoecology*, **170**, 49–57.

Zeebe, R.E. and Caldeira, K. (2008). Close mass balance of long-term carbon fluxes from ice-core CO_2 and ocean chemistry records. *Nature Geoscience*, **1**, 312–15.

Zeebe, R.E. and Marchitto, T.M. (2010). Glacial cycles: atmosphere and ocean chemistry. *Nature Geoscience*, **3**, 386–7.

Zeebe, R.E. and Westbroek, P. (2003). A simple model for the CaCO$_3$ saturation state of the ocean: the 'Strangelove', the 'Neritan', and the 'Cretan' Ocean. *Geochemistry Geophysics Geosystems*, **4**, 1104, doi:10.1029/2003GC000538.

Zeebe, R.E. and Wolf-Gladrow, D.A. (2001). *CO$_2$ in seawater: equilibrium, kinetics, isotopes*. Elsevier, Amsterdam.

Zeebe, R.E., Zachos, J.C., Caldeira, K., and Tyrrell, T. (2008). Oceans: carbon emissions and acidification. *Science*, **321**, 51–2.

Zeebe, R.E., Zachos, J.C., and Dickens, G.R. (2009). Carbon dioxide forcing alone insufficient to explain Palaeocene-Eocene Thermal Maximum warming. *Nature Geoscience*, **2**, 576–80.

Zondervan, I., Zeebe, R.E., Rost, B., and Riebesell, U. (2001). Decreasing marine biogenic calcification: a negative feedback on rising atmospheric pCO$_2$. *Global Biogeochemical Cycles*, **15**, 507–16.

CHAPTER 3

Recent and future changes in ocean carbonate chemistry

James C. Orr

3.1 Introduction

This chapter is about the ongoing human-induced shifts in fundamental ocean carbonate chemistry that are occurring globally and are a growing concern to scientists studying marine organisms. It reviews the current state of ocean pH and related carbonate system variables, how they have changed during the industrial era, and how they are expected to continue to change during this century and beyond.

Surface-ocean pH has been relatively stable for millions of years, until recently. Over the 800 000 years prior to industrialization, average surface-water pH oscillated between 8.3 during cold periods (e.g. during the Last Glacial Maximum, 20 000 yr ago) and 8.2 during warm periods (e.g. just prior to the Industrial Revolution), as reviewed by Zeebe and Ridgwell in Chapter 2. But human activities are upsetting this stability by adding large quantities of a weak acid to the ocean at an ever increasing rate. This anthropogenic problem is referred to as ocean acidification because ocean acidity is increasing (i.e. seawater pH is declining), even though surface-ocean waters are alkaline and will remain so. The cause of the decline in seawater pH is the atmospheric increase in the same gas that is the main driver of climate change, namely carbon dioxide (CO_2).

Due to increasing atmospheric CO_2 concentrations, the ocean takes up large amounts of anthropogenic CO_2, currently at a rate of about 10^6 metric tons of CO_2 per hour (Brewer 2009), which is equivalent to one-fourth of the current global CO_2 emissions from combustion of fossil fuels, cement production, and deforestation (Canadell *et al.* 2007;

Le Quéré *et al.* 2009). If we would partition these emissions equally per capita, each person on the planet would be responsible for 4 kg per day of anthropogenic CO_2 invading the ocean. To grasp the size of the problem, this invisible invasion may be compared with a recent, highly visible environmental disaster. The ocean currently absorbs anthropogenic carbon at a rate that is about a thousand times greater than from when carbon escaped from the BP Deepwater Horizon oil well that exploded on 20 April 2010, releasing 57 000 barrels of petroleum per day into the Gulf of Mexico until it was capped almost 3 months later. Of course, the form of carbon released, the associated impacts, and the duration of the carbon release differ greatly. Anthropogenic ocean acidification is a chronic problem: it has been gradually increasing its intensity for two centuries, and it is continuing.

3.2 Basic chemistry under change

Ocean uptake of anthropogenic CO_2 helps limit the level of CO_2 in the atmosphere but it also changes the ocean's fundamental chemistry. That is, CO_2 is not only a greenhouse gas; it is also an acid gas. It reacts with seawater through a series of well-understood reactions (e.g. Revelle and Suess 1957; Broecker and Takahashi 1966; Stumm and Morgan 1970; Skirrow and Whitfield 1975; Andersen and Malahoff 1977). Like other atmospheric gases, CO_2 exchanges with its dissolved form in surface seawater, $CO_2(g) \leftrightarrow CO_2(aq)$. But CO_2 is exceptionally soluble because its aqueous form reacts with water to form carbonic acid, which dissociates producing hydrogen ions. Most of the additional hydrogen ions that are produced are neutralized when they react with carbonate ions, producing bicarbonate ions. The series of reactions is given by Zeebe and Gattuso in Box 1.1.

For simplicity, the net effect is often summarized as an acid–base neutralization,

$$H_2O + CO_2 + CO_3^{2-} \leftrightarrow 2HCO_3^-, \quad (3.1)$$

but the neutralization reaction is not complete. Excess acid remains because the product above, bicarbonate, also dissociates, producing hydrogen ions (Eq. B1.3 in Box 1.1). The additional hydrogen ions that do remain increase the H^+ concentration $[H^+]$ and lower pH, defined as $-\log_{10}[H^+]$. In summary, as CO_2 is added to seawater, there are increases in $[H^+]$ and bicarbonate ion concentration $[HCO_3^-]$ and simultaneous decreases in pH and carbonate ion concentration $[CO_3^{2-}]$.

Because the concentrations of CO_2, HCO_3^-, and CO_3^{2-} influence one another and are sensitive to changes in temperature, salinity, and pressure, it is fortunate for ocean scientists that the marine carbonate system can be defined in terms of two conservative tracers, namely total dissolved inorganic carbon (C_T) and total alkalinity (A_T), as defined in Box 1.1. Most of the total alkalinity comes from the carbonate alkalinity ($A_C = [HCO_3^-] + 2[CO_3^{2-}]$), and most of the remaining alkalinity comes from borate (Zeebe and Wolf-Gladrow 2001). Linear combinations of C_T and A_T are often used as convenient approximations for concentrations of the individual inorganic carbon species

$$[CO_3^{2-}] \approx A_T - C_T \quad (3.2)$$

$$[HCO_3^-] \approx 2C_T - A_T. \quad (3.3)$$

These approximations are usually good to within about 10% (Sarmiento and Gruber 2006). Inherent in these approximations is the assumption that CO_2 concentrations are relatively low and can be neglected, which works well for the modern ocean. However, in the future as atmospheric CO_2 increases and the $[CO_2]/[CO_3^{2-}]$ ratio approaches 1 at high latitudes (Orr et al. 2005), errors increase dramatically.

An important concept when studying ocean carbonate chemistry and calcification by marine organisms is the level of saturation of a water mass with respect to $CaCO_3$ minerals, defined in Box 1.1 in terms of the saturation state Ω. Another way to describe the level of saturation is by the difference Δ between the actual carbonate ion concentration and the critical carbonate ion concentration $[CO_3^{2-}]_{sat}$, i.e. the threshold below which $CaCO_3$ starts to dissolve:

$$\Delta[CO_3^{2-}] = [CO_3^{2-}] - [CO_3^{2-}]_{sat}. \quad (3.4)$$

Just as for Ω, values of $[CO_3^{2-}]_{sat}$ and $\Delta[CO_3^{2-}]$ differ for each $CaCO_3$ mineral. When $\Delta[CO_3^{2-}]$ is positive ($\Omega > 1$), waters are supersaturated with respect to that $CaCO_3$ mineral. When $\Delta[CO_3^{2-}]$ is negative ($\Omega < 1$), waters are undersaturated and corrosive to the same mineral. To convert between Ω and $\Delta[CO_3^{2-}]$, we only need to use $[CO_3^{2-}]_{sat}=[CO_3^{2-}]/\Omega$, which exploits the definition of the solubility product $K_{sp}=[Ca^{2+}]_{sat}[CO_3^{2-}]_{sat}$ and the 'identity' $[Ca^{2+}]_{sat} = [Ca^{2+}]$, where $[Ca^{2+}]$ is proportional to salinity By 'identity', it is just meant that salinity determines the open-ocean calcium concentration, which is used with K_{sp} to determine the corresponding $[CO_3^{2-}]_{sat}$. Thus $\Delta[CO_3^{2-}]=[CO_3^{2-}](1-1/\Omega)$. Although Ω has the advantage of being non-dimensional, $\Delta[CO_3^{2-}]$ carries the same concentration units as $[CO_3^{2-}]$ and $[CO_3^{2-}]_{sat}$ as well as the measured tracers C_T and A_T, from which it is often computed.

Along with changes in carbonate chemistry variables, it is useful to quantify the changing chemical capacity of the ocean to absorb increases in anthropogenic CO_2. One way to define that capacity is to use what oceanographers typically term the buffer capacity, the inverse of which is the Revelle factor R (Bolin and Eriksson 1959; Broecker et al. 1971; Keeling 1973; Pytkowicz and Small 1977; Sundquist et al. 1979; Wagener 1979; Takahashi et al. 1980), namely the ratio of the relative change in pCO_2 (or CO_2) to the relative change in C_T:

$$R = \frac{\partial pCO_2/pCO_2}{\partial C_T/C_T} = \frac{\partial \ln pCO_2}{\partial \ln C_T} \quad (3.5)$$

The Revelle factor is inversely related to $[CO_3^{2-}]$ (Broecker and Peng 1982) and is useful to help explain why the air–sea equilibration time for CO_2 is much longer than that for other gases such as oxygen (see Box 3.1).

Box 3.1 Future reductions in air–sea CO$_2$ equilibration times

It is known that air–sea equilibration requires many months for CO$_2$ but only a few weeks for most other gases. Here let us consider how air–sea CO$_2$ equilibration times vary regionally and with time. For most gases such as oxygen, a sudden change in gas concentration in the ocean mixed layer equilibrates with the atmosphere (and vice versa) following a simple e-folding time

$$\tau_{O_2} = \frac{z_m}{k_w} \qquad (B3.1)$$

where k_w is the gas transfer or 'piston' velocity (m day^{-1}) and z_m is the mixed layer depth (m). For CO$_2$ though, it is more complicated, because added anthropogenic CO$_2$ does not remain as dissolved gas but reacts with carbonate ions forming bicarbonate ions (see Box 1.1 and Eq. 3.1). As discussed by Broecker and Peng (1974), this reaction increases the equilibration time for CO$_2$. A rigorous development by Zeebe and Wolf-Gladrow (2001) shows that

$$\tau_{CO_2} = \frac{z_m}{k_w}\frac{\partial C_T}{\partial[CO_2]} = \frac{z_m}{k_w}\frac{C_T}{R[CO_2]} \qquad (B3.2)$$

where R is the Revelle factor. A similar development by Sarmiento and Gruber (2006) derived the approximation $\partial C_T/\partial[CO_2] \approx [CO_3^{2-}]/[CO_2]$, first used empirically by Broecker and Peng (1974). All three of these studies used $\partial C_T/\partial[CO_2] \approx 20$, and since τ_{CO_2}/τ_{O_2} has the same ratio (compare Eqs B3.1 and B3.2), air–sea equilibration is much longer for CO$_2$ than for most other gases. Thus a typical 50-m mixed layer equilibrates with atmospheric O$_2$ on a timescale of ~12 days, whereas it requires ~8 months for CO$_2$. But these are only averages for today's ocean.

Indeed the τ_{CO_2}/τ_{O_2} ratio varies regionally and declines as atmospheric CO$_2$ increases. Here, spatiotemporal differences were computed from climatologies for the mixed layer depth and the piston velocity (Fig. B3.1A) and evolving fields of carbonate chemistry variables (Orr et al. 2005). Based on GLODAP gridded data product from the WOCE-era carbon system measurements collected in the 1990s (Key et al. 2004), the τ_{CO_2}/τ_{O_2} ratio (i.e. $\partial C_T/\partial[CO_2]$) varies from 7 to 24, with the lowest values in the Southern Ocean and the highest in the tropics (Fig. B3.1B). Pre-industrial values of that ratio were ~20% higher than in 1994, as computed by subtracting data-based estimates for anthropogenic C_T and recalculating. In the future, the decline continues. Relative to the modern state, at 563 ppmv (2100S) τ_{CO_2}/τ_{O_2} drops by at least 40% everywhere (ranging from 3.6 to 14.2), and at 788 ppmv (2100I) it drops by more than 60% (ranging from 2.5 to 9), based on median changes projected by the OCMIP2 models. Scaling directly with this declining ratio is τ_{CO_2}.

Pre-industrial τ_{CO_2} ranged from 6 to 14 months between the Southern Ocean and the tropics (Fig. B3.1C). At 788 ppmv, the range drops to 1.5 to 4 months. Thus CO$_2$ equilibration times will become more like those of other gases, altering the amplitude and phasing of the annual cycle of surface-layer carbonate system variables, particularly for CO$_2$ and [CO$_3^{2-}$]. These future chemical reductions in τ_{CO_2} will be reinforced by reductions from physical changes driven by climate change, based on projected increased stratification (shallower mixed layers) and reduced sea-ice cover, changes that will reduce equilibration times of all gases.

continues

Box 3.1 *continued*

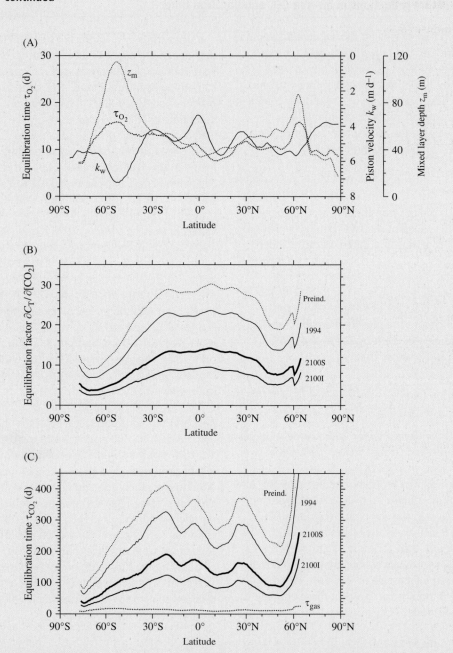

Figure B3.1 Zonal-mean distributions of (A) air–sea equilibration time for oxygen τ_{O_2} and its determining factors, mixed-layer depth z_m and piston velocity k_w, (B) the term $\partial C_T/\partial[CO_2]$ (equivalent to τ_{CO_2}/τ_{O_2}) for the pre-industrial era, the GLODAP central year 1994, 563 ppmv (scenario S650 in year 2100, 2100S), and 788 ppmv (scenario IS92a in 2100, 2100I), and (C) air–sea equilibration time for CO_2, τ_{CO_2}, at the same times. In the top panel, τ_{O_2} was determined from: (1) the global gridded climatologies for the mixed layer depth z_m (de Boyer Montégut *et al.* 2004, variable density criterion, updated with new profiles collected until September 2009) and (2) the gas transfer velocity k_w from OCMIP, including the Schmidt number temperature dependence $(Sc_{O_2}/660)^{-1/2}$ but not fractional sea-ice cover. In the bottom panel, τ_{gas} is the same as τ_{O_2} except that it uses Sc_{CO_2} instead of Sc_{O_2} (making it ~12% greater) and it covers only the GLODAP domain.

3.3   Atmospheric CO_2 emissions, sources, and sinks during the industrial era

Over the industrial era, human activities have released large quantities of CO_2 to the atmosphere. By 1994, the total atmospheric release of anthropogenic carbon amounted to 244 ± 20 Pg C from fossil-fuel combustion and cement production combined with 140 ± 40 Pg C from land-use change (Sabine *et al.* 2004; Denman *et al.* 2007). Out of that total, the atmosphere retained ~43% while the ocean absorbed ~30%. The remainder was taken up by terrestrial plants and soils. For the most recent decades, there is some evidence that the amount of anthropogenic CO_2 remaining in the atmosphere (airborne fraction) may have increased from 40 to 45% from 1959 to 2008 (Le Quéré *et al.* 2009), but the debate remains open (Knorr 2009). The ocean continues to take up a large fraction of the anthropogenic CO_2 emitted to the atmosphere, for example an average of 2.2 ± 0.5 Pg C yr^{-1} during 1990–2005, which is ~27% of the total emissions during that time (Denman *et al.* 2007).

In 2010, the atmospheric CO_2 level reached about 390 ppmv, which is 39% more than the pre-industrial concentration (280 ppmv). Half of that increase has occurred only since 1978. The effect of this large, rapid increase in atmospheric CO_2 is already causing measurable changes in ocean carbonate chemistry, both in the mixed layer, which equilibrates with the atmospheric perturbation on a timescale of roughly 8 months (see Box 3.1), and even in the deep ocean, the ventilation of which typically requires centuries.

3.4   Observed changes in ocean carbonate chemistry during recent decades

From basic marine carbonate chemistry, it is well known that as atmospheric CO_2 increases, surface-ocean pCO_2 will increase, reducing ocean pH and $[CO_3^{2-}]$ (see Section 3.2). Trends and variability in these ocean variables have been quantified and compared with corresponding changes in atmospheric CO_2 through dedicated long-term efforts to maintain three subtropical ocean time-series stations, where surface-ocean carbonate system varia-

bles have been measured with state-of-the-art precision for nearly three decades (Fig. 3.1). In the central North Pacific at station ALOHA of the Hawaii Ocean Time-Series (HOT) program (22.75°N, 158°W) C_T and A_T have been observed since 1989 and the computed surface pH_T (i.e. given on the total scale) exhibits a long-term decline of 0.0019 ± 0.0002 units yr^{-1} (Dore *et al.* 2009). The trend in direct measurements of surface pH_T, made over about half of the period, is not significantly different (Table 3.1). In the western portion of the North Atlantic gyre at the Bermuda Atlantic Time-Series Station (BATS; 31.72°N, 64.17°W), where C_T, A_T, and pCO_2 have been measured since 1983, the trend in the calculated pH_T decline is 0.0017 ± 0.0003 units yr^{-1} (Bates 2007). In the eastern North Atlantic at the European Time Series in the Canary Islands (ESTOC; 24.04°N, 15.50°W), where pH_T and A_T were measured from 1995 to 2004, the decline in surface *in situ* pH_T is 0.0018 ± 0.0003 units yr^{-1} (Santana-Casiano *et al.* 2007; González-Dávila *et al.* 2010). These trends in surface-ocean pH_T are not significantly different between stations. Nor do they differ statistically from the expected decline based on the observed trend in atmospheric CO_2 and the assumption of air–sea equilibrium, a supposition backed up by the similarity of measured trends in atmospheric and oceanic pCO_2 at these stations (Table 3.1). All three stations also exhibit reductions in $[CO_3^{2-}]$ (and thus in the saturation state of seawater with respect to aragonite and calcite, Ω_a and Ω_c respectively), although the reduction is about 80% greater at ESTOC.

There are also substantial subsurface trends in pH_T and other carbonate system variables based on measurements both at time-series stations and along repeated sections. At ALOHA, reductions in pH are significant down to depths of at least 600 m. The maximum 1988–2009 reduction in pH occurs not at the surface but at 250 m, which Dore *et al.* (2009) attribute to a greater increase in C_T. They suggest that this increase is due to subduction and lateral transport of the source waters, located at higher latitude, where the C_T increase has been larger or the location of which may have changed, thus altering its chemical characteristics (Revelle factor). At ESTOC, changes in pH_T and related variables have been measured down to well below 1000 m (González-Dávila *et al.* 2010).

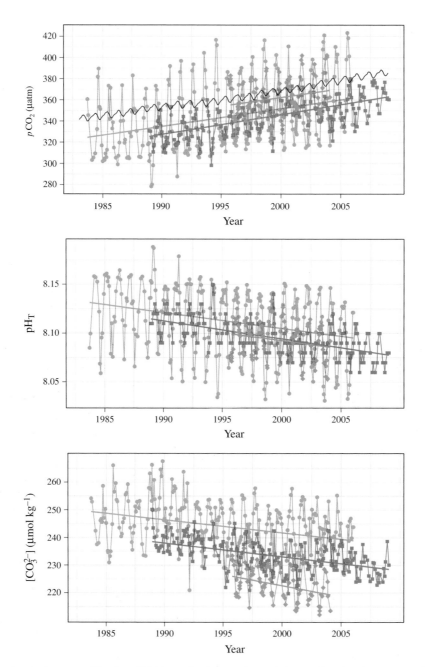

Figure 3.1   Time series of surface-ocean pCO_2, pH_T, and $[CO_3^{2-}]$, as well as for atmospheric mole fraction of CO_2 (xCO_2) at Mauna Loa (black), at three ocean time-series stations ALOHA (green), BATS (red), and ESTOC (blue).

Table 3.1 Published trends (slope ± SE) in atmospheric CO_2, surface pCO_2, pH_T, and $[CO_3^{2-}]$ at three time-series stations: BATS (1983–2005), ESTOC (1995–2004), and ALOHA (1988–2009)

Station	$xCO_{2_{atm}}$ (ppmv yr^{-1})	$pCO_{2_{sea}}$ (µatm yr^{-1})	pH_T (unit yr^{-1})	$[CO_3^{2-}]$ (µmol kg^{-1} yr^{-1})
BATS	1.78 ± 0.02[a]	1.67 ± 0.28[a]	−0.0017 ± 0.0003[a]	−0.47 ± 0.09[a]
	1.80 ± 0.02[b]	1.80 ± 0.13[b]	−0.0017 ± 0.0001[b]	−0.52 ± 0.02[b]
ESTOC	-	1.55 ± 0.43[c]	−0.0017 ± 0.0004[c]	−
	1.7 ± 0.7[d]	1.7 ± 0.7[d]	−0.0018 ± 0.0003[d]	−0.90 ± 0.08[d]
ALOHA	1.68 ± 0.03[e]	1.88 ± 0.16[e]	−0.0019 ± 0.0002[e]	−0.50 ± 0.06[e]
	-	-	−0.0014 ± 0.0002[f]	−

[a]Bates (2007, Table 1), simple linear fit.

[b]Bates (2007, Table 2), seasonally detrended.

[c]Santana-Casiano *et al.* (2007), seasonally detrended.

[d]Gonzalez-Davila *et al.* (2010).

[e]Dore *et al.* (2009), simple linear fit using calculated pH_T (full time series).

[f]Dore *et al.* (2009), simple linear fit using measured pH_T (partial time series).

With more spatial coverage but for only two points in time, ocean pH_T was measured directly on section P16N in the North Pacific, first during the World Ocean Circulation Experiment (WOCE) in 1991 and then again in 2006. During those 15 years, ocean pH_T changed by –0.06 units over the upper 500 m (Byrne *et al.* 2010). Roughly equal contributions were attributed to anthropogenic and non-anthropogenic factors, based on standard separation techniques relying on oxygen measurements. In the surface layer, the anthropogenic decline in pH_T was 0.0018 ± 0.0003 units yr^{-1}, consistent with the observed increase in atmospheric CO_2 as well as results from the three time-series stations.

In the higher latitudes, there is a time-series station in the Iceland Sea where seasonal measurements of C_T and pCO_2 have been made since 1985. The 1985–2008 wintertime trend in computed surface pH_T is 0.0024 units yr^{-1}, one-third greater than at the three lower-latitude time-series stations; simultaneously, surface Ω_a declined by 0.0072 units yr^{-1} while Ω_c declined by 0.0117 units yr^{-1} (Olafsson *et al.* 2009). The decline in pH_T below 1500 m in the Iceland Sea is one-quarter of that at the surface, while Ω_a declines at 0.0009 units yr^{-1}. The latter causes the aragonite saturation horizon (ASH), the interface between supersaturated waters above and undersaturated waters below, to move upward (shoal) at a rate of 4 m yr^{-1}. That shoaling exposes local seafloor previ-

ously covered with supersaturated waters to these newly corrosive conditions at a rate of 2 km^2 d^{-1}.

Although it is not possible to measure anthropogenic C_T directly, data-based techniques have been derived to distinguish anthropogenic C_T from the much larger natural background using measurements of carbon system variables, oxygen, nutrients, and transient tracers (e.g. Gruber *et al.* 1996; Sabine *et al.* 2004; Khatiwala *et al.* 2009). These estimates have been used to evaluate how surface and interior ocean chemistry have changed since the beginning of the industrial era (Feely *et al.* 2004; Orr *et al.* 2005), as detailed in subsequent sections.

3.5 Future scenarios

In 2001, for the Third Assessment Report (TAR) of the Intergovernmental Panel on Climate Change (IPCC), a family of future emissions scenarios was constructed and provided to the scientific community in the IPCC's Special Report on Emissions Scenarios (SRES; Nakićenović and Swart 2000). These SRES scenarios replaced the earlier IS92a family used for the previous IPCC report. They have allowed many different modelling groups to make consistent simulations, under different proposed lines of human behaviour, to project corresponding 21st century changes, and to compare model results within the framework of the IPCC's TAR and the subsequent Fourth Assessment Report (AR4), released in 2007.

For the SRES families of scenarios, less CO_2 is emitted in the more ecologically friendly B scenarios, with B1 being for a more integrated future world and B2 being associated with a more divided future world having higher emissions. More CO_2 is emitted in the non-ecologically friendly A family of scenarios, with A2 being a divided future world and A1 being more integrated. The latter set was further divided by weighting the different energy types: A1T (non-fossil-fuel emphasis), A1B ('balanced'), and A1FI (fossil-fuel intensive). Because modelling groups cannot usually afford to run all scenarios, they often pick a low scenario such as B1 and a high scenario such as A2 in order to bracket the others.

The IPCC has also used concentration scenarios or pathways to investigate processes under a pre-set stabilization goal for atmospheric greenhouse gas concentrations. Further discussion about these and the latest generation of scenarios that will be used in the next IPCC report is provided in Chapter 14.

3.6 Projecting future changes in carbonate chemistry

Under all future scenarios in which atmospheric CO_2 increases, it is well known that ocean acidification will intensify. This basic conceptual understanding is backed up by a simple approach that uses well-known fundamental thermodynamic equations to quantify future changes by assuming equilibrium between atmospheric and surface-ocean CO_2. This equilibrium assumption works well over most of the surface ocean, i.e. where the air–sea CO_2 equilibration timescale (several months; see Box 3.1) is much shorter than the residence time of waters near the surface (Sarmiento and Gruber 2006). A second approach relying on global-scale ocean models confirms this future intensification of ocean acidification and provides a more realistic regional picture by accounting for air–sea CO_2 disequilibrium, such as is found in regions where there is substantial exchange between surface and deep waters. Neither approach considers the effects of eutrophication or atmospheric deposition of anthro-

pogenic nitrogen and sulphur, which can exacerbate acidification of coastal waters (Doney *et al.* 2007). Both approaches also neglect buffering effects from the dissolution of $CaCO_3$ sediments, which would increase alkalinity but is negligible over centuries for shallow sediments and over millennia for deep sediments (see Chapter 7). Below, these approaches are detailed and their projections discussed.

3.6.1 Approaches to project future acidification

The equilibrium approach computes future surface-ocean pH and $[CO_3^{2-}]$ using basic thermodynamic equilibrium equations while varying CO_2 and holding constant another carbonate system variable, typically A_T. Conveniently, this approach does not rely on an ocean model. It is exact when the independent variable is seawater pCO_2, not atmospheric pCO_2. In other words, it assumes thermodynamic equilibrium between CO_2 in the atmosphere and that in surface waters at their *in situ* A_T, temperature, and salinity. The equilibrium approach works well in regions such as the subtropical gyres where waters remain at the surface long enough for C_T to equilibrate with atmospheric CO_2, typically requiring 8 months at present (see Box 3.1); conversely, the equilibrium approach is less accurate in areas such as the tropical Pacific or high latitudes, where waters spend less time at the surface and thus have insufficient time to equilibrate with the atmosphere. The equilibrium approach is also inappropriate for the deep ocean, which is isolated from the atmosphere.

When calculating future changes in surface carbonate system variables, the equilibrium approach is inaccurate wherever the anthropogenic transient of pCO_2 in the surface ocean lags that in the atmosphere. The assumption of equilibrium with the atmosphere leads to the prediction that a given reduction in pH or saturation will occur too soon, at lower atmospheric pCO_2. For example, Orr *et al.* (2005) demonstrated that the equilibrium approach predicts that average surface waters of the Southern Ocean become undersaturated with respect to aragonite, as discussed below, when atmospheric pCO_2 is 550 ppmv (in 2050 under the IS92a scenario); conversely, their disequilibrium approach, relying on a combination of models and data, indicates that such undersaturation will occur at 635 ppmv (in 2070 under IS92a). That 85 ppmv underestimate is substantial, but the difference in the predicted timing of undersaturation is only 20 years, simply because atmospheric CO_2 in the IS92a scenario continues to rise sharply through to the end of this century. With a more conservative scenario, the difference in timing would be larger. Despite their differences, both approaches indicate that under the IS92a scenario, Southern Ocean surface waters will become undersaturated with respect to aragonite during this century (see below).

The disequilibrium approach relies on one or more ocean models. Model projections can be used by themselves, or they can be improved by systematically correcting model results for their present-day biases. The latter approach requires high-quality data with adequate spatial coverage. For this purpose, the discrete bottle data collected in the global CO_2 survey during the Joint Global Ocean Flux Study (JGOFS)/WOCE era in the 1990s (Wallace 2001) has served as the fundamental reference. Key et al. (2004) compiled these data, quality controlled them, and then produced a near global, three-dimensional gridded data product (GLODAP). Two studies have exploited those gridded data to compute a baseline reference for pH and $CaCO_3$ saturation (Caldeira and Wickett 2005; Orr et al. 2005) and then used that reference to improve future model predictions. To that GLODAP data reference, centred around 1994, they added model-simulated changes in C_T relative to the same reference year. Both studies assumed unchanged A_T, then recomputed saturation states and pH (on the seawater scale for Caldeira and Wickett, and on the total scale for Orr et al.).

Caldeira and Wickett (2005) derived the pre-industrial state (by subtracting GLODAP's data-based estimates of anthropogenic C_T from the GLODAP fields for modern C_T) and added to that reference state the simulated changes in C_T from one model (relative to the pre-industrial reference year). They focused on zonal-mean changes during this century under IPCC SRES scenarios and until 2500 under stabilization scenarios and logistic functions (total releases of 1250 to 20 000 Pg C). Orr et al. (2005) also used the modern GLODAP data as the

reference, to which they separately added C_T perturbations from a group of 10 models, each of which participated in phase 2 of the Ocean Carbon Cycle Model Intercomparison Project (OCMIP) (Sarmiento et al. 2000; Orr et al. 2001; Dutay et al. 2002, 2004; Doney et al. 2004; Matsumoto et al. 2004; Najjar et al. 2007). Some related analyses of the Orr et al. (2005) data are presented here for the first time, which for simplicity will be referred to as the OCMIP study. That study focused on regional variations during the 21st century, providing results as the 10-model median $\pm 2\sigma$ for each of two IPCC scenarios, IS92a and S650. The IS92a scenario reaches 788 ppmv in 2100, while the S650 scenario reaches 563 ppmv in 2100 and stabilizes at 650 ppmv before 2200. Projected atmospheric CO_2 and surface-ocean pH from these two scenarios resemble those from IPCC SRES scenarios A2 and B1. Variations in the data–model correction approach have been used to account not only for increasing C_T, but also the effects of climate change by correcting for model biases in other relevant variables (A_T, temperature, salinity, PO_4^{3-}, SiO_2) in climate-change simulations (Orr et al. 2005; Steinacher et al. 2009).

3.6.2 Future trends in open-ocean surface chemistry

As anthropogenic C_T increases in the ocean, it causes shifts in other carbonate system variables, eroding the ocean's capacity to absorb more anthropogenic CO_2 from the atmosphere. That capacity, in terms of the average rate of increase in surface-ocean C_T per unit change in atmospheric CO_2 (i.e. $\partial C_T/\partial pCO_2$ in units of $\mu mol\ kg^{-1}\ ppmv^{-1}$), had already decreased in 1994 to 72% of what it was in the pre-industrial ocean (Fig. 3.2). If atmospheric CO_2 were to reach 563 ppmv then 788 ppmv, that capacity would drop to 40% then 26% of the pre-industrial rate. These changes are well understood (Sarmiento et al. 1995), and for more than 30 years have been accounted for implicitly in ocean models designed to project changes in air–sea CO_2 fluxes. This feedback of ocean acidification on atmospheric CO_2 remains the largest by far, although many much smaller feedbacks have been identified (see Chapter 12). Recently, attention has also focused on projecting associated shifts in other

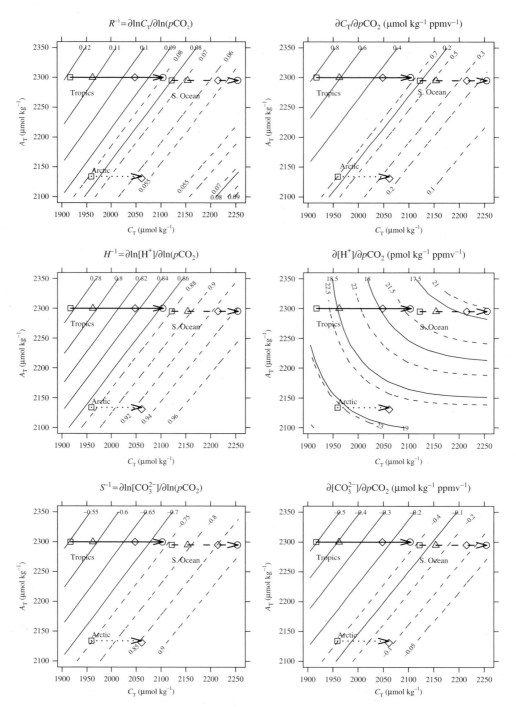

Figure 3.2 A_T–C_T diagrams (Baes 1982) of ratios of anthropogenic changes in C_T (top row), [H$^+$] (middle row), and [CO$_3^{2-}$] (bottom row) relative to those in ocean pCO$_2$ for the case where C_T increases but A_T, temperature, and salinity remain constant. Changes are given in absolute terms (right column) and in relative terms, i.e. as the fractional change in each species (left column). Arrows indicate projections for the evolution of average conditions in the tropics (20°S–20°N) and the Southern Ocean (south of 60°S) based on the OCMIP study. The Arctic average results (north of 70°N) are from Orr *et al.* (2005) based on the model from Institute Pierre Simon Laplace (IPSL). Symbols denote pre-industrial (square) and modern (triangle) conditions as well as projections for 563 ppmv (diamond) and 788 ppmv (circle). Solid contour lines (with horizontal labels) are for tropical conditions (T = 27.01°C, S = 34.92); dashed contour lines (with diagonal labels) are for Southern Ocean conditions (T = −0.49°C, S = 33.96). Contour lines for the Arctic (T = −0.21°C, S = 31.11) are similar to those shown for the Southern Ocean.

Table 3.2 Annual-mean surface pH_T (0–10 m) averaged[†] over the GLODAP domain[‡]

Time (atmospheric CO_2)	pH_T	pH_T change
Pre-industrial[a] (278 ppmv)	8.18	–
1994[b] (360 ppmv)	8.10	−0.08
2050[c] (IS92a, 563 ppmv)	7.95	−0.23
2100[c] (IS92a, 788 ppmv)	7.82	−0.36

[†]Averages given as area-weighted means of pH_T. Identical results are found when pH_T is first converted to $[H^+]$, then averaged and reconverted back to pH_T.

[‡]The near-global GLODAP domain excludes the Arctic Ocean, Indonesian seas, and most other marginal seas.

[a]Pre-industrial pH_T was recomputed from the GLODAP data after subtracting data-based estimates for anthropogenic C_T (Sabine *et al.*, 2004; Key *et al.*, 2004).

[b]The 1994 average is based on the GLODAP data.

[c]The estimates at 563 ppmv and 788 ppmv (nominal years 2050 and 2100 under IS92a) are the medians of the models from the OCMIP study.

carbonate system variables, including pH and $[CO_3^{2-}]$, and how they vary regionally as atmospheric CO_2 continues to increase.

Table 3.2 shows the annual-mean pH_T for the modern ocean based on the 1994 GLODAP data, as well as pre-industrial estimates and future projections. The average change in surface-ocean pH_T, relative to the pre-industrial state, has reached already about –0.1. That change could nearly quadruple by the end of the century under the IS92a or A2 scenarios. To what extent do these changes in surface-ocean pH differ regionally, given the large regional variability for uptake and storage of anthropogenic CO_2 (Sarmiento *et al.* 1992; Orr *et al.* 2001; Sabine *et al.* 2004)?

Reductions in zonal, annual-mean pH_T relative to the pre-industrial distribution vary from 0.33 in the tropics to 0.39 in the Southern Ocean for 2100 under the IS92a scenario (Fig. 3.3). Although pre-industrial $[H^+]$ in the Southern Ocean is on average 13% lower ($pH_T = 8.20$) than in the tropics ($pH_T = 8.14$), the greater anthropogenic $[H^+]$ increase in the Southern Ocean causes both regions to have the same average pH_T of 7.81 when atmospheric CO_2 reaches 788 ppmv. The anthropogenic increase in surface $[H^+]$ is smaller in the tropics, because carbonate-rich surface waters provide greater chemical capacity to take up anthropogenic CO_2 and buffer those changes (they have a lower Revelle factor). Greater carbonate concentrations mean that more carbonate is available to be consumed, neutralizing more of the incoming excess CO_2 and producing less H^+.

Relative to the Southern Ocean, the tropics have an average anthropogenic C_T increase that is 40% greater, whereas the anthropogenic increase in tropical $[H^+]$ is 12% less. This greater tropical buffering necessitates 72% greater $[CO_3^{2-}]$ consumption (–120 µmol kg^{-1}) relative to the Southern Ocean (–70 µmol kg^{-1}).

To better understand regional differences in pH, let us separate the rates at which $[H^+]$ changes with respect to changes in pCO_2 and C_T:

$$\frac{\partial [H^+]}{\partial C_T} = \frac{\partial [H^+]}{\partial pCO_2} \frac{\partial pCO_2}{\partial C_T}. \qquad (3.6)$$

We may expect that spatiotemporal variability in $\partial [H^+]/\partial C_T$ is driven largely by the $\partial pCO_2/\partial C_T$ term, which was shown above to vary greatly. But our goal here is to quantify how $[H^+]$ is directly affected by the increase in pCO_2. To what degree does the less familiar term $\partial [H^+]/\partial pCO_2$ vary, and what drives that variability? Analogous to the familiar Revelle factor, one can frame these questions in terms of the 'relative change' of pCO_2 to that of $[H^+]$. Omta *et al.* (2010) denote that ratio as H and consider it to be constant. Here its variability is quantified to explain the regional differences in anthropogenic pH changes mentioned above. Because the increase in pCO_2 is the driver, let us focus on the inverse of H:

$$H^{-1} = \frac{\partial [H^+]/[H^+]}{\partial pCO_2/pCO_2} = \frac{\partial \ln [H^+]}{\partial \ln pCO_2} = \frac{\gamma_{C_T}}{\beta_{C_T}}. \qquad (3.7)$$

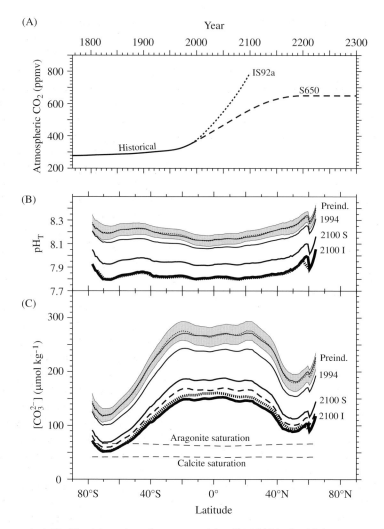

Figure 3.3 Increasing atmospheric CO_2 (A) and decreasing surface-ocean pH_T (B) and $[CO_3^{2-}]$ (C) for the global ocean. In panels B and C, results are given as surface-layer zonal means (global means per band of latitude). Shown are the GLODAP data in 1994 (solid line within grey shading, indicating ± 2σ model range) and the OCMIP median model in 2100 for the IS92a and S650 scenarios (as indicated) as well as year 2300 under S650 (thick dashed line). The effect of future climate change simulated by the Institute Pierre Simon Laplace (IPSL) earth system model (thick dotted line) is shown as a perturbation to IS92a in 2100. The two flat, thin, dashed lines indicate the thresholds where $[CO_3^{2-}]$ in seawater is in equilibrium with aragonite and calcite. From Orr *et al.* (2005).

where the terms $\gamma_{C_T} = (\partial \ln[CO_2] / \partial C_T)^{-1}$ and $\beta_{C_T} = (\partial \ln[H^+] / \partial C_T)^{-1}$ are buffer factors from Egleston *et al.* (2010), both convenient functions of carbonate system variables. Rearranging (3.7), it follows that the corresponding 'absolute change' is

$$\frac{\partial [H^+]}{\partial pCO_2} = \frac{\gamma_{C_T}}{\beta_{C_T}} \frac{[H^+]}{pCO_2}. \qquad (3.8)$$

Figure 3.2 shows these relative and absolute rates of change as a function of C_T and A_T. Also illustrated are regional differences and future changes based on the GLODAP data and the median results from the OCMIP models (Orr *et al.* 2005). Remarkably, there is little variation in either H^{-1} or $\partial[H^+]/\partial pCO_2$. As atmospheric CO_2 increases from 278 to 788 ppmv, the relative rate of change H^{-1} remains nearly constant, increasing by only 7% in the Southern Ocean and in 11% the tropics. The absolute change $\partial[H^+]/\partial pCO_2$ varies by even less (2% in the Southern Ocean and 5% in the tropics). But both the absolute and relative changes are higher in the Southern Ocean than in the tropics. As atmospheric CO_2 increases from 278 to 788 ppmv, the Southern Ocean to tropical regional ratio increases from 1.14 to 1.19 for $\partial[H^+]/\partial pCO_2$, while it decreases from 1.13 to 1.09 for H^{-1}. These slight, opposite trends are explained by the $[H^+]/pCO_2$ ratio, which was identical in both regions during pre-industrial times, but becomes 9% greater in the Southern Ocean when atmospheric pCO_2 reaches 788 ppmv.

At any given time, one may attribute these small regional differences in $\partial[H^+]/\partial pCO_2$ to differences in temperature to the extent that the surface ocean approximates a closed system, i.e. where A_T, pCO_2, and salinity can be considered roughly constant. That is,

$$\frac{\partial[H^+]}{\partial pCO_2} = \frac{\partial[H^+]}{\partial T}\left(\frac{\partial pCO_2}{\partial T}\right)^{-1} =$$
$$\frac{\partial[H^+]}{\partial T}\left(\frac{\partial \ln pCO_2}{\partial T}\right)^{-1}\frac{1}{pCO_2} \qquad (3.9)$$

where T is temperature and $\partial \ln(pCO_2)/\partial T = 0.0423°C^{-1}$ (Takahashi *et al.* 1993). In the real ocean, an open system, temperature still appears to act as the dominant driver of the spatial variability of $\partial[H^+]/\partial pCO_2$, which varies little with increasing atmospheric CO_2 (shown above) and total alkalinity (see Section 3.6.7). In any case, regional and temporal differences in $\partial[H^+]/\partial pCO_2$ are small, i.e. the ocean remains on the flat part of the titration curve, as the acid CO_2 is added and $[H^+]$ is buffered through large reductions in $[CO_3^{2-}]$, our next focus.

Modern surface $[CO_3^{2-}]$ computed from the GLODAP gridded data is naturally lower in colder waters and higher in warmer waters (Caldeira and Wickett 2005; Orr *et al.* 2005). In the OCMIP study, annual-mean $[CO_3^{2-}]$ averages varied from 105 µmol kg^{-1} for the Southern Ocean to 240 µmol kg^{-1} for tropical waters (Fig. 3.3). The polar and subpolar oceans have naturally lower surface $[CO_3^{2-}]$ associated with their higher C_T/A_T ratios. Although variations in A_T are relatively small and generally follow salinity, variations in surface C_T are much larger,

because as temperatures decline toward high latitudes, the solubility of CO_2 increases and more CO_2 invades the ocean from the atmosphere (Sarmiento and Gruber 2006). High C_T/A_T ratios are also found in CO_2-rich subsurface waters, which happen to be cooler. When these waters upwell into high-latitude regions such as the Southern Ocean, they further reduce surface $[CO_3^{2-}]$. These factors combine to make a strong positive correlation between global-scale, annual-mean surface maps of temperature and modern $[CO_3^{2-}]$ ($R^2 = 0.92$; slope of +5 µmol kg^{-1} °C^{-1}), but the link with temperature is indirect. On shorter space- and timescales, one would expect degraded correlations of $[CO_3^{2-}]$ and C_T with temperature. For instance, seasonal variations in C_T cannot keep up with those of temperature, because air–sea CO_2 equilibration requires many months (see Box 3.1), whereas heat transfer is much faster.

Increases in anthropogenic CO_2 have already reduced modern, annual-mean surface $[CO_3^{2-}]$ by more than 10%, relative to pre-industrial conditions based on analysis of the GLODAP data combined with estimates of anthropogenic C_T from data (Feely *et al.* 2004; Sabine *et al.* 2004) and models (Orr *et al.* 2005), assuming no change in A_T. In the OCMIP study, when atmospheric CO_2 reached 788 ppmv in 2100 under the IS92a scenario, annual-mean surface $[CO_3^{2-}]$ declined to levels of 149 ± 14 µmol kg^{-1} in the tropics and 55 ± 5 µmol kg^{-1} in the Southern Ocean, roughly half of pre-industrial values. The latter level is 18% below the threshold where waters become undersaturated with respect to aragonite (~66 µmol kg^{-1}). Thus, well before 2100, typical surface waters of the Southern Ocean become corrosive to aragonite throughout most of the year. Indeed, Southern Ocean surface waters reach these corrosive conditions by 2100 under IPCC SRES scenarios A1, A2, B1, and B2 as well as under any pathway that stabilizes atmospheric CO_2 at 650 ppmv or above (Caldeira and Wickett 2005). Under the same SRES scenarios, the intermediate complexity model from the University of Bern Physics Institute (PIUB) shows similar results (Orr *et al.* 2005). Additionally, surface waters in the subarctic Pacific become slightly undersaturated by 2100 under the IS92a scenario.

For comparison with $[H^+]$ and by analogy with Eq. 3.7, let us evaluate the spatiotemporal variability of $[CO_3^{2-}]$ in terms of the 'inverse saturation factor' (S^{-1}), i.e. the ratio of the relative change of $[CO_3^{2-}]$ to that of pCO_2:

$$S^{-1} = \frac{\partial [CO_3^{2-}]/[CO_3^{2-}]}{\partial pCO_2/pCO_2} = \frac{\partial \ln[CO_3^{2-}]}{\partial \ln pCO_2} = \frac{\gamma_{C_T}}{\omega_{C_T}} \quad (3.10)$$

where γ_{C_T} is defined above and $\omega_{C_T} = (\partial \ln[CO_3^{2-}]/\partial C_T)^{-1}$ is another buffer factor derived by Egleston *et al.* (2010). It follows that the corresponding absolute change with respect to pCO_2 is

$$\frac{\partial [CO_3^{2-}]}{\partial pCO_2} = \frac{\gamma_{DIC}}{\omega_{DIC}} \frac{[CO_3^{2-}]}{pCO_2} \quad (3.11)$$

As atmospheric CO_2 increases from 278 to 788 ppmv, the relative rate of change S^{-1} increases by 32% in the tropics and 17% in the Southern Ocean, while it remains 38% to 22% higher in the latter region relative to the former (Fig. 3.2). In contrast, the absolute change $\partial [CO_3^{2-}]/\partial pCO_2$ declines sharply with increasing atmospheric CO_2. Relative to pre-industrial values of $\partial [CO_3^{2-}]/\partial pCO_2$, those at 788 ppmv are more than three times lower in the tropics and five times lower in the Southern Ocean. Thus temporal changes in $\partial [CO_3^{2-}]/\partial pCO_2$ are much larger than those for $\partial [H^+]/\partial pCO_2$, which was shown above to decline by at most 6% during the same increase in atmospheric CO_2. Regional differences are also about three times larger for $\partial [CO_3^{2-}]/\partial pCO_2$. By far, the dominant factor controlling spatiotemporal variability in the absolute change $\partial [CO_3^{2-}]/\partial pCO_2$ is the $[CO_3^{2-}]/pCO_2$ ratio, which is reduced in all regions by about fivefold as atmospheric CO_2 increases from 278 to 788 ppmv, with tropical ratios about twice those in the Southern Ocean.

Changes in $[CO_3^{2-}]$ are also closely tied to changes in other carbonate chemistry variables (Fig. 3.4). The increase in anthropogenic C_T is largest in the warm tropical waters, where lower C_T/A_T ratios render these waters more chemically suitable to taking up anthropogenic CO_2. This high chemical capacity for taking up anthropogenic CO_2 is linked to its high $[CO_3^{2-}]$ (Fig. 3.3) and thus high buffer capacity and low Revelle factor (Fig. 3.4). Everywhere, the increase in $[HCO_3^-]$ is larger than

the increase in C_T, as required by the overall reduction in $[CO_3^{2-}]$ and the definition of total inorganic carbon (see also Eq. 3.3). Relative to average changes in tropical surface waters, changes in $[CO_2]$ in the Southern Ocean are 2.4 times larger due to enhanced CO_2 solubility (K_0), which increases C_T and drives $[CO_3^{2-}]$ downward. Polar and subpolar regions also have lower buffer capacities (higher Revelle factors) associated with lower $[CO_3^{2-}]$. In the high latitudes, although absolute changes in $[CO_3^{2-}]$ are smallest, changes relative to the pre-industrial level are the largest.

Let us now consider all these changes in terms of the most common denominator, C_T. At 1994 conditions, for every μmol kg^{-1} increase in tropical C_T, 0.65 μmol kg^{-1} of $[CO_3^{2-}]$ is consumed while 1.60 μmol kg^{-1} of $[HCO_3^-]$ is produced. These numbers deviate from the stoichiometric coefficients in Eq. 3.1. Simultaneously, there is only a small increase in $[CO_2]$ of 0.044 μmol kg^{-1} and a minute increase in $[H^+]$ of 0.03×10^{-3} μmol kg^{-1} (pH_T declines by 0.0013 pH units), illustrating the remarkable effectiveness of the seawater buffering system. At 788 ppmv, consumption of $[CO_3^{2-}]$ and production of $[HCO_3^-]$ are at 94% of the 1994 values, while changes in $[CO_2]$, $[H^+]$, and pH_T are about 2.5 times higher. In the Southern Ocean in 1994, consumption of $[CO_3^{2-}]$ and production of $[HCO_3^-]$ are 88% of tropical values in the same year, but the increase in $[CO_2]$ is threefold greater and 1.5 times more $[H^+]$ is produced. By 788 ppmv, Southern Ocean consumption of $[CO_3^{2-}]$ and production of $[HCO_3^-]$ decline to about 72% of 1994 values, while changes in $[CO_2]$, $[H^+]$, and pH_T are 2.7 times greater.

These assessments illustrate the fundamental nature of the GLODAP data for recent evaluations of how ocean acidification is affecting carbonate chemistry and pH. Yet gaps remain. GLODAP is based on 'one-time' survey data that do not cover some key areas, including the Arctic Ocean, marginal seas, and coastal zones. Nor does it account for seasonal variations. These concerns are addressed below.

3.6.3 Future trends in the Arctic Ocean

Recent studies project that undersaturation in the Arctic will occur sooner and be more intense than in the Southern Ocean. A tenth of Arctic surface waters

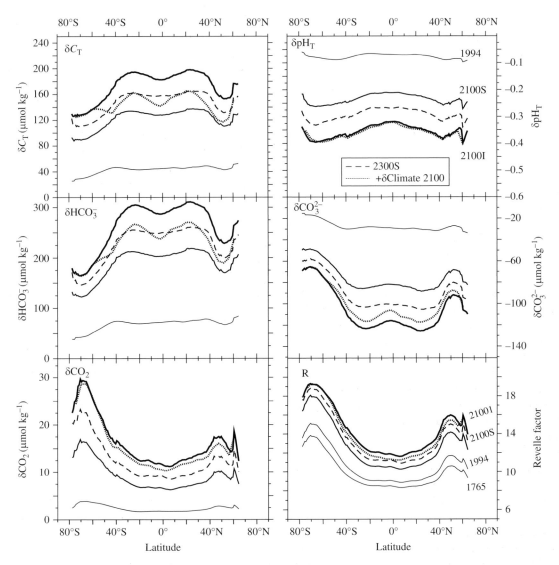

Figure 3.4 Zonal-mean surface changes in C_T, pH_T, [HCO_3^-], [CO_3^{2-}], and [CO_2] during the industrial era until the end of the present century. Snapshots for the data and model results given indicated as in Fig. 3.2. The δ symbol on panels indicates that results are given as perturbations to the pre-industrial state. Conversely, the Revelle factor, R (bottom right panel), is not given as a perturbation but as its absolute value (its pre-industrial model median is indicated by 1765). Line signatures are as in Fig. 3.3. From Orr *et al.* (2005).

will become undersaturated with respect to aragonite (annual-mean $\Omega_a < 1$) by the time that atmospheric CO_2 reaches 428 ppmv (in 2024 ± 1 yr under the A2 and B1 scenarios) based on the combined data–model approach, relying on the CSM1.4 model output with discrete bottle data collected in the Arctic in the 1990s (Steinacher *et al.* 2009). By the

time that atmospheric CO_2 reaches 534 ppmv (in 2050 under A2), annual average Ω_a drops below 1 for half of the surface waters. By 765 ppmv (in 2090 under A2), the same annual-mean undersaturated conditions ($\Omega_a < 1$) occur throughout the water column. Another model study also under the A2 scenario, but using the CCSM3 model without the model-data correction to the Arctic observations, projects that by the end of this century, surface waters with annual average $\Omega_c < 1$ will be found over much of the Arctic (Feely *et al.* 2009). These waters would be chemically corrosive to calcite and all other forms of $CaCO_3$. The main mechanism explaining why undersaturation will occur generally sooner in the Arctic Ocean than the Southern Ocean is the enhanced freshwater input in the Arctic from climate change. In short, enhanced ice melt and increased precipitation dramatically reduce Arctic surface $[CO_3^{2-}]$ (Steinacher *et al.* 2009). This finding is consistent with freshwater dilution suppressing $[CO_3^{2-}]$ as suggested by Salisbury *et al.* (2008), who illustrates large reductions in Ω with declining salinity near river mouths and the generally more acidic pH of river discharge plumes. Indeed, some low-salinity, near-coastal surface waters in the Arctic are already undersaturated with $\Omega_a < 1$ (Yamamoto-Kawai *et al.* 2009). Let us return to these freshwater dilution effects later, when discussing acidification in the context of climate change (Section 3.6.6) and the coastal ocean (Section 3.6.7) after considering natural variability and subsurface changes.

3.6.4 Seasonal and interannual variability

So far, the focus has been on annual-mean surface trends, but surface $[CO_3^{2-}]$ also varies seasonally and interannually, along with natural variations in pCO_2 and related biogeochemical variables, such as nutrients and C_T. The OCMIP study suggests that interannual variability in surface-ocean $[CO_3^{2-}]$ is small everywhere when compared with the magnitude of the anthropogenic transient (trend in annual means). It also suggests that seasonal variability is small at low latitudes, but that at high latitudes the average amplitude of the annual cycle can reach up to ±15 μmol kg^{-1}. Similar or higher seasonal amplitudes have been observed in the subarctic Pacific (Feely *et al.* 1988), the Bering Sea (Merico *et al.* 2006), and the Norwegian Sea (Findlay *et al.* 2008). Comparable amplitudes are also found in the Southern Ocean, based on seasonal variations in $[CO_3^{2-}]$ derived from carbonate system data (McNeil and Matear 2008). In all these cases, $[CO_3^{2-}]$ is highest during summer, when pCO_2 is driven downward mainly by the spring–summer bloom. Secondary factors that also contribute to seasonal variability include wintertime cooling and enhanced wintertime mixing with CO_2-rich deep waters, both of which lower surface $[CO_3^{2-}]$ while raising surface pCO_2. Thus as levels of atmospheric CO_2 continue to increase and $[CO_3^{2-}]$ generally decreases, high-latitude undersaturation ($\Omega_a < 1$) will be reached first during winter, and as years progress, surface waters will remain undersaturated during an increasing number of months per year. Summer conditions will be the most resistant to the advancing undersaturation.

For the Southern Ocean, McNeil and Matear (2008) combined observational estimates of the annual cycle with the future trend from a model and found that those surface waters start to become undersaturated with respect to aragonite in winter when atmospheric CO_2 reaches about 450 ppmv. That undersaturation happens about 100 ppmv sooner than for annual average conditions, which translates to a 30-yr advance for winter undersaturation under the IS92a scenario.

In the Arctic Ocean, seasonal data for the carbonate system are extremely sparse. However, data were collected during spring and summer cruises in 2002 and 2004 over the Chukchi Sea shelf and slope as well as into the Canada Basin during the Shelf-Basin Interactions project. The seasonal amplitude of surface $[CO_3^{2-}]$ in that data reaches up to ±12 μmol kg^{-1} (Bates *et al.* 2009): as found elsewhere, the maximum in surface $[CO_3^{2-}]$ is attained in summer. In contrast, subsurface waters overlying the shelf exhibit a similar magnitude of change but reversed phasing. That is, subsurface $[CO_3^{2-}]$ reaches its minimum in summer due to intense remineralization of organic matter produced from high primary productivity in overlying waters. Elsewhere in the high Arctic, the annual cycle has not been assessed owing to a lack of seasonal observations.

3.6.5 Future changes in interior ocean chemistry

Penetration of anthropogenic CO_2 into the deep ocean also reduces subsurface $[CO_3^{2-}]$ and pH. Feely *et al.* (2004) use observations to demonstrate that the industrial-era invasion of anthropogenic CO_2 has already caused the ASH to shoal. Figure 3.5

illustrates the modern penetration of anthropogenic C_T into the deep Atlantic Ocean and the resulting changes in key carbonate system variables; it also shows corresponding model projections for 2100 from the OCMIP study.

In 1994, the anthropogenic perturbation was mostly confined to the upper 1000 m, except in the North Atlantic. The modern surface perturbation in C_T averages about 50 μmol kg^{-1}. But by 2100 under the IS92a scenario, that same level is projected to penetrate generally beyond 1000 m, while twice that level is reached in North Atlantic bottom waters. Corresponding anthropogenic reductions in pH$_T$ show patterns that are similar to those for the anthropogenic C_T increase, except that the pH$_T$ perturbation appears to penetrate deeper and is more intense in subsurface waters (typically at about 200 m) rather than at the surface. The latter finding is consistent with the subsurface maximum observed at HOT (see Section 3.4). The simulated subsurface maximum must be due to different chemical characteristics of the subsurface waters, because the OCMIP models used here did not account for climate change, which alters ocean circulation. Indeed the spatial pattern of the simulated changes in pH$_T$ is closely matched by the pattern of modelled changes in the Revelle factor (not shown). Changes in C_T have also provoked the global-mean depth of the ASH to shoal from its pre-industrial level of 1090 m to 960 m in 1994. The global-average ASH is projected to shoal to 280 m in 2100 under the IS92a scenario.

There are large regional differences in projected changes in saturation. For example, by 2100 under IS92a, the ASH shoals from 180 m to the surface in the subarctic Pacific, from 1040 m to the surface in the Southern Ocean, and from 2820 m to 110 m in the North Atlantic north of 50°N. Although the average calcite saturation horizon (CSH) in the Southern Ocean remains below 2200 m, by 2100, Weddell Sea surface waters become slightly undersaturated even with respect to calcite. Under both the IS92a and S650 scenarios, the OCMIP models project that during this century there will be large changes in surface and subsurface [CO$_3^{2-}$] due to the invasion of anthropogenic CO$_2$. On longer timescales (a few centuries), stabilization of atmospheric CO$_2$ at even the 450 ppmv target renders most of the

deep ocean volume corrosive to aragonite and calcite (Caldeira and Wickett 2005). With a similar volume analysis but for the CSM1.4 earth system model, Steinacher *et al.* (2009) found that under the A2 scenario, waters with $\Omega_a > 4$ will disappear within three decades, waters with $\Omega_A > 3$ will vanish by 2070, and supersaturated waters ($\Omega_a > 1$) will decrease from 42% during pre-industrial time to 25% by 2100 (see also Chapter 14).

These changes in saturation may affect aragonitic cold-water corals. Nearly all of these corals now live in deep waters where $\Omega_a > 1$, but it is projected that by 2100 under the IS92a scenario, 70% of them will be bathed in waters where $\Omega_a < 1$ (Guinotte *et al.* 2006). Whether or not acidification will dramatically affect cold-water corals is an active area of research (Maier *et al.* 2009). Other deep-sea marine biota, which experience less variation in environmental conditions than surface organisms, may also be affected by deep-ocean increases in CO$_2$ as well as future reductions in deep-sea oxygen, as climate change continues to reduce ventilation of deep waters (Brewer and Peltzer 2009).

3.6.6 Climate change and ocean acidification

In addition to the direct geochemical effect from the CO$_2$ increase, ocean [CO$_3^{2-}$] and related variables are also altered by climate change. The OCMIP study quantified the effects of climate change during this century by analysing results from three atmosphere–ocean climate models, each of which included an ocean carbon cycle module. All models illustrated how climate warming generally results in increased surface-ocean [CO$_3^{2-}$], but that increase typically counteracts less than 10% of the decrease from the increase in anthropogenic CO$_2$ (Figs 3.3 and 3.4). Subsequent studies by McNeil and Matear (2006) and Cao *et al.* (2007) further confirm that 21st-century changes in [CO$_3^{2-}$] due to warming are small compared with chemical changes from invasion of anthropogenic CO$_2$.

But warming is not the only factor, particularly in the Arctic Ocean. The first hint that the Arctic might be different came from an earth system model that projected future reductions in surface [CO$_3^{2-}$] due to climate change (Orr *et al.* 2005). An in-depth analysis

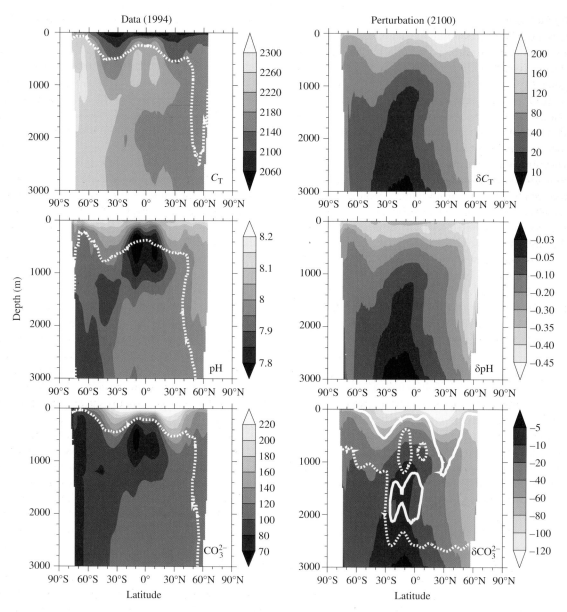

Figure 3.5 Zonal-mean sections in the Atlantic Ocean for C_T (top), pH_T (middle), and $[CO_3^{2-}]$ (bottom). Results are given for the observed state in 1994 (left), based on the GLODAP data, and the pre-industrial to 2100 perturbation (right), indicated as 'δ', based on the IS92a median projection from the OCMIP study. In the left column the dashed white line indicates the depth of penetration of the perturbation in 1994, defined here as the depth where the concentration drops to half of the global-ocean average of the surface perturbation in the same year. These 'half' contours are 22.5 µmol kg^{-1} for δC_T (top), −0.038 units for δpH_T (middle), and −14.5 µmol kg^{-1} for $\delta[CO_3^{2-}]$ (bottom). The lower right panel includes the contour lines for the aragonite saturation horizon in 1994 (dashed) and 2100 (solid).

by Steinacher *et al.* (2009) using another earth system model found that Arctic climate-induced reductions in surface carbonate also exacerbated the decline in carbonate from the invasion of anthropogenic CO_2. Their analysis revealed that reductions in $[CO_3^{2-}]$ from increased freshwater input (from sea-ice melt, more precipitation, and less evaporation) dominated increases from warming and increased primary production. Furthermore, Arctic warming led to reduced summer sea-ice cover and thus greater invasion of anthropogenic CO_2, which reduced $[CO_3^{2-}]$ further. Overall, Steinacher *et al.* (2009) estimated that the net effect of climate change was to enhance the reduction of surface $[CO_3^{2-}]$ in the Arctic by 34% by 2100.

More details concerning the magnitude of the effects of climate change on ocean pH and $CaCO_3$ saturation states can be found in modelling studies from Frölicher and Joos (2010) and those detailed by Joos *et al.* in Chapter 14, with scenarios that go beyond the end of this century, to 2500.

3.6.7 Marginal seas and the coastal ocean

Besides the Arctic Ocean, global projections of future ocean acidification have left out most marginal seas, including the Baltic, Mediterranean, and Black Seas. As a first step, let us estimate how 21st-century acidification of these marginal seas may differ from that of the global ocean by using thermodynamic constants and assuming equilibrium between atmospheric and oceanic pCO_2 at the chemical and hydrographic conditions typical of each sea. Twenty-first century pH and saturation states were computed for each sea by adopting typical local conditions for A_T and salinity at winter and summer temperatures, fixing these variables, and incrementing atmospheric CO_2 each year following IPCC SRES scenarios B1 and A2 (Fig. 3.6). This simple equilibrium approach may produce biased results in areas of deep-water formation (e.g. the Gulf of Lions, Adriatic Sea) and in the low-salinity, low-alkalinity waters of the Baltic Sea, where $[Ca^{2+}]$ may not follow the open-ocean proportionality to salinity and the role of carbonate as the dominant base may be diminished. Nonetheless, this new analysis clearly illustrates the extent to which total alkalinity, which

varies widely between these marginal seas, influences the rate of acidification.

The acidification of the marginal seas surrounding Europe is just starting to be investigated, with some initial studies in the Mediterranean Sea in terms of uptake and storage of CO_2 (Touratier and Goyet 2009; Louanchi *et al.* 2009). No projections have been made as to how the acidification rates of these seas may differ during this century. Yet it has been suggested that the relatively high A_T of the Mediterranean Sea, which drives greater future uptake of anthropogenic CO_2 relative to the open ocean (Touratier and Goyet 2009), thereby implies a greater future reduction in pH (Yilmaz *et al.* 2008). Let us examine this suggestion.

The basic equilibrium calculations made here do show that the average surface pH_T of the Black Sea is substantially higher than that of the Baltic and Mediterranean Seas. Indeed, differences in surface pH_T between these seas are largely explained by differences in carbonate ion concentrations. However, the projected absolute change in surface pH_T over the 21st century is very similar between the global-ocean average and all these seas (Fig. 3.6). For instance under the A2 scenario, the absolute change in pH_T over the 21st century for the Black Sea is identical to that in the global ocean, for the Mediterranean Sea it is 3% less, and for the Baltic it is 9% more. Except for the Baltic, these regional differences are smaller than 'seasonal' differences in the absolute change for a given sea (i.e. the difference in absolute changes computed with summer versus winter temperatures). Under winter conditions, the absolute change in pH_T is 5–7% greater than under summer conditions.

For a greater understanding, let us consider these changes in terms of $\partial[H^+]/\partial pCO_2$, which was shown in Section 3.6.1 to vary by only up to 6% with increasing atmospheric CO_2 (278 to 788 ppmv) and by about 15% across the full range of ocean temperatures. This marginal-sea comparison adds another dimension to our understanding of $\partial[H^+]/\partial pCO_2$. That is, it does not vary substantially even across the large range of A_T found in the global ocean, Black Sea, and Mediterranean Sea. Conversely, in the low-salinity, low-alkalinity Baltic Sea, $\partial[H^+]/\partial pCO_2$ is about 10% larger. This offset for the Baltic Sea, although small, appears linked to its

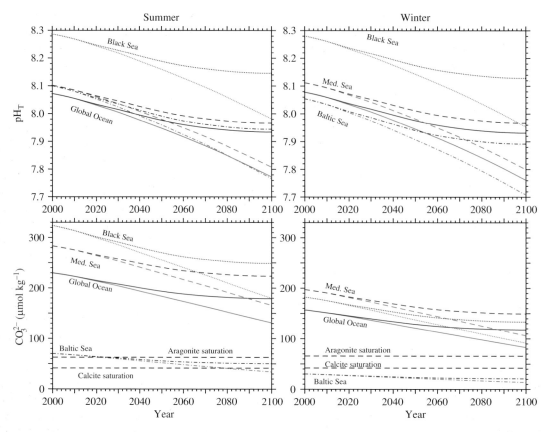

Figure 3.6 Twenty-first century changes in pH$_T$ (top) and [CO$_3^{2-}$] (bottom) for the European seas and the global-ocean average under typical conditions in summer (left) and winter (right). Results are computed using the IPCC A2 scenario (red) and the IPCC B1 scenario (blue) based on observed A_T and assuming thermodynamic equilibrium between atmosphere and ocean. Results differ from the global ocean mean (solid line, Key *et al.* 2004) because of different chemical and physical conditions for each of the European seas: Baltic Sea (dot-dash line, A_T = 1600 µmol kg^{-1}, S = 7, T = 0°C in winter and 20°C in summer); Black Sea (dotted line, A_T = 3256 µmol kg^{-1}, S = 18.075, T = 6.5°C in winter and 24°C in summer); and Mediterranean Sea (dashed line, A_T = 2560 µmol kg^{-1}, S = 38, T = 13°C in winter and 26°C in summer). Calculations were made using the R software package 'seacarb' (Lavigne and Gattuso 2010).

exceptionally low [CO$_3^{2-}$]. Only in the Baltic is the surface [CO$_2$]/[CO$_3^{2-}$] ratio projected to reach and even exceed 1.0 during this century. This ratio reaches unity at pH = (pK_1 + pK_2)/2 when C_T = A_C. As the decline continues to even lower levels of [CO$_3^{2-}$] and higher levels of [CO$_2$], particularly in winter, Baltic surface waters reach the point where C_T = A_T. At that threshold, the traditional buffer capacity −(∂pH/∂A_T)$^{-1}$, which characterizes resistance to changes in pH, reaches its minimum (Egleston *et al.* 2010). Indeed, at that same threshold, all of the six new buffer factors derived by Egleston *et al.* (2010), defined as inversed relative changes in pCO$_2$, [H$^+$], and [CO$_3^{2-}$] with respect to C_T and A_T, also reach their minima.

Although projected changes in pH are largely insensitive to the A_T of seawater, higher A_T does imply a greater uptake of anthropogenic CO$_2$, which is inextricably linked to greater consumption of carbonate ions. These equilibrium calculations project that with the A2 scenario, the mean global ocean reduction in [CO$_3^{2-}$] during the 21st century will be 97 µmol kg^{-1} under summer conditions. Summertime reductions in the more alkaline Mediterranean and Black seas (A_T = 2560 and 3256 µmol kg^{-1}) are 24% and 47% greater, whereas the [CO$_3^{2-}$] reduction in the less alkaline Baltic Sea (A_T = 1600 µmol kg^{-1}) is 62% less. Despite these dramatic differences in absolute changes, the relative change (absolute change divided by the [CO$_3^{2-}$] in 2000) is remarkably constant (0.44 for the Black Sea and 0.42 for both the Mediterranean Sea and the global-ocean mean), except for the Baltic Sea where the proportion is larger (0.53). Therefore over the range of alkalinities found in the open ocean and most marginal seas, the saturation factor S^{-1} (Eq. 3.10) does not vary substantially with total alkalinity for surface waters in equilibrium with atmospheric CO_2.

Present-day carbonate ion concentrations in these marginal seas may already be affecting the abundance of marine calcifying organisms. In the Baltic Sea, very low $[CO_3^{2-}]$ appears to be the factor that prohibits growth of the calcareous phytoplankton *Emiliana huxleyi*; conversely, in the Black Sea, where $[CO_3^{2-}]$ is high, large blooms of the same organism are visible from space (Tyrrell *et al.* 2008). Well before the end of the century, surface waters of the Baltic Sea will become corrosive to all forms of calcium carbonate. In the Black Sea and Mediterranean Sea, there is no danger of surface waters becoming corrosive to $CaCO_3$ before 2100, but they will suffer sharp reductions in $[CO_3^{2-}]$ (–37% in the Mediterranean Sea and –45% in the Black Sea under the A2 scenario). These rapid chemical changes are an added pressure on marine calcifiers and ecosystems of marginal seas already influenced by other anthropogenic factors.

Like the rest of the ocean, coastal waters are affected by acidification from increasing concentrations of anthropogenic CO_2. But they are also affected by acidification from other sources, including: (1) freshwater input (Salisbury *et al.* 2008), (2) atmospheric deposition of anthropogenic nitrogen and sulphur (Doney *et al.* 2007), and (3) delivery of terrestrial organic matter and nutrients that enhance coastal remineralization and redox cycling in adjacent coastal sediments. Freshwater typically has higher $[CO_2]$ and lower pH than does seawater. Surface freshwater dilution also reduces $[CO_3^{2-}]$ since it reduces A_T while surface C_T is partially compensated by input of CO_2 from the atmosphere (see Eq. 3.2). For example, low-salinity surface waters within the Canada Basin have been observed to have $\Omega_a < 1$ (Yamamoto-Kawai *et al.* 2009). In an estuary, Puget Sound in Washington State, Feely *et al.* (2010) observed larger than expected reductions in subsurface pH and estimated that up to half

of that is due to increases in anthropogenic CO_2, the remainder being due to degradation of organic matter produced from both natural and anthropogenic inputs. Unfortunately, it will require centuries for dissolution of coastal $CaCO_3$ sediments to have a significant impact on coastal acidification (see Chapter 7).

Observations in coastal waters off the west coast of the USA illustrate naturally lower pH and $[CO_3^{2-}]$ and large variability due to seasonal upwelling of CO_2-rich subsurface water (Feely *et al.* 2008). Natural seasonality due to this upwelling is now exacerbated by increasing concentrations of anthropogenic CO_2 in subsurface waters. Thus seasonal upwelling now brings with it undersaturated waters ($\Omega_a < 1$) that reach the surface over the Oregon Shelf during spring.

Similar upwelling occurs in other coastal areas, especially along eastern margins (Hauri *et al.* 2009). Yet modelling acidification in coastal zones requires sufficient horizontal resolution to adequately resolve the small-scale coastal features and physical processes such as bottom topography, eddies, and convection that help set local circulation fields and affect carbonate chemistry. Coastal models must also account for river fluxes and closer proximity to the seafloor. New high-resolution regional model configurations have been developed to study the North Sea (Blackford and Gilbert 2007) and the California Coastal Current system (Hauri *et al.* 2009), and these are being extended to include larger coastal areas and other regions, particularly eastern boundary upwelling systems.

3.7   Conclusions

As industrialization continues to drive atmospheric CO_2 concentrations upward, the surface ocean is responding by taking up more of this gas, which reacts with water, reducing surface-ocean pH and carbonate ion concentrations. The basic chemistry is well understood, and the magnitude of these changes is not debated by the scientific community. Indeed, these chemical changes are already measurable. Surface measurements of changes at three subtropical time-series stations agree with what is expected from the atmospheric CO_2 increase, assuming air–sea CO_2 equilibrium.

Future projections are made with models in order to also account for waters that are not in equilibrium with atmospheric CO_2, including high-latitude surface waters and the deep ocean. In the polar oceans, where $[CO_3^{2-}]$ is naturally lower, models project that during the 21st century $[CO_3^{2-}]$ will drop under a critical threshold, below which waters become chemically corrosive to aragonite. Some Arctic surface waters are projected to have annual-mean concentrations that will start falling below this threshold within about a decade under all SRES scenarios. For some surface waters of the

Southern Ocean, this critical saturation concentration is reached later, but still by 640 ppmv for the annual-mean level; in winter, the same critical level is first reached at 450 ppmv. As future $[CO_3^{2-}]$ declines further, a second threshold is reached, below which waters become corrosive to all forms of $CaCO_3$, including the most stable form, calcite. These severe conditions could be prevalent over much of the surface Arctic Ocean by the end of the century. Anthropogenic CO_2 is also penetrating into the deep ocean, altering subsurface carbonate chemistry and causing the ASH to shoal. The greatest

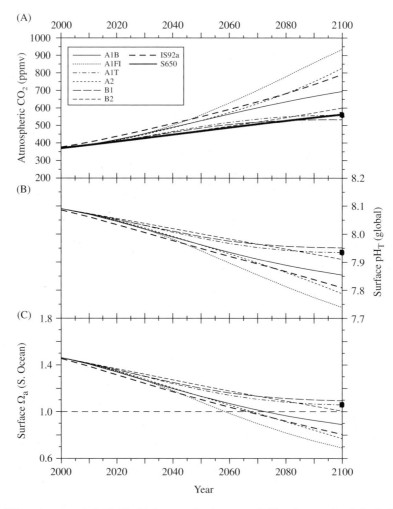

Figure 3.7 Simulated 21st-century atmospheric CO_2 (A), global average of surface-ocean pH_T (B), and mean surface Ω_a for the Southern Ocean (C) projected by the University of Bern Physics Institute (PIUB) ocean model forced under different IPCC emission scenarios. The square symbol indicates the projected result for the S650 scenario in 2099. The older IPCC S650 and IS92a scenarios bracket most of the results as do the newer SRES B1 and A2 scenarios. Adapted and redrawn from Meehl *et al*. (2007, Fig. 10.24).

shoaling occurs in the North Atlantic where the ASH is deepest. By 2100, the ASH shoals all the way to the surface in the Southern Ocean, subarctic Pacific, and Arctic Ocean under the IS92a and A2 scenarios. Beyond this century, stabilizing atmospheric CO_2 even at a 450-ppmv target eventually causes most of the deep ocean to become corrosive to both aragonite and calcite.

Model projections are improved by correcting for the present-day model-data bias. With that approach, differences between model projections for a given scenario grow with time elapsed from the reference year, but they generally remain small throughout this century. The most uncertain aspect of future projections by far is the future atmospheric CO_2 trajectory (Fig. 3.7), for which we can only define a range of scenarios designed to bracket future human behaviour. Although projecting future changes in ocean carbonate chemistry is one of the most certain aspects of ocean acidification research, there remains a need to refine our predictive capacity for anthropogenic changes and variability at high latitudes and in the deep ocean, coastal areas, and marginal seas. Of particular importance will be to establish time-series measurements in these areas, not only to improve our understanding of present-day variability but also to help improve future projections.

3.8   Acknowledgements

I thank N. Bates for providing the BATS data, M. González-Dávila and M. Santana-Casiano for providing the ESTOC data, as well as J. E. Dore and co-authors for making their HOT data publicly available. Thanks also to N. Gruber and M. Steinacher for insightful reviews. C. Sabine provided key data and code used to validate my calculations of the buffer factors defined by Egleston *et al.* (2010). J.-P. Gattuso advised on structuring, documenting, and validating these fundamental calculations, now available as a function 'buffesm' in the 'seacarb' software package, since version 2.3.5 (Lavigne and Gattuso 2010). This work was supported by the 'European Project on Ocean Acidification' (EPOCA) funded by the European Community's Seventh Framework Programme (FP7/2007–2013) under grant agreement no. 211384.

References

Andersen, N. and Malahoff, A. (1977). *The fate of fossil fuel CO_2 in the oceans*, 749 pp. Plenum, New York.

Baes, C.F. (1982). Effects of ocean chemistry and biology on atmospheric carbon dioxide. In: W.C. Clark (ed.), *1982 Carbon dioxide review*, pp. 189–211. Oxford University Press, Oxford.

Bates, N.R. (2007). Interannual variability of the oceanic CO_2 sink in the subtropical gyre of the North Atlantic Ocean over the last 2 decades. *Journal of Geophysical Research*, **112**, C09013, doi:10.1029/2006JC003759.

Bates, N.R., Mathis, J.T., and Cooper, L.W. (2009). Ocean acidification and biologically induced seasonality of carbonate mineral saturation states in the western Arctic Ocean. *Journal of Geophysical Research*, **114**, C11007, doi:10.1029/2008JC004862.

Blackford, J.C. and Gilbert, F.J. (2007). pH variability and CO_2 induced acidification in the North Sea. *Journal of Marine Systems*, **64**, 229–41.

Bolin, B. and Eriksson, E. (1959). Distribution of matter in the sea and the atmosphere, In: B. Bolin (ed.), *The atmosphere and the sea in motion*, pp. 130–43. Rockefeller Institute Press, New York.

de Boyer Montégut, C., Madec, G., Fischer, A. S., Lazar, A., and Iudicone, D. (2004), Mixed layer depth over the global ocean: an examination of profile data and a profile-based climatology. *Journal of Geophysical Research*, **109**, C12003, doi:10.1029/2004JC002378.

Brewer, P.G. (2009). A changing ocean seen with clarity. *Proceedings of the National Academy of Sciences USA*, **106**, 12213–14.

Brewer, P.G. and Peltzer, E.T. (2009). Limits to marine life. *Science*, **324**, 347–8.

Broecker, W.S. and Peng, T.H. (1974). Gas exchange rates between air and sea. *Tellus*, **26**, 21–35.

Broecker, W.S. and Peng, T.-H. (1982). *Tracers in the sea*, 690 pp. Lamont-Doherty Geological Observatory, New York.

Broecker, W.S. and Takahashi, T. (1966). Calcium carbonate precipitation on the Bahama Banks. *Journal of Geophysical Research*, **71**, 1575–602.

Broecker, W.S., Li, Y.-H. and Peng, T.-H. (1971). Carbon dioxide—man's unseen artifact. In: D.W. Hood (ed.), *Impingement of man on the oceans*, pp. 287–324. Wiley, New York.

Byrne, R.H., Mecking, S., Feely, R.A., and Liu, X. (2010). Direct observations of basin-wide acidification of the North Pacific, *Geophysical Research Letters*, **37**, L020601, doi:10.1029/2009GL04099.

Caldeira, K. and Wickett, M.E. (2005). Ocean model predictions of chemistry changes from carbon dioxide emissions to the atmosphere and ocean. *Journal of Geophysical Research*, **110**, C09S04, doi:10.1029/2004JC002671.

Canadell, J.G., Le Quéré, C., Raupach, M.R. *et al.* (2007). Contributions to accelerating atmospheric CO_2 growth from economic activity, carbon intensity, and efficiency of natural sinks. *Proceedings of the National Academy of Sciences USA*, **104**, 18866–70.

Cao, L., Caldeira, K., and Jain, A.K. (2007). Effects of carbon dioxide and climate change on ocean acidification and carbonate mineral saturation, *Geophysical Research Letters*, **34**, L05607, doi:10.1029/2006GL028605.

Denman, K.L., Brasseur, G., Chidthaisong, A. *et al.* (2007). Couplings between changes in the climate system and biogeochemistry. In: S. Solomon, D. Qin, M. Manning *et al.* (eds), *Climate change 2007: the physical science basis. Contribution of Working Group I to the Fourth Assessment Report of the Intergovernmental Panel on Climate Change*, pp. 499–587. Cambridge University Press, Cambridge.

Doney, S.C., Lindsay, K., Caldeira, K. *et al.* (2004). Evaluating global ocean carbon models: the importance of realistic physics. *Global Biogeochemical Cycles*, **18**, GB3017, doi:10.1029/2003GB002150.

Doney, S.C., Mahowald, N., Lima, I. *et al.* (2007). Impact of anthropogenic atmospheric nitrogen and sulfur deposition on ocean acidification and the inorganic carbon system. *Proceedings of the National Academy of Sciences USA*, **104**, 14580–5.

Dore, J.E., Lukas, R., Sadler, D.W., Church, M.J., and Karl, D.M. (2009). Physical and biogeochemical modulation of ocean acidification in the central North Pacific. *Proceedings of the National Academy of Sciences USA*, **106**, 12235–40.

Dutay, J.-C., Bullister, J., Doney, S.C. *et al.* (2002). Evaluation of ocean model ventilation with CFC-11: comparison of 13 global ocean models. *Ocean Modelling*, **4**, 89–120.

Dutay, J.C., Jean-Baptiste, P., Campin, J.M. *et al.* (2004). Evaluation of OCMIP-2 ocean model's deep circulation with mantle helium-3, *Journal of Marine Systems*, **48**, 15–36.

Egleston, E.S., Sabine, C.L., and Morel, F.M.M. (2010). Revelle revisited: buffer factors that quantify the response of ocean chemistry to changes in DIC and alkalinity. *Global Biogeochemical Cycles*, **24**, GB1002, doi:10.1029/2008GB003407.

Feely, R.A., Byrne, R.H., Acker, J.G. *et al.* (1988). Winter-summer variations of calcite and aragonite saturation in the northeast Pacific. *Marine Chemistry*, **25**, 227–41.

Feely, R.A., Sabine, C.L., Lee, K. *et al.* (2004). Impact of anthropogenic CO_2 on the $CaCO_3$ system in the oceans. *Science*, **305**, 362–6.

Feely, R.A., Sabine, C.L., Hernandez-Ayon, J.M., Ianson, D., and Hales, B. (2008). Evidence for upwelling of corrosive 'acidified' water onto the continental shelf. *Science*, **320**, 1490–2.

Feely, R. A., Doney, S.C., and Cooley, S.R. (2009). Ocean acidification: present conditions and future changes in a high-CO_2 world. *Oceanography*, **22**, 36–42.

Feely, R.A., Alin, S.R., Newton, J. *et al.* (2010). The combined effects of ocean acidification, mixing, and respiration on pH and carbonate saturation in an urbanized estuary. *Estuarine, Coastal and Shelf Science*, **88**, 442–9.

Findlay, H. S., Tyrrell, T., Bellerby, R.G.J., Merico, A., and Skjelvan, I. (2008). Carbon and nutrient mixed layer dynamics in the Norwegian Sea. *Biogeosciences*, **5**, 1395–410.

Frölicher, T. L. and Joos, F. (2010). Reversible and irreversible impacts of greenhouse gas emissions in multi-century projections with the NCAR global coupled carbon cycle-climate model. *Climate Dynamics*, **35**, 1–21.

González-Dávila, M., Santana-Casiano, J. M., Rueda, M. J., and Llinás, O. (2010). The water column distribution of carbonate system variables at the ESTOC site from 1995 to 2004. *Biogeosciences*, **7**, 3067–3081.

Gruber, N., Sarmiento, J.L., and Stocker, T. (1996). An improved method for detecting anthropogenic CO_2 in the oceans. *Global Biogeochemical Cycles*, **10**, 809–37.

Guinotte, J.M., Orr, J.C., Cairns, S., Freiwald, A., Morgan, L., and George, R. (2006). Climate change and deep-sea corals: will chemical and physical changes in the world's oceans alter the distribution of deep-sea bioherm forming scleractinians? *Frontiers in Ecology and the Environment*, **4**, 141–6.

Hauri, C., Gruber, N., Plattner, G.-K. *et al.* (2009). Ocean acidification in the California Current System. *Oceanography*, **22**, 60–71.

Keeling, C.D. (1973). The carbon dioxide cycle: reservoir models to depict the exchange of atmospheric carbon dioxide with the oceans and land plants. In: S.J. Rasool (ed.), *Chemistry of the lower atmosphere*, pp. 251–329. Plenum Press, New York.

Key, R.M., Kozyr, A., Sabine, C.L. *et al.* (2004). A global ocean carbon climatology: results from Global Data Analysis Project (GLODAP). *Global Biogeochemical Cycles*, **18**, GB4031, doi:10.1029/2004GB002247.

Khatiwala, S., Primeau, F., and Hall, T. (2009). Reconstruction of the history of anthropogenic CO_2 concentrations in the ocean. *Nature*, **462**, 346–9.

Knorr, W. (2009). Is the airborne fraction of anthropogenic CO_2 emissions increasing? *Geophysical Research Letters*, **36**, L21710, doi:10.1029/2009GL040613.

Lavigne, H. and Gattuso, J.-P. (2010). *seacarb: seawater carbonate chemistry with R*. Available at: http://cran.R-project.org/package=seacarb

Le Quéré, C., Raupach, M.R., Canadell, J.G. *et al.* (2009). Trends in the sources and sinks of carbon dioxide. *Nature Geoscience*, **2**, 831–6.

Louanchi, F., Boudjakdji, M., and Nacef, L. (2009). Decadal changes in surface carbon dioxide and related variables in the Mediterranean Sea as inferred from a coupled data-diagnostic model approach. *ICES Journal of Marine Science*, **66**, 1538–46.

McNeil, B. and Matear, R. (2006). Projected climate change impact on oceanic acidification. *Carbon Balance and Management*, **1**, 10.1186/1750-0680-1-2

McNeil, B. and Matear, R. (2008). Southern Ocean acidification: a tipping point at 450-ppm atmospheric CO_2. *Proceedings of the National Academy of Sciences USA*, **105**, 18860–4.

Maier, C, Hegeman, J., Weinbauer, M.G., and Gattuso, J.-P. (2009). Calcification of the cold-water coral *Lophelia pertusa* under ambient and reduced pH. *Biogeosciences*, **6**, 1671–80.

Matsumoto, K., Sarmiento, J.L., Key, R.M. *et al.* (2004). Evaluation of ocean carbon cycle models with data-based metrics. *Geophysical Research Letters*, **31**, L07303, doi:10.1029/2003GL018970.

Meehl, G.A., Stocker, T.F., Collins, W.D. *et al.* (2007). Global climate projections. In: S. Solomon, D. Qin, M. Manning *et al.* (eds), *Climate change 2007: the physical science basis. Contribution of Working Group I to the Fourth Assessment Report of the Intergovernmental Panel on Climate Change*, pp. 747–845. Cambridge University Press, Cambridge.

Merico, A., Tyrrell, T., and Cokacar, T. (2006). Is there any relationship between phytoplankton seasonal dynamics and the carbonate system? *Journal of Marine Systems*, **59**, 120–42.

Najjar, R.G., Jin, X., Louanchi, F. *et al.* (2007). Impact of circulation on export production, dissolved organic matter, and dissolved oxygen in the ocean: results from phase II of the Ocean Carbon-cycle Model Intercomparison Project (OCMIP-2). *Global Biogeochemical Cycles*, **21**, GB3007, doi:10.1029/2006GB002857.

Nakićenović, N. and Swart, R. (2000). *Special report on emissions scenarios*, 598 pp. Intergovernmental Panel on Climate Change, Cambridge University Press, Cambridge.

Olafsson, J., Olafsdottir, S.R., Benoit-Cattin, A., Danielsen, M., Arnarson, T.S., and Takahashi, T. (2009). Rate of Iceland Sea acidification from time series measurements. *Biogeosciences*, **6**, 2661–8.

Omta, A.W., Goodwin, P., and Follows, M.J. (2010), Multiple regimes of air-sea carbon partitioning identified from constant-alkalinity buffer factors. *Global Biogeochemical Cycles*, **24**, GB3008, doi:10.1029/2009GB003726.

Orr, J.C., Maier-Reimer, E., Mikolajewicz, U. *et al.* (2001). Estimates of anthropogenic carbon uptake for four three-dimensional global ocean models. *Global Biogeochemical Cycles*, **15**, 43–60.

Orr, J.C., Fabry, V.J., Aumont, O. *et al.* (2005) Anthropogenic ocean acidification over the twenty-first century and its impact on calcifying organisms. *Nature*, **437**, 681–6.

Pytkowicz, R.M. and Small, L.F. (1977). Fossil fuel problem and carbon dioxide: an overview. In: N. Anderson and A. Malahoff, A. (eds), *The fate of fossil fuel CO_2 in the oceans*, pp. 7–29, Plenum, New York.

Revelle, R. and Suess, H.E. (1957). Carbon dioxide exchange between atmosphere and ocean and the question of an increase of atmospheric CO_2 during past decades. *Tellus*, **9**, 18–27.

Sabine, C.L., Feely, R.A., Gruber, N. *et al.* (2004). The oceanic sink for anthropogenic CO_2. *Science*, **305**, 367–71.

Salisbury, J., Green, M., Hunt, C., and Campbell, J. (2008). Coastal acidification by rivers: a new threat to shellfish? *EOS Transactions, American Geophysical Union*, **89**, 513.

Santana-Casiano, J. M., González-Dávila, M., Rueda, M.-J., Llinás, O., and González-Dávila, E.-F. (2007). The interannual variability of oceanic CO_2 parameters in the northeast Atlantic subtropical gyre at the ESTOC site. *Global Biogeochemical Cycles*, **21**, GB1015, doi: 10.1029/2006GB002788.

Sarmiento, J.L. and Gruber, N. (2006). *Ocean biogeochemical dynamics*, 503 pp. Princeton University Press, Princeton, NJ.

Sarmiento, J.L., Orr, J.C., and Siegenthaler, U. (1992). A perturbation simulation of CO_2 uptake in an ocean general circulation model. *Journal of Geophysical Research*, **97**, 3621–45.

Sarmiento, J.L., Le Quéré, C., and Pacala, S. (1995). Limiting future atmospheric carbon dioxide. *Global Biogeochemical Cycles*, **9**, 121–37.

Sarmiento, J., Monfray, P., Maier-Reimer, E., Aumont, O., Murnane, R., and Orr, J.C. (2000). Sea-air CO_2 fluxes and carbon transport: a comparison of three ocean general circulation models. *Global Biogeochemical Cycles*, **14**, 1267–81.

Skirrow, G. and Whitfield M. (1975). The effect of increases in the atmospheric carbon dioxide content on the carbonate ion concentration of surface ocean water at 25°C. *Limnology and Oceanography*, **20**, 103–8.

Steinacher, M., Joos, F., Frölicher, T.L., Plattner, G.-K., and Doney, S.C. (2009). Imminent ocean acidification in the Arctic projected with the NCAR global coupled carbon cycle-climate model. *Biogeosciences*, **6**, 515–33.

Stumm, W. and Morgan, J.J. (1970). *Aquatic chemistry*, 583 pp. Wiley, New York.

Sundquist, E.T., Plummer, L.N., and Wigley, T.M.L. (1979). Carbon dioxide in the ocean surface: the homogeneous buffer factor. *Science*, **204**, 1203–5.

Takahashi, T., Broecker, W.S., Werner, S.R., and Bainbridge, A.E. (1980). Carbonate chemistry of the surface waters of the world oceans. In: E.D. Goldberg, Y. Horibe, and K. Saruhashi (eds), *Isotope marine chemistry*, pp. 291–326. Uchinda Rokakuho Publishing Co., Tokyo.

Takahashi, T., Olafsson, J., Goddard, J.G., Chipman, D.W., and Sutherland, S.C. (1993). Seasonal-variation of CO_2 and nutrients in the high-latitude surface oceans: a comparative study. *Global Biogeochemical Cycles*, **7**, 843–78.

Touratier, F. and Goyet, C. (2009). Decadal evolution of anthropogenic CO_2 in the northwestern Mediterranean Sea from the mid-1990s to the mid-2000s. *Deep-Sea Research I*, **56**, 1708–16.

Tyrrell, T., Schneider, B., Charalampopoulou, A., and Riebesell, U. (2008). Coccolithophores and calcite saturation state in the Baltic and Black Seas. *Biogeosciences*, **5**, 485–94.

Wagener, K. (1979). The carbonate system of the ocean. In: B. Bolin, E.T. Degens, S. Kempe, and P. Ketner (eds), *The global carbon cycle. SCOPE 13*, pp. 251–8. Wiley, Chichester.

Wallace, D.W.R. (2001). Introduction to special section: ocean measurements and models of carbon sources and sinks. *Global Biogeochemical Cycles*, **15**, 3–10.

Yamamoto-Kawai, M., McLaughlin, F.A., Carmack, E.C., Nishino, S., and Shimada, K. (2009). Aragonite undersaturation in the Arctic Ocean: effects of ocean acidification and sea ice melt. *Science*, **326**, 1098–100.

Yilmaz, A., De Lange, G., Dupont, S. *et al.* (2008). Executive summary of CIESM Workshop 36 'Impacts of acidification on biological, chemical and physical systems in the Mediterranean and Black Seas'. In: F. Briand (ed.), *CIESM Workshop Monograph 36*, pp. 5–22. CIESM, Monaco.

Zeebe, R.E. and Wolf-Gladrow, D.A. (2001). *CO_2 in seawater: equilibrium, kinetics, isotopes*. Elsevier, Amsterdam.

CHAPTER 4

Skeletons and ocean chemistry: the long view

Andrew H. Knoll and Woodward W. Fischer

4.1 Introduction

In present-day seas, animals, algae, and protozoa are threatened by ocean acidification, amplified in many regions by seawater warming and hypoxia (Doney *et al.* 2009). Many species may be affected adversely by 21st-century environmental change, but a decade of research suggests that the hypercalcifying animals responsible for reef accretion may be especially vulnerable to an acidity-driven decrease in the saturation state (Ω; see Box 1.1) of surface seawater with respect to calcite and aragonite.

The geological record reveals that natural changes in the marine carbonate system have affected the evolution and abundance of calcifying organisms throughout the Phanerozoic Eon (542 million years (Myr) ago to the present). This being the case, we can use our understanding of the dynamic behaviour of the carbon cycle and the stratigraphic comings and goings of reef-building organisms to inform us about what, if any, lessons can be drawn from the long-term past and applied to our near-term future.

4.2 A record of atmospheric $p\text{CO}_2$ and past global change

If there is one thing that geology makes clear it is that the earth and its biota are in a continual state of change. Because of its relationship to climate, the partial pressure of CO_2 ($p\text{CO}_2$) in the atmosphere has been of particular interest to geologists and geochemists, but direct measurement of ancient CO_2 levels is impossible for intervals older than those recorded in glacial ice preserved today near

the poles and at high altitude (Petit *et al.* 1999). Therefore, deep-time estimates of $p\text{CO}_2$ rely on models, broadly constrained by geochemical proxy data. For example, the widely applied models of Berner and colleagues (e.g. GEOCARB III; Berner and Kothavala 2001; Berner 2006; Fig. 4.1C) estimate fluxes of carbon from one reservoir to another, based on geochemical proxies (mainly isotope ratios and abundances of sedimentary carbonate and organic carbon), and then calculate successive steady states of the system through time. Additional parameters are considered, including estimates of carbon fluxes due to erosion, river run-off, plant evolution, volcanic weathering, global CO_2 degassing, and land area; these also influence the model results.

These models suggest that atmospheric $p\text{CO}_2$ was not wildly different from pre-industrial modern levels back into the Miocene (23–5 Myr ago), but was moderately higher earlier in the Cenozoic, and higher yet—perhaps five to eight times the present atmospheric level (PAL)—during the warmest parts of the largely unglaciated Mesozoic Era (252–65 Myr ago). Modelled $p\text{CO}_2$ during the Late Palaeozoic ice age is, as might be predicted, low, but earlier Palaeozoic estimates exceed 10 times PAL, with values in some iterations (Berner and Kothavala 2001) spiking as high as 25 times PAL during the later Cambrian Period (~500 Myr ago; Fig. 4.1). An independent biogeochemical model (COPSE; Bergman *et al.* 2004) suggests a similar history, but with less extreme late Palaeozoic and Mesozoic values. Geochemical proxies for ancient $p\text{CO}_2$ (the C-isotopic composition of alkenones, soil carbonates, and organic matter; the distribution of stomata in the epidermis of fossil leaves; and the stable isotope

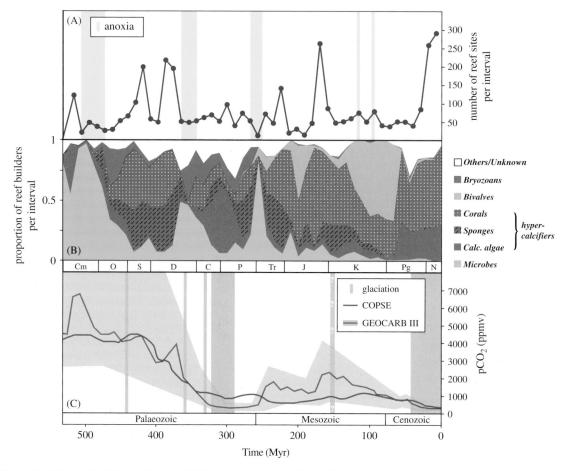

Figure 4.1 Phanerozoic distribution of reefs, reef builders, and environmental variables. A: Reef abundance through time, with intervals of widespread subsurface anoxia in yellow. B: Proportional contribution of reef-building organisms to total reef mass as a function of time. The data are coloured to highlight the comings and going of hypercalcifying taxa (in red). C: Atmospheric pCO_2 estimates and intervals marked by continental ice sheets, as inferred from geological observations, geochemical data, and numerical models. Initials indicate the periods of the Phanerozoic Eon: Cm, Cambrian; O, Ordovician; S, Silurian; D, Devonian; C, Carboniferous; P, Permian; Tr, Triassic; J, Jurassic; K, Cretaceous; Pg, Palaeogene; N, Neogene. Data in A and B are from Kiessling (2009).

ratios of boron) come with their own interpretational challenges (e.g. Royer *et al.* 2001), but generally support model-based hypotheses for Phanerozoic environmental history.

The amount of CO_2 in the atmosphere has clearly varied through geological time, and was often considerably higher than values seen in the atmosphere today. However, when considered alone, estimates of past atmospheric pCO_2 do a poor job in predicting the evolutionary history of skeletal biotas (Fig. 4.1). For example, during the Cambrian and Ordovician periods, when pCO_2 was at its Phanerozoic maximum, skeletal biotas

were radiating throughout the oceans (e.g. Knoll 2003). Clearly, then, pCO_2 is not, in and of itself, a parameter that tracks the evolution of hypercalcifiers. To understand the history of biomineralization, we must place it in the broader context of the expected behaviour of the fluid earth carbonate system as a whole.

Over long timescales ($\geq 10\ 000$ yr) the marine carbonate system operates in a dynamic equilibrium due to feedbacks among fundamental processes operating in the carbon cycle. The carbonate system has six parameters, but because of interdependences can be reduced to two dimensions (see

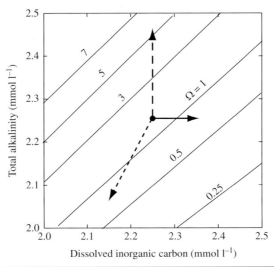

Process	$\Delta A_T : \Delta C_T$
—————— CO_2 outgassing	$0 : 1$
— — · Silicate weathering $CaSiO_3 + 2CO_2 + H_2O \Rightarrow Ca^{2+} + 2HCO_3^- + SiO_2$	$2 : 0$
- - - - · Calcium carbonate precipitation $Ca^{2+} + CO_3^{2-} \Rightarrow CaCO_3$	$-2 : -1$
vector sum	$0 : 0$

Figure 4.2   Three fundamental processes that control the marine carbonate system and influence the saturation state of calcite (Ω_c) of seawater on geological timescales. See text for details.

Chapter 2); here we consider those processes that affect total alkalinity (A_T), total dissolved inorganic carbon (C_T), or both simultaneously. Figure 4.2 depicts one such solution for the marine carbonate system. Given a range of plausible A_T and C_T values, functions of equal saturation state (here calculated for calcite, but a similar reasoning applies to aragonite and magnesian calcite) can be drawn. Three primary processes control the fluid earth carbonate system over long timescales:

(1) CO_2 produced from solid earth sources (volcanoes and metamorphism) and the weathering of sedimentary rocks increases C_T, but does not affect A_T. This process works to lower Ω in seawater and will slow the rate of carbonate precipitation or even begin to promote dissolution of carbonate sediments.

(2) Chemical weathering of silicate minerals consumes protons (derived from CO_2 via carbonic acid) and increases A_T, but not C_T. This process

serves to raise Ω in seawater and both increases the rate of carbonate precipitation and promotes precipitation and preservation of carbonate minerals in areas of the oceans that were previously undersaturated.

(3) Finally, carbonate mineral precipitation provides the mathematical complement to CO_2 outgassing and silicate weathering by consuming A_T and C_T with a slope of –2.

It is not by chance that these fundamental processes have a $\Delta A_T:\Delta C_T$ vector sum that equals zero: they do not operate independently of one another. CO_2 outgassing and silicate weathering are connected via the silicate weathering feedback (e.g. Walker *et al.* 1981), and because the global oceans have a finite and stable water volume they produce carbonate minerals to alleviate inputs of dissolved inorganic carbon and total alkalinity. This forms the basis for a set of negative, or stabilizing, feedbacks on Ω in seawater (a form of 'carbonate compensation'). The result is that, over long timescales, the $CaCO_3$ saturation state of seawater globally is both stable and close to that predicted by thermodynamic equilibrium ($\Omega \sim 1$), despite tropical surface waters being strongly supersaturated. For example, if CO_2 outgassing were to increase, so too would silicate weathering (due to increased temperature). At the same time, the rate of carbonate precipitation from seawater would decrease due to carbonate compensation (lowering Ω) and the system would arrive at a new steady state with a higher atmospheric CO_2, but a similar Ω.

This thought experiment illustrates two important concepts. First, it explains why, in times past, CO_2 can have been far higher than today and yet seawater Ω remained at levels adequate for calcification. Secondly, if we are concerned about ocean acidification events in earth history, we need to look for transient departures from long-term dynamic equilibrium. Global deviations in Ω in seawater cannot last long, in geological terms. Given enough time (> 10 000 yr), carbonate compensation and silicate weathering will work to balance CO_2 outgassing (and inputs of acidity in general). In Chapter 2 Zeebe and Ridgwell provide a detailed discussion of the mechanisms involved.

It is the rapidity of the increase in pCO_2 in present-day oceans that is outstripping the buffering capacity of the earth system and, potentially, the genetic ability of populations to adapt. Thus, if we seek to understand the lessons of the past for our future, we need to identify brief intervals in the past when CO_2 is inferred to have risen too rapidly for the earth system to remain in equilibrium (e.g. Hoegh-Guldberg *et al.* 2007; Knoll *et al.* 2007; and see Section 2.5.3). The geological record does indeed contain several such events, and reveals that perturbations to the marine carbonate system can have complex, and in some cases devastating, effects on populations of calcifying organisms. In addition to these, however, the rock record contains intervals in which the patterns of biological calcification exhibit signal features of stress like those of ocean acidification, but sustained over timescales far longer than those expected from our understanding of ocean acidification and the marine carbonate system (e.g. Knoll *et al.* 2007). These observations highlight an important gap in our understanding and require an

additional class of hypotheses for processes responsible for controlling Ω in surface seawater. The mechanisms of interest are discussed in Section 4.3.3.

4.2.1 The Palaeocene–Eocene Thermal Maximum

Because their calibration depends on assumptions of equilibrium, models such as GEOCARB and COPSE integrate over long intervals of time and cannot be used to identify times of geologically rapid pCO_2 increase in the geological record. We need to find high-resolution geological records in which geochemical data suggest rapid environmental change. Perhaps the best studied example is the so-called Palaeocene–Eocene Thermal Maximum (PETM), a brief interval of pronounced global warming about 55 Myr ago (Kennett and Stott 1991; Zachos *et al.* 1993).

Warming of 5 to 8°C, with larger increases at high latitudes, has been inferred from a sharp excursion of about –1.7‰ in the oxygen isotopic composition of carbonate skeletons (Zachos *et al.* 2003). Other geochemical proxies for sea-surface temperature (Mg/Ca, the relative abundance of unsaturated alkenones, and the structures of archaeal lipids) are consistent with this estimate, as are biogeographical changes among both corals and land plants (reviewed by Scheibner and Speijer 2008). A –2.5 to –3‰ shift in the C-isotopic composition of carbonate skeletons coincides with the temperature excursion, suggesting that increased atmospheric CO_2 supplied from an isotopically light source drove climate change. It has been hypothesized that catastrophic release of methane from shelf/slope clathrate hydrates was involved in the PETM event (Dickens *et al.* 1995), but the inability of clathrate release to supply the quantity of carbon needed to account for recorded C-isotopic change (Zachos *et al.* 2005) suggests that other mechanisms, including thermogenic methane release associated with end-Palaeocene flood basalts, may have played a role (Svensen *et al.* 2004; Higgins and Schrag 2006). In any event, high-resolution stratigraphic and geochemical data indicate that the PETM perturbation was rapid and transient; the decrease in C-isotope values occurred largely in two bursts, each less than 1000 yr in duration, and the system returned to its background state within about 100 000 years (Rohl *et al.* 2000; Nunes and Norris 2006). Shoaling of the calcite compensation depth by as much as 2 km provides empirical evidence of ocean acidification (Zachos *et al.* 2005).

Parallels to the present prompt the question of how earth's biota fared across the PETM event. On land, vascular plants record pronounced but transient species migrations, with only limited extinction (Wing *et al.* 2005). Palaeocene mammals suffered extinctions, but new taxa appeared, including many modern mammalian orders, most at initially small size—marking the PETM as a time of pronounced mammalian *turnover* rather than diversity decline (Gingerich 2006).

Many marine taxa also display a pattern of pronounced turnover but limited extinction (Scheibner and Speijer 2008), with corals (Kiessling 2001) and various microplankton groups (Scheibner and Speijer 2008) showing transient range expansion toward the poles. Major extinction depleted the diversity of deep-sea benthic foraminiferans (Thomas 2007), but corals—a group considered especially vulnerable to present-day ocean acidification (Kleypas *et al.* 1999; Hoegh-Guldberg *et al.* 2007)—show little change in diversity. Diversity, however, does not tell the whole story. In a comprehensive review of carbonate platforms along the Palaeogene margins of the Tethyan Ocean, Scheibner and Speijer (2008) demonstrated that shelf margin reefs built by colonial corals and calcareous algae declined markedly at the PETM. Solitary (but not colonial) scleractinians occur in basal Eocene carbonates, but contribute relatively little to carbonate accumulation. Across the same boundary, larger benthic foraminifera expand dramatically.

Thus, combined warming and ocean acidification 55 Myr ago made only a limited long-term mark on the marine biota. While acidification expanded the volume of undersaturated deep-sea waters, skeleton formers persisted on the shelves. This persistence, however, does not imply strict ecological continuity. Coral reef ecosystems declined widely and did not recover for hundreds of thousands of years—a geological instant but almost impossibly long by the standards of human civilization. Migration may have played an important role in taxonomic persistence on land and in the ocean, but

this required corridors for migration, no longer unimpeded on land or, perhaps, in the sea. In summary, then, the PETM record may be reassuring on an evolutionary timescale, but it raises concerns on the ecological scales relevant to humans. Persistence of coral species, perhaps in isolated populations with little or no calcification (e.g. Fine and Tchernov 2007) may not ensure the continual accretion of reefs, with their attendant ecosystem services.

4.2.2 End-Permian mass extinction

An earlier event interpreted in terms of ocean acidification occurred 252 Myr ago, at the end of the Permian. Estimates of pCO_2 change and global warming coincide broadly with those for the PETM, but the biological consequences were starkly different. On land, a poorly resolved record of vertebrate evolution suggests migration and increased taxonomic turnover across the Permian–Triassic boundary (Smith and Botha 2005), and land plants show both poleward migration and regionally distinct patterns of extinction, most pronounced at high southern latitudes (Rees 2002; Abu Hamad *et al.* 2008). Marine ecosystems, however, were devastated—species loss is estimated at 90% or more, while metazoan reefs and other ecosystems that had long dominated the seafloor disappeared (Erwin 2006).

A reasonable scenario for end-Permian mass extinction invokes rapid, massive influx of CO_2 into the atmosphere and oceans, in association with one of the largest eruptions of flood basalts known from the geological record. At least 1.2×10^6 km^3 of basaltic volcanic rocks were deposited over what is now western Siberia, largely accumulating in a million years or less (Reichow *et al.* 2007). Comparison with modern volcanoes, such as those in Hawaii, suggests that this event might have released 10^{17} to 10^{19} mol CO_2 (equivalent to 10 to 1000 times the amount of CO_2 estimated for the latest Permian atmosphere; Wignall 2001), although integrated into an active carbon cycle over a million years, this would increase atmospheric levels by only twofold or less (Knoll *et al.* 2007). Comparisons with Hawaiian volcanism, however, probably underestimate CO_2 release from Siberian trap volcanism, very likely by a wide margin. The Siberian magmas ascended through thick carbonate and evaporite deposits, adding CO_2 from contact heating and decarbonation (and sulphur dioxide, a second source of acidity; e.g. Knoll *et al.* 2007; Ganino and Arndt 2009; Iacono-Marziano *et al.* 2009). Today, as much as 10% of all CO_2 released from mid-ocean ridges, volcanoes, and convergent plate margins can be attributed to Mount Etna, a volcano developed on extensive platform carbonates (Marty and Tolstikhin 1998). Moreover, Siberian Trap magmas and lavas intruded into and extruded onto extensive late Palaeozoic peat and brown coal deposits, generating large additional fluxes of CO_2 and thermogenic methane (CH_4) to the atmosphere (Retallack and Jahren 2008). Thus, both massive volcanism and the geological context of the volcanism contributed to rapid CO_2 (and SO_2) increase, driving global warming and ocean acidification.

End-Permian extinctions in the oceans were extensive but not random. Knoll *et al.* (1996) documented a strong pattern of selectivity with respect to fundamental physiological and ecological features of the biota. Hypercalcifiers and other animal and algal groups with limited capacity to pump ions across membranes show nearly complete extinction, but groups better able to modulate the composition of fluids from which carbonate skeletons were precipitated survived differentially well. Further, taxa characterized by high rates of exercise metabolism and well developed respiratory and circulatory systems survived better than anatomically simple hypometabolic taxa, and infauna survived better than epifauna. In 1996, the term 'ocean acidification' was not a part of palaeontology's vocabulary, but an extensive physiological literature suggested that observed patterns of extinction and survival matched predictions made on the basis of organismic tolerance to and compensation for hypercapnia (elevated CO_2 in internal fluids).

Stimulated by environmental concerns, a large body of research on marine organisms has accumulated during that past 14 or so years, prompting a number of general statements about vulnerability to hypercapnia and increasing seawater acidity. For example, Widdicombe and Spicer (2008, p. 194) wrote:

> We conclude that there is clear potential for the chemical changes associated with ocean

acidification to impact on individuals at a physiological level particularly through disruption of extracellular acid-base balance. There is some weak evidence that the severity of this impact could be related to an organism's phylogeny suggesting that both species and taxonomic measures of biodiversity could be reduced. However, there is also evidence that potential species extinctions will be more strongly governed by factors related to an organism's lifestyle and activity (e.g. infaunal v epifaunal, deep v shallow, deposit feeder v suspension feeder, large v small) than by its phylogeny. There is also huge uncertainty as to what extent organism adaptation or acclimation will mitigate the long term effects of ocean acidification.

And, in a comparison of marine animals more and less tolerant of hypercapnia, Melzner *et al.* (2009) proposed that 'All more tolerant taxa are characterized by high (specific) metabolic rates and high levels of mobility/activity'. These conclusions about the present recall observed patterns of end-Permian extinction.

In the light of new experimental results, especially those on ocean acidification and calcification, Knoll *et al.* (2007) returned to the Permian–Triassic data, focusing largely on inferred differences in the physiology of skeleton formation. This exercise requires physiological inference from fossil remains. From fossils, we can establish the lifestyles of ancient organisms and, to the extent that phylogeny is a good predictor of anatomy and physiology, those can be inferred, as well. Widdicombe and Spicer (2008) reasonably stress that strict phylogenetic focus in ocean acidification research may be limiting. Nonetheless, in terms of broad physiological attributes important for assessing hypercapnia and ocean acidification, many species within marine classes and phyla share fundamental features of metabolism and (key to interpreting fossils) skeletal biosynthesis. Thus, while individual species may respond variably to increased CO_2 load, groups like corals will have a statistical tendency to respond coherently—and differently from, say, molluscs. And what the fossil record provides is a statistical digest of extinction and survival.

Knoll *et al.*'s (2007) focus on skeletal physiology once again showed evidence of dramatic variations

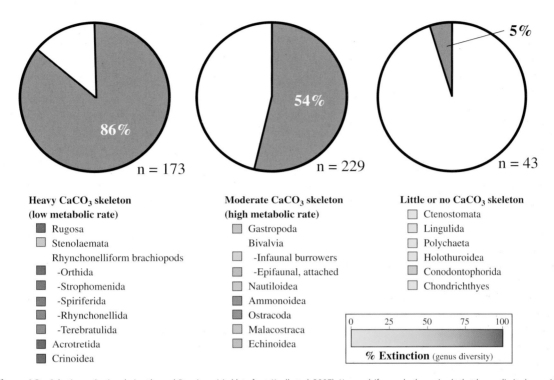

Figure 4.3 Selective extinction during the end-Permian crisis (data from Knoll *et al.* 2007). Hypercalcifers and other animals that have a limited capacity to buffer internal fluids lost 86% of known genera; groups with carbonate skeletons but well-developed physiological mechanisms for buffering internal fluids lost 54% of genera; and groups that use carbonate minerals sparingly or not at all in skeleton formation lost 5% of genera. Colour coding for individual taxa shows how they align along a gradient of increasing extinction severity. The distribution of taxa along this gradient can be predicted from expected variations in vulnerability of these different groups to hypercapnia and ocean acidification as deduced from physiological experiments. See text for discussion and references.

in extinction probability (Fig. 4.3). Hypercalcifiers (corals and massively calcifying sponges) and other groups with minimal capacity to buffer calcifying fluids (e.g. lophophorates and crinoids) lost 86% of their genera during the extinction, whereas genera of animals and protists that made skeletons of materials other than $CaCO_3$ exhibit extinction rates of only about 5%—comparable to or less than background extinction rates for the preceding 50 Myr. Calcifying organisms better able to modulate internal fluids (mostly molluscs and arthropods) show intermediate levels of genera loss (54%), and within this category, groups predicted to be relatively vulnerable to hypercapnic stress based on ecology or anatomy disappeared at rates twice those of groups deemed less vulnerable. Also, for a series of animal, protozoan, and algal taxa, genera characterized by carbonate skeletons showed much higher rates of

extinction than close non-calcifying relatives, providing some control on physiological variability among taxa. All of these observations are consistent with a prominent role for hypercapnia/ocean acidification in generating the selectivity associated with end-Permian mass extinction.

4.2.3 Why the difference?

Clearly, end-Permian fossils record an environmental catastrophe more dire than the PETM, but why did the marine biota respond so differently in the two events? Possibly, end-Permian environmental disruption was simply more pronounced. Certainly, the carbon isotopic excursion across the Permian–Triassic boundary is double that at the PETM. But there is more to the story. In fact, in our summary of environmental triggers for end-Permian mass extinction we revealed only half the story—the other half may be what ensured the unusual severity of this largest mass extinction.

In some ways, the late Permian world into which the Siberian Traps erupted was almost maximally different from the earth we experience today. Continental masses were aggregated into the supercontinent Pangaea, resulting in a Panthalassic ocean more than a hemisphere in extent. With the mid-Permian decay of late Palaeozoic continental ice sheets and climate warming, physical circulation in this ocean may have been relatively sluggish, promoting extensive subsurface hypoxia (e.g. Meyer *et al.* 2008). Both deep-sea black shales, preserved in obducted slivers of late Permian seafloor (Isozaki 1997), and biomarker lipids that document anoxygenic photosynthetic bacteria in latest Permian seas (Cao *et al.* 2009) record oxygen depletion in subsurface water masses. Global warming associated with volatile release from Siberian volcanism appears to have tipped the oceans into a state of widespread anoxia beneath the mixed layer (Wignall and Twitchett 1996), generating additional fluxes of CO_2 from upwelling waters (Knoll *et al.* 1996). In short, as observed at some level today, the end-Permian extinction was not a crisis fomented by hypercapnia/ocean acidification, global warming, *or* subsurface anoxia, it was a crisis in which all three occurred simultaneously (Knoll *et al.* 2007). Given the interconnected nature of the earth system, it could hardly be otherwise.

Subsurface anoxia would have impacted on the biota in several ways. Most obviously, shoaling of the oxycline would have stressed benthic populations, much as it does in seafloor 'dead zones' today. Secondly, the physiological effects of warming, hypercapnia, and hypoxia are not independent, but rather are synergistic, amplifying physiological stress (Pörtner 2008; see Chapter 8). Many organisms in end-Permian oceans probably died of asphyxiation; nonetheless, end-Permian skeletons suggest marked selectivity consistent with the physiological effects of hypercapnia/ocean acidification.

Recently, Higgins *et al.* (2009) explored the consequences of widespread subsurface anoxia for the carbonate system. As discussed in more detail below, anaerobic heterotrophs generate total alkalinity as they remineralize organic matter. Thus, in oceans with widespread subsurface anoxia, subsurface water masses should be expected to have higher Ω than at present, while the Ω of overlying surface waters should be reduced. Reduction of surface-water Ω should, in turn, make skeleton formation by hypercalcifiers more difficult (e.g. Gattuso *et al.* 1999), increasing the physiological stress on latest Permian corals and hypercalcifying sponges and other organisms with limited ability to modulate internal fluid composition.

The punchline for end-Permian extinction, then, is that the ability of marine organisms to precipitate calcium carbonate skeletons was impeded by *two* circumstances. Expanding subsurface anoxia and rapidly rising pCO_2 would both have lowered Ω in the surface ocean; operating in tandem, they appear to have depressed Ω strongly for a biologically protracted interval of time. To the extent that this is correct, it suggests that the end-Permian extinction can inform current research in terms of taxonomic, ecological, and physiological vulnerability to 21st-century global change. Perhaps mercifully, however, the extent of the end-Permian catastrophe appears to rely on concatenated factors only partially in play today.

4.3   Is there a more general historical pattern?

Reefs are a striking component of the carbonate sediments through geological time, and the record of biological calcification is well written in the fossil composition of ancient reefs. The preceding sections examined two historical events wherein ocean acidification occurred due to rapid influx of CO_2 from the solid earth; the effects of these events on the biota were variable, but in one case, at least, devastating for marine hypercalcifiers. Using historical metrics, one can ask a set of broader questions about the processes that have controlled the abundance and diversity of reef-building organisms through time. What conditions are responsible for observed long-term patterns in the evolution of hypercalcifiers? And how might these trends reflect long-term changes in the nature and behaviour of the marine carbonate system that extend beyond short-term ocean acidification events?

4.3.1   Skeletons and surface-water Ω

A large body of experimental research supports the hypothesis that the cost and effectiveness of formation of carbonate skeletons vary inversely with Ω (Gattuso *et al.* 1999; Langdon and Atkinson 2005, and references therein). Skeletal responses to ocean acidification, however, vary across taxa, as might be predicted from the basic features of skeletal physiology. Because hypercalcifiers have only limited physiological capacity to pump ions across membranes to modify the composition of fluids from which skeletal minerals are precipitated, they are particularly vulnerable to decreasing Ω in ambient waters. Ries *et al.* (2009) grew a variety of skeletal invertebrates and algae at a range of Ω. Not surprisingly, a majority of the experimental species showed a decline in skeleton formation with decreasing Ω. Three arthropods, however, actually increased skeletal mass with decreasing Ω, and red and green algae, as well as a limpet and a sea urchin, showed an initial increase in calcification followed by decline as Ω decreased beyond a threshold value. Gooding *et al.* (2009) also observed an increase in skeletal mass in the sea urchin *Pisaster ochraceus* grown at elevated pCO_2 and temperature. These responses are consistent with the pattern of extinction and survival across the Permian–Triassic boundary, and expectations for vulnerability to decreasing Ω. We note, however, that changing calcification rate is only one of many potential physiological responses to ocean acidification—in some cases, skeletal compensation is accompanied by decreased performance in other important aspects of growth or metabolism (Pörtner 2008; see Chapter 10).

4.3.2   Hypercalcification through time

Palaeontologists have long understood that neither the abundance nor the taxonomic composition of reefs has remained constant through time, prompting the generalization that six to eight successive reef biotas have flourished in Phanerozoic oceans, separated by stratigraphic gaps during which metazoans contributed little to reef accretion (e.g. Wood 1999; Copper 2002a; Kiessling 2002; Ezaki 2009). In a recent compilation of reef abundance and diversity, Kiessling (2009) has refined this view. Perhaps surprisingly, Kiessling's data show that widespread reef development, common from early Neogene time through to today, is not generally characteristic of Phanerozoic oceans (Fig. 4.1A).

Reefs can be defined as discrete rigid carbonate structures formed by *in situ* or bound components that develop topographic relief upon the seafloor (Wood 1999). Structures that fit this definition have existed since the evolution of benthic microbial communities more than 3 billion years ago (Grotzinger and Knoll 1999; Allwood *et al.* 2009). With the late Neoproterozoic emergence of complex multicellularity, both animals and algae began to participate in reef accretion, and through the Phanerozoic Eon (the past 542 Myr) a number of major taxa have contributed to reef formation. Kiessling's (2009) compilation highlights several first-order stratigraphic patterns in the composition of reef biotas (Fig. 4.1B). First, microbial accretion did not cease with the evolution of hypercalcifying metazoans, but rather declined slowly and fitfully through time, ceasing to be quantitatively important only in the Cretaceous. Second, bryozoans, bivalves, calcareous algae, and other groups have contributed a moderate volume of reef carbonate through time, with calcareous algae peaking in the late Carboniferous and early Permian periods and again in the Neogene, and rudist bivalves playing a major role in Cretaceous reefs (to the extent that rudist deposits fit Wood's definition of a reef; Gili *et al.* 1995). Throughout the Phanerozoic Eon, however, peaks in reef abundance correspond to times of widespread and diverse hypercalcifying animals, mainly massively calcifying sponges and cnidarians, and calcareous algae. It is principally the episodic waxing and waning of these organisms that gives rise to the widely applied concept of successive reef biotas. Insofar as hypercalcifiers should be sensitive to factors that control Ω, we can ask whether these factors were in play when successive reef biotas expanded and collapsed.

Archaeocyathids, an extinct group of calcareous sponges, were major contributors to reef accretion in Early Cambrian oceans, but a major extinction event near the Cambrian Stage 3–4 (Botomian–Toyonian) boundary essentially wiped out the group, beginning a nearly 50 Myr interval during which metazoans played only a minor role in reef accretion (Rowland and Shapiro 2002). As part of a broader radiation of well-skeletonized animals, sponges, rugose and tabulate corals, and bryozoans renewed metazoan reef accretion during the Ordovician (Harper 2006), and reefs constructed by these organisms persisted with varying abundance until the Frasnian–Famennian boundary in the late Devonian, some 370 Myr ago. At this time, hypercalcifying animals collapsed again, ushering in a brief interval of animal-poor microbial reefs (Copper 2002b). Animal–algal reefs occurred throughout the later Carboniferous and Permian, with hypercalcifying sponges once again playing a particularly important role in later Permian build-ups. Then, as discussed above, hypercalcifiers suffered differentially severe losses during the end-Permian mass extinction.

Early Triassic reefs were microbial. Beginning in the Middle Triassic, however, reef abundance increased with the radiation of scleractinian corals and sponges. Many hypercalcifiers disappeared, once again, at the end of the Triassic, although enough species survived to fuel renewed reef expansion during the Jurassic (Lathuilière and Marchal 2009). Another decline toward the end of the Jurassic was followed by an extended interval dominated by rudist bivalves. Only after the end-Cretaceous mass extinction did modern reef ecosystems begin to take shape.

4.3.3 Mechanisms to explain the pattern of hypercalcification in reefs

Canonically recognized mass extinctions do not fully explain the stratigraphic pattern of hypercalcifier evolution (Kiessling 2009, Kiessling and Simpson 2011). Hypercalcifiers disappeared completely during the end-Permian mass extinction and declined markedly in diversity and extent during Late Devonian and Late Triassic extinctions. On the other hand, mass extinctions at the end of the Ordovician (Sheehan 2001) and Cretaceous do not show the preferential loss of hypercalcifiers observed for end-Permian collapse (Knoll et al. 2007); the proportional extinction of hypercalcifiers was modest during the end-Ordovician and end-Cretaceous mass extinctions, and the loss of metazoan-built reefs was transient.

Although there is good physiological reason to connect the abundance and evolutionary history of hypercalcifiers to state changes in Ω in ambient seawater (e.g. Veron 2008), the long (>10^5 yr) timescales on which reef organisms have waxed and waned introduces a new class of problem. As discussed above, variables such as $p\mathrm{CO_2}$ and temperature do not appear to explain the stratigraphic pattern of hypercalcifier evolution (Kiessling 2009). This should not be surprising, given the flexibility of these parameters within the dynamic equilibrium described in Section 4.2. What we require is a mechanism that is congruent with this dynamic equilibrium and yet can have an impact on marine carbonate chemistry with a characteristic timescale greater than that expected for ocean acidification, bearing in mind the ever-present stabilizing feedbacks.

Hypercalcifying organisms residing in reefs experience the Ω of regional surface seawater. Global Ω is set by the overall marine carbonate system, but this value (which is close to thermodynamic equilibrium) represents a cumulative parameter integrated over the entire volume of global seawater. There are, in spite of this, large gradients in Ω. These gradients result in part from the hydrological cycle (controlling salinity) and inorganic factors controlling the solubility of carbonate polymorphs (e.g. temperature and pressure). An underappreciated process promoting these gradients is the effect of the biological pump (Higgins et al. 2009). $\mathrm{CO_2}$ fixed by primary producers in the surface ocean is aerobically respired in the deep, setting up a gradient in C_T that pushes Ω higher in surface seawater and lower in deep seawater (Fig. 4.4).

We can imagine how the geobiological behaviour of the biological pump differed in times past. The pump could be stronger or weaker. A stronger biological pump means larger gradients in C_T, and would translate into a world characterized by even larger gradients and a higher Ω in surface seawater than we observe today. We can also imagine a world with a reduced depth gradient in Ω, including lower surface-seawater Ω, due to the impact of anaerobic metabolisms. In contrast to aerobic respiration, all anaerobic metabolisms significantly affect A_T in addition to C_T (Soetaert et al. 2007; Higgins et al. 2009). Anoxic environments characterized by

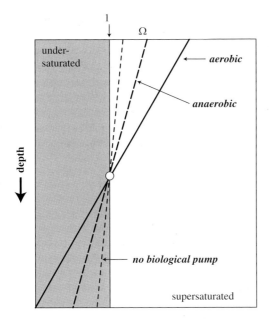

Figure 4.4 Schematic cross-section of gradients in Ω in seawater under different scenarios. Global seawater tends to arrange gradients around a mean value controlled by carbonate compensation ($\Omega \sim 1$). Today, large gradients exist with depth due to the biological pump and aerobic respiration. Surface seawater is strongly supersaturated and deep seawater is undersaturated. A world without a biological pump would still have gradients in Ω in seawater due to the effects of temperature, pressure, and salinity. An idealized world with a biological pump, but anaerobic metabolisms at depth, will have subdued gradients in Ω, due to metabolic gradients in total alkalinity.

anaerobic respiration tend to ameliorate depth gradients in Ω, while aerobic metabolisms tend to promote them. This means that a world in which a significant proportion of biological pump electrons pass through anaerobic metabolisms will tend to have subdued gradients in Ω, even though global seawater may not vary (Fig. 4.4). The ability of the marine carbonate system to accommodate this reorganization is, in principle, intimately tied to the processes that control the long-term and large-scale development of marine anoxia.

It is now possible to develop the logic for hypotheses to explain the abundance and diversity of marine hypercalcifying organisms through time, making reference to both short-term events of rapid CO_2 influx that cause Ω to deviate briefly (in geological terms) from a dynamic equilibrium, and also long-term transitions due to the waxing and waning of ocean basin anoxia. Might multiple ancient

reef crises be related to rapid decreases in Ω associated with ocean acidification, the expansion of subsurface anoxia, or both?

Recent radiometric determinations suggest that the Cambrian extinction of archaeocyathids may coincide with emplacement of the Kalkarindji Large Igneous Province, Australian flood basalts comparable in scale to the end-Permian Siberian Traps (Evins *et al.* 2009). Citing the extensive deposition of black shales, Zhuravlev and Wood (1996) linked Cambrian hypercalcifer extinctions to the expansion of anoxic subsurface waters. In turn, Glass and Phillips (2006) and Hough *et al.* (2006) related anoxia and extinction to the Kalkarindji eruptions. In fact, the palaeobiological particulars of this extinction suggest that we should focus on the modulation of Ω discussed in the previous section. Early Cambrian hypercalcifiers were widespread in shallow shelf and platform environments, and they disappeared despite the limited incursion of anoxic water masses into shallow marine environments. Trilobites, the dominant organisms recorded in Cambrian strata, suffered major extinctions, but because first appearances kept pace, the overall pattern is one of marked turnover, not loss of diversity (Bambach *et al.* 2004). Small shelly fossils of phylogenetically diverse origins declined across this interval, but as their record is closely tied to preservational circumstances that also change, it is difficult to quantify their evolutionary pattern (Porter 2004). The renowned Burgess Shale in British Columbia documents the persistence of diverse animal groups into mid-Cambrian oceans, but few of these made robust carbonate skeletons. In fact, in later Cambrian carbonates, skeletons account for only a few per cent of known carbonate volume—more like the Early Triassic than any other time of Phanerozoic history (Pruss *et al.* 2010). As in the end-Permian example, massive basaltic volcanism appears to have been visited on a planet whose oceans were already characterized by subsurface hypoxia. Volcano-driven warming and expansion of subsurface anoxia caused mass mortality in deeper marine environments, but it may have been the associated decline in surface-water Ω that selectively removed hypercalcifiers from shallow shelves and platforms.

Rapidly accumulating geochemical data (Hurtgen *et al.* 2009; Gill *et al.* 2011) suggest that the ensuing later Cambrian to earliest Ordovician oceans experienced persistent or recurring subsurface hypoxia, at least episodically expanding to widespread anoxia. Thus, the protracted post-extinction interval marked by limited skeletal contributions to accumulating carbonates, a dearth of hypercalcifiers, and few metazoan contributions to reefs may have been governed, at least in part, by the redox-modulated depression of Ω in surface seawater.

Depression of surface Ω may also explain hypercalcifier loss at the boundary between the Frasnian and Famennian stages of the Devonian. Although commonly included as one of the 'big five' mass extinctions, extinction rates were not unusually high during this event. Rather origination rates declined, resulting in a 'mass depletion' of standing diversity (Bambach *et al.* 2004). In fact, the Frasnian–Fammenian loss of reef-building hypercalcifers, especially calcareous sponges, sits in the middle of a protracted interval of diversity decline. As discussed earlier for Permian and Cambrian hypercalcifier extinctions, this entire mid-Devonian to basal Carboniferous interval is characterized by widespread black shales as well as high taxonomic turnover. Once again, then, anoxia at depth may have influenced diversity through a protracted interval, with the Fransian–Famennian boundary representing an extreme perturbation that exceeded the capacity of hypercalcifiers to respond.

Thus, a case can be made that the three Palaeozoic intervals marked by hypercalcifer extinction and subsequent gaps in metazoan reef accretion may share the environmental circumstance of marked reduction of Ω in surface waters. Although in need of geochemical testing, this hypothesis can account for both the timing and taxonomic/physiological selectivity of extinctions at these moments in time.

Transient ocean acidification triggered by eruption of the Central Atlantic Magmatic Province has also been argued to be the trigger mechanism for pronounced extinctions at the end of the Triassic (200 Myr ago; Hautmann 2004; Hautmann *et al.* 2008a). Coral diversity declined strongly (Lathuilière and Marchal 2009), but not completely (Kiessling *et al.* 2009), so that relatively diverse communities, including reefs, became re-established on a million-year timescale (Hautmann *et al.* 2008b). Interestingly, flood basalts and transient subsurface anoxia recur

about 17 Myr later, at the Pliensbachian–Toarcian boundary of the early Jurassic; in this case, corals show elevated turnover rates, but not strong diversity decline (Lathuilière and Marchal 2009). Ocean acidification has been proposed to explain both Mesozoic events (Kiessling and Simpson 2011, and references therein). These events make it clear that while several episodes of hypercalcifier extinction coincide with large igneous eruptions (e.g. Courtillot and Olson 2007), large igneous provinces do not invariably result in mass extinction of reef-building metazoans (Wignall 2001). The hypothesis entertained here suggests that massive volcanism affected hypercalcifiers most strongly when the redox state of the oceans was prone to subsurface anoxia (Fig. 4.1), facilitating a combined influence of ocean acidification and redox-driven redistribution of total alkalinity on surface-water Ω. Of individually moderate effect, ocean acidification and subsurface anoxia in tandem provide a lethal cocktail for organisms with limited physiological capacity to buffer the fluids from which they precipitate carbonate skeletons.

Earth scientists commonly argue the merits of factors X *versus* Y in affecting the history of life; more realistic approximations may occur when we discuss the effects of X *and* Y, occurring together or in series. The point here is not to argue that saturation level was the sole influence on hypercalcifier evolution through time. Additional aspects of seawater chemistry (e.g. Mg/Ca), biological interactions, and other influences may well have affected hypercalcifier evolution (Kiessling 2009). But we do argue that emerging geochemical tools provide a means of exploring both short- and long-term changes in Ω through time and that results to date support the hypothesis that episodic declines in Ω have played a major role in governing the stratigraphic distribution of hypercalcifiers and, hence, metazoan reefs.

4.4 Summary, with lessons for the future

Several events in earth history bear the fingerprints of ocean acidification. The two examples discussed here in detail (the PETM and Permian-Triassic mass extinction) reveal a complex, and sometimes devastating, impact on the abundance, diversity, and evolution of calcifying organisms. More broadly, the geological record indicates that calcifying organisms have been subject to episodically changing Ω throughout the past 542 Myr. Volcanism and associated thermal decomposition of organic matter have generated ocean acidification over intervals that were biologically long if geologically brief. Decreases in surface-seawater Ω sustained for millions of years have also been associated with the expansion of anoxic subsurface waters and, hence, anaerobic heterotrophy. When ocean acidification and the expansion of anoxic waters occur in tandem, the result can be mass extinction that is differentially severe for hypercalcifiers and other animals with limited capacity to modulate the ionic composition of internal fluids. Our understanding of these events remains imperfect, but the perspective they offer can be used to better inform our expectations for the future of reefs during our current anthropogenic experiment.

Several generalities about the past seem relevant to our environmental future:

(1) In assessing the vulnerability of the biota to decreasing Ω, *rate* is key. When the rate of environmental change is fast, the probability of extinction is increased. Times of biological crisis in the past were times, like today, when pCO_2 increased rapidly, not when pCO_2 was high (see also Chapter 2).

(2) There is no clear reason to expect that the coming century will see a 'sixth extinction' comparable to those at the end of the Permian and Cretaceous. Nonetheless, the loss of vulnerable taxa from ocean ecosystems could affect ecological function for many millennia. The past tells us that there will be winners and losers in a changing ocean. Corals, and therefore the coral reef communities that harbour so much of marine diversity, may well be among the losers.

(3) In the past, extinction was more pronounced when several biological challenges were imposed at once. In similar fashion, the consequences of ocean acidification will be amplified by global warming, declining levels of dissolved oxygen (Brewer and Peltzer 2009), habitat loss, overfishing, and the impedance of routes for migration.

(4) The timescale for recovery from ocean acidification is measured in geological time. Thus, even assuming that inputs of CO_2 into the atmosphere and oceans are ameliorated, diversity loss will appear permanent on timescales relevant to the human population.

(5) Physiological experiments and geological history are mutually illuminating. Studies of the past can suggest relative biological vulnerabilities, highlighting candidates for physiological research. Physiology, in turn, provides an important lens through which palaeontological research can be focused.

4.5   Acknowledgements

We thank H. Pörtner, J. Erez, and J. Barry for useful discussions of biomineralization and physiological responses to ocean acidification, and J.-P. Gattuso and R. Zeebe for helpful reviews of our chapter. We also thank W. Kiessling for providing the diversity and abundance data illustrated in Fig. 4.1. Research was supported in part by the NASA Astrobiology Institute (AHK) and the Agouron Institute (WWF).

References

Abu Hamad, A., Kerp, H., Voerding, B., and Bandel, K. (2008). A late Permian flora with *Dicroidium* from the Dead Sea region, Jordan. *Review of Palaeobotany and Palynology*, **149**, 85–130.

Allwood, A.C., Grotzinger, J.P., Knoll, A.H. *et al.* (2009). Controls on development and diversity of Early Archean stromatolites. *Proceedings of the National Academy of Sciences USA*, **106**, 9548–55.

Bambach, R.K., Knoll, A.H., and Wang, S. (2004). Origination, extinction, and mass depletions of marine diversity. *Paleobiology*, **30**, 522–42.

Bergman, N.M., Lenton, T.M., and Watson, A.J. (2004). COPSE: a new model of biogeochemical cycling over Phanerozoic time. *American Journal of Science*, **304**, 397–437.

Berner, R.A. (2006). GEOCARBSULF: a combined model for Phanerozoic atmospheric O_2 and CO_2. *Geochimica et Cosmochimica Acta*, **70**, 5653–64.

Berner, R.A. and Kothavala, Z. (2001). GEOCARB III: a revised model of atmospheric CO_2 over Phanerozoic time. *American Journal of Science*, **301**, 182–204.

Brewer, P.G. and Peltzer, E.T. (2009). Limits to marine life. *Science*, **324**, 347–8.

Cao, C.-Q., Love, G.D., Hays, L.E. *et al.* (2009). Biogeochemical evidence for euxinic oceans and ecological disturbance presaging the end-Permian mass extinction event. *Earth and Planetary Science Letters*, **281**, 188–201.

Copper, P. (2002a). Ancient reef ecosystem expansion and collapse. *Coral Reefs*, **13**, 3–11.

Copper, P. (2002b). Reef development at the Frasnian/Famennian mass extinction boundary. *Palaeogeography, Palaeoclimatology, Palaeoecology*, **181**, 27–65.

Courtillot, V. and Olson, P. (2007). Mantle plumes link magnetic superchrons to Phanerozoic mass depletion events. *Earth and Planetary Science Letters*, **260**, 495–504.

Dickens, G.R., O'Neil, J.R., Rea, D.K., and Owen R.M. (1995). Dissociation of oceanic methane hydrate as a cause of the carbon-isotope excursion at the end of the Paleocene. *Paleoceanography*, **10**, 965–71.

Doney, S.C., Fabry, V.J., Feely, R.A., and Kleypas, J.A. (2009). Ocean acidification: the other CO_2 problem. *Annual Review of Marine Science*, **1**, 169–92.

Erwin, D.H. (2006). *Extinction: how life on earth nearly ended 250 million years ago*, 296 pp. Princeton University Press, Princeton, NJ.

Evins, L.Z., Jourdan, F. and Phillips, D. (2009). The Cambrian Kalkarindji Large Igneous Province: extent and characteristics based on new Ar-40/Ar-39 and geochemical data. *Lithos*, **110**, 294–304.

Ezaki, Y. (2009). Secular fluctuations in Palaeozoic and Mesozoic reef-forming organisms during greenhouse periods: geobiological interrelations and consequences. *Paleontological Research*, **13**, 23–38.

Fine, M. and Tchernov, D. (2007). Scleractinian coral species survive and recover from decalcification. *Science*, **315**, 1811.

Ganino, C. and Arndt, N.T. (2009). Climate changes caused by degassing of sediments during the emplacement of large igneous provinces. *Geology*, **37**, 323–6.

Gattuso, J.-P., Allemand, D., and Frankignoulle, M. (1999). Photosynthesis and calcification at cellular, organismal and community levels in coral reefs: a review on interactions and control by carbonate chemistry. *American Zoologist*, **39**, 160–83.

Gili, E., Masse, J.P., and Skelton, P.W. (1995). Rudists as gregarious sediment dwellers, not reef-builders, on Cretaceous carbonate platforms. *Palaeogeography, Palaeoclimatology, Palaeoecology*, **118**, 245–67.

Gill, B.C., Lyons, T.W., Young, S., Kump, L., Knoll, A.H., and Saltzman, M.R. (2011). Geochemical evidence for widespread euxinia in the later Cambrian ocean. *Nature*, **469**, 80–83.

Gingerich, P.D. (2006). Environment and evolution through the Paleocene–Eocene thermal maximum. *Trends in Ecology and Evolution*, **21**, 246–53.

Glass, L.M. and Phillips, D. (2006). The Kalkarindji continental flood basalt province: a new large igneous province in Australia with possible links to end-Early Cambrian faunal extinctions. *Geology*, **34**, 461–4.

Gooding, R.A., Harley, C.D.G., and Tang, E. (2009). Elevated water temperature and carbon dioxide concentration increase the growth of a keystone echinoderm. *Proceedings of the National Academy of Sciences USA*, **106**, 9316–21.

Grotzinger, J.P. and Knoll, A.H. (1999). Proterozoic stromatolites: evolutionary mileposts or environmental dipsticks? *Annual Review of Earth and Planetary Science*, **27**, 313–58.

Harper, D.A.T. (2006). The Ordovician biodiversification: setting an agenda for marine life. *Palaeogeography, Palaeoclimatology, Palaeoecology*, **232**, 148–66.

Hautmann, M. (2004). Effect of end-Triassic CO_2 maximum on carbonate sedimentation and marine mass extinction. *Facies*, **50**, 257–61.

Hautmann, M., Benton, M.J., and Tomasovych, A. (2008a). Catastrophic ocean acidification at the Triassic–Jurassic boundary. *Neues Jahrbuch für Geologie und Paläontologie, Abhandlungen*, **249**, 119–127.

Hautmann, M., Stiller, F., Cai, H.W., and Sha, J.G. (2008b). Extinction-recovery pattern of level-bottom faunas across the Triassic--Jurassic boundary in Tibet: implications for potential killing mechanisms. *Palaios*, **23**, 711–18.

Higgins, J.A. and Schrag, D.P. (2006). Beyond methane: towards a theory for the Paleocene-Eocene Thermal Maximum. *Earth and Planetary Science Letters*, **245**, 523–37.

Higgins, J.A., Fischer, W.W., and Schrag, D.P. (2009). Oxygenation of the ocean and sediments: consequences for the seafloor carbonate factory. *Earth and Planetary Science Letters*, **284**, 25–33.

Hoegh-Guldberg, O., Mumby, P.J., Hooten, A.J. *et al.* (2007). Coral reefs under rapid climate change and ocean acidification. *Science*, **318**, 1737–42.

Hough, M.L., Shields, G.A., Evins, L.Z., Strauss, H., Henderson, R.A., and Mackenzie, S. (2006). A major sulphur isotope event at c. 510 Ma: a possible anoxia- extinction-volcanism connection during the Early–Middle Cambrian transition? *Terra Nova*, **18**, 257–63.

Hurtgen, M.T., Pruss, S.B., and Knoll, A.H. (2009). Evaluating the relationship between the carbon and sulfur cycles in later Cambrian oceans: an example from the Port au Port Group, western Newfoundland. *Earth and Planetary Science Letters*, **281**, 288–97.

Iacono-Marziano, G., Gaillard, F., Scaillet, B., Pichavant, M., and Chiodini, G. (2009). Role of non-mantle CO_2 in the dynamics of volcano degassing: the Mount Vesuvius example. *Geology*, **37**, 319–22.

Isozaki, Y. (1997). Permo-Triassic boundary superanoxia and stratified superocean: records from lost deep sea. *Science*, **276**, 235–8.

Kennett, J.P. and Stott, L.D. (1991). Abrupt deep-sea warming, palaeoceanographic changes and benthic extinctions at the end of the Palaeocene. *Nature*, **353**, 225–9.

Kiessling, W. (2001). Paleoclimatic significance of Phanerozoic reefs. *Geology*, **29**, 751–4.

Kiessling, W. (2002). Secular variations in the Phanerozoic reef ecosystem. In: W. Kiessling, E. Flügel, and J. Golonka (eds), *Phanerozoic reef patterns*, pp. 625–90. SEPM Special Publication 72. Society of Economic Paleontologists and Mineralogists, Tulsa, OK.

Kiessling, W. (2009). Geologic and biologic controls on the evolution of reefs. *Annual Review of Ecology Evolution and Systematics*, **40**, 173–92.

Kiessling, W. and Simpson, C. (2011). On the potential for ocean acidification to be a general cause of ancient reef crises. *Global Change Biology*, **17**, 56–67.

Kiessling, W., Roniewicz, W., Villier, L., Leonide, P., and Struck, U. (2009). An early Hettangian coral reef in southern France: implications for the end-Triassic reef crisis. *Palaios*, **24**, 657–71.

Kleypas, J.A., Buddemeier, R.W., Archer, D., Gattuso, J.-P., Langdon, C., and Opdyke, B.N. (1999). Geochemical consequences of increased atmospheric carbon dioxide on coral reefs. *Science*, **284**, 118–20.

Knoll, A.H. (2003). Biomineralization and evolutionary history. *Reviews in Mineralogy and Geochemistry*, **54**, 329–56.

Knoll, A.H., Bambach, R., Canfield, D., and Grotzinger, J.P. (1996). Comparative Earth history and late Permian mass extinction. *Science*, **273**, 452–7.

Knoll, A.H., Bambach, R.K., Payne, J., Pruss, S., and Fischer, W. (2007). A paleophysiological perspective on the end-Permian mass extinction and its aftermath. *Earth and Planetary Science Letters*, **256**, 295–313.

Langdon, C. and Atkinson, M.J. (2005). Effect of elevated pCO_2 on photosynthesis and calcification of corals and interactions with seasonal change in temperature/irradiance and nutrient enrichment. *Journal of Geophysical Research–Oceans*, **110**, C09S07, doi:10.1029/2004JC002576.

Lathuilière, B. and Marchal, D. (2009). Extinction, survival and recovery of corals from the Triassic to Middle Jurassic time. *Terra Nova*, **21**, 57–66.

Marty, B. and Tolstikhin, I.N. (1998). CO_2 fluxes from mid-ocean ridges, arcs and plumes. *Chemical Geology*, **145**, 233–48.

Melzner, F., Gutowska, M.A., Langenbuch, M. *et al.* (2009). Physiological basis for high CO_2 tolerance in marine ectothermic animals: pre-adaptation through lifestyle and ontogeny? *Biogeosciences*, **6**, 2313–31.

Meyer, K.M., Kump, L.R., and Ridgwell, A. (2008). Biogeochemical controls on photic- zone euxinia during the end-Permian mass extinction. *Geology*, **36**, 747–50.

Nunes, F. and Norris, R.D. (2006). Abrupt reversal in ocean overturning during the Palaeocene/Eocene warm period. *Nature*, **439**, 60–3.

Petit, J.R., Jouzel, J., Raynaud, D. *et al.* (1999). Climate and atmospheric history of the past 420,000 years from the Vostok ice core, Antarctica. *Nature*, **399**, 429–36.

Pörtner, H.O. (2008). Ecosystem effects of ocean acidification in times of ocean warming: a physiologist's view. *Marine Ecology Progress Series*, **373**, 203–17.

Porter, S.M. (2004). Closing the phosphatization window: testing for the influence of taphonomic megabias on the pattern of small shelly fossil decline. *Palaios*, **19**, 311–14.

Pruss, S., Finnegan, S., Fischer, W.W., and Knoll, A.H. (2010). Carbonates in skeleton- poor seas: new insights from Cambrian and Ordovician strata of Laurentia. *Palaios*, **25**, 73–84.

Rees, P.M. (2002). Land-plant diversity and the end-Permian mass extinction. *Geology*, **30**, 827–30.

Reichow, M.K., Pringle, M.S., Al'Mukhamedov, A.I. *et al.* (2007). The timing and extent of the eruption of the Siberian Traps large igneous province: implications for the end-Permian environmental crisis. *Earth and Planetary Science Letters*, **77**, 9–20.

Retallack, G.J. and Jahren, A.H. (2008). Methane release from igneous intrusion of coal during late Permian extinction events. *Journal of Geology*, **116**, 1–20.

Ries, J.B., Cohen, A.L., and McCorkle, D.C. (2009). Marine calcifiers exhibit mixed responses to CO_2-induced ocean acidification. *Geology*, **37**, 1131–4.

Rohl, U., Bralower, T.J., Norris, R.D., and Wefer, G. (2000). New chronology for the late Paleocene thermal maximum and its environmental implications. *Geology*, **28**, 927–30.

Rowland, S.M. and Shapiro, R.S. (2002). Reef patterns and environmental influences in the Cambrian and earliest Ordovician. In: W. Kiessling and G.I. Flügel (eds), *Phanerozoic reef patterns*, pp. 95–128. SEPM Special Publication 72. Society of Economic Paleontologists and Mineralogists, Tulsa, OK.

Royer, D.L., Berner, R.A., and Beerling, D.J. (2001). Phanerozoic atmospheric CO_2 change: evaluating geochemical and paleobiological approaches. *Earth-Science Reviews*, **54**, 349–92.

Scheibner, C. and Speijer, R.P. (2008). Late Paleocene–early Eocene Tethyan carbonate platform evolution—a response to long- and short-term paleoclimatic change. *Earth-Science Reviews*, **90**, 71–102.

Sheehan, P.M. (2001). The Late Ordovician mass extinction. *Annual Review of Earth and Planetary Sciences*, **29**, 331–64.

Smith, R. and Botha, J. (2005). The recovery of terrestrial vertebrate diversity in the South African Karoo Basin after the end-Permian extinction. *Comptes Rendus Palevol*, **4**, 623–36.

Soetaert K., Hofmann, A.F., Middelburg, J.J., Meysman, F.J.R., and Greenwood J. (2007). The effect of biogeochemical processes on pH. *Marine Chemistry*, **105**, 30–51.

Svensen, H., Planke, S., Malthe-Sørenssen, A. *et al.* (2004). Release of methane from a volcanic basin as a mechanism for initial Eocene global warming. *Nature*, **429**, 542–5.

Thomas, E. (2007). Cenozoic mass extinctions in the deep sea: what perturbs the largest habitat on Earth? *Geological Society of America Special Paper*, **424**, 1–24.

Veron, J.E.N. (2008). Mass extinctions and ocean acidification: biological constraints on geological dilemmas. *Coral Reefs*, **27**, 459–72.

Walker, J.C.G., Hays, P.B., and Kasting, J.F. (1981). A negative feedback mechanism for the long-term stabilization of Earth's surface temperature. *Journal of Geophysical Research*, **86**, 9776–82.

Widdicombe, S. and Spicer, J.I. (2008). Predicting the impact of ocean acidification on benthic biodiversity:

what can animal physiology tell us? *Journal of Experimental Marine Biology and Ecology*, **366**, 187–97.

Wignall, P.B. (2001). Large igneous provinces and mass extinctions. *Earth-Science Reviews*, **53**, 1–33.

Wignall, P.B. and Twitchett, R.J. (1996). Oceanic anoxia and the end Permian mass extinction. *Science*, **272**, 1155–8.

Wing, S.L., Harrington, G.J., Smith, F.A., Bloch, J.I., Boyer, D.M., and Freeman, K.H. (2005). Transient floral change and rapid global warming at the Paleocene–Eocene boundary. *Science*, **310**, 993–6.

Wood, R. (1999). *Reef evolution*, 414 pp. Oxford University Press, Oxford.

Zachos, J.C., Lohmann, K.C., Walker, J.C.G., and Wise, S.W. (1993). Abrupt climate change and transient climates during the Paleogene: a marine perspective. *Journal of Geology*, **101**, 191–213.

Zachos, J.C., Wara, M.W., Bohaty, S. *et al.* (2003). A transient rise in tropical sea surface temperature during the Paleocene-Eocene thermal maximum. *Science*, **302**, 1551–54.

Zachos, J.C., Rohl, U., Schellenberg, S.A. *et al.* (2005). Rapid acidification of the ocean during the Paleocene-Eocene thermal maximum. *Science*, **308**, 1611–15.

Zhuravlev, A.Y. and Wood, R.A. (1996). Anoxia as the cause of the mid-early Cambrian (Botomian) extinction event. *Geology*, **24**, 311–14.

CHAPTER 5

Effects of ocean acidification on the diversity and activity of heterotrophic marine microorganisms

Markus G. Weinbauer, Xavier Mari, and Jean-Pierre Gattuso

5.1 Introduction

Microbe-mediated processes are crucial for biogeochemical cycles and the functioning of marine ecosystems (Azam and Malfatti 2007). If these processes are affected by ocean acidification, major consequences can be expected for the functioning of the global ocean and the systems that it influences, such as the atmosphere. In contrast to phytoplankton, which have been relatively well studied (see Chapter 6), there is comparatively little information on the effect of ocean acidification on heterotrophic microorganisms. Two reviews on the potential effects of ocean acidification on microbial plankton have recently been published (Liu *et al.* 2010; Joint *et al.* 2011) . In a recent perspective paper, Joint *et al.* (2011) concluded that marine microbes possess the flexibility to accommodate pH change and that major changes in marine biogeochemical processes that are driven by microorganisms are unlikely. Narrative reviews, which look at some of the relevant literature, are potentially biased and could lead to misleading conclusions (Gates 2002). Meta-analysis was developed to overcome most biases of narrative reviews. It statistically combines the results (effect size) of several studies that address a shared research hypothesis. Liu *et al.* (2010) used a meta-analytic approach to comprehensively review the current understanding of the effect of ocean acidification on microbes (including phytoplankton) and microbial processes, and to highlight the gaps that need to be addressed in future research. In the following, a brief digest on oceanic microbes and their role is provided for readers unfamiliar with this

topic. Then the research that has been performed to assess the effects of ocean acidification on the diversity and activity of heterotrophic marine microorganisms is reviewed. Finally, scenarios are developed and potential implications are discussed.

5.2 Microbes in the ocean

5.2.1 Structural and functional diversity of microorganisms

Microorganisms are defined as organisms that are microscopic, i.e. too small to be seen by the naked human eye, and mostly comprise single-celled organisms. Viruses are sometimes also included in this definition but it is hotly debated whether viruses are alive or not (Raoult and Forterre 2008). The current phylogeny considers three domains of cellular life, the Bacteria, the Archaea and the Eukarya. Since Archaea are more closely related to Eukarya than to Bacteria, the old term 'Prokaryota' is phylogenetically invalid for Bacteria and Archaea (Pace 2009). However, the term 'prokaryotes' can still be used in a non-phylogenetic context for these two domains (Whitman 2009) as is done in this chapter.

The development of molecular tools was an important step for assessing marine microbial diversity. Previously, the assessment of species diversity was done by the characterization of isolated bacteria. However, it is well known that only a small fraction of the cells (often less than 1%) can be isolated (great plate count anomaly; Staley and Konopka 1985). The advent of molecular tools has considerably changed the perception of the diversity of the microbial food web. Rapid finger-printing methods such as denaturing gradient gel electrophoresis (DGGE; Muyzer *et al.* 1993), terminal fragment length polymorphism (TRFLP; Moyer *et al.* 1994), and automated ribosomal intergenic spacer analysis (ARISA; Fisher and Triplett 1999), as well as single-cell techniques such as fluorescence *in situ* hybridization (FISH; Amann *et al.* 1990) are available to assess the community structure of Bacteria, Archaea, and Eukarya. Large-scale sequencing techniques, such as pyrosequencing based on 16S rRNA genes, now enable the study of 'true' species diversity. It is noteworthy that these techniques do not allow for a full species identification, for which an isolation is obligatory. However, they allow for the distinction of phylogenetic types (phylotypes), which are sometimes also called operational taxonomic units (OTUs). For example, a large number of highly diverse, low-abundance OTUs was found in bacterioplankton using large-scale sequencing. This constitutes a 'rare biosphere' that is largely unexplored (Sogin *et al.* 2006). Some of its members might serve as keystone species within complex consortia; others might simply be the products of historical ecological change with the potential to become dominant in response to shifts in environmental conditions (e.g. when local or global change favours their growth). The use of such techniques has provided new answers for old questions (what is the diversity of microorganisms in a system?) and provoked new questions (what is the role of a rare biosphere?).

Large-scale sequencing techniques and progress in bioinformatics now allow us to study the genomics (organization of genes), transcriptomics (set of all RNA molecules, i.e. expressed genes), and proteomics (structure and function of proteins) of microbial communities. Such metagenomics and metatranscriptomics approaches have been successfully used to detect new metabolic pathways and assess environment-specific activities of microorganisms (Shi *et al.* 2009). New pathways have been found such as aerobic anoxygenic phototrophs (AAnP) mediated by bacteriorhodopsin (Béjà *et al.* 2000). Such studies now allow for assessing the functional diversity of microorganisms. Characterization of the metaproteome is expected to provide a link to genetic and functional diversity of microbial communities.

Studies on the metaproteome together with those on the metagenome and the metatranscriptome will contribute to progress in our knowledge of microbial communities and their role in ecosystem functioning. More specifically, the analysis of the metaproteome in contrasting environmental situations should allow tracking new functional genes and metabolic pathways (Maron *et al.* 2007). These advancements have started to revolutionize our understanding of the role of microbes in the ocean and their response to environmental change.

5.2.2 Microbial food webs

In the late 1970s and early 1980s, it became obvious that microorganisms play a crucial role in microbial food webs and biogeochemical cycles (Pomeroy 1974; Azam *et al.* 1983). The classical grazing food chain composed of phytoplankton and herbivorous and carnivorous zooplankton (see Fig. 5.1) was extended by several concepts on microorganisms. Phytoplankton, even when healthy, release or 'exude' dissolved organic matter (DOM). Another form of release of DOM is the sloppy feeding of zooplankton. Prokaryotes are the main users of this DOM, and are themselves grazed upon by flagellates and ciliates, which are eaten in turn by meso- and macrozooplankton. Thus, prokaryotic assimilation and subsequent grazing on prokaryotes return carbon that would otherwise be lost, back to the food web. These pathways are collectively referred to as the microbial loop (see Fig. 5.1). Related to this concept is the finding that the grazing activity is an important factor for the remineralization of nutrients and that prokaryotes are competing with phytoplankton for nutrients. It was subsequently shown that viral lysis is another important cause of prokaryotic (and phytoplankton) mortality (Proctor and Fuhrman 1990). Viral lysis not only kills cells and releases new viral particles, but it also sets free the cell content and converts cell walls into small debris. This material is taken up by prokaryotes generating a viral loop or viral shunt (Wilhelm and Suttle 1999; Fig. 5.1), which stimulates bacterial production and respiration (Fuhrman 1999). Thus, viral lysis catalyses nutrient generation and lubricates the microbial food web (Suttle 2007). Coagulation of DOM can result in the formation of transparent

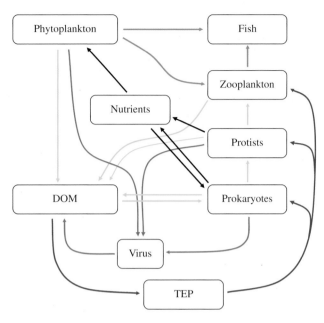

Figure 5.1 Simplified scheme of the pelagic food web. Four major conceptual pathways are shown, the classical food chain (green), the microbial loop (yellow), the viral shunt (red) and the abiotic loop (blue). Nutrient pathways are shown in black. See the main text for details.

exopolymeric particles (TEP) in seawater (Alldredge *et al.* 1993), which can be used or ingested by prokaryotes, protists, and zooplankton; this is sometimes called the abiotic loop (Fig. 5.1). In the dark ocean, i.e. below ~200 m, the food web has a reduced complexity and consists of sinking particles, prokaryotes, protists, and (in the upper layers) zooplankton. The dark ocean is the largest habitat in the biosphere and harbours 75% of the prokaryotic biomass and 50% of the prokaryotic production of the global ocean (Arístegui *et al.* 2009).

The concept of the microbial loop has been developed for planktonic systems; however, a microbial food web also exists in sediments. The functioning of sedimentary systems mainly relies on carbon supplied by particles that sink from the surface or are transported horizontally, except in shallow waters where microphytes that include cyanobacteria and algae provide autochthonous carbon. Other carbon sources can be locally important for sediments (and systems close to sediments) in the dark ocean. Methane can be used (Boetius *et al.* 2000) and chemoautotrophic microorganisms are the base of the food web in environments such as hot vents (Nakagawa and Takai 2008). The abundance

of heterotrophic microorganisms is high in marine sediments; they populate the porewater and are key components of biofilms. Meiofauna, such as nematodes, are the trophic equivalent to zooplankton in the sediment. The digging and burrowing activity of macrofauna such as enteropneusts or crustaceans can change the strong chemical gradients typically found in sediments and also influence microbial activity (see Chapter 9).

Benthic and pelagic animals can host viral and microbial pathogens as well as symbionts. They can also be associated with non-pathogenic and non-symbiotic microbial communities, for example in sponges (Taylor *et al.* 2007). Many sponges harbour a dense and diverse community of Bacteria and Archaea that are involved in element cycles within the animals or at their surface, i.e. in their mucus. Recently, major progress has been made for corals of tropical or cold-water communities (Ainsworth *et al.* 2009). It has been suggested that microorganisms associated with corals function as an equivalent to an immune system (the coral probiotic hypothesis; Reshef *et al.* 2006). The coral probiotic hypothesis proposes that there is a dynamic relationship between microorganisms and corals that selects for the coral holobiont (coral plus microorganisms) that is best suited to the prevailing environmental conditions. A generalization of the coral probiotic hypothesis has led to the proposition of the hologenome theory of evolution (Rosenberg *et al.* 2007). Much of this theory applies to sponges. Despite this progress, little is known about the interactions between microorganisms, i.e. on the microbial food web within invertebrates.

5.2.3   Microbe-mediated ecosystem functions and biogeochemical cycles

The carbon pool in DOM is approximately as large as the carbon pool in atmospheric CO_2 (Hedges 2002). Thus, the fate of dissolved organic carbon (DOC) and particulate organic carbon (POC), which is strongly mediated by microorganisms, is an important factor influencing climate. Organic matter in pelagic systems is separated operationally into DOM and particulate organic matter (POM). Non-living organic matter occurs in a size continuum from dissolved molecules to marine snow (Nagata and Kirchman 1997) and several types of non-living organic particles have been described in marine environments (Nagata 2008). Seawater is considered as a diluted medium with gels or gel-like structures of different size, that can be formed in minutes to hours from DOM or polymer chains released by phyto- or bacterioplankton, hence bridging the DOM–POM continuum (Verdugo *et al.* 2004).

At the microscale level, the microbial community is operationally structured into free and particle-attached microorganisms. Attached microorganisms are embedded in a nutrient-rich organic matrix, thus particles can be considered as a physical refuge for microorganisms (Alldredge and Cohen 1987). Particles can be characterized by strong chemical microgradients. This patchiness of chemical properties within the particles at the microscale level can modify the processes which rely upon the concentration of organic carbon and nutrients, such as bacterial activity (Simon *et al.* 2002). Marine snow and the plume of sinking (or rising) particles are considered to be hot spots of biogeochemical transformations mediated by microorganisms (Azam and Long 2001; Kiørboe and Jackson 2001).

Heterotrophic microorganisms are among the main drivers of the carbon and nutrient cycles in the ocean (Azam and Malfatti 2007). For example, some metabolic pathways, such as ammonium oxidation or denitrification, are only performed by heterotrophic organisms. Most of the organic carbon generated by primary production in the upper ocean is remineralized to CO_2 which accumulates in deep waters until it is eventually ventilated again at the sea surface. However, a small fraction of the fixed carbon is not mineralized but stored for millennia as recalcitrant DOM. The processes and mechanisms involved in the generation of this large carbon reservoir are poorly understood. Recently, the microbial carbon pump has been proposed as a conceptual framework to address this important, multifaceted biogeochemical problem (Jiao *et al.* 2010).

Prokaryotes are major remineralizers of nutrients such as nitrogen, phosphorus, iron, and sulphur in the water column and sediments by degrading organic matter. However, they also use nutrients and compete with phytoplankton for nutrients such as nitrate or phosphate. Another important group of remineralizers are the protists, which egest organic matter as well as inorganic nutrients. Viral lysis is also considered as a catalyst of nutrient regeneration (see above and Fig. 5.1). The link between microbial food webs and microbe-mediated ecosystem functions and biogeochemical cycles (and the potential consequences of ocean acidification) in the pelagic realm is shown in Figure 5.2.

Tropical coral reefs exhibit important ecological and biogeochemical functions (Gattuso *et al.* 1998). Corals and coralline algae are ecosystem engineers which provide the physical structure of coral reefs and thus provide a plethora of ecological niches explaining the high biodiversity found in reefs. The build-up of coral skeletons is possible due to symbiosis between the animal and symbiotic dinoflagellates. The loss of the symbionts ('bleaching') can jeopardize the ecosystem function of reef corals. Bleaching can be caused by bacteria such as *Vibrio coralliilyticus* which can become pathogenic at elevated temperatures (Rosenberg *et al.* 2007). It has been argued that changes in the microbial community associated with corals can prevent new *Vibrio* infections (Ainsworth *et al.* 2009), thus sustaining

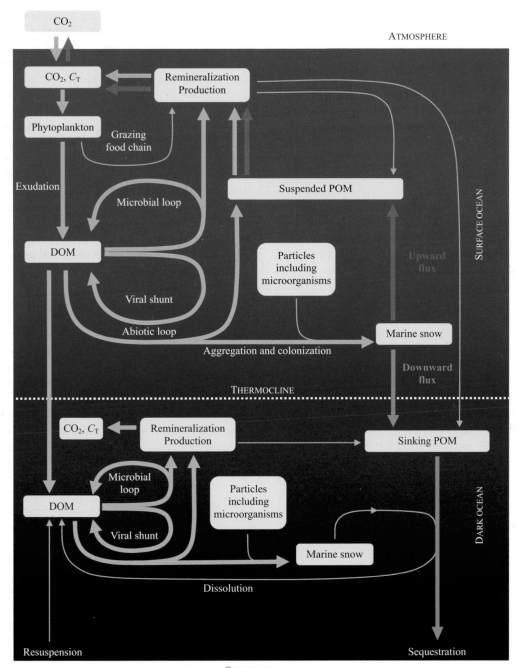

Figure 5.2   Potential consequences of ocean acidification for organic matter cycling and microbial activity. Thick grey arrows indicate processes that will probably be affected by ocean acidification, whereas thin lines show processes which are likely to be less affected or not affected at all. Red arrows show processes enhanced by ocean acidification, which will result in an increase in aggregates and remineralization of organic matter in the euphotic zone. Green arrows show processes enhanced by ocean acidification, which result in higher vertical export rates and higher remineralization of organic matter in the dark ocean. Note that these two types of scenarios are not mutually exclusive but can occur simultaneously. The net effect of ocean acidification remains an open question. The surface and deep oceans are not to scale.

the ecosystem functions and biogeochemical roles of corals.

5.3 Ocean acidification: approaches and evidence

Table 5.1 summarizes the studies and reports on the response of microorganisms to ocean acidification and provides details on parameters such as experimental set-ups, type of manipulations, and pCO_2 and pH levels used. The focus is mainly on *in situ* observations and on studies which tested parameters of the carbonate chemistry within the range of values projected for this century as a result of ocean acidification driven by the uptake of CO_2 through the air–sea interface. Unless mentioned otherwise, pH values are reported on the total scale. If pCO_2 levels were not reported in publications, they were calculated from pH and other ancillary data with the R software package 'seacarb' as described by Nisumaa *et al.* (2010). Data collected at pCO_2 and pH levels that are not environmentally relevant are nevertheless reported when no other data are available on a particular process. Four types of responses are considered: (1) positive, when ocean acidification stimulates the parameter or process; (2) negative, when ocean acidification decreases the parameter or process; (3) change, when ocean acidification affects the community composition (whether it is a positive or negative effect cannot be determined); and (4) neutral, when ocean acidification has no significant effect.

5.3.1 Approaches to study the effect of ocean acidification on microbes

The effect of ocean acidification on heterotrophic microorganisms has been investigated using perturbation experiments with isolates of natural communities over short periods of time (days to weeks). So far, only a few bacterial and viral isolates have been tested, and to the best of our knowledge, no study has been performed on heterotrophic protistan isolates. Open-water CO_2 fertilization experiments still seem unrealistic, but it is potentially rewarding to take advantage of systems naturally enriched with CO_2 such as shallow-water CO_2 vents (Hall-Spencer *et al.* 2008), deep-sea vents

(Inagaki *et al.* 2006), cold-eddies, and upwelling systems which have lower pH and higher pCO_2 levels than ambient water and are thus suitable for studying the effects of ocean acidification. Nevertheless, the experimental set-up is challenging due to factors such as advection or migration of microorganisms. The data interpretation is also challenging, given the temporal and spatial variability in pH/pCO_2 and hence the uncertainty on the dose–response relationship.

In order to study the effects of ocean acidification, seawater pH must be lowered while maintaining total alkalinity constant (Gattuso and Lavigne 2009). Bubbling seawater with gas mixtures is often used to reach the desired level of pCO_2. However, this approach can be a problem when addressing the question of the effects of ocean acidification on the microbial ecosystem, since bubbling can promote aggregation of organic matter via surface aggregation and by increasing turbulence (Kepkay 1994), which is known to affect microbial processes (Kepkay and Johnson 1989). Thus, bubbling should be avoided whenever possible or kept as short as possible. If bubbling is used, it must be carried out in the same way in all treatments, including the control. Addition of acid avoids bubbling but has other disadvantages; for example it does not mimic the natural process as it decreases total alkalinity (see Gattuso and Lavigne 2009).

5.3.2 Direct versus indirect effects of ocean acidification

Heterotrophic marine microorganisms can be affected either directly by the chemical changes generated by ocean acidification or indirectly through effects on other levels of the community. The direct effects of elevated pCO_2 on the physiology and metabolism of microorganisms can be studied using isolates. Microbial species have growth optima at different pH values and many physiological functions such as hydrolytic enzyme activities are dependent on pH (Yamada and Suzumura 2010). Therefore, it is critical that the internal pH (pH_i) is controlled, which is achieved by a pH homeostatic system (Booth 1985). Studies at pH and pCO_2 levels relevant to ocean acidification have not been performed yet. The only results

Table 5.1 Effect of elevated pCO_2 or decreased pH on heterotrophic microbes and parameters relevant for heterotrophic microbes. When pCO_2 was not mentioned and two parameters of the carbonate system were available, it was calculated as described by Nisumaa et al. (2010).

Parameter	Effect	Method	pCO_2 (µatm) and pH levels	Comments	Environment, isolate	Ref.
Dissolved and particulate organic carbon						
[DOC]	ns	pCO_2	pCO_2: 190, 410, 710	Mesocosm	Bergen, Norway	4
	ns	pCO_2	pCO_2: 190, 410, 710	Mesocosm	Bergen, Norway	12
	ns	pCO_2	pCO_2: 190, 370, 700	Mesocosm	Bergen, Norway	5
	ns	pCO_2	pCO_2: 350, 700, 1050	Mesocosm	Bergen, Norway	14
[cDOM]	ns	pCO_2	pCO_2: 190, 410, 710	Mesocosm	Bergen, Norway	12
[TEP] and TEP production	+	Acid–base	pCO_2: 53–2686	On-board incubation	Baltic Sea	3
[TEP] and TEP production	+	pCO_2	pCO_2: 190, 410, 710	Mesocosm	Bergen, Norway	4
[TEP] and TEP production	ns	pCO_2	pCO_2: 350, 700, 1050	Mesocosm	Bergen, Norway	2
[TEP] and TEP production	+	acid	pH$_{NBS}$: 7.36, 7.76, 7.96, 8.16	Batch cultures	Lagoon of New Caledonia	9
TEP buoyancy	+	acid	pH$_{NBS}$: 7.36, 7.76, 7.96, 8.16	Batch cultures	Lagoon of New Caledonia	9
Prokaryotes						
Bacterial abundance	ns	pCO_2	pCO_2: 190, 414, 714	Mesocosm	Bergen, Norway	5
	ns	pCO_2	pCO_2: 190, 370, 700	Total community, mesocosm	Bergen, Norway	1
	ns	pCO_2	pCO_2: 350, 700, 1050	Mesocosm	Bergen, Norway	10
	ns	pCO_2	pCO_2: 350, 700, 1050	Mesocosm	Bergen, Norway	7
Bacterial production (BP)	+	pCO_2	pCO_2: 190, 370, 700	Mesocosm	Bergen, Norway	1
	ns	pCO_2	pCO_2: 350, 700, 1050	Mesocosm	Bergen, Norway	5
Specific BP	+	pCO_2	pCO_2: 190, 370, 700	Total community, mesocosm	Bergen, Norway	5
	+	pCO_2	pCO_2: 190, 370, 700	Attached community, mesocosm	Bergen, Norway	5
Protease activity	+	pCO_2	pCO_2: 190, 370, 700	Mesocosm	Bergen, Norway	5
α-Glucosidase activity	ns	pCO_2	pCO_2: 190, 370, 700	Mesocosm	Bergen, Norway	5
β-Glucosidase activity	ns	pCO_2	pCO_2: 190, 370, 700	Mesocosm	Bergen, Norway	5
α + β-Glucosidase activity	+	Acid, pCO_2	pCO_2: 380, 750; pH$_{NBS}$: 7.7–8.3	Cultures and communities	North Sea	11
Alkaline phosphatase activity	ns	pCO_2	pCO_2: 350, 700, 1050	Mesocosm	Bergen, Norway	16
Nitrification	–	pCO_2	pH$_{NBS}$: 6–8	CO_2 disposal study	Strait of Juan de Fuca, WA, USA	6
Leucine aminopeptidase + lipase	–	buffer	pH$_{NBS}$: 7.8–8.2	Batch cultures	Tokyo Bay and offshore, Japan	18
β-Glucosidase + phosphatase	ns	buffer	pH$_{NBS}$: 7.8–8.2	Batch cultures	Tokyo Bay and offshore, Japan	18
PO_4 turnover	ns	pCO_2	pCO_2: 350, 700, 1050	Mesocosm	Bergen, Norway	16
Community composition	Change	pCO_2	pCO_2: 350, 700, 1050	Free-living community, mesocosm	Bergen, Norway	1
Community composition	Change	pCO_2	pCO_2: 350, 700, 1050	Attached community, mesocosm	Bergen, Norway	1
Community composition	Change	Acid	pH$_{NBS}$: 6.7, 8.1	Coral associated	Hawaii	17

(continued)

Table 5.1 Continued

Parameter	Effect	Method	pCO_2 (µatm) and pH levels	Comments	Environment, isolate	Ref.
Viruses						
Viral abundance	ns	pCO_2	pCO_2: 190, 370, 700	Mesocosm	Bergen, Norway	12
Viral abundance	ns	pCO_2	pCO_2: 350, 700, 1050	Mesocosm	Bergen, Norway	8
Low-fluorescence viruses	ns	pCO_2	pCO_2: 350, 700, 1050	Mesocosm	Bergen, Norway	8
Middle-fluorescence viruses	ns	pCO_2	pCO_2: 350, 700, 1050	Mesocosm	Bergen, Norway	8
High-fluorescence viruses	+	pCO_2	pCO_2: 350, 700, 1050	Mesocosm	Bergen, Norway	8
EhV + unknown dsDNA virus	–	pCO_2	pCO_2: 350,700, 1050	Mesocosm	Bergen, Norway	8
Grazers						
Grazing rate, grazer abundance	+	pCO_2	pCO_2: 390, 690	On-board incubation	North Atlantic spring bloom	13
Grazing rate, grazer abundance	ns	pCO_2	pCO_2: 350, 700, 1150	Mesocosm	Bergen, Norway	5
Community composition	ns	pCO_2	pCO_2: 350, 700, 1150	Mesocosm	Bergen, Norway	5

DOC, dissolved organic carbon; CDOM, chromophoric dissolved organic matter; TEP, transparent exopolymeric particles; EhV, virus infecting *Emiliania huxleyi*; dsDNA, double-stranded DNA; +, positive effect; –, negative effect; ns, no significant effect; change, change in community composition (distinction between positive and negative not possible).

References: 1, Allgaier *et al.* (2008); 2, Egge *et al.* (2009); 3, Engel (2002); 4, Engel *et al.* (2004); 5, Grossart *et al.* (2006); 6, Huesemann *et al.* (2002); 7, Inagaki *et al.* (2006); 8, Larsen *et al.* (2008); 9, Mari (2008); 10, Paulino *et al.* 2008; 11, Piontek *et al.* (2010); 12, Rochelle-Newall *et al.* (2004); 13, Rose *et al.* (2009); 14, Schulz *et al.* (2008); 15, Suffrian *et al.* (2008); 16, Tanaka *et al.* (2008); 17, Vega Thurber *et al.* (2009); 18, Yamada and Suzumura (2010).

available were obtained in the context of CO_2 disposal at pH and pCO_2 levels well outside the range of values expected in the next decades (Takeuchi et al. 1997; Labare et al. 2010). A Vibrio sp. isolated from the deep sea and grown at a pH of 5.2 responded with a change in morphology and temporary growth inhibition. At a pH of 6 the growth recovered. As pH is usually around 8 in natural marine systems (see Chapter 1), this suggests a strong capacity for pH homeostasis or high tolerance to low pH_i. If the direct effects differ between species, this could change microbial diversity, and thus associated ecosystem functions. Direct effects on viruses seem unlikely as marine phage isolates remain quite stable and infective over a range of pH much wider than that found in marine systems (Børsheim 1993).

Indirect effects of ocean acidification are likely to occur in complex food webs. For example, a change in pCO_2 could affect phytoplankton community composition and metabolism, e.g. by influencing primary production (see Chapter 6), leading to changes in the quantity and quality of DOM released, which in turn would influence bacterial growth. Indirect effects could also occur in zooxanthellate corals. For example, if rates of primary production are influenced by changes in pCO_2, this could also influence metabolism and mucus production and thus the associated microorganisms. The studies discussed in the following sections of this chapter were not designed to separate direct from indirect effects; they rather report net effects.

5.3.3   Effect of ocean acidification on microbial diversity and community composition

Only a few studies have been performed on the effects of ocean acidification on microbial diversity (Table 5.1). Allgaier et al. (2008) found, in a mesocosm study and using genetic fingerprints (DGGE), that pCO_2 induced changes in the composition of the free-living community but not in the attached community. Vega Thurber et al. (2009) used a metagenomic analysis to study the effect of strong decreases in pH_{NBS} from 8.1 to 6.7 on bacteria associated with the zooxanthellate coral Porites compressa. Corals stressed by a change in pH showed strong shifts in bacterial community composition and the

occurrence of bacterial groups often found in diseased corals such as Bacteroidetes and Fusobacteria. Larsen et al. (2008) investigated, in a mesocosm study, the effect of elevated pCO_2 on the viral community composition using flow cytometry, and showed that the 'high-fluorescence viruses' were stimulated and that no other viral subgroups were altered. Viral groups identified by flow cytometry and nucleic acid staining belong to different 'species' or types of viruses. Using molecular markers, Larsen et al. (2008) could also show that the abundance of EhV, a virus infecting Emiliania huxleyi, and of an unidentified virus decreased at elevated pCO_2 (1050 µatm). This indicates changes in the viral community composition. Suffrian et al. (2008) did not find any effect of elevated pCO_2 (700 and 1150 versus 350 µatm) on the composition of microzooplankton (mainly dinoflagellates and ciliates) in a mesocosm study. Overall, there is no consistent effect of ocean acidification on microbial biodiversity and community composition.

5.3.4   Effect of ocean acidification on prokaryotic abundance, production, and enzyme activity

No or only a small effect of elevated pCO_2 levels was found on bacterial abundance in mesocosm studies (Rochelle-Newall et al. 2004; Grossart et al. 2006; Allgaier et al. 2008; Paulino et al. 2008; Yamada et al. 2008). Grossart et al. (2006) found that the total prokaryotic production and the cell-specific total and attached production were enhanced under elevated pCO_2 (750 µatm). In contrast, Allgaier et al. (2008) did not detect a significant difference in bacterial production at pCO_2 levels of 350, 700, and 1050 µatm. However, they found that regression lines between bacterial production and the C:N ratio of suspended matter varied depending on pCO_2. This suggests an indirect effect via modification of the quality of the organic matter rather than a direct effect on bacterial production. Overall, to date no study has shown a clear negative effect of elevated pCO_2 on prokaryotic production at pH levels relevant in the context of ocean acidification. Thus, the few data collected so far suggest a neutral or positive effect on total and cell-specific prokaryotic production.

Several perturbation experiments performed in mesocosms investigated the effect of ocean acidification on enzymatic activity. Grossart *et al.* (2006) found higher enzyme (α- and β-glucosidase and protease) activity in prokaryotes at elevated pCO_2, with significant differences only found for protease activity. Piontek *et al.* (2010) reported higher rates of extracellular glucosidase at higher pCO_2, which also resulted in higher rates of polysaccharide degradation. This might result in an enhanced processing of organic matter at higher pCO_2. However, Tanaka *et al.* (2008) reported that there was no effect of elevated pCO_2 on glucose affinity. In a mesocosm experiment, there was no effect of a pCO_2 level of 1050 µatm on the uptake of phosphate by bacteria, whereas the uptake of phosphate was stimulated in the >10 µm fraction (Tanaka *et al.* 2008). A recent study using a buffer to adjust pH showed evidence for a negative effect of elevated pCO_2 on enzyme activity, with a rapid decrease of some enzyme activities, such as leucine aminopeptidase or lipase, for a decrease of pH_{NBS} from 8.2 to 7.8 (Yamada and Suzumura 2010). Thus, elevated pCO_2 levels seem to influence the activity of some enzymes.

5.3.5 Viral lysis and grazing

The lysis of cells by viruses as well as the ingestion and digestion of cells by protistan predators are the two main factors involved in the mortality of prokaryotes (see Section 5.2.2). To the best of our knowledge, no information is available on the effect of ocean acidification on rates of viral lysis. Thus, the effects of ocean acidification on viral lysis can only be inferred from changes in viral abundance. In a mesocosm experiment, total viral abundance did not change between the three pCO_2 levels that were investigated (190, 414, and 714 µatm; Rochelle-Newall *et al.* 2004). In a subsequent study, viral abundance was followed by flow cytometry in mesocosms subject to pCO_2 levels of 350, 700, and 1050 µatm (Larsen *et al.* 2008). Viral abundance and the abundance of three viral groups detected by flow cytometry did not show any significant change between pCO_2 levels, whereas for the 'high-fluorescence viruses' maximum abundances were higher at a pCO_2 of 700 and 1050 µatm than at 350 µatm. Thus, the findings published so far suggest

either higher lysis rates of cells, which can be infected and lysed by this type of virus, or higher viral production rates of these specific viruses at elevated pCO_2 levels (e.g. by an increased number of viruses released per cell).

A single study reported a decrease in the abundance of the *E. huxleyi* virus EhV at elevated (700 and 1050 compared to 350 µatm) pCO_2 (Larsen *et al.* 2008), but there was no significant difference in the abundance of *E. huxleyi* across treatments (see Paulino *et al.* 2008). This could suggest that a potentially negative effect of elevated pCO_2 on the host was counterbalanced by a negative effect on viral infection and lysis, thus reducing mortality.

It is known that metabolically active cells are often infected at higher rates, for example due to a greater number of receptors on the cell surface (Weinbauer 2004). Thus, if ocean acidification reduces the metabolism of the host, this could lead to a decrease in the infection rate. Also, metabolically active cells often produce more progeny, i.e. they have a higher burst size (number of viruses released upon cell lysis), than less active ones (Parada *et al.* 2006). In contrast, one could argue that a weakening of the defence system by ocean acidification could increase viral infection as happens in metazoans. However, there seems to be no evidence for that in microorganisms.

Viruses, i.e. viral DNA, can also remain within the host cells, usually integrated as provirus or prophage into the host DNA (Weinbauer 2004). This virus–host relationship is called lysogeny for prokaryotes and latency for eukaryotes. In such lysogenic cells and cells with latent infections, the viral DNA replicates along with the host. This occurs until either so-called prophage curing occurs spontaneously, which excludes the viral DNA, or until a factor induces the lytic cycle in the viral DNA, which results in the formation of viral particles, the lysis of the cell, and the release of the viral progeny. Inducing agents can be stress factors such as DNA damage. It is unknown whether changes in pCO_2 levels can act as an inducing agent.

The effect of elevated pCO_2 on grazing by microzooplankton (including ciliates) on phytoplankton such as cyanobacteria has been investigated in only two studies. One experiment reported higher prey abundances and grazing rates in an on-board continuous culture experiment (Rose *et al.* 2009), whereas the other experiment did not detect any significant effect in a mesocosm experiment (Suffrian *et al.* 2008). Rose *et al.* (2009) suggested that the grazing rates were influenced by a change in phytoplankton community composition rather than by physiological effects. Ocean acidification could also change the food quality and thus affect grazing. For example, it is known that the formation of filamentous cells, filamentous cell colonies, or aggregation of cells prevents or reduces grazing (Jürgens and Güde 1994). Since Takeuchi *et al.* (1997) have reported on morphological changes of microorganisms due to increased pCO_2, it is possible that the changes in grazing rates are due to changes in the appearance of morphologies resistant to grazing.

5.3.6 Effect of ocean acidification on microbial nutrient cycling and organic matter dynamics

Nitrification is the biological oxidation of ammonia into nitrite followed by the oxidation of nitrite to nitrate. Huesemann *et al.* (2002) investigated the effects of CO_2-induced pH changes on marine nitrification in the context of deep-sea CO_2 disposal. They found that the rate of nitrification drops drastically with decreasing pH. Relative to the rates at pH 8 (presumably on the NBS scale), nitrification decreased by ~50% at pH 7 and by more than 90% at pH 6.5, while it was completely inhibited at pH 6. Despite the fact that this study was not aimed at projected ocean acidification conditions of the future ocean, it indicates a potentially large sensitivity of nitrification to changes in pCO_2.

No significant effects of ocean acidification were found on the concentrations of chromophoric DOM (cDOM) and DOC (Engel *et al.* 2004; Rochelle-Newall *et al.* 2004; Grossart *et al.* 2006; Schulz *et al.* 2008). However, it must be noted that the fraction of readily bioavailable DOC is small. Thus, effects of pCO_2 on the bioavailability of DOM might not be detectable with the available or applied methods. Data on the effect of ocean acidification on the processing of organic matter by bacteria and the heterotrophic microbial food web are not available. For example, there has been no study of the effect of ocean acidification on bacterial respiration and growth efficiency which

are key parameters for assessing the efficiency of energy transfer to higher trophic levels via the microbial loop.

In on-board experiments under increased pCO_2, the production of TEP increased as a function of CO_2 uptake (Engel 2002). In a mesocosm experiment which induced a phytoplankton bloom dominated by *E. huxleyi*, the TEP production per cell was highest at 710 µatm, lowest at 190 µatm, and intermediate at 410 µatm, whereas the total TEP concentration and production were not affected by the highest and lowest pCO_2 levels (Engel *et al.* 2004). Egge *et al.* (2009) did not find a significant effect of pCO_2 (350, 700, and 1050 µatm) on TEP concentration in a subsequent mesocosm study. It was concluded that the increase in TEP concentration (when observed) was due to the increased production of TEP precursors. In a batch culture study, an increase in TEP concentration and production was observed in the absence of cells (Mari 2008). This supports the idea that the observed increase is due to a modification of the TEP structure linked to an alteration of aggregation processes, rather than to increased production. Overall, a negative effect of ocean acidification on TEP formation has never been detected. An increased TEP aggregation often results in higher abundances and production of bacteria by attracting these cells to the microbial hot spots (Simon *et al.* 2002). Therefore, increased TEP aggregation could result in higher bacterial abundance and production.

5.4 Implications

5.4.1 Potential implications for microbial food webs and biogeochemical cycles

Since most effects of elevated pCO_2 on microorganisms are poorly known and evidence is sometimes contradictory, the implications are highly speculative and must be viewed with caution. Among the effects of ocean acidification are changes in the composition of viral and microbial communities at elevated pCO_2. Changes in diversity, i.e. changes in the relative abundance of species, can also mean changes in the specific functions of single species. Since changes in microbial diversity can be linked to changes in ecosystem functions (Bell *et al.* 2005), there is the possibility that a rise in $p\text{CO}_2$ will affect ecosystem functions and biogeochemical cycles by altering the relative abundance of species. Alternatively, changes in the bacterial community composition of species with similar functions could be a mechanism of adaptation. The net outcome of ocean acidification will depend on whether rare species will serve as a seed bank of species or whether they are the target of local extinctions.

It is known that the degradation of organic matter by bacteria in diatom frustules releases silica and could thus control silica regeneration and diatom production (Nagata 2008). This mechanism could be enhanced by the increased bacterial production in a high-CO_2 ocean. In marine snow, there are also inorganic particles, e.g. calcified cell parts of phytoplankton (Nagata 2008). Since ocean acidification enhances the dissolution of CaCO_3 (see Chapter 3), it is also likely that the access of prokaryotes to the organic matter matrix should be facilitated and result in faster degradation rates, and a decrease of export to the deep sea.

One of the few relatively consistent findings is that TEP formation is often stimulated at higher $p\text{CO}_2$ levels (Table 5.1). One can therefore speculate that ocean acidification directly increases TEP formation (or changes its spatial structure). It has been argued that this will increase the downward flux of organic matter (Riebesell *et al.* 2007; Schulz *et al.* 2008). The reason for postulating this process is inferred from the imbalance between the build-up of organic matter and the drawdown of inorganic nutrients. Consequently, the biological pump and thus carbon export to the deep sea could be enhanced. The pathways of this potential scenario are shown by hatched arrows in Fig. 5.2. This could also stimulate remineralization in the dark ocean and increase the occurrence of oxygen minimum zones (Riebesell *et al.* 2007).

An additional and not exclusive scenario is possible based on the finding of increased TEP formation (Table 5.1). One can also speculate that this could stimulate bacterial production and enzymatic activity, since it is well known that aggregation stimulates these processes (Simon *et al.* 2002). Indeed, in ocean acidification experiments only neutral and positive effects on bacterial production and enzyme activity were found whereas no nega-

tive effects were reported (Table 5.1). Particles are hot spots of microbial activity and biogeochemical transformation (Azam and Long 2001; Kiørboe and Jackson 2001; Azam and Malfatti 2007). Ocean acidification could therefore enhance this transformation and lead to a faster recycling of elements and increased respiration. If this occurs, then the export of carbon by the biological pump or the microbial carbon pump could be reduced. Mari (2008) has shown that the buoyancy of aggregates may increase at lower pH due to lowering of the sticking properties of TEP (i.e. a lower ability to aggregate dense particles and, thus, to form fast-sinking aggregates) and concluded that marine aggregates may ascend to the sea surface under elevated $p\text{CO}_2$. This could reduce the efficiency of the biological pump by partly inverting the carbon flow to the deep ocean. Since a significant number of bacteria, viruses, and protists are attached to TEP (Simon *et al.* 2002; Weinbauer *et al.* 2009), ocean acidification might play a role as a microbial elevator. In the surface microlayer, TEP and associated microorganisms could become even more concentrated and eventually be aerosolized (Wurl and Holmes 2008). As a result, one can propose a scheme in which ocean acidification will increase the ascent of organic material and aggregate-attached microorganisms. The pathways of this potential scenario are shown by open arrows in Fig. 5.2. This could enhance remineralization in the euphotic layer (the surface ocean). Since microorganisms, cell debris, and small particles are transported by spray from the surface microlayer into the atmosphere (Kuznetsova *et al.* 2005) and can serve as nuclei for cloud formation (Leck and Bigg 2005), there could be an impact on climate. Both scenarios are likely to operate in parallel; the net outcome of ocean acidification, i.e. whether it primes or short-circuits the biological pump, remains unknown.

Ocean acidification reduces the calcification rate of many coral species (see Chapter 7). This could be even more important for deep-water coral communities, which can be found at depths close to the aragonite saturation level, which will become increasingly shallow due to ocean acidification (see Chapter 3). Such a decrease in skeleton formation rate could have other consequences which have not been investigated yet. According to the holobiont theory of evolution (Rosenberg *et al.* 2007), changes in the microbial community composition could be a means to counterbalance environmental changes. Although data are missing, there is increasing evidence that microorganisms are strongly involved in carbon and nutrient cycling in the holobiont. Thus, corals (and sponges) may be able to partly adapt to ocean acidification via changes in the structural and functional diversity of associated microorganisms. The microbial community associated with a zooxanthellate coral showed shifts towards a bacterial community composition characteristic of dead corals when exposed to higher pCO_2 levels (Vega Thurber *et al.* 2009). This seems to contradict the probiotic coral hypothesis of Reshef *et al.* (2006), but one must keep in mind that the experiment was performed using a strong decrease in pH and ran over a short time frame, which might have prevented the detection of adaptation.

5.4.2   Acclimation, adaptation, and perspectives

Potential evidence for acclimation (i.e. the adjustment to environmental change by physiological and morphological change) of bacteria to ocean acidification comes from pH_i regulation in a *Vibrio* strain (Labare *et al.* 2010). Adaptation, the adjustment to environmental change by genetic change, is probably faster in microbes than for multicellular marine organisms. This is due to their short generation time of only a few days which allows for thousands of generations within the projected ocean acidification scenarios, hence increasing the potential for accumulation of mutations, and (at least for prokaryotes) due to the efficient mechanisms of lateral gene transfer. Genomics, transcriptomics, proteomics, and assessment of the expression of specific marker genes for crucial functions are among the most promising methods that are available, or are on the verge of development, for investigating potential of acclimation and adaptation. In this context, it is necessary to note that no data are available for long-term experiments, which could shed light on potential acclimation and adaptation. Since the currently used experimental approaches involve perturbation experiments with a pCO_2 or pH 'shock' which does not mimic the gradual increase in pCO_2 that occurs

in the real world, studies on acclimation and adaptation will be crucial for an evaluation of the performance of microorganisms in a high-CO_2 ocean.

An overall conclusion of microbe-related ocean acidification studies is that only a few environments have been studied. Among the systems not yet studied or barely studied with respect to microbe interactions are the dark ocean, sediments, as well as microorganisms associated with benthic animals, macrophytes, zooplankton, and fish. To the best of our knowledge, no study has been performed on microbial pathogens and (heterotrophic) symbionts. Also, studies on some crucial processes such as prokaryotic respiration and growth efficiency are missing. Data from in-depth sequencing studies are not available yet. Thus, many potential effects of ocean acidification on viruses and heterotrophic microbes remain to be investigated.

5.5   Acknowledgements

This work is a contribution to the 'European Project on Ocean Acidification' (EPOCA) which received funding from the European Community's Seventh Framework Programme (FP7/2007–2013) under grant agreement no. 211384. It was also supported by the projects ANR-AQUAPHAGE (no. ANR 07 BDIV 015–06) and ANR-MAORY (no. ANR 07 BLAN 0116) from the French Science Ministry. The comments of Carol Turley and Ulf Riebesell on an early draft of this chapter are much appreciated.

References

Ainsworth, T.D., Vega Thurber, R.V., and Gates, R.D. (2009). The future of coral reefs: a microbial perspective. *Trends in Ecology and Evolution*, **25**, 233–40.

Alldredge, A.L. and Cohen, Y. (1987). Can microscale chemical patches persist in the sea? Microelectrode study of marine snow and fecal pellets. *Science*, **235**, 689–91.

Alldredge, A.L., Passow, U., and Logan, B.E. (1993). The abundance and significance of a class of large, transparent organic particles in the ocean. *Deep-Sea Research*, **40**, 1131–40.

Allgaier, M., Riebesell, U., Vogt, M., Thyrhaug, R., and Grossart, H.P. (2008). Coupling of heterotrophic bacteria to phytoplankton bloom development at different levels: a mesocosm study. *Biogeosciences*, **5**, 1007–22.

Amann, R., Krumholz, I., and Stahl, D.A. (1990). Fluorescent-oligonucleotide probing of whole cells for determinative, phylogenetic, and environmental studies in microbiology. *Journal of Bacteriology*, **1972**, 762–70.

Arıstegui, J., Gasol, J.M., Duarte, C.M., and Herndl, G.J. (2009). Microbial oceanography of the dark ocean's pelagic realm. *Limnology and Oceanography*, **54**, 1501–29.

Azam, F., Fenchel, T., Field, J.G., Gray, J.S., Meyer-Reil, L.A., and Thingstad, F. (1983). The ecological role of water-column microbes in the sea. *Marine Ecology Progress Series*, **10**, 257–63.

Azam, F. and Long, R.A. (2001). Sea snow microcosms. *Nature*, **414**, 495–8.

Azam, F. and Malfatti, F. (2007). Microbial structuring of marine ecosystems. *Nature Reviews Microbiology*, **5**, 966.

Béjà, O., Aravind, L., Koonin, E.V. *et al.* (2000). Bacterial rhodopsin: evidence for a new type of phototrophy in the sea. *Science*, **289**, 1902–6.

Bell, T., Newman, J.A., Silverman, B.W., Turner, S.L., and Lilley, A.K. (2005). The contribution of species richness and composition to bacterial services. *Nature*, **436**, 1157–60.

Boetius, A., Ravenschlag, K., Schubert, C.J., *et al.* (2000). A marine microbial consortium apparently mediating anaerobic oxidation of methane. *Nature*, **407**, 623–6.

Booth, I.R. (1985). Regulation of cytoplasmic pH in bacteria. *Microbiology and Molecular Biology Reviews*, **49**, 359–78.

Børsheim, K.Y. (1993) Native marine bacteriophages. *FEMS Microbial Ecology*, **102**, 141–59.

Egge, J.K., Thingstad, T.F., Larsen, A. *et al.* (2009). Primary production during nutrient-induced blooms at elevated CO_2 concentrations. *Biogeosciences*, **6**, 877–85.

Engel, A. (2002). Direct relationship between CO_2 uptake and transparent exopolymer particles production in natural phytoplankton. *Journal of Plankton Research*, **24**, 49–53.

Engel, A., Delille, B., Jacquet, S. *et al.* (2004). Transparent exopolymer particles and dissolved organic carbon production by *Emiliania huxleyi* exposed to different CO_2 concentrations: a mesocosm experiment. *Aquatic Microbial Ecology*, **34**, 93–104.

Fisher, M.M. and Triplett, E.W. (1999). Automated approach for ribosomal intergenic spacer analysis of microbial diversity and its application to freshwater bacterial communities. *Applied Environmental Microbiology*, **65**, 4630–36.

Fuhrman, J.A. (1999). Marine viruses and their biogeochemical and ecological effects. *Nature*, **399**, 541–8.

Gates, S. (2002). Review of methodology of quantitative reviews using meta-analysis in ecology. *Journal of Animal Ecology*, **71**, 547–57.

Gattuso, J.-P. and Lavigne, H. (2009). Technical note: approaches and software tools to investigate the impact of ocean acidification. *Biogeosciences*, **6**, 2121–33.

Gattuso, J.-P., Frankignoulle, M., and Wollast, R., (1998). Carbon and carbonate metabolism in coastal aquatic ecosystems. *Annual Review of Ecology and Systematics*, **29**, 405–34.

Grossart, H.P., Allgaier, M., Passow, U., and Riebesell, U. (2006). Testing the effect of CO_2 concentration on the dynamics of marine heterotrophic bacterioplankton. *Limnology and Oceanography*, **51**, 1–11.

Hall-Spencer, J.M., Rodolfo-Metalpa, R., Martin, S. *et al.* (2008). Volcanic carbon dioxide vents show ecosystem effects of ocean acidification. *Nature*, **454**, 96–9.

Hedges, J.I. (2002). Why dissolved organics matter. In: D.A. Hansell and C.A. Carlson (eds) *Biogeochemistry of marine dissolved organic matter*, pp. 1–33. Academic Press, San Diego.

Huesemann, M.H., Skilmann, A.D., and Crecelius, E.A. (2002). The inhibition of marine nitrification by ocean disposal of carbon dioxide. *Marine Pollution Bulletin*, **44**, 142–8.

Inagaki, F., Kuypers, M.M.M., Tsunogai, U. *et al.* (2006). Microbial community in a sediment-hosted CO_2 lake of the southern Okinawa Trough hydrothermal system. *Proceedings of the National Academy of Sciences USA*, **103**, 14164–9.

Jiao, N., Herndl, G.J., Hansell, D.A. *et al.* (2010). Microbial production of recalcitrant dissolved organic matter: long-term carbon storage in the global ocean. *Nature Reviews Microbiology*, **8**, 593–9.

Joint, I., Doney, S.C., and Karl, D.M. (2011). Will ocean acidification affect marine microbes? *The ISME Journal*, **5**, 1–7.

Jürgens, K. and Güde, H. (1994). The potential importance of grazing-resistant bacteria in planktonic systems. *Marine Ecology Progress Series*, **112**, 169–88.

Kepkay, P.E. (1994). Particle aggregation and biological reactivity of colloids. *Marine Ecology Progress Series*, **109**, 293–304.

Kepkay, P.E. and Johnson, B.D. (1989). Coagulation on bubbles allows the microbial respiration of oceanic dissolved organic carbon. *Nature*, **385**, 63–5.

Kiørboe, T. and Jackson, G. (2001). Marine snow, organic solute plumes, and optimal chemosensory behavior of bacteria. *Limnology and Oceanography*, **43**, 1309–18.

Kuznetsova, M., Lee, C., and Aller, J. (2005). Characterization of the proteinaceous matter in marine aerosols. *Marine Chemistry*, **96**, 359–77.

Labare, M.P., Bays, J.P., Butkus, B.A. *et al.* (2010). The effects of elevated carbon dioxide levels on a *Vibrio* sp. isolated from the deep-sea. *Environmental Science and Pollution Research*, **17**, 1009–15.

Larsen, J.B., Larsen, A., Thyrrhaug, R., Bratbak, G., and Sandaa, R.-A. (2008). Response of marine viral populations to a nutrient induced phytoplankton bloom at different pCO_2 levels. *Biogeosciences*, **5**, 523–33.

Leck, C. and Bigg, E.K. (2005). Biogenic particles in the surface microlayer and overlaying atmosphere in the central Arctic Ocean during summer. *Tellus B*, **57**, 305–16.

Liu, J., Weinbauer, M.G., Maier, C., Dai, M., and Gattuso, J.-P. (2010) Effect of ocean acidification on microbial diversity, and on microbe-driven biogeochemistry and ecosystem functioning. *Aquatic Microbial Ecology*, **61**, 291–305.

Mari, X. (2008). Does ocean acidification induce an upward flux of marine particles? *Biogeosciences*, **5**, 1023–31.

Maron, P.A., Ranjard, L., Mougel, C., and Lemanceau, P. (2007). Metaproteomics: a new approach for studying functional microbial ecology. *Microbial Ecology*, **53**, 486–93.

Moyer, C.L., Dobbs, F.C., and Karl, D.M. (1994). Estimation of diversity and community structure through restriction fragment length polymorphism distribution analysis of bacterial 16S rRNA genes from a microbial mat at an active, hydrothermal vent system, Loihi Seamount, Hawaii. *Applied Environmental Microbiology*, **60**, 871–9.

Muyzer, G., de Waal, E.C., and Uitterlinden, A.G. (1993). Profiling of complex microbial populations by denaturing gradient gel electrophoresis analysis of polymerase chain reaction-amplified genes coding for 16S rRNA. *Applied Environmental Microbiology*, **59**, 695–700.

Nagata, T. (2008). Organic matter–bacteria interactions in seawater. In: D.L. Kirchman (ed.) *Microbial ecology of the oceans*, pp. 207–41. John Wiley and Sons, New York.

Nagata, T. and Kirchman, D.L. (1997). Roles of submicron particles and colloids in microbial food webs and biogeochemical cycles within marine environments. *Advances in Microbial Ecology*, **15**, 81–103.

Nakagawa, S. and Takai, K. (2008). Deep-sea vent chemoautotrophs: diversity, biochemistry and ecological significance. *FEMS Microbial Ecology*, **65**, 1–14.

Nisumaa, A.-M., Pesant, S., Bellerby, R.G.J., *et al.* (2010). EPOCA, EUR-OCEANS data compilation on the biological and biogeochemical responses to ocean acidification. *Earth System Science Data*, **2**, 167–75.

Pace, N.R. (2009). Problems with 'procaryote'. *Journal of Bacteriology*, **191**, 2008–10.

Parada, V., Herndl, G.J. and Weinbauer, M.G. (2006). Viral burst size of heterotrophic prokaryotes in aquatic systems. *Journal of the Marine Biology Association UK*, **86**, 613–21.

Paulino, A.I., Egge, J.K., and Larsen, A. (2008). Effects of increased atmospheric CO_2 on small and intermediate sized osmotrophs during a nutrient induced phytoplankton bloom. *Biogeosciences*, **5**, 739–48.

Piontek, J., Lunau, M., Händel, N., Borchard, C., Wurst, M., and Engel, A. (2010). Acidification increases microbial polysaccharide degradation in the ocean. *Biogeosciences*, **7**, 1615–24.

Pomeroy, L.R. (1974). The ocean's food web, a changing paradigm. *BioScience*, **24**, 499–504.

Proctor, L.M. and Fuhrman, J.A. (1990). Viral mortality of marine bacteria and cyanobacteria. *Nature*, **343**, 60–2.

Raoult, D. and Forterre, P. (2008). Redefining viruses: lessons from Mimivirus. *Nature Reviews Microbiology*, **6**, 315–19.

Reshef, L., Koren, O., Loya, Y., Zilber-Rosenberg, I., and Rosenberg, E. (2006). The coral probiotic hypothesis. *Environmental Microbiology*, **8**, 2068–73.

Riebesell, U., Schulz, K.G., Bellerby, R.G. *et al.* (2007). Enhanced biological carbon consumption in a high CO_2 ocean. *Nature*, **450**, 545–8.

Rochelle-Newall, E., Delille, B., Frankignoulle, M. *et al.* (2004). Chromophoric dissolved organic matter in experimental mesocosms maintained under different pCO_2 levels. *Marine Ecology Progress Series*, **272**, 25–31.

Rose, J.M., Feng, Y.Y., Gobler, C.J. *et al.* (2009). Effects of increased pCO_2 and temperature on the North Atlantic spring bloom. II. Microzooplankton abundance and grazing. *Marine Ecology Progress Series*, **388**, 27–40.

Rosenberg, E., Koren, O., Reshef, L., Efrony, R., and Zilber-Rosenberg, I. (2007). The role of microorganisms in coral health, disease, and evolution. *Nature Reviews Microbiology*, **5**, 355–62.

Schulz, K.G., Riebesell, U., Bellerby, R.G.J. *et al.* (2008). Build-up and decline of organic matter during PeECE III. *Biogeosciences*, **5**, 707–18.

Shi, Y., Tyson, G.W., and DeLong, E.F. (2009). Metatranscriptomics reveals unique microbial small RNAs in the ocean's water column. *Nature*, **459**, 266–9.

Simon, M., Grossart, H.-P., Schweitzer, B., and Ploug, H. (2002). Microbial ecology of organic aggregates in aquatic ecosystems. *Aquatic Microbial Ecology*, **28**, 175–211.

Sogin, M.L., Morrison, H.G., Huber, J.A. *et al.* (2006). Microbial diversity in the deep sea and the underexplored 'rare biosphere'. *Proceedings of the National Academy of Sciences USA*, **103**, 12115–20.

Staley, J.T. and Konopka, A. (1985). Measurement of *in situ* activities of nonphotosynthetic microorganisms in aquatic and terrestrial habitats. *Annual Review of Microbiology*, **39**, 321–46.

Suffrian, K., Simonelli, P., Nejstgaard, J.C., Putzeys, S., Carotenuto, Y., and Antia, A.N. (2008). Microzooplankton grazing and phytoplankton growth in marine mesocosms with increased CO_2 levels. *Biogeosciences*, **5**, 1145–56.

Suttle, C.A. (2007). Marine viruses-major players in the global ecosystem. *Nature Reviews Microbiology*, **5**, 801–12.

Takeuchi, K., Fujioka, Y., Kawasaki, Y., and Shirayama, Y. (1997). Impacts of high concentration of CO_2 on marine organisms; a modification of CO_2 ocean sequestration. *Energy Conversion and Management*, **38**, 5337–41.

Tanaka, T., Thingstad, T.F., Lovdal, T. *et al.* (2008). Availability of phosphate for phytoplankton and bacteria and of glucose for bacteria at different pCO_2 levels in a mesocosm study. *Biogeosciences*, **5**, 669–78.

Taylor, M.W., Radax, R., Steger, D., and Wagner, M. (2007). Sponge-associated microorganisms: evolution, ecology, and biotechnological potential. *Microbiology and Molecular Biology Reviews*, **71**, 295–347.

Vega Thurber, R.V., Willner-Hall, D., Rodriguez-Mueller, B. *et al.* (2009). Metagenomic analysis of stressed coral holobionts. *Environmental Microbiology*, **11**, 2148–63.

Verdugo, P., Alldredge, A.L., Azam, F., Kirchman, D.L., Passow, U., and Santschi, P.H. (2004). The oceanic gel phase: a bridge in the DOM–POM continuum. *Marine Chemistry*, **92**, 67–85.

Weinbauer, M.G. (2004). Ecology of prokaryotic viruses. *FEMS Microbial Ecology*, **28**, 127–81.

Weinbauer, M.G., Bettarel, Y., Cattaneao, R. *et al.* (2009). Viral ecology of organic and inorganic particles in aquatic systems: avenues for further research. *Aquatic Microbial Ecology*, **57**, 321–41.

Whitman, W.B. (2009). The modern concept of the procaryote. *Journal of Bacteriology*, **191**, 2000–5.

Wilhelm, S.W. and Suttle, C.A. (1999). Viruses and nutrient cycles in the sea. *Bioscience*, **49**, 781–8.

Wurl, O. and Holmes, M. (2008). The gelatinous nature of sea-surface microlayer, *Marine Chemistry*, **110**, 89–97.

Yamada, N., Suzumura, A., Tsurushima, N., and Harada K. (2008). Impact on bacterial activities of ocean sequestration of carbon dioxide into bathypelagic layers. *Oceans 2008— MTS/IEEE Kobe Techno-Ocean*, 1–3.

Yamada, N. and Suzumura, M. (2010). Effects of seawater acidification on hydrolytic enzyme activities. *Journal of Oceanography*, **66**, 233–41.

CHAPTER 6

Effects of ocean acidification on pelagic organisms and ecosystems

Ulf Riebesell and Philippe D. Tortell

6.1 Introduction

Over the past decade there has been rapidly growing interest in the potential effects of ocean acidification and perturbations of the carbonate system on marine organisms. While early studies focused on a handful of phytoplankton and calcifying invertebrates, an increasing number of investigators have begun to examine the sensitivity to ocean acidification of various planktonic and benthic organisms across the marine food web. Several excellent review articles have recently summarized the rapidly expanding literature on this topic (Fabry *et al.* 2008; Doney *et al.* 2009; Joint *et al.* 2011). The focus of this chapter is on the potential ecosystem-level effects of ocean acidification. Starting with a brief review of the basic physical, chemical, and biological processes which structure pelagic marine ecosystems, the chapter explores how organismal responses to perturbations of the carbonate system could scale up in both time and space to affect ecosystem functions and biogeochemical processes. As with many chapters in this volume, and indeed much of the ocean acidification literature at present, our review raises more questions than it answers. It is hoped that these questions will prove useful for articulating and addressing key areas of future research.

6.2 Planktonic processes and the marine carbon cycle

6.2.1 Planktonic organisms, pelagic food webs, and ecosystems

Complexity in marine pelagic food webs results from the interactions of multiple trophic levels across a range of temporal and spatial scales. The traditional view of marine food webs (Steele 1974) involved a relatively short trophic system in which large phytoplankton (e.g. net plankton such as diatoms) were grazed by a variety of mesozooplankton (e.g. copepods), which were in turn consumed by second-level predators, including many economically important fish and invertebrate species. This 'classic' marine food web is typical of high-productivity regions such as coastal upwelling regimes (Lassiter *et al.* 2006). A characteristic feature of these systems is a strong decoupling between primary production and grazing, which results from the different metabolic rates of consumers and producers and, in many cases, ontogenetic and seasonal delays in the emergence of feeding predators. The uncoupling between phytoplankton and their consumers leads to significant export of organic material out of the euphotic zone, the so-called biological carbon pump (discussed further below). In a steady-state system, export of organic material from the euphotic zone is balanced by external nutrient inputs (e.g. NO_3^- supply from vertical mixing; Eppley and Peterson 1979), and the primary production that is fuelled by these nutrient supplies is referred to as 'new' or 'export' production (Dugdale and Goering 1967).

In addition to the classical food web, there exists a parallel 'microbial loop' (Azam *et al.* 1983; see Chapter 5) which contributes to total marine primary productivity and plays an important role in nutrient recycling. This microbial food web is driven by small photosynthetic algae, heterotrophic bacteria, and microzooplankton (which are actually protists). While many of these small organisms were unknown or greatly undersampled before the latter half of the 20th century, it is now clear that they are amongst the most abundant cells in seawater, dominating autotrophic and hetero-trophic metabolism in many oceanic systems (Sherr *et al.* 2007). The microbial loop is particularly important in regions with limited 'new' nutrient supplies where small primary producers have a distinct advantage in nutrient uptake based on their high surface area to volume ratio. In addition to the well-known photosynthetic bacteria *Synechococcus* and *Prochlorococcus*, recent work has documented the apparent abundance of picoeu-karyotic phytoplankton (Sherr *et al.* 2007). It has been shown that this group can contribute signifi-cantly to primary productivity and biogeochemical cycles in marine waters (Worden *et al.* 2004), but the taxonomic identity of these picoeukaryotes has only been examined in a few locations and their physiological capabilities remain poorly known.

Microbial food webs are characterized by a very tight coupling between organic carbon production and consumption. While some controversy exists about the extent to which these microbially domi-nated ecosystems are net autotrophic or net hetero-trophic (Williams 1998; Duarte and Prairie 2005), it is clear that small primary producers are grazed voraciously by a variety of taxonomically diverse single-celled microzooplankton (Sherr and Sherr 1994), including many different ciliated and flagel-lated species (this latter group includes mixotrophic dinoflagellates). Unlike copepods and other larger zooplankton, micrograzers have metabolic rates that are similar to those of their prey, and they are thus able to graze small phytoplankton at rates close to those at which the phytoplankton grow (Calbet and Landry 2004). Moreover, predator and prey in the microbial loop are not subject to seasonal timing 'mismatches', which can occur in the classic food web. The close coupling between production and grazing results in high rates of nutrient reminer-alization, with inorganic nitrogen recycled in the form of NH_4^+ during grazer excretion (Glibert 1982). In the absence of strong external nutrient inputs, primary production in these systems is mainly fuelled by such 'regenerated' nutrients, in a tight cycle of production and grazing which limits bio-mass accumulation.

In both the classic and microbial food webs, a sig-nificant fraction of phytoplankton-derived organic carbon accumulates in a large pool of dissolved organic carbon (DOC), much of which is refractory and long-lived, with residence times of the order of thousands of years (Hansell and Carlson 2002). This DOC can be released directly by phytoplankton through exudation and from zooplankton as a result of egestion and excretion. Marine viruses also play a role in DOC production through the lysis (i.e. infection and subsequent rupture) of phytoplank-ton and heterotrophic bacteria, which releases cellular constituents into surface waters (see Chapter 5). Some fraction of the DOC pool is highly labile and is rapidly taken up by heterotrophic bac-teria. Bacterial growth efficiency determines the fraction of the assimilated DOC that can be reincor-porated into the food web through microzooplank-ton grazing versus the fraction that is remineralized back to CO_2 and inorganic nutrients. The latter proc-ess contributes to the recycling of nutrients that fuels regenerated production. Given the relatively low growth efficiencies of many natural bacterial and microzooplankton assemblages (del Giorgio and Cole 1998), a large fraction of the organic car-bon cycled through the microbial loop will eventu-ally be 'lost' as CO_2, with only relatively small amounts recycled back into the classic food chain.

6.2.2 The marine carbon cycle

The marine carbon cycle is driven by two independ-ent processes, the 'solubility pump' and the 'bio-logical carbon pump' (see also Chapter 12). As the solubility of gases increases with decreasing seawa-ter temperature, the cold waters sinking to depths during deep-water formation at high latitudes are CO_2-rich relative to average oceanic surface waters. As the newly formed deep waters flow towards lower latitudes, they carry a high CO_2 load, spread-ing it throughout the deep ocean. About one-third of the surface-to-depth gradient of dissolved inor-ganic carbon (C_T) is generated by this solubility pump. The other two-thirds of the vertical carbon gradient is caused by the biological carbon pump; the sinking of biogenic material from the sunlit surface layer to the deep ocean. Integrated over the global ocean, the biotically mediated oceanic surface-to-depth C_T gradient corresponds to a car-bon pool 3.5 times larger than the total amount of atmospheric CO_2 (Gruber and Sarmiento 2002).

Small changes in this carbon pool, caused for example by biological responses to ocean change, therefore have the potential to cause large changes in atmospheric CO_2 concentration.

Based on the biological processes responsible for carbon fixation, two biological carbon pumps can be distinguished: (1) the organic carbon pump, driven by photosynthetic CO_2 fixation, and (2) the carbonate counter pump, generated by the formation of calcium carbonate ($CaCO_3$) shell material by calcifying plankton. While photosynthetic carbon fixation lowers the CO_2 partial pressure in the euphotic zone, causing a net flux of CO_2 from the atmosphere to the surface ocean, $CaCO_3$ precipitation lowers C_T and total alkalinity in the surface ocean, causing an increase in CO_2 partial pressure (Chapter 2). Thus, the two biological carbon pumps reinforce each other in terms of maintaining a vertical C_T gradient whereas they counteract each other with respect to their impact on air–sea CO_2 exchange (Heinze et al. 1991). The latter aspect has led to the term 'counter pump'. With a global vertical flux of particulate organic carbon of approximately 10 Pg C yr^{-1} compared with a $CaCO_3$ flux of about 1 Pg C yr^{-1} the organic carbon pump clearly dominates over the carbonate counter pump (Milliman 1993). It is worth noting that dissolved organic carbon produced in the surface layer and transported to depth during deep-water formation, with an estimated flux of 2 Pg C yr^{-1} (Hansell and Carlson 2002), also contributes to the organic carbon pump.

6.2.3 Biogeochemical provinces

Longhurst (1998) was the first to utilize satellite-based observations of phytoplankton biomass (based on inferred chlorophyll *a* distributions) to characterize the broad-scale patterns of marine 'ecological geography'. This work provided a conceptual and practical framework for classifying oceanic regions into domains with spatially and temporally coherent biological dynamics. The distribution of these ecological domains is coupled with underlying physical processes that drive chemical and biological gradients in the upper mixed layer. For example, a clear distinction can be made between the highly stratified subtropical gyre systems (the 'trades biome' *sensu* Longhurst 1998) and the

equatorial upwelling regimes, where nutrient-rich deep waters are transported into the mixed layer promoting phytoplankton growth. Similarly, distinct biogeochemical provinces occur in coastal upwelling regimes and along shallow continental margins that are influenced by riverine input and/or tidal forcing. The classical phytoplankton 'spring bloom' is found in the North Atlantic, but not in other subpolar waters (e.g. the subarctic north-east Pacific) where iron limitation prevents the complete utilization of excess macronutrients (Martin and Fitzwater 1988). In high-latitude polar systems, seasonal light limitation restricts productivity over much of the growing season, with large pulses of productivity typically occurring over a single period of the year (Arrigo et al. 2008; Pabi et al. 2008). Despite some gross similarities between Arctic and Antarctic marine ecosystems (e.g. the importance of seasonal ice cover), large differences in physical circulation and nutrient supply between these regions (both macronutrients and trace metals) result in very different ecological and biogeochemical dynamics.

While a complete review of oceanic biomes is well beyond the scope of this chapter (readers are referred to the original work of Longhurst 1998 for details), the broad concept is important for understanding the potential ecological effects of ocean acidification. Productivity in the different ecological domains is controlled by a unique combination of physical, chemical, and biological factors (e.g. light, macronutrients, trace metals, grazing, etc.), and these factors will probably influence CO_2-dependent responses. For example, CO_2-dependent growth of phytoplankton is likely to differ in nutrient replete versus nutrient-limited regions, and under conditions of strong stratification versus deep vertical mixing (see discussion below). Moreover, the unique attributes of each ecological domain should be considered when designing manipulative experiments (see Section 6.4). A different approach should be taken, for example, in designing manipulative experiments in subtropical regions versus subpolar waters. In the subtropics, experiments should aim to mimic a tightly coupled, nutrient-limited environment, since enrichment of macronutrients in bottles would perturb the system into a new ecological state. In contrast, steady-state approaches, based on chemostat experiments (e.g. Sciandra et al.

2003), may not provide the best simulation of the ecological dynamics observed in many subpolar regimes, which are characterized by strong non-steady-state behaviour.

6.3 Direct effects of ocean acidification on planktonic organisms

6.3.1 Photosynthesis and carbon fixation

In the oceans, photosynthesis, the formation of organic matter using energy from sunlight, is carried out chiefly by microscopic phytoplankton. For this purpose, these single-cell organisms must acquire, from surface seawater, inorganic carbon and a suite of major and trace nutrients, including nitrogen, phosphorus, and trace metals such as iron. CO_2 rather than the much more abundant bicarbonate ion, HCO_3^-, is the substrate used in the 'carbon fixation' step of photosynthesis that is catalyzed by the enzyme ribulose-1,5-bisphosphate carboxylase oxygenase (RubisCO). This enzyme has an intrinsically low affinity for CO_2, achieving half saturation of carbon fixation at CO_2 concentrations well above those present in seawater (Badger *et al.* 1998). To overcome RubisCO's low affinity, CO_2 must be concentrated at the site of fixation, an energy-consuming process. Because CO_2 diffuses readily through biological membranes and leaks out of the cell it is expected that an increase in the CO_2 concentration of surface seawater will reduce CO_2 leakage, which in some phytoplankton groups facilitates photosynthesis and can lead to an increase in primary production (i.e. the rate of synthesis of organic matter per unit time and unit area of the ocean). This theoretical expectation is supported by both laboratory and field experiments (see below).

The extent to which phytoplankton may respond to increased CO_2 (decreased pH) is likely to depend, to a significant extent, on the physiological mechanisms of inorganic carbon uptake and intracellular assimilation. Primary producers in the marine realm encompass phylogenetically very diverse groups (Falkowski *et al.* 2004), from prokaryotes to angiosperms, differing widely in their photosynthetic apparatus and carbon enrichment systems (Giordano *et al.* 2005). Species with effective carbon-concentrating mechanisms (CCMs) are less sensitive

to increased CO_2 levels than those with less efficient CCMs (Burkhardt *et al.* 2001; Rost *et al.* 2003). Because the sensitivity to carbon enrichment differs widely among taxa, rising CO_2 levels will alter competitive relationships and result in shifts of plankton species composition (see below; Rost *et al.* 2008).

Stimulating effects of elevated CO_2 on photosynthesis and carbon fixation have been observed in a variety of phytoplankton taxa (Table 6.1), including diatoms, coccolithophores, cyanobacteria, and dinoflagellates. No effect of elevated pCO_2 on organic matter production was observed in the two coccolithophore species *Coccolithus pelagicus* and *Calcidiscus leptoporus* (Langer *et al.* 2006). The only study showing decreasing organic matter production at elevated pCO_2 in *Emiliania huxleyi* was conducted in chemostats under nitrate limitation (Sciandra *et al.* 2003). Modest increases (~10%) in growth and productivity in response to elevated pCO_2 were also observed in natural phytoplankton assemblages (Hein and Sand-Jensen 1997; Tortell *et al.* 2002, 2008). In a mesocosm study the net drawdown of inorganic carbon by primary production was 27% and 39% higher at 700 μatm and 1050 μatm compared with 350 μatm (Riebesell *et al.* 2007; Bellerby *et al.* 2008), whereas increases in net community production of only 8% and 15% were estimated based on day-time ^{14}C primary production measurements from the same mesocosm experiment (Egge *et al.* 2009). In this study no CO_2/pH effect was observed on community respiration rate.

In summary, investigations on both individual species and natural assemblages of phytoplankton reveal CO_2 fertilization of photosynthetic carbon fixation and rates of production of organic matter. While few individual phytoplankton species appear to be insensitive to ocean acidification, studies on natural phytoplankton assemblages consistently show increased carbon fixation in response to elevated pCO_2. The magnitude of this stimulation is, however, relatively small compared with the stimulation of plankton growth by iron enrichment in iron-limited waters.

6.3.2 The cell division rate

No consistent response to ocean acidification has been obtained with respect to phytoplankton cell

Table 6.1   Effects of ocean acidification on photosynthesis and carbon fixation in planktonic organisms, comparing rates at pre-industrial/present-day pCO_2 levels with those at pCO_2 projected for the end of this century.

Group	Response	References
Diatoms	↑	Riebesell *et al.* (1993), Burkhardt and Riebesell (1997), Burkhardt *et al.* (1999), Gervais and Riebesell (2001), Wu *et al.* (2010)
Coccolithophores	↑	Buitenhuis *et al.* (1999), Riebesell *et al.* (2000), Rost *et al.* (2002), Zondervan *et al.* (2002), Leonardos and Geider (2005), Feng *et al.* (2008), Barcelos e Ramos *et al.* (2010), Shi *et al.* (2009), De Bodt *et al.* (2010), Müller *et al.* (2010), Rickaby *et al.* (2010)
	↓	Sciandra *et al.* (2003)
	↔	Langer *et al.* (2006), Gao *et al.* (2009)
Dinoflagellates	↑	Burkhardt *et al.* (1999), Rost *et al.* (2006)
Cyanobacteria	↑	Barcelos e Ramos *et al.* (2007), Hutchins *et al.* (2007, 2009), Levitan *et al.* (2007), Fu *et al.* (2008a), Kranz *et al.* (2009)
	↔	Czerny *et al.* (2009)
Natural assemblages	↑	Hein and Sand-Jensen (1997), Tortell *et al.* (2002, 2008), Riebesell *et al.* (2007), Bellerby *et al.* (2008), Egge *et al.* (2009)

↑, enhanced; ↓, slowed down; ↔, unaffected/inconclusive.

Table 6.2   Effects of ocean acidification on cell division rate in planktonic organisms, comparing rates at pre-industrial/present-day pCO_2 levels with those at pCO_2 projected for the end of this century.

Group	Response	References
Diatoms	↑	Riebesell *et al.* (1993), Burkhardt and Riebesell (1997), Burkhardt *et al.* (1999), Gervais and Riebesell (2001)
Coccolithophores	↑	Shi *et al.* (2009), Rickaby *et al.* (2010; *Gephyrocapsa oceanica*)
	↓	Iglesias-Rodriguez *et al.* (2008), Langer *et al.* (2009), Barcelos e Ramos *et al.* (2010), Müller *et al.* (2010), Rickaby *et al.* (2010; *Coccolithus braarudii*)
	↔	Buitenhuis *et al.* (1999), Riebesell *et al.* (2000), Rost *et al.* (2002), Zondervan *et al.* (2002)
Dinoflagellates	↑	Burkhardt *et al.* (1999)
Cyanobacteria	↑	*Trichodesmium* sp.: Barcelos e Ramos *et al.* (2007), Hutchins *et al.* (2007), Levitan *et al.* (2007), Kranz *et al.* (2009, 2010)
	↓	*Nodularia spumigena*: Czerny *et al.* 2009

↑, enhanced; ↓, slowed down; ↔, unaffected/inconclusive.

division rate (also referred to as growth rate). While some studies reported enhanced growth rates at elevated CO_2 concentrations for diatoms, the cyanobacterium *Trichodesmium*, the coccolithophores *E. huxleyi* (only one study) and *Gephyrocapsa oceanica*, and the dinoflagellate *Scrippsiella trochoidea* (Table 6.2), most studies on coccolithophores show no change or a slight decrease in cell division rate in response to elevated pCO_2. In coccolithophores, a pCO_2/pH-induced decrease in cell division rate generally concurs with increased rates of production of organic carbon (see Section 6.3.1), resulting in higher carbon cell quota and larger cell sizes at elevated pCO_2 (e.g. Zondervan *et al.* 2002).

The CO_2 sensitivity of growth and the cell division rate is generally strongest at pCO_2 levels lower than present-day values, with only small changes in cell division rate when pCO_2 is elevated above present-day levels (e.g. see Burkhardt *et al.* 1999). A decline in growth rate, possibly due to adverse effects of decreasing pH, has been observed at pCO_2 levels higher than projected for the end of this century (i.e. >1200 µatm; e.g. Burkhardt *et al.* 1999). This effect also becomes evident in the review of pH sensitivities of a wide range of coastal marine phytoplankton by Hinga (2002), showing declining growth rates as pH is shifted from present values to much higher and much lower levels. As most of the early studies reviewed by Hinga (2002) used extremely broad pH ranges with often poorly constrained carbonate chemistry, it is difficult to interpret them in the context of projected future ocean acidification.

In summary, the rather uniform response in carbon fixation rate to elevated $p\mathrm{CO_2}$ is not mirrored in phytoplankton cell division rates. This is explained by the flexible carbon quota in most phytoplankton taxa and is consistent with the notion that cell division is triggered by cell constituents other than cellular carbon content. The observed differences in the $\mathrm{CO_2}$/pH sensitivity of cell division rate between phytoplankton groups and species suggests that ocean acidification has an impact on phytoplankton community composition and succession.

6.3.3   Nitrogen fixation

Recent studies with the dominant diazotrophic cyanobacterium *Trichodesmium* revealed enhanced rates of both carbon and nitrogen fixation at elevated $p\mathrm{CO_2}$ (Table 6.3). These laboratory results are confirmed by bioassay experiments with natural populations of *Trichodesmium* showing similar responses when exposed to high $\mathrm{CO_2}$ (Hutchins *et al.* 2009). A stimulation of nitrogen fixation by elevated $\mathrm{CO_2}$ was also observed for the unicellular cyanobacterium *Crocosphaera* under iron-replete conditions (Fu *et al.* 2008a). The opposite trend, a decrease in growth and nitrogen fixation rates at elevated $\mathrm{CO_2}$, was observed in *Nodularia spumigena*, a heterocystous cyanobacterium regularly blooming in the open Baltic Sea during the summer months (Czerny *et al.* 2009). While a mechanistic understanding of the effects of ocean acidification on diazotrophic nitrogen fixation is still lacking,

Czerny *et al.* (2009) speculate that structural and regulatory differences between heterocystous and non-heterocystous cyanobacteria may be a reason for the different responses. It should be noted that heterocystous cyanobacteria are largely confined to freshwater and brackish water environments. Even within marine environments, there appears to be a high diversity of potentially nitrogen-fixing organisms (Zehr *et al.* 1998) whose sensitivity to ocean acidification is, at present, entirely unknown. Experimental studies with these organisms are hampered by the difficulty in culturing them under controlled laboratory conditions.

While all studies to date, with the exception of that of Czerny *et al.* (2009), suggest a stimulating effect of ocean carbonation on nitrogen fixation (Table 6.3), a note of caution should be applied when extrapolating these findings to the global ocean. Nitrogen fixers have a high demand for iron and their ecological niche requires excess phosphate after nitrate depletion. Whether $\mathrm{CO_2}$ fertilization of nitrogen fixation also occurs under iron- and phosphorus-limited conditions is largely unknown. In a comparison of iron-limited and iron-replete *Crocosphaera*, only the iron-replete cultures responded to elevated $p\mathrm{CO_2}$ with increased nitrogen fixation (Fu *et al.* 2008a). When exposing phosphorus-depleted and phosphorus-replete cultures of *Trichodesmium* to elevated $p\mathrm{CO_2}$ Hutchins *et al.* (2007) found that $\mathrm{CO_2}$ fertilization of nitrogen fixation still persisted in severely phosphorus-limited cultures. Clearly, more work needs to be done on the interacting effects of $p\mathrm{CO_2}$ with other nutrients and temperature and on the $\mathrm{CO_2}$-sensitivity of natural diazotroph communities before reliable predictions about the impacts of ocean acidification on nitrogen fixation can be made (see discussion below).

Table 6.3   Observed effects of ocean acidification on nitrogen fixation in planktonic organisms, comparing rates at pre-industrial/present-day $p\mathrm{CO_2}$ levels with those at $p\mathrm{CO_2}$ projected for the end of this century.

Species	Response	References
Trichodesmium erythraeum	↑	Barcelos e Ramos *et al.* (2007), Hutchins *et al.* (2007), Levitan *et al.* (2007), Kranz *et al.* (2009, 2010)
Natural colonies of *Trichodesmium*	↑	Preliminary data reported in Hutchins *et al.* (2009)
Crocosphaera watsonii	↑ ↔	Fu *et al.* 2008a
Nodularia spumigena	↓	Czerny *et al.* 2009

↑, enhanced; ↓, slowed down; ↔, unaffected/inconclusive.

6.3.4   Calcification

$CaCO_3$ is one of the most common building materials used in the formation of skeletons, shells, and other protective structures in the marine biota. Organisms exploit the supersaturation with respect to $CaCO_3$ in the surface ocean, which prevents crystallized $CaCO_3$ from dissolving. Calcification, the precipitation of $CaCO_3$, is facilitated by high pH and high carbonate ion (CO_3^{2-}) concentration (see Box 1.1 in Chapter 1). In calcifying organisms these conditions are achieved at the site of calcification through energy-consuming ion transport processes (Mackinder *et al.* 2010). With ocean acidification causing a decrease in pH and [CO_3^{2-}], the energetic cost of calcification is thought to increase. The extra energy needed to compensate for these changes in seawater carbonate chemistry depends on the specific pathways employed in $CaCO_3$ precipitation, the details of which are currently poorly understood but which are likely to differ between taxonomic groups.

Most planktonic calcifying organisms tested so far show a decrease in calcification in response to elevated CO_2/reduced pH (Table 6.4), such as foraminifera, pteropods, and planktonic larvae of echinoderms. A wide range of responses to ocean acidification was obtained for coccolithophores (Table 6.4). Whereas calcification in *E. huxleyi*, *G. oceanica*, and *Calcidiscus quadriperforatus* decreases to varying degrees with increasing pCO_2, *Calcidiscus leptoporus* shows an optimum curve with reduced calcification at pCO_2 levels below and above present conditions and *Coccolithus pelagicus/braarudii* appears to be insensitive to elevated pCO_2. In a comparison of different strains of *E. huxleyi* Langer *et al.* (2009) observed either no change or a decrease in calcification rate with increasing pCO_2. In all studies on coccolithophores the ratio of $CaCO_3$ to organic matter production (PIC:POC; PIC being particulate inorganic carbon and POC particulate organic carbon) decreases or remains unchanged with elevated pCO_2. No interacting effects of pCO_2 and temperature were observed on calcite production, coccolith morphology, or on coccosphere size by De Bodt *et al.* (2010).

Two recent studies on coccolithophores appear to suggest a stimulating affect of ocean acidification on calcification. Iglesias-Rodriguez *et al.* (2008)

reported *E. huxleyi* to double its calcification in response to pCO_2 increasing from 280 to 750 μatm. However, the difference can be explained by a size difference in the cells incubated in the different CO_2 treatments. Cells grown at high CO_2 had an initial biomass two to three times greater than low CO_2-grown cells (possibly due to differences in pre-cultures). Due to the difference in cell size, however, a comparison between CO_2 treatments on a per cell basis is meaningless, but can only be done on a per biomass basis. In fact, when normalized to algal biomass, the trends in calcification and primary production with increasing pCO_2 disappear (see Riebesell *et al.* 2008a). A re-analysis of the same strain of *E. huxleyi* (NZEH) used by Iglesias-Rodriguez *et al.* (2008) by Hoppe *et al.* (pers. comm.) revealed no effect on growth rate and a moderate decrease in calcification with increasing pCO_2.

In a study by Shi *et al.* (2009) both growth rate (cell division rate) and the cellular POC and PIC content of *E. huxleyi* (strain NZEH) was higher at pH_T 7.8 compared with pH_T 8.1, yielding higher rates of organic carbon production and calcification at elevated pCO_2. The PIC:POC ratio was slightly lower in cultures maintained at lower pH levels. As discussed above, increased carbon cell quota and cell size are frequently observed in coccolithophores at elevated pCO_2. However, the results reported by Shi *et al.* (2009) differ from all other studies on coccolithophores in showing an increased cell division rate at elevated pCO_2. The significance of this finding is difficult to assess, partly because that study was based on only two pCO_2 levels and was observed in only two out of the three CO_2 manipulation approaches, with the opposite trend when carbonate chemistry manipulation was done through CO_2 bubbling. Using the same strain of *E. huxleyi*, Hoppe *et al.* (pers. comm.) found no effect on growth rate.

In summary, although much of the work on CO_2/pH sensitivity focuses on coccolithophores, evidence currently available suggests that ocean acidification will cause a decline in $CaCO_3$ production in most planktonic calcifiers. It is currently unknown whether a decreased calcification rate will affect the competitive fitness of calcifying organisms relative to their non-calcifying competitors and to what extent CO_2-sensitive calcifying organisms will

Table 6.4 Effects of ocean acidification on calcification in planktonic organisms, comparing rates at pre-industrial/present-day pCO_2 levels with those at pCO_2 projected for the end of this century.

Group/species		Response	References
Foraminifera		↓	Lombard et al. (2010)
Pteropods		↓	Comeau et al. (2009)
Echinoderm larvae		↓	Clark et al. (2009), Sheppard Brennand et al. (2010)
Coccolithophores			
Emiliania huxleyi	Calcification	↑	Iglesias-Rodriguez et al. (2008), Shi et al. (2009)
		↓	Riebesell et al. (2000), Zondervan et al. (2002), Sciandra et al. (2003), Delille et al. (2005), Engel et al. (2005), Langer et al. (2006), Feng et al. (2008), Gao et al. 2009, De Bodt et al. (2010), Hoppe et al. (pers. comm.)
		↔	Buitenhuis et al. (1999)
	PIC:POC	↓	Buitenhuis et al. (1999), Riebesell et al. (2000), Zondervan et al. (2002), Delille et al. (2005), Engel et al. (2005), Langer et al. (2006, 2009), Feng et al. (2008), Iglesias-Rodriguez et al. (2008), Gao et al. (2009), Shi et al. (2009), De Bodt et al. (2010), Müller et al. (2010), Hoppe et al. (pers. comm.)
		↔	Sciandra et al. (2003), Langer et al. (2009)
Gephyrocapsa oceanica	Calcification	↓	Riebesell et al. (2000)
		↔	Rickaby et al. (2010)
	PIC:POC	↓	Riebesell et al. (2000), Rickaby et al. (2010)
Calcidiscus leptoporus	Calcification	↓	Langer et al. (2006)
	PIC:POC	↓	Langer et al. (2006)
Coccolithus pelagicus/ braarudii	Calcification	↓	Müller et al. (2010)
		↔	Langer et al. (2006), Rickaby et al. (2010)
	PIC:POC	↓	Müller et al. (2010)
		↔	Langer et al. (2006), Rickaby et al. (2010)

↑, enhanced; ↓, slowed down; ↔, unaffected/inconclusive.
PIC, particulate inorganic carbon; POC, particulate organic carbon.

be replaced by other groups and/or more CO_2/pH-tolerant calcifying species. It is also unknown whether adaptation will allow calcifiers to overcome the adverse effects of ocean acidification.

6.3.5 Cell stoichiometry

In view of the multiple effects of ocean acidification on key metabolic processes such as carbon and nitrogen fixation, respiration, and calcification, it is not surprising to also find strong changes in the chemical composition of primary producers with changing carbonate chemistry (see Hutchins et al. 2009 for a detailed review). A common response to elevated pCO_2 is an increase in cellular carbon to nitrogen ratios (Table 6.5). A comparatively uniform response in this respect is obtained for coccolithophores and cyanobacteria, which with few exceptions either show increased C:N and N:P ratios or

no effect of elevated pCO_2. A diverse set of responses is seen in diatoms, which in some cases show opposite trends even in closely related species (e.g. Burkhardt et al. 1999).

A trend towards higher C:N and C:P ratios also emerges from studies on natural phytoplankton assemblages in mesocosm CO_2 perturbation studies, as observed in suspended particulate organic matter (POM) (Engel et al. 2005), in the drawdown of dissolved inorganic carbon relative to nitrogen (Riebesell et al. 2007; Bellerby et al. 2008), and in sedimented organic matter (Schulz et al. 2008). No effect of CO_2 on natural community C:N ratios was observed by Kim et al. (2006) or Feng et al. (2008). In contrast, no consistent response was obtained for N:P ratios (Tortell et al. 2002; Engel et al. 2005; Bellerby et al. 2008; Feng et al. 2010). Elevated CO_2 has also been observed leading to enhanced production of extracellular organic matter (Engel

Table 6.5   Effects of ocean acidification on the elemental composition of planktonic organisms, comparing rates at pre-industrial/present-day pCO_2 levels with those at pCO_2 projected for the end of this century.

Group		Response	References
Diatoms	C:N	↑	*Thalassiosira punctigera*: Burkhardt et al. (1999)
			Asterionella: Burkhardt et al. (1999)
		↓	*Skeletonema costatum*: Burkhardt and Riebesell (1997), Burkhardt et al. (1999), Gervais and Riebesell (2001)
		↔	*Phaeodactylum tricornutum*: Burkhardt et al. (1999)
			Coscinodiscus wailesii: Burkhardt et al. (1999)
			Thalassiosira weissflogii: Burkhardt et al. (1999)
	N:P	↑	*Skeletonema costatum*: Burkhardt and Riebesell (1997), Burkhardt et al. (1999)
		↓	*Thalassiosira weissflogii*: Burkhardt et al. (1999)
			Coscinodiscus wailesii: Burkhardt et al. (1999)
		↔	*Thalassiosira punctigera*: Burkhardt et al. (1999)
Coccolithophores	C:N	↑	*Emiliania huxleyi*: Leonardos and Geider (2005), Feng et al. (2008), Iglesias-Rodriguez et al. (2008), De Bodt et al. (2010)
			Gephyrocapsa oceanica: Rickaby et al. (2010)
		↓	*Coccolithus braarudii*: Rickaby et al. (2010)
		↔	*Emiliania huxleyi*: Buitenhuis et al. (1999), Müller et al. (2010)
			Coccolithus braarudii: Müller et al. (2010)
	N:P	↔	*Emiliania huxleyi*: Leonardos and Geider (2005), Feng et al. (2008), Müller et al. (2010)
Dinoflagellates	C:N	↔	*Scrippsiella trochoidea*: Burkhardt et al. (1999)
	N:P	↑	*Scrippsiella trochoidea*: Burkhardt et al. (1999)
Cyanobacteria	C:N	↑	*Synechococcus*: Fu et al. (2007)
			Trichodesmium: Levitan et al. (2007)
			Nodularia: Czerny et al. (2009)
		↔	*Trichodesmium*: Barcelos e Ramos et al. (2007), Hutchins et al. (2007), Kranz et al. (2009)
			Prochlorococcus: Fu et al. (2007)
			Crocosphaera: Fu et al. (2007)
	N:P	↑	*Trichodesmium*: Barcelos e Ramos et al. (2007), Hutchins et al. (2007)
			Synechococcus: Fu et al. (2007)
		↓	*Nodularia*: Czerny et al. (2009)
		↔	*Prochlorococcus*: Fu et al. (2007)
			Crocosphaera: Fu et al. (2007)
Other taxa	C:N	↑	*Heterosigma*: Fu et al. (2008b)
		↓	*Phaeocystis antarctica*: Hutchins et al. (2009)
		↔	*Prorocentrum*: Fu et al. (2008b)
	N:P	↑	*Heterosigma*: Fu et al. (2008b)
			Phaeocystis antarctica: Hutchins et al. (2008)
		↔	*Prorocentrum*: Fu et al. (2008b)
Natural assemblages	C:N	↑	Engel et al. (2005), Riebesell et al. (2007), Bellerby et al. (2008), Schulz et al. (2008)
		↔	Kim et al. (2006), Feng et al. (2009)
	N:P	↑	Bellerby et al. (2008)
		↓	Torteli et al. (2002), Engel et al. (2005)
		↔	Feng et al. (2009)

↑, increased; ↓, decreased; ↔, unaffected/inconclusive.

2002; see also Chapter 5), as well as an increase in carbohydrate to protein ratios of particulate matter (Tortell *et al.* 2002).

There is some evidence for a direct pCO_2 effect on diatom silicification and silicon quotas, with higher quotas under low CO_2 conditions (Milligan *et al.* 2004). Indirect effects may also arise from CO_2-induced shifts in species composition. Lower Si:N consumption in response to elevated pCO_2 was observed by Tortell *et al.* (2002) in a natural community of the equatorial Pacific, associated with a shift in community composition from diatom- to *Phaeocystis*-dominated. Increasing Si:N ratios due to a shift from lightly to more heavily silicified diatoms in response to elevated pCO_2 was observed by Tortell *et al.* (2008) and Feng *et al.* (2008). No change in silicate drawdown was found in a natural community exposed to 350, 700, and 1050 µatm CO_2 in a mesocosm experiment (Bellerby *et al.* 2008). The reasons for CO_2-induced changes in phytoplankton stoichiometry remain poorly understood but are critical for understanding trophic- and ecosystem-level consequences of ocean acidification. Changes in the C:N and C:P ratios of primary produced organic matter alter its nutritional value and may adversely affect the growth and reproduction of herbivorous consumers, for example, as seen in copepods and daphnids (Sterner and Elser 2002). A change in the stoichiometry of export production is also a powerful mechanism by which biology can alter ocean carbon storage (Boyd and Doney 2003; Riebesell *et al.* 2009; see also Chapter 12).

6.3.6   Species composition and succession

The process of phytoplankton species succession is relatively well understood for classic food webs, for example in the North Atlantic bloom. In such systems, there is a recurrent and predictable transition from diatoms to coccolithophores, dinoflagellates, and cyanobacteria as nutrient concentrations and mixing depths decrease from early spring to summer. This ecological succession was explained by Margalef (1958) as a result of changing energy input into the upper mixed layer. Early successional species such as diatoms thrive under high-nutrient conditions (associated with significant vertical mixing) by virtue of their rapid growth rates and large

capacity for nutrient storage in intracellular vacuoles (Sarthou *et al.* 2005). As the phytoplankton bloom progresses, decreased nutrient availability and an increased biomass of grazers (e.g. copepods) lead to a demise of diatom populations, shifting phytoplankton assemblages towards a greater fraction of smaller cells with high surface area to volume ratios and, correspondingly, more efficient nutrient uptake systems. Some large dinoflagellates can also thrive under these conditions through mixotrophic metabolism (i.e. facultative photoautotrophy supplemented by grazing) and by their ability to migrate into nutrient-rich deeper waters.

In the context of ocean acidification research, the critical question is how decreasing pH (and the associated increase in pCO_2) might alter ecological dynamics in planktonic ecosystems. Early work by Hinga (1992) demonstrated changes in phytoplankton succession in mesocosms induced by pH manipulations. Chen and Durbin (1994) demonstrated large differences in the composition of phytoplankton assemblages across a broad range of natural pH variation. Results from the latter study are difficult to interpret, since natural pH gradients across systems are typically associated with changes in multiple variables (e.g. macronutrient concentrations and mixing depth) that are also likely to influence phytoplankton productivity and species composition. More recent studies have focused on controlled pCO_2 manipulations in various oceanic regimes. A number of authors have reported an increase in the relative abundance of diatoms under elevated CO_2 (decreased pH) at the expense of haptophytes and other nanoflagellates (Tortell *et al.* 2002, 2008), and a shift within diatom assemblages to large centric species (Tortell *et al.* 2008; Feng *et al.* 2010). However, other studies have not observed the same CO_2-dependent increases in relative diatom biomass (Hare *et al.* 2007; Feng *et al.* 2009). This discrepancy may result in part from differences in experimental methodology (chemostats versus semi-continuous cultures) and/or from region-specific differences attributable to background oceanographic conditions.

To understand, mechanistically, how elevated pCO_2 and ocean acidification could influence phytoplankton species succession, fundamental information is needed on the taxonomic diversity of carbon uptake mechanisms and CO_2 sensitivity in phytoplankton, as well as information on the direct effects of pH on various species. Early work established upper and lower pH limits for a variety of phytoplankton species, as summarized by Hinga (2002), but these studies examined pH gradients much larger than those expected to occur naturally over the next century, and seawater carbonate chemistry was often poorly constrained. More recent work has focused on taxonomic differences in carbon uptake in marine phytoplankton (e.g. Rost et al. 2003). As discussed above (Section 6.3.1), phytoplankton possess physiological/biochemical carbon concentrating mechanisms that impose energy and mineral resource demands on cells. Taxonomic differences in the sensitivity to ocean acidification could thus result from differences in carbon uptake among species (Raven and Johnston 1991). In this respect, perhaps the most fundamental distinction among phytoplankton groups is the difference in the substrate specificity factor of RubisCO. The poor catalytic properties of this enzyme (low CO_2 affinity, competitive O_2 inhibition, and slow catalytic rate) determine an intrinsic sensitivity to increased CO_2 concentrations in the absence of any physiological carbon-concentrating mechanism (CCM).

Meta-analysis of existing biochemical data demonstrates that there is a statistically significant difference among phytoplankton groups in the CO_2 affinity of RubisCO, and an apparent trend of improved CO_2 affinity over the course of phytoplankton evolution as atmospheric CO_2 has decreased and O_2 increased over the past 500 Myr (Tortell 2000). Differences in RubisCO specificity appear to be related to the intrinsic capacity of various phytoplankton to concentrate inorganic carbon intracellularly. For example, diatoms which have high-affinity RubisCO tend to have much lower cellular carbon concentration factors than other taxonomic groups such as cyanobacteria whose RubisCO has a low selectivity for CO_2 binding over O_2 (Tortell 2000). The implications of such differences for the CO_2 sensitivity of diatoms relative to other phytoplankton groups remain unclear. It has been suggested, however, that the rapid growth of diatoms may be attributable to the relatively low energetic costs of a CCM in cells with more efficient isoforms of RubisCO (Tortell 2000). Conversely, cyanobacte-ria, which have low-affinity RubisCO (half saturation constant for $CO_2 > 200$ µmol l^{-1}), appear to invest heavily in cellular carbon accumulation. This may explain the large apparent CO_2 stimulation of productivity and nitrogen fixation observed in *Trichodesmium* (see Table 6.3) as energetic resources become reallocated from carbon uptake to other biochemical and physiological processes (Levitan et al. 2010). These hypotheses should be examined experimentally in order to gain a more mechanistic understanding of how ocean acidification could influence phytoplankton species composition in the oceans. Such a process-based understanding is critical for predicting ecosystem-level responses.

Our discussion thus far has focused on the bottom-up effects structuring phytoplankton assemblage composition. Yet grazers can play an important role in this process through, for example, size-selective grazing. An excellent example of this occurs in iron-limited open-ocean waters where iron availability controls the biomass of large phytoplankton, while rapid grazing by microzooplankton maintains a low standing crop of small phytoplankton (Landry et al. 1997). Under various ocean acidification scenarios, changes in relative grazing pressure across both the classic food chain and the microbial loop could thus act to influence the relative abundance of certain phytoplankton species (see Chapter 5). Moreover, changes in zooplankton assimilation efficiencies and excretion rates could strongly influence nutrient remineralization. As discussed below, very little information is currently available to examine these potential ecological feedbacks.

6.3.7 Zooplankton growth, reproduction, and grazing

Relative to work on phytoplankton, much less information is available about the effects of ocean acidification on zooplankton. This reflects, in part, the greater experimental complexity inherent in working with these organisms. Nonetheless, there has been a significant effort in recent years to begin documenting the potential CO_2/pH sensitivity of various zooplankton. Much of this work has focused on groups including pteropods and foraminifera that produce $CaCO_3$ shells. These organisms clearly show decreased calcification under more acidic conditions (see above), but the implications of this for growth, reproduction, and grazing rates remain unclear. Less environmentally relevant information is available for non-calcifying zooplankton species. Much of the work that has been done with copepods, for example, has utilized extreme CO_2 levels (e.g. >10 000 µatm) that induce juvenile mortality in a number of species (see Fabry *et al.* 2008 for a recent and comprehensive review). In a mesocosm CO_2 perturbation study with pCO_2 levels of 350, 700, and 1050 µatm no effect of ocean acidification was observed on microzooplankton grazing (Suffrian *et al.* 2008) and copepod feeding and egg production (Carotenuto *et al.* 2007), but differences between CO_2 treatments were observed with respect to copepod nauplii recruitment (Carotenuto *et al.* 2007; Riebesell *et al.* 2008b). Experimental data are urgently needed on the potential effects of environmentally relevant CO_2/pH levels on feeding rates, growth, and assimilation efficiencies of both micro- and mesozooplankton. In a recent field experiment, Rose *et al.* (2009) reported no consistent CO_2 or temperature effects on microzooplankton biomass during a late North Atlantic spring bloom. Such effects, if present, will result from a combination of direct physiological influences on grazers, and indirect ecological effects associated with altered phytoplankton assemblage composition.

6.4 Synergistic effects of ocean acidification with other environmental changes

6.4.1 Ocean warming

Temperature exerts a fundamental control on the metabolic rates of planktonic organisms, and warming of surface-ocean waters is thus expected to influence marine metabolic cycles. The classic temperature-dependent growth characteristics of phytoplankton were described by Eppley (1972), and have been incorporated in many ecosystem models and productivity algorithms (e.g. Sarmiento *et al.* 2004; Schmittner *et al.* 2008). In general, growth rates of individual phytoplankton species have been observed to increase exponentially with increasing temperature. However, each individual species has a temperature optimum and its growth is inhibited above a certain threshold value. As a result, observed temperature responses of mixed phytoplankton assemblages reflect the aggregate response of many interacting species. Recent experiments have examined the influence of temperature on phytoplankton productivity in bottle experiments, demonstrating significantly increased rates of carbon fixation under elevated temperature (Hare *et al.* 2007; Feng *et al.* 2009). This temperature stimulation occurred irrespective of CO_2/pH levels, but the extent of any temperature response should depend on other limiting factors (e.g. light, nutrients, trace metals; see below). Moreover, the metabolic rates of grazers also increase with temperature (e.g. Rose *et al.* 2009), potentially reducing the impacts of increased phytoplankton productivity.

6.4.2 Shoaling of the mixed-layer depth

While the direct ecological impacts of increased sea-surface temperature remain unclear, the effects of warming on physical circulation have been examined by a number of authors. As surface waters warm, mixed-layer stratification (i.e. vertical density gradients) increases and this effect is enhanced in some regions by increased sea-ice melt (Boyd and Doney 2003; Sarmiento *et al.* 2004). Increased stratification has several important implications for biological processes, which could offset or exacerbate the effects of elevated pCO_2 alone. First, increased stratification reduces vertical water mass exchange and hence the flux of nutrients from the subsurface into the mixed layer. This, in turn, decreases the capacity of an ecosystem for new production. Second, enhanced stratification increases the mean depth-integrated irradiance and decreases the range (i.e. variability) of irradiance levels experienced by phytoplankton mixed to a shallower depth. The potential decrease in nutrient concentrations and increased irradiance levels could exert opposing effects on primary productivity. As CO_2 increases (pH decreases), phytoplankton may have lower nitrogen requirements associated with reduced cellular RubisCO content, as well as lower energy requirements associated with down-regulated carbon concentration (Raven and Johnston 1991).

Similarly, phytoplankton may have higher light-use efficiencies under elevated CO_2 as a result of the lowered metabolic costs of inorganic carbon assimilation. Such a response has been observed in culture studies with freshwater dinoflagellates (Berman-Frank *et al.* 1998). Future experiments should thus carefully consider how best to manipulate temperature, light, and nutrients in conjunction with pCO_2. With respect to potential changes in mixed-layer depth, it will be important to simulate not only changes in mean irradiance levels, but also decreased variability in the light field. This latter effect could have important implications for the ability of phytoplankton to photo-acclimate, and could thus impose a selective pressure on phytoplankton species.

6.4.3 Nutrient speciation and availability

In different oceanic regions, primary production is limited by the availability of various key nutrients, most commonly nitrogen, phosphorus, or iron. The acquisition of such nutrients depends on their chemical form(s)—the chemical 'species'—present in the water. For example, free phosphate in seawater (HPO_4^{2-}) can be readily taken up by phytoplankton but, in most cases, phosphate bound in organic compounds must first be cleaved enzymatically from the organic moiety before being utilized. In the same way, nitrate (NO_3^-) and ammonium (NH_4^+) are readily bioavailable, while organic nitrogen compounds, including urea, amino acids, or amines can only be utilized through specialized enzymatic processes. Changing the pH of seawater is expected to affect the efficiency of the enzymatic processes involved in acquiring organic forms of nitrogen and phosphorus (Millero *et al.* 2009). A similar situation applies to essential trace metals such as iron (Fe), which are readily taken up when present as free ions or ions bound to chloride, hydroxide, or other inorganic ligands, but require specialized uptake machinery when bound to organic complexing agents (Maldonado and Price 2001). As ocean acidification decreases the concentrations of hydroxyl (OH^-) and carbonate (CO_3^{2-}) ions, which both form strong complexes with divalent and trivalent metals, these metals will have a higher fraction in their free forms at lower pH (Millero *et al.* 2009). These

changes in speciation will also increase the thermodynamic and kinetic activity of the metals. Because most organic particles in seawater are negatively charged, the surface sites will become less available to adsorb metals as pH decreases. These changes are expected to alter the availability and toxicity of metals for marine organisms.

Contrary to expectations, reduced uptake of iron in response to seawater acidification was observed in monospecific cultures of diatoms and coccolithophores by Shi *et al.* (2010) and was attributed to decreased bioavailability of dissolved iron under ocean acidification. The iron requirement of the phytoplankton remained unchanged with increasing CO_2. Decreased iron and zinc requirements and increased growth and carbon fixation rates at elevated CO_2 were observed in a centric diatom by King *et al.* (pers. comm.). In a mesocosm CO_2 enrichment study Breitbarth *et al.* (2010) found significantly higher concentrations of dissolved iron and higher Fe(II) turnover rates in high- compared with the mid- and low-CO_2 treatments, suggesting enhanced iron bioavailability in response to ocean acidification. This agrees with the expectation of Millero et al. (2009), who predicted that declining pH will increase the half-life of Fe(II) in seawater, making it more available for biological consumption. Clearly, more work is needed to entangle the effects of ocean acidification on trace metal bioavailability and the synergistic effects with changing chemistry of other essential nutrients to assess their impacts on biogeochemical cycling. Coastal upwelling and oxygen minimum zones, which already have low pH, would be useful areas to study these processes in the modern ocean (Millero *et al.* 2009).

6.5 Ecological processes and biogeochemical feedbacks

Responses of pelagic ecosystems to ocean change have the potential to affect marine biogeochemical cycles and carbon sequestration, leading to a series of climate feedback mechanisms (Fig. 6.1, Table 6.6). These potential biogeochemical feedbacks, i.e. changes in the ocean's elemental cycling with repercussions on the climate system, are discussed in detail by Riebesell *et al.* (2009) and in Chapter 12. The discussion below aims at highlighting the

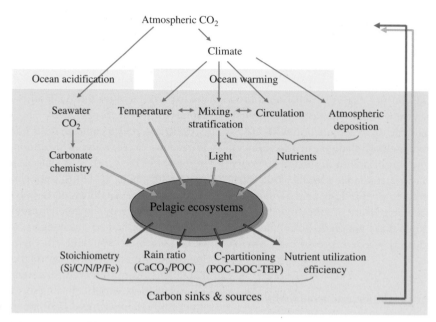

Figure 6.1 Schematic representation of the environmental changes affecting pelagic ecosystems and possible feedbacks to the global carbon cycle (POC, particulate organic carbon; DOC, dissolved organic carbon; TEP, transparent exopolymeric particles).

Table 6.6 Biotic responses to ocean acidification and their feedback potential to the global carbon cycle. Responses are characterized with regard to feedback sign, sensitivity, capacity, and longevity using best guesses (±0 for negligible, + for low, ++ for moderate, and +++ for high). Empty boxes indicate missing information/understanding (adapted from Riebesell *et al.* 2009).

Process	Sign of feedback	Sensitivity	Capacity	Longevity
Calcification	Negative	+[1]	+	+[1]
Ballast effect	Positive		+++	+++
Extracellular organic matter prod.	Negative	++[1]	+++	?[2]
Stoichiometry	Negative	++[1]	++	++
Nitrogen fixation	Negative	++[1]	+	?[2]

[1] Available information mainly based on short-term perturbation experiments.
[2] Potential for adaptation currently unknown.

CO_2- (and pH)-dependent ecological processes, which may drive these biogeochemical feedbacks. In Chapter 12 Gehlen *et al.* use a variety of experimental data and ocean models to explore the potential magnitude of these feedbacks and their quantitative impacts on the climate system.

As discussed in the sections above, changes in the carbonate chemistry of the surface ocean have the potential to influence the physiology and growth of various members of the pelagic ecosystem, and to affect the ecological and trophic dynamics which govern the exchange of energy and cycling of nutri-

ents through the marine food web. For the purposes of the present discussion, the following changes will be considered: changes in cellular carbon processing, changes in phytoplankton species assemblage composition, and changes in biological nitrogen cycling.

6.5.1 Changes in cellular carbon processing

Marine pelagic organisms mediate the transformation of inorganic carbon into organic and inorganic particulate forms, and into a large dissolved organic pool. All these transformations have been shown to be potentially sensitive to CO_2. The production of POC through pelagic photosynthesis is the first step in the biological carbon cycle, providing organic carbon that is either consumed by heterotrophic organisms in surface waters or exported into the deep sea. When exported to depth it is slowly remineralized and remains out of contact with the atmosphere over timescales of hundreds to thousands of years or is buried in sediments over much longer timescales. Results of laboratory and field studies suggest that primary production is likely to increase slightly in response to increased CO_2, and all else being equal, this should act to enhance the biological carbon pump and oceanic CO_2 sequestration—providing a negative climate feedback. This potential CO_2-dependent increase in primary productivity could, however, be mitigated by changes in ocean nutrient supply related to changes in surface-water stratification (Boyd and Doney 2003). Moreover, changes in the composition of phytoplankton assemblages could either amplify or diminish the impacts of altered photosynthetic rates on carbon export (see below).

The extent to which primary production is consumed versus exported depends on what fraction of the inorganic carbon is converted into dissolved organic and particulate inorganic forms (DOC and PIC, respectively). The presence of both DOC and PIC can enhance the downward transport of POC into the deep sea, but the formation of these two carbon pools appears to have different CO_2 sensitivities. Decreased pelagic calcification and the resulting decline in the strength of the carbonate pump lower the drawdown of total alkalinity in the surface layer, thereby increasing the uptake capacity for atmospheric CO_2 in the surface ocean. Assuming a rate of $CaCO_3$ export of 1 Gt C yr^{-1} (Milliman 1993), the capacity of this negative feedback is relatively small (Chapter 12; Table 6.6). The sensitivity of this response in coccolithophores appears to be species-dependent (Langer *et al.* 2006), permitting changes in species composition to dampen or eliminate the response at the community level. Shifting species composition and the potential for adaptation (currently unknown) could make this a transient feedback. In view of the scarcity of information on the pH sensitivity of

foraminifera and pteropods it is premature to speculate on the significance of this feedback process.

CO_2-dependent changes in calcification can also have an impact on marine biogeochemical cycles by influencing POC fluxes, since $CaCO_3$ may act as ballast in particle aggregates, accelerating the flux of particulate material to depth (Armstrong *et al.* 2002; Klaas and Archer 2002; but see also Passow 2004). Reduced $CaCO_3$ production could therefore slow down the vertical flux of biogenic matter to depth, shoaling the remineralization depth of organic carbon and decreasing carbon sequestration (positive feedback; see Chapter 12). The capacity of this feedback depends on the pH sensitivity of pelagic calcifiers and the quantitative importance of $CaCO_3$ (versus opaline silicate) as ballast for particle export, both of which are poorly understood. If $CaCO_3$ is a prerequisite for deep transport of POM and if pelagic calcification remains sensitive to high CO_2, this feedback could have a high capacity and extended duration (Table 6.6). A decrease in the efficiency of the biological pump would lead to a change in upper-ocean nutrient and oxygen status with likely impacts on the pelagic community structure.

Potential changes in $CaCO_3$ ballasting could be offset by CO_2-dependent changes in DOC production. Increased production of extracellular organic matter under high CO_2 levels (Engel 2002) may enhance the formation of particle aggregates (Engel *et al.* 2004; Schartau *et al.* 2007) and thereby increase the vertical flux of organic matter (negative feedback; Table 6.6; see also Chapter 5). This response may in fact have been responsible for the increase in carbon drawdown observed at elevated pCO_2 in some mesocosm experiments with natural phytoplankton assemblages (Arrigo 2007). Provided that the response is of a general nature in bloom-forming phytoplankton, this process may have significant biogeochemical implications by influencing the remineralization depth for organic carbon. Again, not knowing whether evolutionary adaptation will select against this response prevents us judging the longevity of this process.

Substantial changes in marine primary production and nutrient cycling must be expected if ocean acidification causes a systematic shift in phytoplankton stoichiometry. No clear trend with pCO_2 is obtained for N:P ratios, which show large differences within and between phytoplankton functional groups. While single-species laboratory experiments also revealed mixed responses in C:N ratio with rising pCO_2, a more consistent pattern is observed in natural communities in field assays and mesocosm experiments, generally showing stable or higher C:N (and higher C:P) at elevated pCO_2 (Table 6.5). An increase in the community C:N ratio would increase the carbon drawdown for a given amount of nutrient supply to the sunlit surface layer, with the potential of accelerating the biological pump. This would provide a negative feedback to rising atmospheric pCO_2 (Riebesell *et al.* 2007), but could also greatly expand the volume of oxygen-depleted waters (Oschlies *et al.* 2008). In view of the scarcity of information, global extrapolations of the observed responses should be viewed merely as sensitivity tests. Open questions concern the longevity of the observed changes in elemental ratios, possible dampening effects from adaptation and shifts in species composition and the interacting effects from changes in temperature, light, and nutrient availability.

6.5.2 Changes in biological nitrogen cycling

Nitrogen fixation, carried out by a few specialized microorganisms, represents a major input of 'new' nitrogen to oligotrophic marine ecosystems and is key in controlling primary production in large regions of the world's oceans. Nitrogen fixation is an 'expensive' biochemical process that requires synthesis of a complex iron-rich enzyme and uses large amounts of energy. This process is thus only ecologically viable under conditions of nitrogen deficiency in sunlit surface waters. pH-dependent changes in the availability of iron and, possibly, other nutrients may thus change the rate of nitrogen fixation in the oceans. Enhanced nitrogen fixation at elevated pCO_2, reported for the diazotrophic cyanobacterium *Trichodesmium* (Table 6.3) has the potential to increase the reservoir of bioavailable nitrogen in the surface layer which, in a predominantly nitrogen-limited ocean, would increase primary production and carbon fixation (negative feedback). However, since the biomass produced by cyanobacteria is generally thought not to be exported to great

depth, it is uncertain to what extent the additional bioavailable nitrogen would lead to enhanced carbon sequestration or would remain suspended in the surface layer. In the latter scenario, an accumulation of biologically available nitrogen in the surface ocean would eventually eliminate the ecological niche for nitrogen-fixing cyanobacteria, in a self-limiting process. As CO_2-sensitivity of nitrogen fixation has so far been reported for only two species, it is too early to speculate about the sensitivity and longevity of this response. Indeed, one still lacks fundamental information on the identity and metabolic characteristics of key nitrogen fixers in the oceans, and significant certainty still exists on the rate of global nitrogen fixation (Gruber and Galloway 2008). Future experiments should consider the potential phylogenetic and metabolic diversity of marine nitrogen fixers, while modelling studies should take into account the selective pressures favouring (or limiting) the abundance of these organisms.

6.5.3 Changes in the composition of phytoplankton assemblages

The qualitative nature of phytoplankton and zooplankton assemblages exerts a strong mitigating influence on the biogeochemical impacts of primary productivity. This was demonstrated by Boyd and Newton (1995) in a study of the 1989–1990 North Atlantic spring bloom. Despite similar rates of total primary productivity and nutrient drawdown in both years, very different rates of vertical carbon export were observed, in conjunction with significant differences in the species composition of phytoplankton and zooplankton. Higher carbon fluxes were observed when phytoplankton assemblages were dominated by larger diatom species. Such a shift has been observed to occur under high-CO_2 conditions in several experiments (Tortell *et al.* 2002, 2008). In contrast, other studies have observed reductions in diatom biomass under high-CO_2 conditions (Hare *et al.* 2007). In general, a CO_2 (pH)-dependent shift in phytoplankton species composition can influence biogeochemical cycles in several ways. First, larger cells (e.g. diatoms) should have intrinsically higher sinking rates and thus promote greater carbon fluxes out of the surface layer.

Moreover, species shifts may change the abundance of biogeochemically important functional classes such as calcifiers or groups which produce large amounts of climate-active gases including dimethyl-sulphide (DMS; see Chapter 11). In experiments conducted in the equatorial Pacific, Tortell *et al.* (2002) observed a shift from diatoms to *Phaeocystis pouchetii* under low-CO_2 conditions. This taxonomic shift is biogeochemically significant since *Phaeocystis* is a prolific producer of DMS (Stefels and Vanboekel 1993). DMS levels appear to have been high during the Last Glacial Maximum (Legrand *et al.* 1991), and it has been hypothesized that these high DMS values may have resulted from a CO_2-dependent enhancement of *Phaeocystis* growth during this time (Tortell *et al.* 2002). Conversely, if high CO_2 favours the growth of large diatoms relative to coccolithophores or *Phaeocystis*, oceanic DMS concentrations could be expected to decrease in the future. In contrast to these expectations, recent mesocosm experiments have demonstrated increased DMS production under high-CO_2 conditions (Chapter 11). Other bottle experiments have demonstrated CO_2-dependent shifts in dissolved and particulate DMSP, the precursor molecule to DMS. Given the complexity of the oceanic DMS cycle (Stefels *et al.* 2007), CO_2-dependent effects on the production of this gas could result from either direct physiological effects on phytoplankton or indirect ecological effects involving bacteria, viruses, and/or grazers. Fully understanding the CO_2 sensitivity of the DMS cycle will thus require a much deeper ecosystem-level understanding than is currently available.

The focus of this section has been on phytoplankton-dependent processes since these are best studied in the context of ocean acidification. However, changes in phytoplankton community structure have ecological implications for higher trophic levels, which can in turn influence biogeochemical feedbacks. In particular, CO_2-dependent shifts from small to large phytoplankton species would influence the composition of zooplankton communities, and thereby affect the degree of coupling between producers and consumers (see Section 6.2.2). This would affect the partitioning of carbon between its various dissolved and particulate phases in the upper ocean, and thus affect the marine carbon cycle. Understanding the nature and quantitative significance of such multitrophic-level effects should be considered as a priority for future studies.

6.6 Critical information gaps

Our present knowledge of the pH/CO_2 sensitivities of marine organisms is almost entirely based on short-term perturbation experiments, neglecting the possibility of evolutionary adaptation. With generation times of about 1 day, unicellular algae and bacteria will go through tens of thousands of generations as pCO_2 increases to projected maximum levels, which may be sufficient for adaptive processes to become relevant when environmental change occurs over decadal timescales or longer. Also, little is currently known regarding effects from multiple and interacting stressors, such as sea-surface warming, enhanced stratification, and changes in nutrient availability and speciation. Moreover, there is a complete lack of information on the transfer of responses from the organism to the community and ecosystem level and the replacement of sensitive species by those tolerant to ocean acidification.

Given the lack of fundamental information on the biological impacts of ocean acidification on organisms and communities, experimental manipulations will continue to play an important role in informing our understanding of this topic. Laboratory studies have the advantage of isolating individual variables under well-controlled conditions, thus facilitating the interpretation of cause and effect. However, such studies may be limited in the extent to which they can be extrapolated to natural plankton assemblages in the oceans. One of the main problems is the inherent biological variability observed among individual species, or even within different strains of the same species. Phylogenetic analysis has demonstrated very high microdiversity and unique ecologically distinct phytoplankton clones (Iglesias-Rodríguez *et al.* 2006; Rynearson *et al.* 2006). Genomics, transcriptomics, proteomics, and the expression of specific marker genes for crucial functions are promising methods for addressing this issue. Even in the case where individual isolates of a given species show strong and consistent responses to ocean acidification (e.g. *Trichodesmium*'s

CO_2-dependent increase in nitrogen fixation), community-level responses may be significantly different from those observed in pure culture. Such discrepancies could result, for example, from 'ecological buffering' processes, whereby species replacement and selection leads to functional stability and resilience. These processes have not been studied with respect to ocean acidification but have been documented experimentally in terrestrial ecosystems (Tilman and Downing 1996).

A full understanding of ecosystem responses will require knowledge of how multiple trophic levels respond. For example, hypothetical CO_2-dependent increases in the growth rates and productivity of picoeukaryotes could be counteracted by a compensatory increase in microzooplankton grazing rates. For this reason, it is important to consider not only the temporal scale of the experiments, but also the spatial scale. Typical ship-board manipulation experiments use incubation bottles ranging in volume from about 100 ml to 4 l, which may be too small to adequately sample larger zooplankton species. In this regard, mesocosm studies provide a much better ecological approximation of natural plankton assemblages. Yet the results of even these large and logistically challenging experiments may not scale up easily to a basin scale. For example, enclosed mesocosms do not allow for the horizontal advection of organisms, which may be better adapted to newly created ecophysiological niches.

Given all the uncertainties in the ecological and biogeochemical responses of marine systems to increased pCO_2 and decreased pH, it may seem premature to begin parameterizing quantitative models. Yet these models can provide significant insight into the potential sensitivities of biogeochemical systems to various perturbations, for example a change in the C:N stoichiometry of sinking POC. In addition, models provide help in determining the magnitude of potential physical and chemical changes used in manipulative experiments (Boyd and Doney 2003; Sarmiento et al. 2004). Through a concerted and collaborative effort over the coming decade, it should be possible for ocean modellers and experimentalists to begin understanding some of the consequences of increased oceanic CO_2 on marine ecosystems and biogeochemical cycles. This effort will require a sophisticated approach which takes into account the tremendous complexity of marine ecosystems and the interaction of multiple driving forces which shape biological responses over a range of spatial and temporal scales.

6.7 Acknowledgements

U.R. gratefully acknowledges funding received as part of the EU integrated project EPOCA (European Project on OCean Acidification) and the coordinated project BIOACID (Biological Impacts of Ocean ACIDificaton) by the German Ministry for Education and Research (BMBF).

References

Armstrong, R.A., Lee, C., Hedges, J.I., Honjo, S., and Wakeham, S.G. (2002). A new, mechanistic model for organic carbon fluxes in the ocean based on the quantitative association of POC with ballast minerals. *Deep-Sea Research, Part II: Topical Studies in Oceanography*, **49**, 219–36.

Arrigo, K.R. (2007). Marine manipulations. *Nature*, **450**, 491–2.

Arrigo, K.R., van Dijken, G.L., and Bushinsky, S. (2008). Primary production in the Southern Ocean, 1997–2006. *Journal of Geophysical Research*, **113**, C08004, doi:10.1029/2007JC004551.

Azam, F., Fenchel, T., Field, J.G., Gray, J.S., Meyer-Reil, L.A., and Thingstad, F. (1983). The ecological role of water-column microbes in the sea. *Marine Ecology Progress Series*, **10**, 257–63.

Badger, M.R., Andrews, T.J., Whitney, S.M. et al. (1998). The diversity and coevolution of Rubisco, plastids, pyrenoids, and chloroplast-based CO_2-concentrating mechanisms in algae. *Botany*, **76**, 1052–71.

Barcelos e Ramos, J., Biswas, H., Schulz, K.G., LaRoche, J., and Riebesell, U. (2007). Effect of rising atmospheric carbon dioxide on the marine nitrogen fixer *Trichodesmium*. *Global Biogeochemical Cycles*, **21**, GB2028, doi:10.1029/2006GB002898.

Barcelos e Ramos, J., Müller, M.N., and Riebesell, U. (2010). Short-term response of the coccolithophore *Emiliania huxleyi* to an abrupt change in seawater carbon dioxide concentrations. *Biogeosciences*, **7**, 177–86.

Bellerby, R.G.J., Schulz, K.G., Riebesell, U. et al. (2008). Marine ecosystem community carbon and nutrient uptake stoichiometry under varying ocean acidification during the PeECE III experiment. *Biogeosciences*, **5**,1517–27.

Berman-Frank, I., Erez, J., and Kaplan, A. (1998). Changes in inorganic carbon uptake during the progression of a dinoflagellate bloom in a lake ecosystem. *Botany*, **76**, 1043–51.

Boyd, P. and Newton, P. (1995). Evidence of the potential influence of planktonic community structure on the interannual variability of particulate organic carbon flux. *Deep-Sea Research Part I: Oceanographic Research Papers*, **42**, 619–39.

Boyd, P.W. and Doney, S.C. (2003). The impact of climate change and feedback processes on the ocean carbon cycle. In: M.J.R. Fasham (ed.), *Ocean biogeochemistry: the role of the ocean carbon cycle in global change*, pp. 157–93, Springer-Verlag, Berlin.

Breitbarth, E., Bellerby, R.J., Neill, C.C. *et al.* (2010). Ocean acidification affects iron speciation during a coastal seawater mesocosm experiment. *Biogeosciences*, **7**, 1065–73.

Buitenhuis, E.T., De Baar, H.J.W., and Veldhuis, M.J.W. (1999). Photosynthesis and calcification by *Emiliania huxleyi* (Prymnesiophyceae) as a function of inorganic carbon species. *Journal of Phycology*, **35**, 949–59.

Burkhardt, S. and Riebesell, U. (1997). CO_2 availability affects elemental composition (C:N:P) of the marine diatom *Skeletonema costatum*. *Marine Ecology Progress Series*, **155**, 67–76.

Burkhardt, S., Amoroso, G., Riebesell, U., and Sültemeyer, D. (2001). CO_2 and HCO_3^- uptake in marine diatoms acclimated to different CO_2 concentrations. *Limnology and Oceanography*, **46,** 1378–91.

Burkhardt, S., Zondervan, I., and Riebesell, U. (1999). Effect of CO_2 concentration on the C:N:P ratio in marine phytoplankton: a species comparison. *Limnology and Oceanography*, **44**, 683–90.

Calbet, A. and Landry, M.R. (2004). Phytoplankton growth, microzooplankton grazing, and carbon cycling in marine systems. *Limnology and Oceanography*, **49**, 51–7.

Carotenuto, Y., Putzeys, S., Simonelli, P. *et al.* (2007). Copepod feeding and reproduction in relation to phytoplankton development during the PeECE III mesocosm experiment. *Biogeosciences Discussions*, **4**, 3913–36.

Chen, C.Y. and Durbin, E.G. (1994). Effects of pH on the growth and carbon uptake of marine phytoplankton. *Marine Ecology Progress Series*, **109**, 83–83.

Clark, D., Lamare, M., and Barker, M. (2009). Response of sea urchin pluteus larvae (Echinodermata: Echinoidea) to reduced seawater pH: a comparison among a tropical, temperate, and a polar species. *Marine Biology*, **156**, 1125–37.

Comeau, S., Gorsky, G., Jeffree, R., Teyssié, J.-L., and Gattuso, J.-P. (2009). Impact of ocean acidification on a key Arctic pelagic mollusc (*Limacina helicina*). *Biogeosciences*, **6**, 1877–82.

Czerny, J., Barcelos e Ramos, J., and Riebesell, U. (2009). Influence of elevated CO_2 concentrations on cell division and nitrogen fixation rates in the bloom-forming cyanobacterium *Nodularia spumigena*. *Biogeosciences*, **6**, 1865–75.

De Bodt, C., Van Oostende, N., Harlay, J., Sabbe, K., and Chou, L. (2010). Individual and interacting effects of pCO_2 and temperature on *Emiliania huxleyi* calcification: study of the calcite production, the coccolith morphology and the coccosphere size. *Biogeosciences*, **7**, 1401–12.

Delille, B. Harlay, J., Zondervan, I. *et al.* (2005). Response of primary production and calcification to changes of pCO_2 during experimental blooms of the coccolithophorid *Emiliania huxleyi*. *Global Biogeochemical Cycles*, **19**, GB2023, doi:10.1029/2004GB002318.

Doney, S.C., Fabry, V.J., Feely, R.A., and Kleypas, J.A. (2009). Ocean acidification: the other CO_2 problem. *Annual Review of Marine Science*, **1**, 169–92.

Duarte, C.M. and Prairie, Y.T. (2005). Prevalence of heterotrophy and atmospheric CO_2 emissions from aquatic ecosystems. *Ecosystems*, **8**, 862–70.

Dugdale, R.C. and Goering, J.J. (1967). Uptake of new and regenerated forms of nitrogen in primary productivity. *Limnology and Oceanography*, **12**, 196–206.

Egge, J.K., Thingstad, T.F., Engel, A., Bellerby R.G.J., and Riebesell, U. (2009). Primary production during nutrient-induced blooms at elevated CO_2 concentrations. *Biogeosciences*, **6**, 877–85.

Engel, A. (2002). Direct relationship between CO_2 uptake and transparent exopolymer particles production in natural phytoplankton. *Journal of Plankton Research*, **24**, 49–53.

Engel, A., Thoms, S., Riebesell, U., Rochelle-Newall, E., and Zondervan, I. (2004). Polysaccharide aggregation as a potential sink of marine dissolved organic carbon. *Nature*, **428**, 929–32.

Engel, A., Zondervan, I., Aerts K. *et al.* (2005). Testing the direct effect of CO_2 concentration on a bloom of the coccolithophorid *Emiliania huxleyi* in mesocosm experiments. *Limnology and Oceanography*, **50**, 493–504.

Eppley, R.W. (1972). Temperature and phytoplankton growth in the sea. *Fish Bulletin*, **70**, 1063–85.

Eppley, R.W. and Peterson, B.J. (1979). Particulate organic matter flux and planktonic new production in the deep ocean. *Nature*, **282**, 677–80.

Fabry, V.J., Seibel, B.A., Feely, R.A., and Orr, J.C. (2008). Impacts of ocean acidification on marine fauna and ecosystem processes. *ICES Journal of Marine Science*, **65**, 414–32.

Falkowski, P.G., Katz, M.E., Knoll, A.H. *et al.* (2004). The evolution of modern eukaryotic phytoplankton. *Science*, **305**, 354–60.

Feng, Y., Warner, M., Zhang, Y. *et al.* (2008). Interactive effects of increased pCO_2, temperature and irradiance on the marine coccolithophore *Emiliania huxleyi* (Prymnesiophyceae). *European Journal of Phycology*, **43**, 87–98.

Feng, Y., Hare, C.E., Leblanc, K. *et al.* (2009). The effects of increased pCO_2 and temperature on the North Atlantic spring bloom: I. phytoplankton community and biogeochemical response. *Marine Ecology Progress Series*, **388**, 13–25.

Feng, Y., Hare, C.E., Rose, J.M. *et al.* (2010). Interactive effects of iron, irradiance and CO_2 on Ross Sea phytoplankton. *Deep-Sea Research Part I: Oceanographic Research Papers*, **57**, 368–83.

Fu, F.-X., Mulholland, M.R., Garcia, N.S. *et al.* (2008a). Interactions between changing pCO_2, N_2 fixation, and Fe limitation in the marine unicellular cyanobacterium *Crocosphaera*. *Limnology and Oceanography*, **53**, 2472–84.

Fu, F., Zhang, Y., Warner, M.E., Feng, Y., and Hutchins, D.A. (2008b). A comparison of future increased CO_2 and temperature effects on sympatric *Heterosigma akashiwo* and *Prorocentrum minimum*. *Harmful Algae*, **7**, 76–90.

Gao K., Ruan Z., Villafane V.E., Gattuso J.-P., and Helbling, W. (2009). Ocean acidification exacerbates the effect of UV radiation on the calcifying phytoplankter *Emiliania huxleyi*. *Limnology and Oceanography*, **54**, 1855–62.

Gervais, F. and Riebesell, U. (2001). Effect of phosphorus limitation on elemental composition and stable carbon isotope fractionation in a marine diatom growing under different CO_2 concentrations. *Limnology and Oceanography*, **46**, 497–504.

Giordano, M., Beardall, J., and Raven, J.A. (2005). CO_2 concentrating mechanisms in algae: mechanisms, environmental modulation, and evolution. *Annual Review of Plant Biology*, **56**, 99–131.

del Giorgio, P.A. and Cole, J.J. (1998). Bacterial growth efficiency in natural aquatic systems. *Annual Review of Ecology and Systematics*, **29**, 503–41.

Glibert, P.M. (1982). Regional studies of daily, seasonal and size fraction variability in ammonium remineralization. *Marine Biology*, **70**, 209–22.

Gruber, N. and Galloway, J.N. (2008). An Earth-system perspective of the global nitrogen cycle. *Nature*, **451**, 293–6.

Gruber, N. and Sarmiento, J.L. (2002). Large-scale biogeochemical/physical interactions in elemental cycles. In: A.R. Robinson, J.J. McCarthy, and B.J. Rothschild (eds), *The sea: biological–physical interactions in the oceans*, pp. 337–99. John Wiley and Sons, New York.

Hansell, D.A. and Carlson, C.A. (2002). *Biogeochemistry of marine dissolved organic matter*, 745 pp. Academic Press, New York.

Hare, C.E., Leblanc, K., DiTullio, G.R. *et al.* (2007). Consequences of increased temperature and CO_2 for phytoplankton community structure in the Bering Sea. *Marine Ecology Progress Series*, **352**, 9–16.

Hein, M. and Sand-Jensen, K. (1997). CO_2 increases oceanic primary production. *Nature*, **388**, 526–7.

Heinze, C., Maier-Reimer, E. and Winn, K. (1991). Glacial pCO_2 reduction by the world ocean: experiments with the Hamburg carbon cycle model. *Paleoceanography*, **6**, 395–430.

Hinga, K.R. (1992). Co-occurrence of dinoflagellate blooms and high pH in marine enclosures. *Marine Ecology Progress Series*, **86**, 181–7.

Hinga, K.R. (2002). Effects of pH on coastal marine phytoplankton. *Marine Ecology Progress Series*, **238**, 281–300.

Hutchins, D.A., Fu, F.-X., Zhang, Y. *et al.* (2007). CO_2 control of *Trichodesmium* N_2 fixation, photosynthesis, growth rates, and elemental ratios: implications for past, present and future ocean biogeochemistry. *Limnology and Oceanography*, **52**, 1293–304.

Hutchins, D.A., Mulholland, M.R., and Fu, F. (2009). Nutrient cycles and marine microbes in a CO_2-enriched ocean. *Oceanography*, **22**, 128–45.

Iglesias-Rodríguez, M.D., Schofield, O.M., Batley, J., Medlin, L.K., and Hayes, P.K. (2006). Intraspecific genetic diversity in the marine coccolithophore *Emiliania huxleyi* (Prymnesiophyceae): the use of microsatellite analysis in marine phytoplankton population studies. *Journal of Phycology*, **42**, 526–36.

Iglesias-Rodriguez, M.D., Halloran, P.R., Rickaby, R.E.M. *et al.* (2008). Phytoplankton calcification in a high-CO_2 world. *Science*, **320**, 336–40.

Joint, I., Doney, S.C., and Karl, D.M. (2011). Will ocean acidification affect marine microbes? *The ISME Journal*, **5**, 1–7.

Kim, J.-M., Lee, K., Shin, K. *et al.* (2006). The effect of seawater CO_2 concentration on growth of a natural phytoplankton assemblage in a controlled mesocosm experiment. *Limnology and Oceanography*, **51**, 1629–36.

Klaas, C. and Archer, D.A. (2002). Association of sinking organic matter with various types of mineral ballast in the deep sea: implications for the rain ratio. *Global Biogeochemical Cycles*, **16**, 1116, doi:10.1029/2001GB001765.

Kranz, S.A., Sültemeyer, D., Richter, K.U., and Rost, B. (2009). Carbon acquisition by *Trichodesmium*: the effect of pCO_2 and diurnal changes. *Limnology and Oceanography*, **54**, 548–59.

Kranz, S.A., Levitan, O., Richter, K.-U., Prášil, O., Berman-Frank, I., and Rost, B. (2010). Combined effects of CO_2 and light on the N_2-fixing cyanobacterium *Trichodesmium*

IMS101: physiological responses. *Plant Physiology*, **154**, 334–45.

Landry, M.R., Barber, R.T., Bidigare, R.R. *et al.* (1997). Iron and grazing constraints on primary production in the central equatorial Pacific: an EqPac synthesis. *Limnology and Oceanography*, **42**, 405–18.

Langer, G., Geisen, M., Baumann, K. *et al.* (2006). Species-specific responses of calcifying algae to changing seawater carbonate chemistry. *Geochemistry, Geophysics, Geosystems*, **7**, Q09006, doi:10.1029/2005GC001227.

Langer, G., Nehrke, G., Probert, I., Ly, J., and Ziveri, P. (2009). Strainspecific responses of *Emiliania huxleyi* to changing seawater carbonate chemistry. *Biogeosciences*, **6**, 2637–46.

Lassiter, A.M., Wilkerson, F.P., Dugdale, R.C., and Hogue, V.E. (2006). Phytoplankton assemblages in the CoOP-WEST coastal upwelling area. *Deep Sea Research Part II: Topical Studies in Oceanography*, **53**, 3063–77.

Legrand, M., Fenietsaigne, C., Saltzman, E.S., Germain, C., Barkov, N.I., and Petrov, V.N. (1991). Ice-core record of oceanic emissions of dimethylsulfide during the last climate cycle. *Nature*, **350**, 144–6.

Leonardos, N. and Geider, R.J. (2005). Elevated atmospheric CO_2 increases organic carbon fixation by *Emiliania huxleyi* (Haptophyta) under nutrient-limited, high-light conditions. *Journal of Phycology*, **41**, 1196–203.

Levitan, O., Rosenberg, G., Setlik, I. *et al.* (2007). Elevated CO_2 enhances nitrogen fixation and growth in the marine cyanobacterium *Trichodesmium*. *Global Change Biology*, **13**, 531–8.

Levitan, O., Kranz, S.A., Spungin, D., Prášil, O., Rost, B., and Berman-Frank, I. (2010). Combined effects of CO_2 and light on the N_2-fixing cyanobacterium *Trichodesmium* IMS101: a mechanistic view. *Plant Physiology*, **154**, 346–56.

Lombard, F., da Rocha, R.E., Bijma, J., and Gattuso, J.-P. (2010). Effect of carbonate ion concentration and irradiance on calcification in planktonic foraminifera. *Biogeosciences*, **7**, 247–55.

Longhurst, A. (1998). *Ecological geography of the sea*, 398 pp. Academic Press, San Diego, CA.

Mackinder, L., Wheeler, G., Schroeder, D., Riebesell, U., and Brownlee, C. (2010). Molecular mechanisms underlying calcification in coccolithophores. *Geomicrobiology*, **27**, 585–95.

Maldonado, M.T. and Price, N.M. (2001). Reduction and transport of organically bound iron by *Thalassiosira oceanica* (Bacillariophyceae). *Journal of Phycology*, **37**, 298–310.

Margalef, R. (1958). Temporal succession and spatial heterogeneity in phytoplankton. In: A.A. Buzzato-Traverso (ed.), *Perspectives in marine biology*, pp. 323–49. University of California Press, Berkeley.

Martin, J.H. and Fitzwater, S.E. (1988). Iron deficiency limits phytoplankton growth in the north-east Pacific subarctic. *Nature*, **331**, 341–3.

Millero, F.J., Woosley, R., DiTrolio, B. and Waters, J. (2009). Effect of ocean acidification on the speciation of metals in seawater. *Oceanography*, **22**, 72–85.

Milligan, A.J., Varela, D.E., Brzezinski, M.A., and Morel, F.M.M. (2004). Dynamics of silicon metabolism and silicon isotopic discrimination in a marine diatom as a function of pCO_2. *Limnology and Oceanography*, **49**, 322–9.

Milliman, J.D. (1993). Production and accumulation of calcium carbonate in the ocean: budget of a nonsteady state. *Global Biogeochemical Cycles*, **7**, 927–57.

Müller, M.N., Schulz, K.G., and Riebesell, U. (2010). Effects of long-term high CO_2 exposure on two species of coccolithophores. *Biogeosciences*, **7**, 1109–16.

Oschlies, A., Schulz, K.G., Riebesell, U., and Schmittner, A. (2008). Simulated 21st century's increase in oceanic suboxia by CO_2-enhanced biotic carbon export. *Global Biogeochemical Cycles*, **22**, GB4008, doi: 10.1029/2007GB003147.

Pabi, S., van Dijken, G.L., and Arrigo, K.R. (2008). Primary production in the Arctic Ocean, 1998–2006. *Journal of Geophysical Research*, **113**, C08005, doi:10.1029/2007JC004578.

Passow, U. (2004). Switching perspectives: do mineral fluxes determine particulate organic carbon fluxes or vice versa? *Geochemistry, Geophysics, Geosystems*, **5**, 1–5, doi: 10.1029/2003GC000670.

Raven, J.A. and Johnston, A.M. (1991). Mechanisms of inorganic-carbon acquisition in marine phytoplankton and their implications for the use of other resources. *Limnology and Oceanography*, **36**, 1701–14.

Rickaby, R.E.M., Henderiks, J., and Young, J.N. (2010). Perturbing phytoplankton: response and isotopic fractionation with changing carbonate chemistry in two coccolithophore species. *Climate of the Past*, **6**, 771–85.

Riebesell, U., Wolf-Gladrow, D.A., and Smetacek, V. (1993). Carbon dioxide limitation of marine phytoplankton growth rates. *Nature*, **361**, 249–51.

Riebesell, U., Zondervan, I., Rost, B., Tortell, P.D., Zeebe, R.E., and Morel, F.M.M. (2000). Reduced calcification in marine plankton in response to increased atmospheric CO_2. *Nature*, **407**, 634–7.

Riebesell, U., Schulz, K.G., Bellerby, R.G.J., *et al.* (2007). Enhanced biological carbon consumption in a high CO_2 ocean. *Nature*, **450**, 545–9.

Riebesell, U., Bellerby, R.G.J., Engel, A. *et al.* (2008a). Comment on 'Phytoplankton calcification in a high-CO_2 world'. *Science*, **322**, 1466b.

Riebesell, U., Bellerby, R., Grossart, H.-P. and Thingstad, F. (2008b). Mesocosm CO_2 perturbation studies: from organism to community level. *Biogeosciences*, **5**, 1157–64.

Riebesell, U., Körtzinger, A. and Oschlies, A. (2009). Sensitivities of marine carbon fluxes to ocean change. *Proceedings of the National Academy of Sciences USA*, **106**, 20602–9.

Rose, J.M., Feng, Y., Gobler, C.J. *et al.* (2009). The effects of increased pCO_2 and temperature on the North Atlantic Spring Bloom. II. Microzooplankton abundance and grazing. *Marine Ecology Progress Series*, **388**, 27–40.

Rost, B., Zondervan, I., and Riebesell, U. (2002). Light-dependent carbon isotope fractionation in the coccolithophorid *Emiliania huxleyi*. *Limnology and Oceanography*, **47**, 120–8.

Rost, B., Riebesell, U., Burkhardt, S., and Sültemeyer, D. (2003). Carbon acquisition of bloom-forming marine phytoplankton. *Limnology and Oceanography*, **48**, 55–67.

Rost, B., Richter, K.-U., Riebesell, U., and Hansen, P.J. (2006). Inorganic carbon acquisition in red-tide dinoflagellates. *Plant Cell and Environment*, **29**, 810–22.

Rost, B., Zondervan, I., and Wolf-Gladrow, D. (2008). Sensitivity of phytoplankton to future changes in ocean carbonate chemistry: current knowledge, contradictions and research directions. *Marine Ecology Progress Series*, **373**, 227–37.

Rynearson, T.A., Newton, J.A., and Armbrust, E.V. (2006). Spring bloom development, genetic variation, and population succession in the planktonic diatom *Ditylum brightwellii*. *Limnology and Oceanography*, **51**, 1249–61.

Sarmiento, J.L., Slater, R., Barber, R. *et al.* (2004). Response of ocean ecosystems to climate warming. *Global Biogeochemical Cycles*, **18**, GB3003, doi: 10.1029/2003GB002134.

Sarthou, G., Timmermans, K.R., Blain, S., and Tréguer, P. (2005). Growth physiology and fate of diatoms in the ocean: a review. *Journal of Sea Research*, **53**, 25–42.

Schartau, M., Engel, A., Schröter, J., Thoms, S., Völker, C., and Wolf-Gladrow, D. (2007). Modelling carbon overconsumption and the formation of extracellular particulate organic carbon. *Biogeosciences*, **4**, 433–54.

Schmittner, A., Oschlies, A., Matthews, H.D., and Galbraith, E.D. (2008). Future changes in climate, ocean circulation, ecosystems and biogeochemical cycling simulated for a business-as-usual CO_2 emission scenario until 4000 AD. *Global Biogeochemical Cycles*, **22**, GB1013, doi:10.1029/2007GB002953.

Schulz, K.G., Riebesell, U., Bellerby, R.G.J. *et al.* (2008). Build-up and decline of organic matter during PeECE III. *Biogeosciences*, **5**, 707–18.

Sciandra, A., Harlay, J., Lefèvre, D. *et al.* (2003) Response of coccolithophorid *Emiliania huxleyi* to elevated partial pressure of CO_2 under nitrogen limitation. *Marine Ecology Progress Series*, **261**, 111–22.

Sheppard Brennand, H., Soars, N., Dworjanyn, S.A., Davis, A.R., and Byrne, M. (2010). Impact of ocean warming and ocean acidification on larval development and calcification in the sea urchin *Tripneustes gratilla*. *PLoS ONE*, **5**, e11372, doi:10.1371/journal.pone.0011372.

Sherr, B.F., Sherr, E.B., Caron, D.A., Vaulot, D., and Worden, A.Z. (2007). Oceanic protists. *Oceanography*, **20**, 130–4.

Sherr, E.B. and Sherr, B.F. (1994). Bacterivory and herbivory: key roles of phagotrophic protists in pelagic food webs. *Microbial Ecology*, **28**, 223–35.

Shi, D.L., Xu, Y., and Morel, F.M.M. (2009). Effects of the pH/pCO_2 control method in the growth medium of phytoplankton. *Biogeosciences*, **6**, 1199–207.

Shi D.L., Xu Y., Hopkinson B.M., and Morel F.M.M. (2010). Effect of ocean acidification on iron availability to marine phytoplankton. *Science*, **327**, 676–9.

Steele, J.H. (1974). *The structure of marine ecosystems*, 128 pp. Harvard University Press, Cambridge, MA.

Stefels, J. and Vanboekel, W.H.M. (1993). Production of DMS from dissolved DMSP in axenic cultures of the marine phytoplankton species *Phaeocystis* sp. *Marine Ecology Progress Series*, **97**, 11–18.

Stefels, J., Steinke, M., Turner, S., Malin, G., and Belviso, S. (2007). Environmental constraints on the production and removal of the climatically active gas dimethylsulphide (DMS) and implications for ecosystem modelling. *Biogeochemistry*, **83**, 245–75.

Sterner, R.W. and Elser, J.J. (2002). *Ecological stoichiometry: the biology of elements from molecules to the biosphere*, 584 pp. Princeton University Press, Oxford, UK.

Suffrian, K., Simonelli, P., Nejstgaard, J.C., Putzeys, S., Carotenuto, Y., and Antia, A.N. (2008). Microzooplankton grazing and phytoplankton growth in marine mesocosms with increased CO_2 levels. *Biogeosciences*, **5**, 1145–56.

Tilman, D. and Downing, J.A. (1996). Biodiversity and stability in grasslands. In: F.B. Samson and F.L. Knopf (eds), *Ecosystem management: selected readings*, pp. 3–7. Springer Verlag, New York.

Tortell, P.D. (2000). Evolutionary and ecological perspectives on carbon acquisition in phytoplankton. *Limnology and Oceanography*, **45**, 744–50.

Tortell, P.D., DiTullio, G.R., Sigman, D.M. and Morel, F.M.M. (2002). CO_2 effects on taxonomic composition and nutrient utilization in an Equatorial Pacific phytoplankton assemblage. *Marine Ecology Progress Series*, **236**, 37–43.

Tortell, P.D., Payne, C.D., Li, Y., *et al.* (2008). CO_2 sensitivity of Southern Ocean phytoplankton, *Geophysical Research Letters*, **35**, L04605, doi:10.1029/2007GL032583.

Williams, P.J.leB. (1998). The balance of plankton respiration and photosynthesis in the open oceans. *Nature*, **394**, 55–7.

Worden, A.Z., Nolan, J.K., and Palenik, B. (2004). Assessing the dynamics and ecology of marine picophytoplankton: the importance of the eukaryotic component. *Limnology and Oceanography*, **49**, 168–79.

Wu, Y., Gao, K., and Riebesell, U. (2010). CO_2-induced seawater acidification affects physiological performance of the marine diatom *Phaeodactylum tricornutum*. *Biogeosciences*, **7**, 2915–23.

Zehr, J.P., Mellon, M.T., and Zani, S. (1998). New nitrogen-fixing microorganisms detected in oligotrophic oceans by amplification of nitrogenase (*nifH*) genes. *Applied and Environmental Microbiology*, **64**, 3444–50.

Zondervan, I., Rost, B., and Riebesell, U. (2002). Effect of CO_2 concentration on the PIC/POC ratio in the coccolithophore *Emiliania huxleyi* grown under light-limiting conditions and different daylengths. *Journal of Experimental Marine Biological and Ecology*, **272**, 55–70.

CHAPTER 7

Effects of ocean acidification on benthic processes, organisms, and ecosystems

Andreas J. Andersson, Fred T. Mackenzie, and Jean-Pierre Gattuso

7.1 Introduction

The benthic environment refers to the region defined by the interface between a body of water and the bottom substrate, including the upper part of the sediments, regardless of the depth and geographical location. Hence, benthic environments, their organisms, and their ecosystems are highly variable as they encompass the full depth range of the oceans with associated changes in physical and chemical properties as well as differences linked to latitudinal and geographical variation. The effects of ocean acidification on the full range of different benthic organisms and ecosystems are poorly known and difficult to ascertain. Nevertheless, by integrating our current knowledge on the effects of ocean acidification on major benthic biogeochemical processes, individual benthic organisms, and observed characteristics of benthic environments as a function of seawater carbonate chemistry, it is possible to draw conclusions regarding the response of benthic organisms and ecosystems to a world of increasingly higher atmospheric CO_2 levels. The fact that there are large-scale geographical and spatial differences in seawater carbonate system chemistry (see Chapter 3), owing to both natural and anthropogenic processes, provides a powerful means to evaluate the effect of ocean acidification on marine benthic systems. In addition, there are local and regional environments that experience high-CO_2 and low-pH conditions owing to special circumstances such as, for example, volcanic vents (Hall-Spencer *et al.* 2008; Martin *et al.* 2008; Rodolfo-Metalpa *et al.* 2010), seasonal stratification (Andersson *et al.* 2007), and upwelling (Feely *et al.* 2008; Manzello *et al.* 2008) that may provide important clues to the impacts of ocean acidification on benthic processes, organisms, and ecosystems.

The objective of this chapter is to provide an overview of the potential consequences of ocean acidification on marine benthic organisms, communities, and ecosystems, and the major biogeochemical processes governing the cycling of carbon in the marine benthic environment, including primary production, respiration, calcification, and $CaCO_3$ dissolution.

7.2 The effect of ocean acidification on major biogeochemical processes

7.2.1 Photosynthesis and primary production

The depth of the euphotic zone, i.e. the depth of water exposed to sufficient sunlight to support photosynthesis, varies depending on a range of factors affecting the clarity of seawater, including river input and run-off to the coastal ocean, upwelling, mixing, and planktonic production. Benthic primary producers include micro- and macroalgae and seagrasses (Gattuso *et al.* 2006). Some organisms, such as corals and foraminifera, can host within their tissues symbiotic algae that contribute to benthic primary production in certain environments.

Photosynthesis reduces CO_2 using energy captured from the sun in order to produce organic material in the form of sugars. Most, but not all, autotrophs produce oxygen as a by-product of the process of photosynthesis. Because dissolved CO_2

exists in very low concentrations in seawater (~10 to 15 µmol kg^{-1} at current typical surface-seawater conditions), marine organisms have mechanisms to catalyse the uptake of CO_2 from this relatively scarce source of carbon, including enzymatic activity (e.g. carbonate anhydrase) and the ability to utilize the bicarbonate ion (HCO_3^-) as a source of carbon for photosynthesis (Raven 1997). HCO_3^- is approximately two orders of magnitude more abundant than CO_2(aq) (CO_2(aq) includes dissolved CO_2 and H_2CO_3 in the approximate ratio of 400:1), but dehydrating HCO_3^- to CO_2 has a high energetic cost. Regardless of the source of carbon for photosynthesis, ocean acidification results in increasing concentrations of both of these dissolved carbon species, although the relative increase in CO_2 is much greater than the relative increase in HCO_3^- (see Chapter 1). However, other constituents such as nitrogen, phosphorus, and iron may limit photosynthesis, so increasing CO_2 does not automatically result in increased production of organic matter. Although studies are limited, some experimental results suggest that photosynthesis as well as net primary production (= gross primary production – respiration) for certain benthic autotrophs will increase in a high-CO_2 world. For example, seagrasses, which appear limited by the availability of CO_2, respond positively to conditions of increasing seawater CO_2 (Table 7.1; Palacios and Zimmerman 2007; Hall-Spencer et al. 2008). Seagrasses might also benefit from a reduction in calcareous epibiont organisms that currently foul their blades and decrease their photosynthetic area. On the other hand, calcification by the same epibionts could benefit seagrasses by supplying CO_2 that could be used for photosynthesis (Barrón et al. 2006). Similar to seagrasses, experiments conducted with non-calcifying algae have shown increased production and growth in response to elevated CO_2 conditions. For example, primary production was observed to increase in an Arctic specimen of the brown alga *Laminaria saccharina* exposed to elevated CO_2 (S. Martin, pers. comm.), and the relative growth rate of the red seaweed *Lomentaria articulata* was observed to increase in response to the CO_2 levels anticipated by the end of this century (Kübler et al. 1999). Experiments investigating the effect of elevated CO_2 on photosynthesis and/or carbon production of calcifying

algae and corals show complex and species-specific responses with variable results. Anthony et al. (2008) reported a decrease in net productivity of a coralline alga as a function of increasing CO_2 and decreasing pH. A number of experiments conducted with corals have reported increased zooxanthellae density and/or chlorophyll *a* content with increasing CO_2 (e.g. Reynaud et al. 2003; Crawley et al. 2009), but both gross photosynthesis and respiration showed variable responses with no clear trend in terms of the resulting net production. Schneider and Erez (2006) observed no effect on the net productivity in the coral *Acropora eurystoma* as a function of CO_2.

Increased benthic primary production could act as a small negative feedback to rising atmospheric CO_2 and ocean acidification if the organic material produced were exported to the deep sea or permanently buried within the sediments. However, this assumes that decomposition and remineralization of organic material remain unchanged or that any increase in these processes is smaller relative to the amount of carbon buried in the sediments; that is, gross productivity exceeds gross respiration (net ecosystem production is positive). The magnitude of this negative feedback in the early 21st century is of the order of 0.1 Gt C yr^{-1} (Mackenzie and Lerman 2006).

7.2.2 Oxidation of organic matter

Living organisms break down organic material in order to extract energy, typically using oxygen as a terminal electron acceptor and releasing CO_2 as a waste product. Under suboxic or anoxic conditions, many microbes can use alternative electron acceptors such as NO_3^-, SO_4^{2-}, CH_4, and CO_2 (see Chapter 9). The decomposition of organic material causes acidification of seawater. This natural acidification is evident from the general trends of increasing dissolved inorganic carbon and decreasing pH as a function of depth in the oceans (see Chapter 3). The same trends are also observed as a function of age of the water mass, i.e. the time since the water mass was last in contact with the atmosphere. For example, bottom waters in the Pacific are more acidic than Atlantic bottom waters, which have been in contact with the atmosphere more recently than waters in the Pacific. The acidification of seawater

Table 7.1 Summary of the effects of CO_2, pH, and ocean acidification on major benthic organisms

CCA, crustose coralline algae; pH_U, pH scale unknown; pH_T, total pH scale; pH_{SWS}, pH, seawater scale; pH_{NBS}, pH, National Bureau of Standards scale; N/A, not available; T, temperature; S, salinity; Q_{10}, the relative change in a rate based on a 10°C temperature increase; Ref., reference. The complete reference list for this table is available as Appendix 7.1 at http://ukcatalogue.oup.com/product/9780199591091.do

Species	$CO_2(g)$ (ppmv); pCO_2 (μatm); pH; Ω; $[CO_3^{2-}]$ (μmol kg^{-1})	Other variables/ comments	Duration	Observed effect(s)	Ref.
Algae					
Non-calcifying algae					
Feldmania spp.	CO_2: 380; 550	Temperature	14 weeks	High CO_2 had no effect on the cover of turfs at ambient T (17°C), but combined with elevated T (20°C), turfs occupied 25% more space than predicted from CO_2 and T independently. Both high CO_2 and T yielded an increase in the dry mass of turfs.	[13]
Feldmannia spp.	CO_2: 380; 550	Nutrients	76 days	Turf biomass increased in response to high CO_2 while nutrients had no detectable effect. High CO_2 had no effect on the new recruitment of turf at ambient nutrient conditions. Combined with elevated nutrients, turf recruited on to 34% more substratum than predicted from CO_2 and nutrients independently. Photosynthetic yield increased in response to high CO_2 and high nutrients.	[68]
Hizikia fusiforme	CO_2: 360; 700 pH_U: 8.4–9.2; 8.0–8.4 (diurnal)		6–8 days	Mean relative growth rate and nitrogen assimilation increased in high-CO_2 conditions. There were no significant differences in light-saturated photosynthesis rates, dark respiratory rates, and apparent photosynthetic efficiency.	[80]
Lomentaria articulata	CO_2: 235; 350; 700; 1750	Oxygen	3 weeks	Relative growth rate based on carbon production increased by 52% in 700 ppmv and 25% in 1750 ppmv treatments compared with control (350 ppmv). The tissue content of C per unit dry weight was greater in cultures grown at low CO_2. Growth rates were unaffected by $[O_2]$.	[35]
Non-calcareous algae	pCO_2: 334; 957; 20812 pH_T: 8.14; 7.83; 6.57 Ω_a: 3.91; 2.43; 0.19	Volcanic vent	N/A	Algal cover significantly increased from near 0 at ambient pH to >60% at the highest CO_2, and lowest pH conditions.	[27]
Calcifying algae					
CCA	pCO_2: 311; 306; 365; 549; 1564 pH_T: 8.17; 8.17; 8.11; 8.00; 7.66 Ω_a: 4.16; 4.10; 3.74; 3.17; 1.89	Volcanic vent 2 weeks aquarium expt	N/A	CCA dominated the epiphytic community on seagrass blades in pH_T 8.0 to 8.2 (18 to 69% cover), but were absent in pH_T 7.7. Additional controlled laboratory experiments showed that epiphytic coralline algae completely dissolved at pH_T 7.0.	[46]
Hydrolithon boreale, Hydrolithon cruciatum, Hydrolithon farinosum, Pneophyllum confervicola, Pneophyllum fragile, Pneophyllum zonale					

Species	Stressor	Duration	Carbonate chemistry	Observations	Ref.
CCA		51 days	pCO_2: 400; 765 (midday) pH_{NBS}: 8.17; 7.91	Recruitment of CCA decreased by 78% and percentage cover by 92% under high CO_2.	[36]
Corallinaceae	Volcanic vent	N/A	pCO_2: 334; 957; 20812 pH_T: 8.14; 7.83; 6.57 Ω_a: 3.91; 2.43; 0.19	Coralline algae cover significantly decreased from >60% at ambient pH to 0% at the highest CO_2 and lowest pH conditions.	[27]
Corallina sessilis	UV radiation	28 days	CO_2: 380; 1000 pH_{NBS}: 8.3–8.4; 7.8–7.9	Synergistic effects of UV radiation and CO_2 decreased growth by 13%, photosynthesis by 6%, and calcification by 3% in the low-CO_2 treatment. In the high-CO_2 treatment the same parameters decreased by 47%, 20%, and 8%, respectively. Both UVA and UVB had negative effects on photosynthesis and calcification although the inhibition by UVB was 2.5 times that caused by UVA.	[20]
Halimeda incrassata		60 days	CO_2: 409; 606; 903; 2856 Ω_a: 2.5; 2.0; 1.5; 0.7	Net calcification rate (wt% day^{-1}) increased linearly between Ω_a 0.7 to 2.0 but decreased in the high-Ω_a treatment.	[64]
Halimeda opuntia		3 weeks	pH_T: 8.1; 7.5	$CaCO_3$ crystals were 43% smaller and the number of crystals per unit area was 66% greater in low pH compared with ambient pH.	[65]
Halimeda renshii		2.5 h	pH_U: 8.3–8.9	Calcification increased 1.6-fold in response to an increase in pH_U from ~8.3 to ~8.9 induced by photosynthesis by seagrass.	[70]
Halimeda tuna		3 weeks	pH_T: 8.1; 7.5	$CaCO_3$ crystals were 18% smaller and the number of crystals per unit area was 11% greater in low pH compared to ambient pH.	[65]
Halimeda tuna		2 h	pH_{NBS}: 6.0–9.0	In light conditions, photosynthetic rate increased as a function of decreasing pH while calcification showed a variable response with an initial decrease between pH_{NBS} 9 to 7, an increase between pH_{NBS} 7 to 6.5, and a radical decrease between pH_{NBS} 6.5 to 6.	[8]
Hydrolithon sp.		1 h; 5 days	pCO_2: ~395; ~790; ~1–1382 pH_U: 8.1; 7.8; 7.6–9.8 Ω_a: 4; 2; 0.8–21	Overall, calcification and photosynthesis, respectively, were positively and inversely correlated to pH. However, changes in pH_U between 7.6 and 8.6 had no significant effect on calcification. pH_U >8.6 significantly enhanced calcification, but resulted in decreased photosynthesis. A reduction in pH_U from 8.1 to 7.8 caused a reduction in calcification by 20% and an increase in photosynthetic rates by 13% after 5 days' exposure.	[71]
Hydrolithon sp.		2.5 h	pH_U: 8.3–8.9	Calcification increased 5.8-fold in response to an increase in pH_U from ~8.3 to ~8.9 induced by photosynthesis by seagrass.	[70]

(continued)

Table 7.1 Continued

Species	$CO_2(g)$ (ppmv); pCO_2 (µatm); pH; Ω; $[CO_3^{2-}]$ (µmol kg⁻¹)	Other variables/comments	Duration	Observed effect(s)	Ref.
Lithophyllum cabiochae	pCO_2: 397; 436; 703; 753 pH$_T$: 8.08; 8.05; 7.87; 7.85 Ω_a: 5.26; 5.43; 3.54; 3.72	Temperature	392 days	Algal necrosis was observed at the end of summer at elevated T of 3°C above ambient T (60% necrosis at high CO_2 and 30% at ambient CO_2). Calcification was initially reduced under high CO_2, but during summer, calcification was only reduced in response to high CO_2 and T combined. Independently of CO_2 treatment, calcification increased during autumn and winter with increasing T. Dissolution of dead algae was two- to fourfold higher under high CO_2 compared with ambient CO_2.	[45]
Lithophyllum sp.	CO_2: 380; 550	Nutrients	76 days	High CO_2 and high nutrients both independently decreased biomass of coralline crusts. The largest decrease was observed when these treatments were combined. No new recruitment occurred under high-CO_2 conditions. The photosynthetic yield decreased owing to high CO_2 but was unaffected by high nutrients.	[68]
Mesophyllum sp.	pH$_U$: 8.3–8.9		2.5 h	Calcification increased 1.6-fold in response to an increase in pH$_U$ from ~8.3 to ~8.9 induced by photosynthesis by seagrass.	[70]
Neogoniolithon spp.	CO_2: 409; 606; 903; 2856 Ω_a: 2.5; 2.0; 1.5; 0.7		60 d	Net calcification rate (wt% day⁻¹) increased linearly between Ω_a 0.7 to 2.0 but decreased in the high-Ω_a treatment.	[64]
Porolithon gardineri	Ω_c: 1–8.6		3–6 months	Extension rate and skeletal $MgCO_3$ content decreased linearly as a function of decreasing Ω_a.	[1], [42]
Porolithon onkodes	pCO_2: 130–465; 520–705; 1010–1360 pH$_{SWS}$: 8.0–8.4; 7.85–7.95; 7.60–7.70 Ω_a: 3.3–7.1; 2.5–3.5; 1.5–2.2	Temperature	8 weeks	Calcification (wt% month⁻¹) and net productivity decreased by 130 to 190% and 45 to 160%, respectively, in response to high CO_2. These conditions also led to bleaching by 40 to 50%. The negative response was exacerbated by warming. Net dissolution was observed in the high CO_2 treatment.	[3]
Rhodolith (*Lithophyllum* cf. *pallescens*, *Hydrolithon* sp., *Porolithon* sp.)	pCO_2: 414; 670 (midday) pH$_{NBS}$: 8.17; 7.97 Ω_a: 2.91; 1.95		268 days	Algae lost weight at a rate of −0.9 g yr⁻¹ in the high-CO_2 treatment compared with gaining 0.6 g yr⁻¹ in ambient conditions.	[33]
Annelida					
Hydroides crucigera	CO_2: 409; 606; 903; 2856 Ω_a: 2.5; 2.0; 1.5; 0.7		60 days	Marginal increase in net calcification rate (wt% day⁻¹) as a function of increasing Ω_a, although the correlation coefficient was very low.	[64]
Nereis virens	pH$_U$: eight treatments between pH 5.07 and 8.12	Temperature	10–30 days	No negative effects at pH levels expected as a result of anthropogenic ocean acidification. Negative effects in terms of mortality, burrowing activity, and dry weight were not observed until pH$_U$ < 6.5.	[5]
Vermetids	pCO_2: 414; 670 (midday) pH$_{NBS}$: 8.17; 7.97 Ω_a: 2.91; 1.95		268 days	No significant difference in recruitment.	[33]

Bivalves/gastropods

Species	Chemistry	Additional stressor	Duration	Findings	Ref
Argopecten irradians	CO_2: 409; 606; 903; 2856 Ω_a: 2.5; 2.0; 1.5; 0.7		60 days	Increased net calcification rate (wt% day^{-1}) as a function of increasing Ω_a.	[64]
Crassostrea ariakensis	pCO_2: 291; 386; 581; 823 pH_{NBS}: 8.17; 8.08; 7.92; 7.79 Ω_a: 1.3; 1.1; 0.8; 0.6		28 days	No significant changes in growth, calcification, and shell thickness were observed in response to increasing CO_2.	[54]
Crassostrea gigas	pCO_2: 698–2774 pH_{NBS}: 8.07–7.55 Ω_a: 3.1–1.1		2 h	Significant decreased calcification as a function of increasing CO_2 and decreasing pH.	[22]
Crassostrea gigas	pCO_2: 348; 2268 pH_{NBS}: 8.21; 7.42 Ω_a: 3.0; 0.68		48 h	Inhibition of shell synthesis and reduced larval size during early development in high-CO_2 treatment.	[39]
Crassostrea gigas	pH_{NBS}: 8.08–8.21; 7.81–7.86		1 h	No significant effect of low pH on sperm swimming speed, sperm motility, or fertilization kinetics.	[28]
Crassostrea virginica	CO_2: 409; 606; 903; 2856 Ω_a: 2.5; 2.0; 1.5; 0.7		60 days	Increased net calcification rate (wt% day^{-1}) as a function of increasing Ω_a.	[64]
Crassostrea virginica	pCO_2: 284; 389; 572; 840 pH_{NBS}: 8.16; 8.06; 7.91; 7.76 Ω_a: 1.2; 1.0; 0.8; 0.6		28 days	Shell area decreased by 16% and calcium content 42% in the highest CO_2 treatment relative to the lowest. No significant difference was observed in shell thickness.	[54]
Crassostrea virginica	pH_{NBS}: 12 treatments between pH 7.41 and 8.29 Ω_c: 0.31–4.62	Temperature Salinity	<15 h	In general, calcification rates decreased as a function of decreasing pH. The interaction of T, S, and pH was not significant but the two-way interaction of T or S with pH was significant.	[76]
Crepidula fornicata	CO_2: 409; 606; 903; 2856 Ω_a: 2.5; 2.0; 1.5; 0.7		60 days	Net calcification rate (wt% day^{-1}) increased linearly between Ω_a 0.7 and 1.5 but decreased in the high-Ω_a treatment.	[64]
Haliotis coccoradiata	pCO_2: 324–335; 801–851; 1033–1104; 1695–1828 pH_U: 8.25; 7.9; 7.8; 7.6 Ω_a: 3.20–3.99; 1.64–2.10; 1.33–1.72; 0.87–1.13	Temperature	15 min	Per cent fertilization was not significantly affected by CO_2 or T treatments ranging from 18 to 26°C.	[9]
Littorina littorea	CO_2: 409; 606; 903; 2856 Ω_a: 2.5; 2.0; 1.5; 0.7		60 days	Calcification rate increased linearly (wt% day^{-1}) as a function of increasing Ω_a.	[64]
Mercenaria mercenaria	CO_2: 409; 606; 903; 2856 Ω_a: 2.5; 2.0; 1.5; 0.7		60 days	Net calcification rate (wt% day^{-1}) increased between Ω_a 0.7 and 1.5 but remained relatively constant between Ω_a 1.5 and 2.5.	[64]
Mercenaria mercenaria	pH_U: 7.8; 7.1 Ω_a: 1.5; 0.3		21 days	Per cent mortality of bivalves of varying sizes (0.2, 0.3, 1.0, 2.0 mm) ranged from 11.8, 4.8, 1.9, and 1.1% day^{-1} in low-Ω_a conditions.	[25]

(continued)

Table 7.1 Continued

Species	$CO_2(g)$ (ppmv); pCO_2 (µatm); pH; Ω; $[CO_3^{2-}]$ (µmol kg⁻¹)	Other variables/ comments	Duration	Observed effect(s)	Ref.
Mercenaria mercenaria	pH_U: 7.9; 7.3; 7.0 Ω_a: 1.6; 0.6; 0.4		25 days	Per cent mortality of bivalves of varying sizes (0.2, 0.4, 0.6 mm) ranged from 14, 9.6, and 2.8% day⁻¹ at Ω_a=0.4 compared to 3.9, 0.0, and 0.27% at Ω_a=1.6, respectively.	[26]
Mya arenaria	CO_2: 409; 606; 903; 2856 Ω_a: 2.5; 2.0; 1.5; 0.7		60 days	Calcification rate increased linearly (wt% day⁻¹) as a function of increasing Ω_a.	[64]
Mytilus edulis	pH_U: 8.1; 7.6; 7.4; 7.1; 6.7		44 days	Shell growth was not significantly affected between pH_U 7.4 and 8.1 but was significantly reduced at pH_U 7.1 and 6.7. No net growth occurred at pH_U 6.7. From day 23 mortality was observed in this treatment.	[6]
Mytilus edulis	pCO_2: 421–2351 pH_{NBS}: 8.13–7.46 Ω_a: 3.4–1.0		2 h	Significant decreased calcification as a function of increasing CO_2, and decreasing pH.	[22]
Mytilus edulis	CO_2: 409; 606; 903; 2856 Ω_a: 2.5; 2.0; 1.5; 0.7		60 days	No significant effect.	[64]
Mytilus edulis	pCO_2: 385–2309 (field); 464–4254 (lab) pH_{NBS}: 7.49–8.23 (field); 7.08–8.13 (lab) Ω_a: 0.35–0.96 (field) 0.12–1.01 (lab)		2 months	Mussels thrived in natural conditions with pCO_2 exceeding pre-industrial pCO_2 by a factor of 3 to 5. Experimental results showed that growth rates were not significantly different in mussels exposed to pCO_2 of 385 and 1400 µatm, but significantly decreased under pCO_2 of 4000 µatm. Although all mussels gained weight in all treatments, external shell dissolution increased with increasing pCO_2.	[74]
Mytilus edulis	pCO_2: 665.7; 1160.7; 1435.2; 3316.2 pH_{NBS}: 7.83; 7.65; 7.49; 6.70		32 days	Over time, exposure to elevated CO_2 disrupted the ability to express an immune response by suppressing levels of phagocytosis. The predominant mechanism of internal defence in bivalves involves phagocytosis by circulating haemocytes followed by a range of physiological processes. In controls, phagocytosis increased >800% in 32 days compared with <200% in the lowest pH treatment.	[7]
Mytilus galloprovincialis	CO_2: 380; 2000 pH_{NBS}: 8.13; 7.42 Ω_a: 2.23; 0.49		6 days	Embryogenesis was unaffected, but delayed development, morphological abnormalities, shell malformation, and reduced larval height (26 ± 1.9%) and length (20 ± 1.1%) were observed in the high-CO_2 treatment.	[40]
Mytilus galloprovincialis	pH_U: 8.05; 7.3		3 months	Exposure to low pH caused slower growth, lower metabolic rate, and a permanent reduction in haemolymph pH. An observed increase in haemolymph HCO_3^- was attributed to be a result of shell dissolution to some extent counterbalancing the effect of low pH.	[53]
Nucula annulata	pH_U: 8.1; 7.5 Ω_a: 0.95; 0.29		70 days	No statistical differences in bivalve density as a function of time in the two treatments.	[24]

Species	Perturbation	Chemistry	Duration	Effect	Ref.
Nucella lamellosa		CO_2: 385; 785; 1585 pH_{NBS}: 7.98; 7.80; 7.54	6 days	Gain in shell weight in live snails decreased linearly with increasing CO_2. This trend was paralleled by a loss in weight in empty shells suggesting that the observed decline may not be due to a decrease in rates of deposition, but rather an increase in dissolution.	[57]
Osilinus turbinata	Volcanic vent	pCO_2: 334; 957; 20812 pH_T: 8.14; 7.83; 6.57 Ω_a: 3.91; 2.43; 0.19	N/A	Specimen were absent in areas where minimum $pH_T < 7.4$.	[27]
Patella caerulea	Volcanic vent	pCO_2: 334; 957; 20812 pH_T: 8.14; 7.83; 6.57 Ω_a: 3.91; 2.43; 0.19	N/A	Specimen were absent in areas where minimum $pH_T < 7.4$.	[27]
Saccostrea glomerata	Temperature	CO_2: 375; 600; 750; 1000	24–48 h	Smaller sizes, lower fertilization, fewer and more abnormal larvae (D-veliger stage) increased with prolonged increase of high pCO_2.	[61]
Saccostrea glomerata		pCO_2: 220; 509; 776 pH_{NBS}: 8.10–8.14; 7.78–7.84; 7.59–7.67 Ω_a: 1.15; 0.64; 0.45	8 days	Larval survival decreased by 43% at pH_{NBS} 7.8 and by 72% at pH_{NBS} 7.6 relative to control pH. No significant difference in mass was observed, but morphology and shell-surface properties were significantly affected. Per cent of empty shells remaining from dead larvae also decreased with decreasing pH indicating dissolution.	[78]
Strombus luhuanus		pH_U: 7.936–7.945; 7.897–7.899	6 months	In the initial stages of the experiment, growth rates were not significantly different between different pH treatments, but were significantly different at the end of the experiment. For one experimental run, mortality was higher in treatment conditions, but for a second run, there were no significant differences in mortality.	[72]
Strombus alatus		CO_2: 409; 606; 903; 2856 Ω_a: 2.5; 2.0; 1.5; 0.7	60 days	Net calcification rate (wt% day^{-1}) increased between Ω_a 0.7 and 1.5 but remained relatively constant between Ω_a 1.5 and 2.5.	[64]
Tellina agilis		pH_U: 8.1; 7.5 Ω_a: 0.95±0.05; 0.29±0.013	68 days	No statistical differences in bivalve density as a function of time in the two treatments.	[24]
Urosalpinx cinerea		CO_2: 409; 606; 903; 2856 Ω_a: 2.5; 2.0; 1.5; 0.7	60 days	Calcification rate increased linearly (wt% day^{-1}) as a function of increasing Ω_a.	[64]
Bryozoans					
Bryozoans (*Callopora lineata, Electra posidoniae, Microporella ciliata, Tubulipora* spp.)	Volcanic vent	pCO_2: 311; 306; 365; 549; 1564 pH_T: 8.17; 8.17; 8.11; 8.00; 7.66 Ω_a: 4.16; 4.10; 3.74; 3.17; 1.89	N/A	No significant effect was observed in per cent cover of bryozoans on seagrass blades as a function of a natural gradient in CO_2, in a natural seagrass meadow.	[46]

(continued)

Table 7.1 Continued

Species	$CO_2(g)$ (ppmv); pCO_2 (µatm); pH; Ω; $[CO_3^{2-}]$ (µmol kg^{-1})	Other variables/comments	Duration	Observed effect(s)	Ref.
Corals					
Tropical and subtropical corals					
Acropora sp.	Ω_a: 0.98; 1.95; 2.93; 3.90; 5.85	Ca^{2+}	20 days	Nonlinear decrease in calcification rate as a function of decreasing aragonite saturation state. Saturation state was altered by manipulating calcium concentration.	[21]
Acropora cervicornis	pCO_2: 352–373; 714–771 pH_{SWS}: 8.00–8.10; 7.66–7.73	Nutrients	16 weeks	Significant decrease in growth rate as a result of both nutrient and high-CO_2 treatments. The negative effect was exacerbated when these parameters were combined.	[62]
Acropora digitifera	pCO_2: 400–475; 775–1005; 930–1260; 905–1660; 2115–3585; 12600–21100 pH_T: 8.03; 7.77; 7.69; 7.64; 7.31; 6.55 Ω_a: 3.2–3.5; 1.9–2.3; 1.6–2.0; 1.2–2.0; 0.6–1.0; 0.1–0.2		N/A	Sperm flagellar motility decreased significantly in response to decreasing pH. 69% of sperm were motile at pH_T 8.0, 46% at pH_T 7.8, and fewer than 20% at pH_T <7.7.	[55]
Acropora digitifera	pCO_2: 400–475; 905–1660; 2115–3585 pH_T: 8.03; 7.64; 7.31 Ω_a: 3.2–3.5; 1.2–2.0; 0.6–1.0		4–14 days	Survival of larvae was not significantly affected by pH, but polyps were significantly smaller under conditions of lower pH. The infection rate by zooxanthellae was significantly delayed in low-pH conditions, but all polyps in all treatments acquired zooxanthellae by day 4 of the experiment.	[73]
Acropora eurystoma	pCO_2: 175.65–889.43 pH_{SWS}: 7.87–8.5 Ω_a: 1.5–6.4	Different combinations of keeping C_T, pH or pCO_2 constant	1–2 h	Calcification rate was positively correlated to $[CO_3^{2-}]$ and Ω_a. A reduction in $[CO_3^{2-}]$ by 30% reduced calcification by 50%. Photosynthesis and respiration did not show any significant response to changes in seawater CO_2.	[69]
Acropora formosa	pCO_2: 260–460; 600–790; 1160–1500 pH_{SWS}: 8.0–8.2; 7.8–7.9; 7.55–7.65		4 days	Chlorophyll a per cell increased while photosynthetic capacity per chlorophyll decreased with increasing CO_2. Dark respiration remained constant whereas light-enhanced dark respiration increased with increasing CO_2.	[14]
Acropora intermedia	pCO_2: 130–465; 520–705; 1010–1360 pH_{SWS}: 8.0–8.4; 7.85–7.95; 7.60–7.70 Ω_a: 3.3–7.1; 2.5–3.5; 1.5–2.2	Temperature	8 weeks	Net productivity was unaffected by intermediate CO_2 levels, but increased combined with a temperature increase of 3°C. At the high CO_2 treatment, productivity decreased radically. Elevated CO_2 and T caused bleaching by 40–50%. Calcification (wt% month^{-1}) decreased by 60% at the high-CO_2 and high-T treatment.	[3]

Species	Conditions	Duration	Observation	Reference
Acropora tenuis	CO_2: ambient; 1000 pH_U: ambient; 7.6	14 days	Corals were unaffected until the larval stage in high-CO_2 treatment. After settlement, the polyp endoskeleton was observed to be disturbed and malformed compared with the control.	[37]
Acropora tenuis	pCO_2: 400–475; 905–1660; 2115–3585 pH_T: 8.03; 7.64; 7.31 Ω_a: 3.2–3.5; 1.2–2.0; 0.6–1.0	4–14 days	Survival of larvae was higher in the lowest pH treatment (84.8%) compared with the intermediate pH (62%) and control (~70%) treatments.	[73]
Acropora verweyi	pCO_2: 407–416; 857–882 pH_{SWS}: 8.06; 7.75–7.76 Ω_a: 4.37–4.43; 2.26–2.33	8 days	Statistically significant reduction in calcification rate by 18% under high CO_2.	[50]
Favia fragum	pH_{NBS}: 8.17; 8.04; 7.87; 7.54 Ω_a: 3.71; 2.40; 1.03; 0.22	8 days	Significant delays in initiation, calcification, and growth of the primary corallite of new recruits as a function of decreasing Ω_a. Crystal morphology changed from densely packed bundles of aragonite needles to a disordered aggregate of rhombs with decreasing Ω_a.	[12]
Fungia sp.	pCO_2: 226; 399; 887 pH_{NBS}: 8.417; 8.207; 7.855 Ω_a: ~5.18; ~3.43; ~1.56	3–6 h	Positive linear correlation between calcification rate and seawater Ω_a.	[30]
Galaxea fascicularis	pCO_2: 407–416; 857–882 pH_{SWS}: 8.06; 7.75–7.76 Ω_a: 4.37–4.43; 2.26–2.33	8 days	Statistically significant reduction in calcification rate by 16% under high CO_2.	[50]
Madracis auretenra	pCO_2: 171–1480 pH_T: 7.60–8.07 Ω_a: 1.74–4.30	Different combinations of manipulating either C_T or A_T / 2 h	Calcification rates correlated strongly with $[HCO_3]$ ranging from 777 to 3579 μmol kg^{-1} and less consistently with $[CO_3^{2-}]$, Ω_a, and pH.	[34]
Montipora capitata	pCO_2: 414; 670 (midday) pH_{NBS}: 8.17; 7.97 Ω_a: 2.91; 1.95	59–263 days	Reduction in coral colony calcification by 15% and 20% in long- and short-term experiments, respectively, under high CO_2. No significant effect was observed on coral nubbins.	[33]
Pavona cactus	pCO_2: 407–416; 857–882 pH_{SWS}: 8.06; 7.75–7.76 Ω_a: 4.37–4.43; 2.26–2.33	8 days	Statistically significant reduction in calcification rate by 18% under high CO_2.	[50]

(continued)

Table 7.1 Continued

Species	CO_2(g) (ppmv); pCO_2 (µatm); pH; Ω; $[CO_3^{2-}]$ (µmol kg^{-1})	Other variables/comments	Duration	Observed effect(s)	Ref.
Pocillopora damicornis	pCO_2: 414; 670 (midday) pH_{NBS}: 8.17; 7.97 Ω_a: 2.91; 1.95		268 days	No significant difference in recruitment.	[34]
Pocillopora meandrina	pCO_2: 395; 720 pH_{NBS}: 8.18; 7.80 Ω_a: 3.8; 2.0	Temperature	14 days	Calcification was not significantly affected by CO_2 or T although a significant positive interaction was observed between these parameters. Dark adapted maximum quantum yield of photosystem II was unaffected at all treatments.	[56]
Porites asteroides	pH_{SWS}: 7.95–7.99; 7.88–7.91; 7.80–7.82 Ω_a: 3.1–3.2; 2.61–2.62; 2.2–2.23		1 month	No significant effect on larval settlement. Reduction in skeletal extension by 50% and 78% in the intermediate and high-CO_2 treatments, respectively.	[2]
Porites compressa	pCO_2: 186; 440 pH_{SWS}: 8.31; 7.97 Ω_a: 5.05; 2.48	Light	6 weeks	Calcification rate increased with increasing light and Ω_a.	[49]
Porites compressa	pCO_2: 186; 336; 641 pH_{SWS}: 8.31; 8.08; 7.82 Ω_a: 5.05; 3.64; 2.25	Light (depth)	30 weeks	Calcification rate decreased as a function of decreasing Ω_a, but also as a function of increasing depth, i.e. lower light level.	[49]
Porites compressa	pCO_2: 757; 3982 pH_{NBS}: 7.96; 7.17 Ω_a: 1.81; 0.28	Nutrients	5 weeks	Calcification rate in the low-pH treatment was half the rate observed in the high-pH treatment. Calcification rates of corals exposed to low pH recovered to normal rates after 2 days in ambient seawater.	[47]
Porites lobata	pCO_2: 130–465; 520–705; 1010–1360 pH_{SWS}: 8.0–8.4; 7.85–7.95; 7.60–7.70 Ω_a: 3.3–7.1; 2.5–3.5; 1.5–2.2	Temperature	8 weeks	Calcification (wt% month^{-1}) and net productivity decreased as a function of decreasing pH under ambient T. Intermediate CO_2 and 3°C warming caused an increase in calcification. High T and CO_2 led to a maximum of 20% bleaching.	[3]
Porites lutea	pH_{NBS}: ~8.6; ~8.4; ~8.2; ~7.9 Ω_a: 7.2; 5.5; 3.6; 1.6		3–6 h	Positive linear correlation between coral calcification rate and Ω_a. Evidence of positive nighttime calcification observed at ~Ω_a > 4.	[59]
Porites lutea	pCO_2: 226; 399; 887 pH_{NBS}: 8.417; 8.207; 7.855 Ω_a: ~5.18; ~3.43; ~1.56			Positive linear correlation between rate of calcification and seawater Ω_a.	[30]
Porites porites	pCO_2: 461; 605 pH_U: 8.1; 8.27 $[CO_3^{2-}]$: 153; 442	Nutrients	32 days	Addition of 2 mM bicarbonate caused an approximate doubling of seawater C_T and A_T, and also a doubling of the calcification rate. Addition of nitrate or ammonium caused a significant reduction in coral growth.	[48]
Porites rus	pCO_2: 395; 720 pH_{NBS}: 8.18; 7.80 Ω_a: 3.8; 2.0	Temperature	14 days	Calcification was reduced by 70% at high CO_2 and low T (27°C), but was unaffected by CO_2 at high T (29°C). Dark adapted maximum quantum yield of photosystem II was unaffected at all treatments.	[56]

Species		Duration		Observations	Reference
Stylophora pistillata	Bicarbonate	8 days	pCO_2: 293–2367; pH_{SWS}: 7.56–8.19; Ω_a: 1.54–9.39	Coral calcification decreased by 0.1 mg $CaCO_3$ g^{-1} day^{-1} per 0.1 unit decrease in pH. The rate of calcification was most strongly correlated with pH, $[CO_3^{2-}]$, and Ω_a. A hyperbolic relationship was observed between calcification and $[CO_3^{2-}]$ (and Ω_a) suggesting a Michaelis–Menten dependence of calcification on $[CO_3^{2-}]$. Photosynthesis was not affected by changes in pH and pCO_2 alone, but significantly increased owing to a 2 mM bicarbonate addition.	[51]
Stylophora pistillata	Temperature	5 weeks	pCO_2: 450–470; 734–798 pH_{SWS}: 8.02–8.04; 7.83–7.86	Net photosynthesis and respiration normalized per unit protein decreased and remained unchanged, respectively, in response to elevated pCO_2. Photosynthesis increased owing to a warming of 3°C. Calcification decreased by 50% under high CO_2, combined with high-T treatment, but was not affected by CO_2 under ambient temperature.	[63]
Stylophora pistillata	Ca^{2+}	20 days	Ω_a: 0.98; 1.95; 2.93; 3.90; 5.85	Nonlinear decrease in calcification rate as a function of decreasing aragonite saturation state. Saturation state was altered by manipulating calcium concentration.	[21]
Tubinaria reniformis		8 days	pCO_2: 407–416; 857–882 pH_{SWS}: 8.06; 7.75–7.76 Ω_a: 4.37–4.43; 2.26–2.33	Statistically significant reduction in calcification rate by 13% under high CO_2.	[50]
Temperate and deep-sea corals					
Astrangia poculata	Nutrients	6 months	CO_2: 390; 780 pH_T: 8.03; 7.78 Ω_a: 3.0; 1.8	Calcification rates of corals exposed to elevated CO_2 were lower than for corals exposed to ambient conditions. However, no significant difference was observed when high CO_2 was combined with elevated nutrient conditions.	[31]
Cladocora caespitosa	Temperature	1 yr	pCO_2: 381–423; 407–475; 693–713; 733–779 pH_T: 8.06–8.10; 8.01–8.07; 7.87–7.88; 7.84–7.86 Ω_a: 3.04–3.77; 3.17–3.86; 1.96–2.68; 2.10–2.78	High-pCO_2 conditions did not have a significant effect on calcification and photosynthetic rates, which seemed to be more dependent on temperature.	[66, 67]
Lophelia pertusa		24 h	pCO_2: 352; 386; 568; 827 pH_{NBS}: 8.1; 8.1; 7.91; 7.76 Ω_a: 2.25; 1.89; 1.40; 1.02	Intermediate- and low-pH conditions resulted in a reduction in calcification by 30 and 56%, respectively. Fast-growing, young polyps were affected more than older, slower-growing polyps (~59% compared with ~40%). Positive net calcification was observed at $\Omega_a < 1$.	[43]
Madracis pharencis		12 months	pH_U: 8.0–8.3; 7.3–7.6	Complete dissolution of skeleton at low pH, but polyps maintained basic life functions skeleton-less. Corals started calcifying and forming colonies when returned to normal seawater conditions.	[19]

(continued)

Table 7.1 Continued

Species	$CO_2(g)$ (ppmv); pCO_2 (µatm); pH; Ω; $[CO_3^{2-}]$ (µmol kg⁻¹)	Other variables/ comments	Duration	Observed effect(s)	Ref.
Oculina arbuscula	CO_2: 409; 606; 903; 2856 Ω_a: 2.5; 2.0; 1.5; 0.7		60 days	Net calcification rate (wt% d⁻¹) increased between Ω_a 0.7 and 1.5 and remained relatively constant between Ω_a 1.5 and 2.5.	[64]
Oculina patagonica	pH_U: 8.0–8.3; 7.3–7.6		12 months	Complete dissolution of skeleton at low pH, but polyps maintained basic life functions skeleton-less. Corals started calcifying and forming colonies when returned to normal seawater conditions.	[19]
Crustaceans					
Amphibalanus amphitrite	pH_{NBS}: 8.2; 7.4		11 weeks	No significant effects were observed on larval condition, cyprid size, cyprid attachment and metamorphosis, juvenile to adult growth, or egg production. Barnacles developed in low-pH conditions showed larger basal diameter and enhanced calcification. Nonetheless, central cell walls of these barnacles required less force to penetrate than those developed in ambient conditions.	[52]
Callinectes sapidus	CO_2: 409; 606; 903; 2856 Ω_a: 2.5; 2.0; 1.5; 0.7		60 days	Linear increase in net calcification rate (wt% d⁻¹) as a function of decreasing Ω_a	[64]
Chthamalus stellatus	pCO_2: 334; 957; 20812 pH_T: 8.14; 7.83; 6.57 Ω_a: 3.91; 2.43; 0.19	Volcanic vent	N/A	Abundance was not significantly different between pH_T 7.83 to 8.14, but was significantly reduced at the lowest pH condition.	[27]
Elminus modestus	CO_2: 412–413; 1076–1075 pH_{NBS}: 7.96–7.98; 7.73 Ω_c: 1.9–2.4; 1.4–1.5	Temperature	30 days	Growth rate was significantly reduced in high-CO_2 and elevated T conditions (~20°C vs. 14°C). No other significant effect was observed in response to CO_2 and/or T. Elevated temperature increased mortality, but there was no significant effect of CO_2 on the survival of this species.	[18]
Homarus americanus	CO_2: 409; 606; 903; 2856 Ω_a: 2.5; 2.0; 1.5; 0.7		60 days	Maximum rate of calcification was observed at lowest Ω_a. The net calcification rate (wt% day⁻¹) decreased between Ω_a 0.7 and 1.5 and remained relatively constant between Ω_a 1.5 and 2.5.	[64]
Homarus gammarus	CO_2: 315; 1200 pH_U: 8.39; 8.10 Ω_a: 4.33; 4.38		28 days	The experimental conditions did not significantly affect carapace length, but there was a reduction in carapace mass during the final stage of larval development with a concurrent reduction in exoskeletal mineral content. Caution is advised in interpreting the results as there was an unexplained elevation of A_T in the high-CO_2 treatment.	[4]
Hyas araneus	CO_2: 380; 710; 3000 pH_U: 8.0; 7.8; 7.3	Temperature	24 h	Thermal sensitivity in terms of Q_{10} values of heart rate increased with increasing CO_2 concentration, suggesting a potential narrowing of the thermal window for these crabs as a result of ocean acidification.	[77]
Semibalanus balanoides	CO_2: 346; 922 pH_{NBS}: 8.07; 7.70		104 days	Survival of adults was reduced by 22% in high-CO_2 conditions compared with control conditions. Embryonic development was significantly slower in high CO_2 and the time to hatch was delayed by 19 days compared with the control conditions.	[17]

Species		Stressor	Duration	Results	Reference
Semibalanus balanoides	CO_2: 409–423; 1109–1132; pH_{NBS}: 8.05–8.07; 7.71–7.73; Ω_c: 2.4–2.9; 1.5	Temperature	30 days	No significant effect of CO_2 and/or T was observed in mean growth rates. Elevated temperature increased mortality, but there was no significant effect of CO_2 on the survival of this species.	[18]

Echinoderms

Species		Stressor	Duration	Results	Reference
Arbacia lixula, Paracentrotus lividus	pCO_2: 334; 957; 20812; pH_T: 8.14; 7.83; 6.57; Ω_a: 3.91; 2.43; 0.19	Volcanic vent	N/A	Abundance was significantly reduced at average intermediate pH levels which reach pH_T minima of 7.4–7.5.	[27]
Arbacia punctulata	CO_2: 409; 606; 903; 2856; Ω_a: 2.5; 2.0; 1.5; 0.7		60 days	Net calcification rate (wt% d^{-1}) increased linearly between Ω_a 0.7 to 1.3, but decreased at higher Ω_a treatments.	[64]
Centrostephanus rodgersii	pCO_2: 324–335; 801–851; 1033–1104; 1695–1828; pH_U: 8.25; 7.9; 7.8; 7.6; Ω_a: 3.20–3.99; 1.64–2.10; 1.33–1.72; 0.87–1.13	Temperature	15 min	Percentage fertilization was not significantly affected by CO_2 or temperature treatments ranging from 18 to 26°C.	[9]
Crossaster papposus	CO_2: 372; 930; pH_U: 8.1; 7.7; Ω_a: 2.0; 1.0		38 days	Lavae and juveniles grew faster under elevated CO_2 compared with ambient conditions with no apparent negative effects.	[16]
Echinometra mathaei	CO_2: ambient; 500; 1000; 2000; 5000; 10000; pH_U: 8.11; 7.80; 7.71; 7.33; 7.12; 6.79		5 months	No significant effect was observed in the percentage of fertilized eggs in the pH_U range 7.33–8.11. At pH_U < 7.2 a significant reduction was observed.	[38]
Echinometra mathaei	pH_U: 7.936–7.945; 7.897–7.899		6 months	In the initial stages of the experiment, growth rates were not significantly different between different pH treatments, but diverged in weeks 12 to 16. For one experimental run, sea urchins started to lose weight. Mortality rates were variable in different experimental runs ranging from increased mortality under low pH to no significant difference between treatment and control conditions.	[72]
Eucidaris tribuloides	CO_2: 409; 606; 903; 2856; Ω_a: 2.5; 2.0; 1.5; 0.7		60 days	Net calcification rate (wt% day^{-1}) increased between Ω_a 0.7 and 1.3 and remained relatively constant between Ω_a 1.3 and 2.5.	[64]
Evechinus chloroticus	pH_{NBS}: 7 treatments of pH between 6.0 and 8.1 (survival expt); pH_{NBS}: 8.1; 7.7 (growth expt)		9–13 days	Survival of larvae ranged from 71 to 88% between pH_{NBS} 6.5 and 8.1. 100% mortality was observed in pH_{NBS} 6.0. Growth and calcification of larvae were reduced by 4.2% and 30.6%, respectively, in pH_{NBS} 7.7 compared with pH_{NBS} 8.1. The morphology of larvae appeared unaffected.	[11]

(continued)

Table 7.1 Continued

Species	$CO_2(g)$ (ppmv); pCO_2 (µatm); pH; Ω; $[CO_3^{2-}]$ (µmol kg^{-1})	Other variables/ comments	Duration	Observed effect(s)	Ref.
Heliocidaris erythrogramma	pCO_2: 324–335; 801–851; 1033–1104; 1695–1828 pH$_{NBS}$: 8.25; 7.9; 7.8; 7.6 Ω_a: 3.20–3.99; 1.64–2.10; 1.33–1.72; 0.87–1.13	Temperature	15 min	Percentage fertilization was not significantly affected by CO_2 or temperature treatments ranging from 18 to 26°C.	[9]
Heliocidaris erythrogramma	pCO_2: 367–373; 1105–1142; 1823–1892 pH$_{NBS}$: 8.17; 7.8; 7.6 Ω_a: 3.5–3.9; 2.4–2.7; 1.6–1.8	Temperature	2 h	Elevated CO_2 and lower pH had no effect on fertilization. Similarly, elevated T did not affect fertilization.	[10]
Heliocidaris erythrogramma	pH$_U$: 8.1; 7.7		3–24 h	Sperm swimming speed and motility were reduced by 11.7% and 16.3%, respectively, in low-pH conditions. Based on these data, model simulations predicted a decrease in fertilization success by 24.9% which was close to an observed decrease of 20.4%.	[29]
Heliocidaris tuberculata	pCO_2: 324–335; 801–851; 1033–1104; 1695–1828 pH$_U$: 8.25; 7.9; 7.8; 7.6 Ω_a: 3.20–3.99; 1.64–2.10; 1.33–1.72; 0.87–1.13	Temperature	15 min	Percentage fertilization was not significantly affected by CO_2 or temperature treatments ranging from 18 to 26°C.	[9]
Hemicentrotus pulcherrimus	pH$_U$: 7.936–7.945; 7.897–7.899		6 months	In the initial stages of the experiment, growth rates were not significantly different between different pH treatments, but diverged in weeks 12 to 16. Mortality rates were variable in different experimental runs ranging from increased mortality under low pH to no significant difference between treatment and control conditions.	[72]
Hemicentrotus pulcherrimus	CO_2: ambient; 500; 1000; 2000; 5000; 10 000 pH$_U$: 8.11; 7.80; 7.71; 7.33; 7.12; 6.79		3 months	Overall, fertilization rate, cleavage rate, developmental speed, and pluteus larval size decreased in the highest-CO_2 treatments. However, no significant effect was observed in the percentage fertilized eggs in the pH$_U$ range 7.0 to 8.11. At pH$_U$ < 7.0 a significant reduction was observed. Pluteus larvae size was significantly different in all pH treatments compared with the control.	[38]
Holothuria spp.	pCO_2: 400–475; 775–1005; 930–1260; 905–1660; 2115–3585; 12600–21100 pH$_T$: 8.03; 7.77; 7.69; 7.64; 7.31; 6.55 Ω_a: 3.2–3.5; 1.9–2.3; 1.6–2.0; 1.2–2.0; 0.6–1.0; 0.1–0.2		N/A	Sperm flagellar motility decreased significantly in response to decreasing pH. 73% of sperm were motile at pH$_T$ 8.0, 72% at pH$_T$ 7.8, and fewer than 30% at pH$_T$ < 7.7.	[55]

Species	Parameters	Additional stressor	Duration	Observations	Ref.
Ophiothrix fragilis	pH_U: 8.1; 7.9; 7.7		8 days	Intermediate pH resulted in 100% larval mortality after 8 days compared with 30% mortality in control pH. Low pH also resulted in a temporal decrease in larval size and abnormal development and skeletogenesis.	[15]
Ophiura ophiura	pCO_2: 553–594; 1282–1400; 2275–2546; pH_U: 7.95–7.99; 7.62–7.66; 7.38–7.42; Ω_a: 1.35–1.36; 0.63–0.77; 0.38–0.44	Temperature	40 days	No difference in net calcification rates was observed between treatments, but a 30% reduction in the rate of arm regeneration was observed in the low-pH treatment.	[79]
Patiriella regularis	pCO_2: 324–335; 801–851; 1033–1104; 1695–1828; pH_U: 8.25; 7.9; 7.8; 7.6; Ω_a: 3.20–3.99; 1.64–2.10; 1.33–1.72; 0.87–1.13	Temperature	15 min	Percentage fertilization was not significantly affected by CO_2 or temperature treatments ranging from 18 to 26°C.	[9]
Pisaster ochraceus	pCO_2: 380; 780	Temperature	70 days	Growth and feeding rates increased with increasing T. Growth rate also increased with increasing CO_2, while the relative calcified mass of the total wet mass decreased from a mean of 11.5% at ambient CO_2 to 10.9% at high CO_2 conditions.	[23]
Pseudechinus huttoni	pH_{NBS}: 7 treatments of pH between 6.0–8.1 (survival exp); pH_{NBS}: 8.1; 7.7 (growth expt)		9–13 days	Survival of larvae ranged from 76 to 104% between pH_{NBS} 7.0 and 8.1. 100% mortality was observed in pH_{NBS} 6.0 and 34% survival was observed in pH_{NBS} 6.5. Growth and calcification of larvae were reduced by 18.3% and 36.9%, respectively, in pH_{NBS} 7.7 compared with pH_{NBS} 8.1. The morphology of larvae appeared unaffected.	[11]
Sterechinus neumayeri	pH_{NBS}: 7 treatments of pH between 6.0 and 8.0 (survival expt); pH_{NBS}: 8.0; 7.6 (growth expt)		7–17 days	Survival of larvae ranged from 89 to 110% between pH_{NBS} 6.5 and 8.1. 100% mortality was observed in pH_{NBS} 6.0. Growth and calcification of larvae were reduced by 1.8% and 3.9%, respectively, in pH_{NBS} 7.6 compared with pH_{NBS} 8.1. The morphology of larvae appeared unaffected.	[11]
Strongylocentrotus franciscanus	pCO_2: 380; 540; 970; pH_U: 8.04; 7.98; 7.87	Temperature	96 h	Sea urchin larvae subjected to 1-h acute heat stress showed significantly reduced ability to activate a gene for heat-stress inducible heat shock protein under elevated CO_2 conditions.	[58]
Strongylocentrotus purpuratus	pCO_2: 380; 540; 1020; pH_U: 8.01; 7.96; 7.88		<72 h	Sea urchin larvae exposed to elevated CO_2 showed significant decreases in gene expression associated with major cellular processes including biomineralization, cellular stress response, metabolism, and apoptosis.	[75]

(continued)

Table 7.1 Continued

Species	CO_2(g) (ppmv); pCO_2 (µatm); pH; Ω; $[CO_3^{2-}]$ (µmol kg^{-1})	Other variables/ comments	Duration	Observed effect(s)	Ref.
Tripneustes gratilla	pCO_2: 324–335; 801–851; 1033–1104; 1695–1828 pH$_U$: 8.25; 7.9; 7.8; 7.6 Ω_a: 3.20–3.99; 1.64–2.10; 1.33–1.72; 0.87–1.13		15 min	Percentage fertilization was not significantly affected by CO_2 or temperature treatments ranging from 18 to 26°C.	[9]
Tripneustes gratilla	pH$_{NBS}$: 7 treatments of pH between 6.0–8.2 (survival expt) pH$_{NBS}$: 8.2; 7.8 (growth expt)		4 days	Survival of larvae ranged from 52 to 78% between pH$_{NBS}$ 7.0 and 8.2. 98% mortality was observed in pH$_{NBS}$ 6.0 and 11% survival was observed in pH$_{NBS}$ 6.5. Growth and calcification of larvae were reduced by 3.2% and 13.8%, respectively, in pH$_{NBS}$ 7.8 compared with pH$_{NBS}$ 8.2. The morphology of larvae appeared unaffected.	[11]
Foraminifera					
Elphidium clavatum (exclavatum)	pH$_{NBS}$: 8.1; 7.5 Ω_a: 1.0; 0.4		70 days	Mortality rates of live foraminifera were three times greater in treatment than in control conditions. Visual evidence of dissolution of discarded tests was clearly evident after 14 days in the low-pH treatment.	[24]
Marginopora kudakajimensis	pH$_{NBS}$: 8.3; 8.2; 7.9; 7.7		71 days	Growth rate, shell weight, and the number of chambers added generally decreased with lower pH. However, growth rates at pH$_{NBS}$ 8.2 and 7.9 were not significantly different and a significant decrease was not seen until pH$_{NBS}$ 7.7.	[41]
Seagrasses					
Cymodocea nodosa	pH$_U$: 8.0–9.0		1 h	Significant decrease in photosynthetic rates with increasing pH observed in both field and laboratory experiments.	[32]
Posidonia oceanica	pH$_U$: 8.0–9.0		1 h	Significant decrease in photosynthetic rates with increasing pH observed in both field and laboratory experiments.	[32]
Posidonia oceanica	pCO_2: 309; 304; 542; 1824 pH$_T$: 8.17; 8.17; 8.00; 7.60 Ω_a: 4.10; 4.07; 3.13; 1.71	Volcanic vent	N/A	No significant difference in photosynthetic rate was observed as a function of pH, but seagrass production and shoot density were significantly higher at pH$_T$ 7.60. Leaves at pH$_T$ 8.17 had 75% cover of calcified epiphytes compared with 2% at pH$_T$ 7.60.	[27]
Zostera marina	pH$_U$: 8.1; 7.75; 7.5; 6.4 $[CO_3^{2-}]$: 204; 108; 55; 10	Light	1 yr	High-CO_2 conditions did not affect biomass-specific growth rates, leaf size, or leaf sugar content of above-ground shoots, but resulted in higher reproductive output, below-ground biomass and vegetative proliferation of new shoots.	[60]
Zostera noltii	pH$_U$: 8.0–9.0		1 h	In general, photosynthetic rates were not affected by pH, although one incubation at pH$_U$ 9.0 showed a significant decrease.	[32]

as a result of decomposition of organic material is also evident in sediment porewaters, which in the top few millimetres to several decimetres of the sediments typically have much higher C_T and lower pH (Fig. 7.1), and, with increasing depth, generally lower SO_4^{2-} and higher sulphide concentrations compared with the overlying bottom waters. These trends are the direct result of microbial activity in the sediments, but also the activity of large deposit feeders that directly consume organic material and indirectly affect microbial populations and geo-chemical distributions of dissolved porewater chemical species by mixing the sediments (see Chapter 9).

Decomposition of organic material can be a prominent process in the benthic boundary layer and at shallow depth in sediments, since all particulate organic material sedimented from the sunlit surface and not decomposed during transit to the bottom is either decomposed or buried here. In the coastal ocean, this accumulation of organic matter may be equivalent to most of the particulate organic material produced in the surface water or deposited via rivers and terrestrial run-off. In the open ocean, only a few per cent of the particulate organic material produced in the surface makes it to the benthos. For the oceans as a whole, less than 1% of marine net carbon production is preserved in sediment accumulations (Mackenzie and Lerman 2006).

As a result of rising atmospheric CO_2 and the enhanced greenhouse effect, air and seawater temperatures are increasing. Warmer seawater temperatures may lead to increased metabolic rates, resulting in higher rates of respiration and decomposition of organic material, and consequently a greater flux of CO_2 from this process. Furthermore, if increasing CO_2 is fertilizing net carbon production in surface seawater (e.g. Riebesell *et al.* 2007), the amount of organic material deposited in both shallow and deep-sea sediments, and subsequent

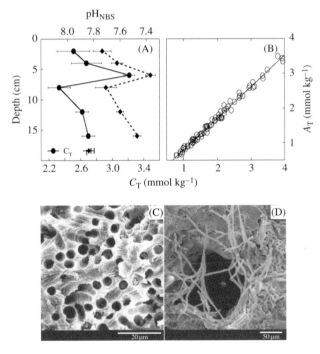

Figure 7.1   (A) Changes in daily average porewater pH_{NBS} and total dissolved inorganic carbon (C_T) as a function of sediment depth in shallow-water (<1 m) carbonate sediments in Bermuda. Data were collected every 2 h at each depth for a complete diel cycle. Error bars represent one standard deviation. (B) Changes in total alkalinity (A_T) as a function of C_T in shallow-water carbonate sediments in Mangrove Bay, Bermuda. The slope of a best fit line is close to 1, illustrating that the observed changes are due to metabolic dissolution of carbonate sediments. (C), (D) Visual evidence of microbioerosion in dead coral substrate owing to microendoliths (from Tribollet 2008, with kind permission of Springer Science + Business Media).

remineralization of this material and production of CO_2, could increase. On the other hand, if the production and sinking of biogenic $CaCO_3$ particles from the surface ocean decreased, less organic material may make it down to the bottom of the deep sea because $CaCO_3$ acts as important ballast (Barker *et al.* 2003; see Chapters 6 and 12). In recent years, declining oxygen concentrations have been detected in large regions of the oceans, and it has been proposed that the oxygen minimum zones—or 'dead zones', void of aerobic life—of the oceans could increase in response to global warming (O_2 becomes less soluble) as well as increased export and breakdown of organic material (e.g. Oschlies *et al.* 2008; Stramma *et al.* 2008; Brewer and Peltzer 2009). Oxygen starvation affects decomposition and recycling of organic material in benthic environments and has drastic consequences for benthic ecosystems (Stramma *et al.* 2008; Brewer and Peltzer 2009). Coastal ocean environments, particularly those with seasonal thermoclines, are particularly vulnerable to warming and the development of regions of hypoxia and anoxia.

7.2.3 Calcification

Many benthic organisms deposit skeletal hard parts made of $CaCO_3$. One hypothesis suggests that the calcification process in marine organisms originally evolved under conditions of high calcium concentration in the ocean as a detoxification mechanism (Brennan *et al.* 2004), but the explanation may be much more complex than this. Regardless, calcareous hard parts provide myriad advantages to marine benthic calcifiers, including protection from predators, a refuge for intertidal organisms to avoid desiccation when exposed to air during low tide, structural support, increased surface area, a mechanism to maintain elevation above the sediment–water interface, a way of maintaining close proximity to high light levels, and as a means of keeping up with sea-level rise. There are three commonly occurring carbonate mineral phases deposited by benthic marine calcifiers, namely aragonite, calcite, and magnesian calcite (Mg-calcite). Aragonite and calcite have the same chemical composition but a different mineral structure (orthorhombic versus rhombohedral), whereas calcite and Mg-calcite have the same basic

mineral structure but calcium ions have been randomly replaced by magnesium ions in the latter (up to ~30 mol%; Morse and Mackenzie 1990). These differences result in somewhat different chemical and physical properties. For example, Mg-calcite with a significant mol% magnesium in calcite is more soluble than aragonite, which is more soluble than calcite. It is not known in detail why some organisms and not others favour a certain mineral phase, but it is most certainly linked to the mechanism and control of the calcification process, which is different in different organisms. Some organisms deposit different mineralogies at different life stages and some even have multiple mineralogies in different parts of their calcareous hard parts. Evidence exists from palaeo-oceanographic records and controlled laboratory experiments that the mineralogy being deposited may change as a function of temperature and seawater chemical composition including changes in the Mg-to-Ca ratio and the distribution of inorganic carbon species (e.g. Mackenzie *et al.* 1983; Agegian 1985; Ries 2010; Stanley *et al.* 2010).

Based on thermodynamic and kinetic principles, as seawater carbonate ion concentration ($[CO_3^{2-}]$) and carbonate mineral saturation state (Ω) decrease as a result of ocean acidification (see Chapter 1), one would expect that the rate of calcification of benthic marine calcifiers as well as other calcifiers would decrease. Indeed the majority of studies show a consistent decline in the rate of benthic calcification as a result of increasing CO_2 and ocean acidification (Fig. 7.2 and Table 7.1; e.g. Marubini *et al.* 2003; Langdon and Atkinson 2005; Schneider and Erez 2006; Gazeau *et al.* 2007; Anthony *et al.* 2008; Jokiel *et al.* 2008), although a few recent studies show no response or an increase in calcification in a range of different benthic calcifiers exposed to moderately elevated CO_2 conditions (Ries *et al.* 2009; Rodolfo-Metalpa *et al.* 2010). However, one has to be cautious in interpreting these results. Regardless of whether calcification in marine organisms has been observed to increase or decrease in response to elevated CO_2 and lower Ω, deposition of $CaCO_3$ is thermodynamically less favourable under such conditions. Wood *et al.* (2008) proposed that some organisms may be able to up-regulate their metabolism and calcification to compensate for increased acidity of seawater.

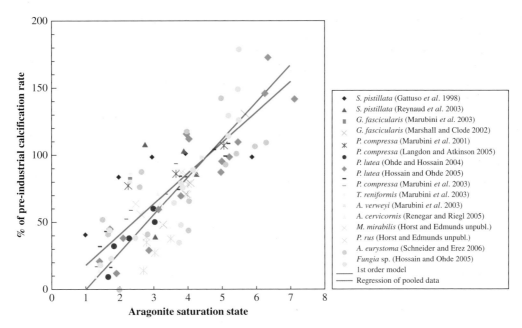

Figure 7.2   Relative rate of calcification for a number of coral species as a function of seawater aragonite saturation state (Ω_a; from C. Langdon, pers. comm.). References can be found in Appendix 7.1 available online at http://ukcatalogue.oup.com/product/9780199591091.do

Nonetheless, this comes at a substantial energetic cost, energy that would otherwise have been spent on other essential processes such as protein synthesis, growth, and/or reproduction, and this up-regulation is probably not sustainable in the long run. However, if organisms have sufficient energy resources in terms of food, nutrients, and light (for those organisms dependent on photosynthesis), they may be able to compensate for the additional energy demand required to calcify under conditions of elevated CO_2. Overall, despite different and even opposing responses between different marine calcifiers and different responses between the same species in different experiments, the majority of experimental results to date strongly suggest that ocean acidification will result in a slowdown in calcification by benthic organisms. Observations of linear extension and/or calcification rates in coral colonies from the Atlantic, Pacific, and Indian Oceans, including *Porites* colonies on the Great Barrier Reef (Cooper *et al.* 2008; De'ath *et al.* 2009) and in Phuket, Thailand (Tanzil *et al.* 2009), *Pocillopora* colonies in Panama (Manzello 2010), and *Diploria* colonies in Bermuda

(A. Cohen, pers. comm.), show decreasing trends in their rates for the past several decades. However, it is not possible to conclude unequivocally that this is a result of ocean acidification. A recent study in the central Red Sea demonstrated a 30% reduction in the growth of *Diploastrea heliopora* since 1998, which was directly related to rising sea-surface temperature (Cantin *et al.* 2010).

A small fraction of benthic $CaCO_3$ production is chemically precipitated $CaCO_3$, often as a direct consequence of biological processes changing the seawater chemistry and resulting in precipitation, or dissolution and re-precipitation of a carbonate mineral phase. Such carbonate cements serve an important role in consolidating reef structures. Ocean acidification could result in slower rates of abiotic carbonate precipitation, changes in the average mineral composition of the cements, and weaker carbonate reefs and structures (Andersson *et al.* 2005). Manzello *et al.* (2008) observed that cements in intraskeletal pores were almost absent on coral reefs in the eastern tropical Pacific. These reefs experience high CO_2 and low Ω as a result of upwelling and are, in general, poorly developed and subject to high rates of bioerosion. The abundance of cement appeared correlated to the seawater aragonite saturation state and inversely related to measured rates of bioerosion.

A slowdown in benthic calcification would result in less CO_2 being released to the atmosphere from this process, and thus acts as a negative feedback to rising atmospheric CO_2 on a decadal to centennial timescale (e.g. Mackenzie and Lerman 2006; see Chapter 12). If all marine calcification stopped, the amount of CO_2 that would otherwise have been released to the atmosphere from this process corresponds to about 4 to 6% of the current total anthropogenic CO_2 emissions of 9.3 Gt C yr^{-1} (Le Quéré et al. 2009) assuming a global $CaCO_3$ production of 0.64 to 1 Gt C yr^{-1} (Milliman 1993; Milliman and Droxler 1996) and a release of 0.6 mol of CO_2 to the atmosphere for every mole of $CaCO_3$ precipitated (Frankignoulle et al. 1994). Note, though, that this is a rough approximation. In reality the proportion of CO_2 released to the atmosphere from calcification will increase with increasing seawater acidification. To some extent this will compensate for the reduction in CO_2 production caused by reduced calcification. Benthic calcification probably does not correspond to more than 6 to 10% of the annual global total calcification (Milliman and Droxler 1996). Nonetheless, the majority of this calcification occurs within the global coastal area, which only makes up approximately 7% of the global ocean area. In addition, as much as 60% of the $CaCO_3$ produced within this area may actually accumulate as carbonate structures or sediments (Milliman and Droxler 1996; Mackenzie et al. 2005).

7.2.4 Calcium carbonate dissolution

Dissolution of $CaCO_3$ minerals is the reverse process of calcification, and results in the chemical disintegration of the solid mineral phase into its individual components of calcium and carbonate ions. From a thermodynamic perspective, dissolution is expected if $\Omega < 1$. In simple terms, the farther away from equilibrium, the faster the rate of mineral dissolution (ignoring diffusional limitations). Surface seawater is typically supersaturated with respect to calcite and aragonite globally but

could become undersaturated with respect to aragonite in high-latitude regions within a few decades as a result of ocean acidification (Orr et al. 2005; Steinacher et al. 2009; Chapter 3). Since Mg-calcite minerals with a magnesium content greater than 8 to 12 mol% $MgCO_3$ are more soluble than aragonite, seawater will become undersaturated with respect to these mineral phases before aragonite. Thus, Mg-calcite minerals are the first responders to ocean acidification and declining carbonate saturation states (Morse et al. 2006; Andersson et al. 2008). The exact magnesium content of the Mg-calcite phase with the same solubility as aragonite is somewhat uncertain and dependent on the experimental solubility curve adopted, which is currently poorly constrained (e.g. Plummer and Mackenzie 1974; Walter and Morse 1984; Bischoff et al. 1993; Morse et al. 2006; Andersson et al. 2008).

As a result of increasing pressure, decreasing temperature, and natural acidification of seawater from decomposition of organic material, the saturation state with respect to carbonate minerals decreases as a function of depth. The majority of the benthic environment of the open ocean is immersed in waters undersaturated with respect to all commonly occurring carbonate phases. Because of the difference in age of water masses between the Atlantic and the Pacific Oceans, and thus the amount of dissolved inorganic carbon that has accumulated in these water masses, the saturation horizons with respect to carbonate minerals in the Pacific are located at much shallower depths than in the Atlantic (Morse and Mackenzie 1990; Chapter 3). These differences may be responsible for ecological and mineralogical differences in the benthic environment between the two ocean basins (see Section 7.3.4).

Although most surface seawaters are currently supersaturated with respect to the majority of carbonate mineral phases, carbonate dissolution is an ongoing process in all environments as a result of microbial metabolic activity causing corrosive conditions in sediment porewaters and microenvironments, defined as small specific areas isolated from their immediate surroundings. There are also many macro- and microorganisms (e.g. endolithic autotrophic or heterotrophic organisms) that actively chew, rasp, break, or penetrate carbonate substrates by releasing CO_2 or other acids (Fig. 7.1; e.g. Alexandersson 1975; Tribollet 2008). This process is referred to as bioerosion. The absolute extent and rate of $CaCO_3$ dissolution from any given shallow environment is poorly quantified owing to the fact that most studies are only able to characterize the net effect of calcification minus dissolution (Langdon *et al.* 2010). The relative magnitude and importance of metabolic dissolution and bioerosion are not well known. Nonetheless, assuming that the biological processes generating bioerosion and metabolic dissolution remain unaffected by ocean acidification and climate change, carbonate dissolution as a result of these processes is likely to increase because the initial seawater carbonate saturation state will become progressively lower as a result of ocean acidification. Recent experimental results indicate that bioerosion may indeed become increasingly efficient in breaking down carbonate material under higher CO_2 conditions (Tribollet *et al.* 2009).

It is highly likely that carbonate dissolution will increase due to ocean acidification, which is important because this consumes CO_2 and acts as a sink of anthropogenic CO_2, albeit a small one on the decadal to centennial timescale. It also increases total alkalinity, which increases the ability of seawater to absorb CO_2. Hypothetically, if dissolution of shallow-water carbonate minerals could keep up with the oceanic uptake of anthropogenic CO_2, this process could act as a buffer and prevent major changes in surface-seawater pH and carbonate saturation state resulting from this process. However, it has been demonstrated that the rate of dissolution is too slow relative to the rate of uptake of anthropogenic CO_2 and the time seawater resides in shallow regions in contact with carbonate minerals to produce a significant buffer effect on timescales of decades to centuries (Andersson *et al.* 2003, 2005; Morse *et al.* 2006). In addition, the size of the reactive coastal ocean carbonate reservoir is too small to enhance the buffer capacity substantially (Morse *et al.* 2006). On longer timescales of several thousands of years, dissolution of carbonate sediments, particularly in the deep sea, will be the ultimate sink of anthropogenic CO_2 (Archer *et al.* 1998; see Chapter 2).

7.3 Effect of ocean acidification on benthic organisms, communities, and ecosystems

7.3.1 Effect on major benthic organisms

Until the late 1990s, only a few studies had been conducted to investigate the response of benthic organisms such as corals and algae to seawater CO_2 conditions anticipated as a result of anthropogenic ocean acidification (e.g. Smith and Roth 1979; Agegian 1985; Gao *et al.* 1993). For the past decade, and concurrent with the rising awareness and concern about the problem of ocean acidification, numerous studies and experiments have been conducted with a range of different taxa of benthic organisms (see Chapter 1). Nonetheless, although some observed trends appear relatively consistent for some organisms, such as the dependence of coral calcification rates on seawater Ω (Fig. 7.2), there are still inconsistencies and substantial variations between results, and there are many important groups of organisms for which we have a poor understanding or a lack of data on how they might respond to rising CO_2. Table 7.1 summarizes some of the major results reported to date for a range of marine benthic organisms including algae, bivalves, corals, crustaceans, echinoderms, foraminifera, and seagrasses exposed to elevated CO_2 conditions under different experimental settings and durations. The focus is mainly on results from studies conducted under pCO_2 and pH conditions anticipated as a result of present and future anthropogenic ocean acidification, although some of these studies have extended their observations and treatments well beyond these conditions. Nonetheless, Table 7.1 is not an exhaustive list of studies and contains a subset of the published results available. For additional discussion on the effects of ocean acidification on benthic organisms and physiological effects in general see also Chapters 9 and 8, respectively.

7.3.2 Effect on shallow benthic communities

A limited number of ocean acidification studies have been conducted at the community scale. The majority of these investigations have been conducted on subtropical or tropical calcifying communities in the natural environment (e.g. Yates and

Halley 2006; Silverman *et al.* 2007; Bates *et al.* 2010) or in experimental mesocosms (e.g. Leclercq *et al.* 2000, 2002; Langdon and Atkinson 2005; Jokiel *et al.* 2008; Andersson *et al.* 2009) as well as in the Biosphere 2 facility (Langdon *et al.* 2000, 2003). In general, studies on benthic communities have shown that there is a strong positive coupling between seawater carbonate saturation state and the net community calcification rate (*G*). That is, under elevated CO_2 and lower pH conditions, the daily net community calcification rate and deposition of $CaCO_3$ were significantly lower relative to the rate and deposition at ambient conditions in these experiments (Fig. 7.3). Based on observed relationships between net community calcification and seawater aragonite saturation state, a reduction in average Ω_a from 3.5 to 2.5 could result in a decline in net community calcification of about 15 to 130% (–51% to –73%, Langdon *et al.* 2000; –18%, Leclercq

et al. 2000; –15%, Leclercq *et al.* 2002; –97 to –102%, Yates and Halley 2006; –133%, Silverman *et al.* 2007; –50 to –64%, Andersson *et al.* 2009).

In some experiments, negative net community calcification has been observed (i.e. carbonate dissolution exceeded gross calcification) at times of very low carbonate saturation state occurring predominantly during the night in the absence of photosynthesis (Leclercq *et al.* 2000; Langdon *et al.* 2003; Yates and Halley 2006; Andersson *et al.* 2009). However, the chemical threshold in terms of the seawater pCO_2 and carbonate saturation state at which different systems become subject to net dissolution varies significantly and is a function of a range of properties such as community composition, amount of reactive organic material, and sediment mineral composition. Nevertheless, as the surface-seawater carbonate saturation state continues to decrease as a result of anthropogenic

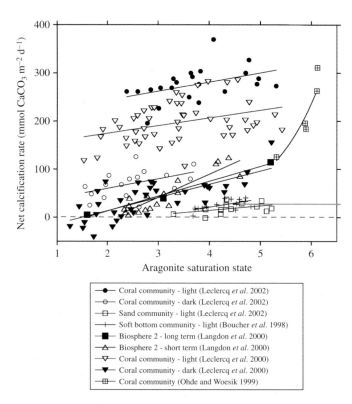

Figure 7.3 Net community calcification as a function of seawater aragonite saturation state in a range of different experimental settings and environments. Positive rates imply net calcification and net accretion of $CaCO_3$ whereas negative rates imply net dissolution and net loss of $CaCO_3$ (from Andersson *et al.* 2005, reprinted by permission of the *American Journal of Science*).

emissions of CO_2, calcification will decrease and carbonate dissolution increase, potentially pushing calcifying systems past their critical thresholds and towards a condition of net dissolution. Andersson et al. (2009) observed marginal daily net dissolution (–0.1 mmol $CaCO_3$ m^{-2} d^{-1}) in replicated subtropical coral reef communities incubated in continuous-flow mesocosms exposed to a daily average pCO_2 of ~1147 µatm and Ω_a of 1.4. However, despite the low seawater carbonate saturation state, individual corals (*Montipora capitata*) appeared healthy and deposited significant amounts of $CaCO_3$, albeit at lower rates by a few per cent compared with corals incubated under ambient conditions (pCO_2 = 568 µatm; Ω_a = 2.8). Other marine calcifiers were not as successful in dealing with the high CO_2 conditions and were very negatively affected, including crustose coralline algae for which recruitment was reduced by 85% (Kuffner et al. 2008).

If benthic gross community photosynthesis were to increase in response to increasing CO_2 and respiration remained unchanged, net community production (NCP) of benthic communities would increase. In contrast, if respiration were to increase in proportion to the increase in photosynthesis, NCP would remain unchanged, but the turnover of carbon would increase. Both Leclercq et al. (2002) and Langdon et al. (2003) reported that NCP did not change significantly in their experiments in response to increasing CO_2. However, Langdon et al. (2003) observed a significant increase in gross primary production that was accompanied by a similar increase in light respiration in response to higher CO_2 conditions (dark respiration remained unchanged), meaning that the rate of cycling of carbon through the organic carbon pool increased under these conditions. Looking strictly at the response of an assemblage of corals, Langdon and Atkinson (2005) observed a significant increase in net community production at a rate of 3 ± 2% per µmol CO_2 kg^{-1} based on measurements of dissolved inorganic carbon (C_T), but did not observe a significant increase based on changes in dissolved oxygen concentrations. Results from other experiments conducted on individual corals and other benthic primary producers show variable responses ranging from no effect (e.g. Schneider and Erez 2006) to positive and even negative effects on some factors relevant to the pro-

duction of organic carbon (e.g. zooxanthellae density, photosynthetic rate; Borowitzka and Larkum 1976; Reynaud et al. 2003; Crawley et al. 2009) in response to increasing CO_2 (Table 7.1). Note that the community-scale experiments discussed here fail to capture any changes in NCP that could arise from changes in the community structure in nature over longer time periods as a result of ocean acidification and from changes in other variables such as temperature.

7.3.3   Effect on benthic ecosystems

Although ocean acidification experiments conducted with individual benthic organisms and communities provide critical information regarding the effects on their function and role under these conditions, it is challenging to extrapolate and make strong conclusions as to what this means from an ecological perspective. In the context of ecological changes that may result from ocean acidification, it is important to realize that there will be both organisms that benefit from higher CO_2 and organisms that are negatively affected by higher CO_2. It is also important to consider the potential effects on all life stages of organisms. Many organisms may be more vulnerable during early developmental and reproductive stages, and many benthic organisms have planktonic larval stages, which may make them more vulnerable (Kurihara 2008; Findlay et al. 2009). Furthermore, it is important to realize that the structure and diversity of benthic communities are strongly controlled by interactions between species (see Chapter 10). Thus, benthic organisms may not only be affected by ocean acidification directly, but also by the consequences of this process for other organisms including competitors, prey, predators, and the quantity and quality of food supply. For individual species, this effect could be either positive or negative. As a result, ecosystems are not expected to suddenly crash as a result of ocean acidification, but rather to undergo successive changes as organisms are affected differently (directly or indirectly) by the increasing concentration of CO_2 and decreasing pH. From an ecological perspective, it also becomes important whether or not organisms are able to acclimatize in the short term or adapt over generations to the predicted changes in

CO_2 chemistry. Naturally, it is very difficult to conduct rigorous ecological experiments to address these questions as these changes occur over relatively long periods of time. Hence, in addressing the question about the effects of ocean acidification on benthic ecosystems, it is useful to look at environments that naturally experience high-CO_2 conditions at the present time or did so in the past. One way is to compare the benthic environments of the Atlantic and Pacific oceans, which experience significantly different CO_2 chemistry as a function of depth (Section 7.3.4). There are also local or regional shallow environments exposed to high-CO_2 and low-pH conditions resulting from volcanic vents (Hall-Spencer *et al.* 2008; Martin *et al.* 2008; Rodolfo-Metalpa *et al.* 2010), seasonal stratification (Andersson *et al.* 2007), or upwelling of deep corrosive water on to the continental shelf (Feely *et al.* 2008; Manzello *et al.* 2008; Section 7.3.5) that may provide important clues in terms of the effects of ocean acidification on benthic ecosystems.

7.3.4 Deep-ocean environments

As illustrated in Chapter 3, seawater of the Pacific Ocean becomes increasingly less alkaline (decreasing pH) and increasingly corrosive with respect to carbonate minerals at much shallower depths than seawater in the Atlantic Ocean. This raises the question of whether there are differences between the benthic ecosystems of the Atlantic and Pacific oceans that reflect this observed difference in seawater acidity and may provide clues in terms of the effects of future ocean acidification on benthic ecosystems?

It is well known that the aragonite and the calcite compensation depths (ACD, CCD), i.e. the depths at which the rates of sedimentation of these phases equal their rates of dissolution and thus where aragonite and calcite generally are no longer found in the sediments, are located much deeper in the Atlantic than in the Pacific (e.g. Morse and Mackenzie 1990). The compensation depths are located below the saturation horizons ($\Omega = 1$), which are thermodynamic boundaries (see Chapter 1), and below the lysoclines (a kinetic boundary) where these phases initially begin to undergo significant dissolution in the water column. Nevertheless, the observed differences in compensation depths between the ocean basins are mostly a result of different seawater chemistry and the greater extent of carbonate dissolution in the Pacific than in the Atlantic (Morse and Mackenzie 1990). Thus, as the saturation horizons continue to shoal throughout the global ocean as a result of anthropogenic CO_2 emissions (Feely *et al.* 2004; see Chapter 3), carbonate sediment dissolution will increase and the CCD and ACD are likely to shoal assuming that the influx of carbonate particles does not increase. However, this process is slow and there is likely to be a substantial time lag relative to the shoaling of the saturation horizon before significant shoaling of the compensation depths can be detected in most places (Chapter 2). The rate of shoaling of the saturation horizon is also different between different regions, depending on the uptake and subduction of anthropogenic CO_2 (Feely *et al.* 2004). As a result of the formation of North Atlantic Deep Water (NADW) and the associated uptake and subduction of CO_2 from the atmosphere (solubility pump) in this region, anthropogenic CO_2 can be detected at depths exceeding 2500 m in the North Atlantic Ocean compared with ~1500 m in the South Atlantic (Feely *et al.* 2004). Thus, this region may serve as an important indicator of the effect of ocean acidification on deep-sea benthic ecosystems. For example, in the waters surrounding Iceland, Olafsson *et al.* (2009) reported that the aragonite saturation horizon (ASH) between 1985 and 2008 shoaled by 4 m yr^{-1}, immersing 800 km^2 of seafloor and associated benthic communities in seawater undersaturated with respect to aragonite every year. No data currently exist on how this might have affected the communities or biogenic carbonate material present in the sediments.

Many benthic marine calcifiers found in shallow seas are also found in the deep sea, including echinoderms, molluscs, crustaceans, foraminifera, and scleractinian corals. Deep-sea coral ecosystems, sometimes improperly called 'deep-sea coral reefs', are found in shallow to intermediate-depth waters (50–1000 m) at high latitudes and at greater depths at low latitudes (Roberts *et al.* 2006). They are probably the most three-dimensionally complex habitat in the deep sea, providing numerous ecological niches and hosting a high biodiversity (Clark *et al.*

2010). Their geographical distribution seems to be constrained in part by the ASH. The present-day ASH ranges from 50 to 600 m in the North Pacific where corals are few, mostly calcitic, and do not build bioherms (Guinotte *et al.* 2006; Tittensor *et al.* 2009). In contrast, the ASH is much deeper (> 2000 m) in the North Atlantic, which harbours a much larger coral diversity, including the aragonitic scleractinian coral *Lophelia pertusa* that builds spectacular bioherms tens of metres high and kilometres long off the Atlantic coasts of northern Europe. The shoaling of the ASH implies that deep-sea coral ecosystems could soon become immersed in seawater undersaturated with respect to aragonite. Guinotte *et al.* (2006) showed that more than 95% of the deep-sea, bioherm-forming corals were located in areas that were supersaturated with respect to aragonite in 1765 and that only about 30% of coral locations will remain in supersaturated waters in 2099 (Fig. 7.4). Even though *L. pertusa* seems to be able to calcify in moderately undersaturated seawater (Maier *et al.* 2009), it is likely that undersaturation and increased carbonate dissolution will weaken bioherms, decrease their structural complexity, and greatly reduce their biodiversity. In response to the shoaling of the ASH, it is likely that the maximum depth where these corals and other deep-sea calcifiers are found will transition to shallower depths. It is not possible to conclude unequivocally that the observed difference of deep-sea corals and

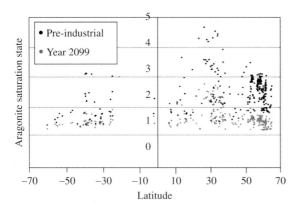

Figure 7.4 Model-projected depth of the aragonite saturation horizon for globally distributed deep-sea coral locations in pre-industrial time and in the year 2099 (from Guinotte *et al.* 2006, reprinted by permission of the Ecological Society of America).

bioherms in the Atlantic and Pacific Oceans are strictly due to the differences in seawater CO_2 chemistry, but based on theory, it is reasonable to assume that the chemistry is a key factor.

7.3.5   Shallow-ocean environments

Organisms and ecosystems in the surface ocean are the first to experience the direct effects of anthropogenic ocean acidification as there is a time lag before the anthropogenic CO_2 reaches the deep ocean. Nonetheless, anthropogenic CO_2 has already penetrated to depths exceeding 1000 m in all major ocean basins causing shoaling of carbonate saturation horizons (Feely *et al.* 2004; see Chapter 3). Even shallow regions of the oceans that are currently supersaturated with respect to atmospheric pCO_2 and act as sources of CO_2 to the atmosphere due to net heterotrophy and/or calcification (e.g. many estuaries and coral reefs frequently experience pCO_2 well above the pCO_2 expected from equilibrium with the atmosphere) will experience acidification as the pH of the open-ocean source waters continuously decreases.

In considering the effects of ocean acidification on shallow benthic ecosystems, it is important to bear in mind that a whole new pattern of benthic communities emerged as sea level rose from the Last Glacial Maximum (LGM) 18 000 years ago to late pre-industrial time. It is during this time that extensive carbonate reef, bank, platform, and shelf ecosystems were established amounting to at least 4500 Gt of $CaCO_3$ accumulation (Milliman 1993; Vecsei and Berger 2004). These ecosystems today account for 26% of the total $CaCO_3$ produced and 45% of the $CaCO_3$ accumulated, annually, in the ocean (Wollast 1994; Milliman and Droxler 1996; Mackenzie *et al.* 2005). These deposits represent the establishment of important benthic communities that were not present to a significant degree for 80 000 years and the accumulation of the calcifying shoal-water ecosystems may have contributed in part to the natural rise in atmospheric CO_2 from the LGM to late pre-industrial times of approximately 100 µatm (the 'coral reef hypothesis'; Berger 1982; Opdyke and Walker 1992; Vecsei and Berger 2004). Importantly, these coastal carbonate systems are dominated by highly soluble benthic skeletal mineralogies with aragonite and high-magnesian calcites comprising ~63% and ~24%, respectively, of the calcifiers and carbonate sediments that could be reactive on the decadal to centennial timescale (Chave 1967; Morse and Mackenzie 1990; Morse *et al.* 2006). The less reactive calcite phase makes up ~13% of the sediments in these coastal systems. Not only did benthic calcifiers like corals spread across the increasing accommodation space as sea level rose but benthic calcifying algae, such as rhodolith beds in polar and temperate areas and *Halimeda* beds in the tropics, became important calcifying ecosystems (Nelson 2009).

As these shoal-water benthic carbonate deposits accumulated with rising sea level and atmospheric CO_2, global-mean sea-surface temperature also rose 4 to 5°C and the carbonate system of coastal ocean water changed significantly. Despite the fact that the global shoal-water carbonate mass was growing extensively, the pH_T of global coastal waters actually declined from ~8.35 to ~8.18 and the carbonate ion concentration (and consequently Ω) declined by ~19% from the LGM to late pre-industrial time (F. Mackenzie, pers. comm.). The latter represents a rate of decline of about 0.028 µmol CO_3^{2-} per decade. For comparison, the decline in coastal water pH from the year 1900 to 2000 was about 8.18 to 8.08 and projected from the year 2000 to 2100 about 8.08 to 7.85, using the IS92a business-as-usual scenario of CO_2 emissions (e.g. Andersson *et al.* 2005). Over these 200 years, the carbonate ion concentration will fall by ~120 µmol kg^{-1} or 6 µmol kg^{-1} $decade^{-1}$. This decadal rate of decline of CO_3^{2-} in the Anthropocene is 214 times the average rate of decline for the entire Holocene. Hence when viewed against the timescale of geological change in the coastal ocean marine carbon system of millennia to several millennia, one can easily appreciate why ocean acidification is the 'other CO_2 problem' (see also Chapter 2). The problem is magnified in the calcifying ecosystems of the shoal-water ocean, where mainly aragonite- and Mg-calcite-secreting benthic calcifiers have proliferated in a relatively slowly changing marine carbonate chemistry for thousands of years. Many of the skeletal fragments and inorganic cements made of Mg-calcite have solubilities greater than that of aragonite and are viewed as the 'first responders' to ocean acidification (Morse *et al.* 2006). Despite the complications arising from vital effects, their

Mg-calcite compositions appear to be very close to those predicted for the phases in metastable equilibrium with the ambient composition and temperature of their seawater environment (Mackenzie *et al.* 1983; Morse *et al.* 2006; Andersson *et al.* 2008). Hence, even small changes in pH and CO_3^{2-} concentration brought about by modern ocean acidification will displace this metastable equilibrium and hence can significantly affect the rate of calcification and composition of Mg-calcite skeletal organisms as the organisms attempt to acclimate or adapt to the new environment (e.g. Mackenzie *et al.* 1983; Agegian 1985). Metastable mineral phases are thermodynamically unstable under the earth's surface temperature and pressure conditions, but persist owing to kinetic constraints. This means that metastable mineral phases in seawater are expected to dissolve based on thermodynamic principles, but they do not because of slow reaction rates or inhibition of the reaction by other constituents.

The important point is that the benthic carbonate system of the whole coastal ocean was evolving and diversifying as sea level and accommodation space slowly rose from the LGM to late pre-industrial times and the coastal system was becoming a major player in the carbon cycle. This evolution took place under relatively slowly rising atmospheric CO_2 and temperature and declining pH and CO_3^{2-} relative to the rate of these changes in the Anthropocene. The question is whether the modern rates are such as to overwhelm one or more coastal ocean carbonate ecosystems, or whether one or more of these systems will be able to acclimate or adapt to the changes. So far the answer seems to be, at least for coral reefs, that in this century or the early 22nd century at the latest, the net ecosystem calcification of many reef systems could become negative because the production of $CaCO_3$ will be exceeded by its dissolution. Reef accretion rates will slow, cease, or decline, and the reefs degrade in quality (e.g. Andersson *et al.* 2005, 2009; Silverman *et al.* 2009). Silverman *et al.* (2009) predicted that all coral reef ecosystems will become subject to net dissolution and net loss of $CaCO_3$ at an atmospheric CO_2 concentration of 560 ppmv. In contrast, Andersson *et al.* (2005) made a more modest projection based on the shallow-water ocean carbonate model (SOCM) and suggested that the global coastal ocean will become subject to net dissolution at an approximate seawater pCO_2 close to 1000 µatm. This value is surprisingly close to the daily average seawater pCO_2 observed in replicated flow-through mesocosms containing subtropical calcifying communities that were marginally subject to net dissolution (Andersson *et al.* 2009). Regardless of the timing of this transition in nature, it represents a reversal of a major geochemical process that has been in effect for thousands of years. It is also important to make the distinction and to realize that coral reef ecosystems will become subject to a condition of net dissolution well before individual live coral colonies are likely to experience this condition.

Manzello *et al.* (2008) proposed that the coral reef ecosystems of the eastern tropical Pacific, which naturally experience high CO_2, low pH, and low Ω_a as a result of upwelling, could serve as an example of how coral reefs will be affected in a high-CO_2 world. As previously discussed, Manzello *et al.* (2008) showed that reef cementation is limited in this region and the reported rates of bioerosion are high. Furthermore, in general, $CaCO_3$ accumulations of the reefs in this region are thin deposits relative to those of Indo-Pacific and Caribbean reefs, small in areal extent and limited to shallow depths.

According to the current experimental results using individual organisms exposed to elevated CO_2 conditions (Table 7.1), it appears likely that benthic marine ecosystems will undergo a gradual transition in the relative composition (and success) of calcifying and non-calcifying organisms, favouring the latter group of organisms. This transition may be specifically noticeable on coral reefs, where a combination of changes in community structure with decreasing, and perhaps even negative, reef accretion could result in a drastically different ecosystem (e.g. Hoegh-Guldberg *et al.* 2007). Similar changes may occur elsewhere because calcifying organisms are ubiquitous components of almost all marine ecosystems (Chave 1967). Observations from a region in the Mediterranean Sea subject to a natural gradient in seawater pH and CO_2 as a result of volcanic vent activity show convincing evidence of a gradual decrease in the abundance of marine calcifiers including sea urchins, coralline algae, gastropods, limpets, and barnacles as a function of decreasing seawater pH (e.g. Hall-Spencer *et al.* 2008; Martin *et al.* 2008). Some gastro-

pod shells observed near the vent site even showed direct evidence of dissolution. In contrast, non-calcifying algae and seagrasses thrived and flourished under the high-CO_2, low-pH conditions observed near the vent site (see also Chapter 10).

Feely *et al.* (2009) recently described how upwelling along the California Shelf occasionally exposes the benthic systems on the continental shelf to undersaturated and corrosive seawater with respect to aragonite. As the upwelling phenomenon is not a new occurrence, the benthic ecosystems on the California Shelf may provide important information regarding an ecosystem subject to high CO_2 and low pH. However, the frequency, extent, and duration of these upwelling events are currently poorly known. Smith (1971) carefully quantified the carbonate budget on the southern California continental shelf. Despite high abundance and production of $CaCO_3$ by marine benthic calcifiers present on hard-bottom portions of the shelf as well as encrusting kelps and other plants, negligible quantities of carbonate material were found to accumulate within this region. Smith (1971) suggested that the material was transported to adjacent basins and subsequently dissolved. Upwelling of corrosive seawater would greatly accelerate this process and may provide an explanation (in addition to export) of why only negligible amounts of $CaCO_3$ relative to its production are found on the California Shelf.

7.4 Conclusions and final remarks

As the carbonate chemistry of the oceans changes as a result of anthropogenic CO_2 emissions, benthic processes, organisms, communities, and ecosystems will most certainly undergo changes reflecting this major change in ocean chemistry. Many benthic ecosystems already experience high seawater CO_2, low pH, and a low saturation state with respect to $CaCO_3$ minerals and may serve as real-world examples of how other systems will be affected in a high-CO_2 world. Based on observations and trends from these systems combined with current experimental results, it is likely that marine benthic ecosystems will change in a predictable sense, although the rate and extent of change are more difficult to predict. Certainly, there could also be unexpected consequences that our current knowledge fails to recognize. Nevertheless, based on our current understanding of experimental, field, and modelling research, some changes that will probably occur as seawater CO_2 increases and pH decreases are:

(1) Benthic ecosystems are likely to undergo a transition where benthic non-calcifying organisms are favoured over benthic calcifiers because several important life stages of calcifiers and processes such as recruitment and growth are negatively affected by ocean acidification.

(2) Net calcification and $CaCO_3$ accretion of coral reefs will decrease and, at some point in time, a transition to net dissolution and net loss of $CaCO_3$ will occur. Over time, this will result in decreased structural complexity and decreased biodiversity, but it is important to recognize that the transition itself is not the same as the 'end' to corals and other calcifiers, which will continue to build their shells and skeletons even once this threshold has been crossed.

(3) As the saturation state with respect to carbonate minerals is shoaling throughout the global ocean, the carbonate compensation depth and the maximum depths where deep-sea corals, bioherms, and other calcifiers could develop are likely to move to shallower depths in all ocean basins.

(4) It is important to keep in mind that although tropical and subtropical coral reefs may be the most visible and important benthic calcifying ecosystems to be affected by ocean acidification, other carbonate systems are vulnerable to the same process and represent approximately 60% of benthic carbonate production and 50% of carbonate accumulation in the shallow ocean.

7.5 Acknowledgements

This work is a contribution to the 'European Project on Ocean Acidification' (EPOCA) which received funding from the European Community's Seventh Framework Programme (FP7/2007–2013) under grant agreement no. 211384. A.J.A. gratefully acknowledges support from NSF Grant OCE 09–28406 and NOAA NA10AR4310094. F.T.M. gratefully acknowledges support from NSF Grant OCE 07–49401 for his part in the preparation of this chapter. The authors are grateful to Ms. Lica Krug for assistance compiling Table 1.

References

Agegian, C.R. (1985). *The biogeochemical ecology of Porolithon gardineri (foslie)*, 178 pp. PhD Thesis. University of Hawaii, Honolulu, HI.

Alexandersson, E.T. (1975). Marks of unknown carbonate-decomposing organelles in cyanophyte borings. *Nature*, **254**, 212–38.

Andersson, A.J., Bates, N.R., and Mackenzie, F.T. (2007). Dissolution of carbonate sediments under rising pCO_2 and ocean acidification: observations from Devil's Hole, Bermuda. *Aquatic Geochemistry*, **13**, 237–64.

Andersson, A.J., Mackenzie, F.T., and Ver, L.M. (2003). Solution of shallow-water carbonates: an insignificant buffer against rising atmospheric CO_2. *Geology*, **31**, 513–16.

Andersson, A.J., Mackenzie, F.T., and Lerman, A. (2005). Coastal ocean and carbonate systems in the high CO_2 world of the Anthropocene. *American Journal of Science*, **305**, 875–918.

Andersson, A.J., Mackenzie, F.T., and Bates, N.R. (2008). Life on the margin: implications of ocean acidification on Mg-calcite, high latitude and cold-water marine calcifiers. *Marine Ecology Progress Series*, **373**, 265–73.

Andersson, A.J., Kuffner, I.B., Mackenzie F.T., Tan, A., Jokiel, P.L., and Rodgers, K.S. (2009). Net loss of $CaCO_3$ from a subtropical calcifying community due to seawater acidification: mesocosm-scale experimental evidence. *Biogeosciences*, **6**, 1811–23.

Anthony, K.R.N., Kline, D.I., Diaz-Pulido, G., Dove, S., and Hoegh-Guldberg, O. (2008). Ocean acidification causes bleaching and productivity loss in coral reef builders. *Proceedings of the National Academy of Sciences USA*, **105**, 17442–6.

Archer, D., Kheshgi, H., and Maier-Reimer, E. (1998). Dynamics of fossil fuel CO_2 neutralization by marine $CaCO_3$. *Global Biogeochemical Cycles*, **12**, 259–76.

Barker, S., Higgins, J.A., and Elderfield, H. (2003). The future of the carbon cycle: review, calcification response, ballast and feedback on atmospheric CO_2. *Philosophical Transactions of the Royal Society A: Mathematical, Physical and Engineering Sciences*, **361**, 1977–99.

Barrón, C., Duarte, C.M., Frankignoulle, M., and Borges A.V. (2006). Organic carbon metabolism and carbonate dynamics in a Mediterranean seagrass (*Posidonia oceanica*) meadow. *Estuaries and Coasts*, **29**, 417–26.

Bates, N.R., Amat, A., and Andersson, A.J. (2010). Feedbacks and responses of coral calcification on the Bermuda reef system to seasonal changes in biological processes and ocean acidification. *Biogeosciences*, **7**, 2509–30.

Berger, W.H. (1982). Increase of carbon dioxide in the atmosphere during deglaciation: the coral reef hypothesis. *Naturwissenschaften*, **69**, 87–8.

Bischoff, W.D., Bertram, M.A., Mackenzie, F.T., and Bishop, F.C. (1993). Diagenetic stabilization pathways of magnesian calcites. *Carbonates and Evaporites*, **8**, 82–9.

Borowitzka, M.A. and Larkum, A.W.D. (1976). Calcification in the green alga *Halimeda*. 3. The sources of inorganic carbon for photosynthesis and calcification and a model of the mechanism of calcification. *Journal of Experimental Botany*, **27**, 879–93.

Boucher, G., Clavier, J., Hily, C., and Gattuso, J.-P. (1998). Contribution of soft-bottoms to the community metabolism (primary production and calcification) of a barrier reef flat (Moorea, French Polynesia). *Journal of Experimental Marine Biology and Ecology*, **225**, 269–83.

Brennan, S.T., Lowenstein, T.K., and Horita, J. (2004). Seawater chemistry and the advent of biocalcification. *Geology*, **32**, 473–6.

Brewer, P.G. and Peltzer, E. (2009). Limits to marine life. *Science*, **324**, 347–8.

Cantin, N.E., Cohen, A.L., Karnauskus, K.B., Tarrant, A.M., and McCorkle, D.C. (2010). Ocean warming slows coral growth in the central Red Sea. *Science*, **329**, 322–5.

Chave, K.E. (1967). Recent carbonate sediments: an unconventional view. *Journal of Geological Education*, **15**, 200–4.

Clark, M.R., Rowden, A.A., Schlacher, T. *et al.* (2010). The ecology of seamounts: structure, function and human impacts. *Annual Review of Marine Science*, **2**, 253–78.

Cooper, T.F., De'ath, G., Fabricius, K.E., and Lough, J.M. (2008). Declining coral calcification in massive *Porites* in two nearshore regions of the northern Great Barrier Reef. *Global Change Biology*, **14**, 529–38.

Crawley, A., Kline, D., Dunn, S., Anthony, K., and Dove, S. (2009). The effect of ocean acidification on symbiont photorespiration and productivity in *Acropora formosa*. *Global Change Biology*, **16**, 851–63.

De'ath, G., Lough, J.M., and Fabricius, K.E. (2009). Declining coral calcification on the Great Barrier Reef. *Science*, **323**, 116–19.

Feely, R.A., Sabine, C.L., Lee, K. *et al.* (2004). Impact of anthropogenic CO_2 on the $CaCO_3$ system in the oceans. *Science*, **305**, 362–6.

Feely, R.A., Sabine, C.L., Hernandez-Ayon, J.M., Ianson, D., and Hales, B. (2008). Evidence for upwelling of corrosive acidified water onto the continental shelf. *Science*, **320**, 1490–2.

Feely, R.A., Doney, S.C., and Cooley, S.R. (2009). Present conditions and future changes in a high-CO_2 world. *Oceanography*, **22**, 36–47.

Findlay, H.S., Kendall, M.A., Spicer, J.I., and Widdicombe, S. (2009). Future high CO_2 in the intertidal may compromise adult barnacle *Semibalanus balanoides* survival and embryonic development rate. *Marine Ecology Progress Series*, **389**, 193–202.

Frankignoulle, M., Canon, C., and Gattuso, J.-P. (1994). Marine calcification as a source of carbon dioxide: positive feedback of increasing atmospheric CO_2. *Limnology and Oceanography*, **39**, 458–62.

Gao, K., Aruga, Y., Asada, K., Ishihara, T., Akano, T., and Kiyohara, M. (1993). Calcification in the articulated coralline alga *Corallina pilulifera*, with special reference to the effect of elevated CO_2 concentration. *Marine Biology*, **117**, 129–32.

Gattuso, J.-P., Gentili, B., Duarte, C.M., Kleypas, J.A., Middelburg, J.J., and Antoine, D. (2006). Light availability in the coastal ocean: impact on the distribution of benthic photosynthetic organisms and their contribution to primary production. *Biogeosciences*, **3**, 489–513.

Gazeau, F., Quiblier, C., Jansen, J.M., Gattuso, J.-P., Middelburg, J.J., and Heip, C.H.R. (2007). Impact of elevated CO_2 on shellfish calcification. *Geophysical Research Letters*, **34**, L07603, doi:10.1029/2006GL028554.

Guinotte, J.M., Orr, J., Cairns, S., Freiwald, A., Morgan, L., and George, R. (2006). Will human-induced changes in seawater chemistry alter the distribution of deep-sea scleractinian corals? *Frontiers in Ecology and the Environment*, **4**, 141–6.

Hall-Spencer, J.M, Rodolfo-Metalpa, R., Martin, S. *et al.* (2008). Volcanic carbon dioxide vents show ecosystem effects of ocean acidification. *Nature*, **454**, 96–9.

Hoegh-Guldberg, O., Mumby, P.J., Hooten, A.J. *et al.* (2007). Coral reefs under rapid climate change and ocean acidification. *Science*, **318**, 1737–42.

Jokiel, P.L., Rodgers, K.S., Kuffner, I.B., Andersson, A.J., Cox, E.F., and Mackenzie, F.T. (2008). Ocean acidification and calcifying reef organisms: a mesocosm investigation. *Coral Reefs*, **27**, 473–83.

Kübler, J.E., Johnston, A.M., and Raven, J.A. (1999). The effects of reduced and elevated CO_2 and O_2 on the seaweed *Lomentaria articulate*. *Plant, Cell and Environment*, **22**, 1303–10.

Kuffner, I.B., Andersson, A.J., Jokiel, P., Rodgers, K.S., and Mackenzie, F.T. (2008). Decreased abundance of crustose coralline algae due to ocean acidification. *Nature Geoscience*, **1**, 114–17.

Kurihara, H. (2008). Effects of CO_2-driven ocean acidification on the early developmental stages of invertebrates. *Marine Ecology Progress Series*, **373**, 275–84.

Langdon, C. and Atkinson, M.J. (2005). Effect of elevated pCO_2 on photosynthesis and calcification of corals and interactions with seasonal change in temperature/irradiance and nutrient enrichment. *Journal of Geophysical Research*, **110**, C09S07, doi:10.1029/2004JC002576.

Langdon, C., Takahashi, T., Sweeney, C. *et al.* (2000). Effect of calcium carbonate saturation state on the calcification rate of an experimental coral reef. *Global Biogeochemical Cycles*, **14**, 639–54.

Langdon, C., Broecker, W.S., Hammond, D.E. *et al.* (2003). Effect of elevated CO_2 on the community metabolism of an experimental coral reef. *Global Biogeochemical Cycles*, **17**, 1011, doi:10.1029/2002GB001941.

Langdon, C.R., Gattuso, J.-P., and Andersson, A.J. (2010). Measurements of calcification and dissolution of benthic organisms and communities. In: U. Riebesell, V.J. Fabry, L. Hansson, and J.-P. Gattuso (eds), *Guide to best practices in ocean acidification research and data reporting*, pp. 213–34. Publications Office of the European Union, Luxembourg.

Leclercq, N., Gattuso, J.-P., and Jaubert, J. (2000). CO_2 partial pressure controls the calcification rate of a coral community. *Global Change Biology*, **6**, 329–34.

Leclercq, N., Gattuso, J.-P., and Jaubert, J. (2002). Primary production, respiration, and calcification of a coral reef mesocosm under increased CO_2 partial pressure. *Limnology and Oceanography*, **47**, 558–64.

Le Quéré, C., Raupach, M.R., Canadell, J.G. *et al.* (2009). Trends in the sources and sinks of carbon dioxide. *Nature Geoscience*, **2**, 831–6.

Mackenzie, F.T., and Lerman, A. (2006). *Carbon in the geobiosphere: earth's outer shell*. Springer, Dordrecht.

Mackenzie, F.T., Bischoff, W.D., Bishop, F.C., Loijens, M., Schoonmaker, J., and Wollast, R. (1983). Magnesian calcites: low temperature occurrence, solubility and solid-solution behavior. In: R.J. Reeder (ed.), *Carbonates: mineralogy and chemistry*. Reviews in Mineralogy, Vol. 11, pp. 97–143. Mineralogical Society of America, Washington, DC.

Mackenzie, F.T., Andersson, A.J., Lerman, A., and Ver, L.M. (2005). Boundary exchanges in the global coastal margin: implications for the organic and inorganic carbon cycles. In: A.R. Robinson, J. McCarthy, and B.J. Rothschild (eds), *The sea*, Vol. 13, pp. 193–225. Harvard University Press, Cambridge, MA.

Maier, C., Hegeman, J., Weinbauer, M.G., and Gattuso, J.-P. (2009). Calcification of the cold-water coral *Lophelia pertusa* under ambient and reduced pH. *Biogeosciences*, **6**, 1671–80.

Manzello, D.P. (2010). Coral growth with thermal stress and ocean acidification: lessons from the eastern tropical Pacific. *Coral Reefs*, **29**, 749–58.

Manzello, D.P., Kleypas, J.A., Budd, D.A., Eakin, C.M., Glynn, P.W., and Langdon, C. (2008). Poorly cemented coral reefs of the eastern tropical Pacific: possible insights into reef development in a high-CO_2 world. *Proceedings of the National Academy of Sciences USA*, **105**, 10450–5.

Martin, S., Rodolfo-Metalpa, R., Ransome, E. *et al.* (2008). Effects of naturally acidified seawater on seagrass calcareous epibionts. *Biology Letters*, **4**, 689–92.

Marubini, F., Ferrier-Pagès, C., and Cuif, J.-P. (2003). Suppression of skeletal growth in scleractinian corals by decreasing ambient carbonate-ion concentration: a cross-family comparison. *Proceedings of the Royal Society B: Biological Sciences*, **270**, 179–84.

Milliman, J.D. (1993). Production and accumulation of calcium carbonate in the ocean: budget of a nonsteady state. *Global Biogeochemical Cycles*, **7**, 927–57.

Milliman, J.D. and Droxler, A.W. (1996). Neritic and pelagic carbonate sedimentation in the marine environment: ignorance is not bliss. *Geologische Rundschau*, **85**, 496–504.

Morse, J.W. and Mackenzie, F.T. (1990). *Geochemistry of sedimentary carbonates*. Elsevier Science, Amsterdam.

Morse, J.W., Andersson, A.J., and Mackenzie, F.T. (2006). Initial responses of carbonate-rich shelf sediments to rising atmospheric pCO_2 and ocean acidification: role of high Mg-calcites. *Geochimica et Cosmochimica Acta*, **70**, 5814–30.

Nelson, W.A. (2009). Calcified macroalgae—critical to coastal ecosystems and vulnerable to change: a review. *Marine and Freshwater Research*, **60**, 787–801.

Ohde, S. and van Woesik, R. (1999). Carbon dioxide flux and metabolic processes of a coral reef, Okinawa. *Bulletin of Marine Science*, **65**, 559–76.

Olafsson, J., Olafsdottir, S.R., Benoit-Cattin, A., Danielsen, M., Arnarson, T.S., and Takahashi, T. (2009). Rate of Iceland Sea acidification from time series measurements. *Biogeosciences*, **6**, 2661–8.

Opdyke, B.N. and Walker, J.C.G. (1992). Return of the coral reef hypothesis: basin to shelf partitioning of $CaCO_3$ and its effect on atmospheric CO_2. *Geology*, **20**, 733–6.

Orr, J.C., Fabry, V.J., Aumont, O. *et al.* (2005). Anthropogenic ocean acidification over the twenty-first century and its impacts on calcifying organisms. *Nature*, **437**, 681–6.

Oschlies, A., Schulz, K.G., Riebesell, U., and Schmittner, A. (2008). Simulated 21st century's increase in oceanic suboxia by CO_2-enhanced biotic carbon export. *Global Biogeochemical Cycles*, **22**, GB4008, doi:10.1029/2007GB003147.

Palacios, S.L. and Zimmerman, R.C. (2007). Response of eelgrass *Zostera marina* to CO_2 enrichment: possible impacts of climate change and potential for remediation of coastal habitats. *Marine Ecology Progress Series*, **344**, 1–13.

Plummer, L.N. and Mackenzie, F.T. (1974). Predicting mineral solubility from rate data: application to the dissolution of magnesian calcites. *American Journal of Science*, **274**, 61–83.

Raven, J.A. (1997). Inorganic carbon acquisition by marine autotrophs. *Advances of Botanical Research*, **27**, 85–209.

Reynaud, S., Leclercq, N., Romaine-Lioud, S., Ferrier-Pagès, C., Jaubert, J., and Gattuso, J.-P. (2003). Interacting effects of CO_2 partial pressure and temperature on photosynthesis and calcification in a scleractinian coral. *Global Change Biology*, **9**, 1660–8.

Riebesell, U., Schulz, K.G., Bellerby, R.G.J. *et al.* (2007). Enhanced biological carbon consumption in a high CO_2 ocean. *Nature*, **450**, 545–8.

Ries, J.B. (2010). Review: geological and experimental evidence for secular variation in seawater Mg/Ca (calcite-aragonite seas) and its effects on marine biological calcification. *Biogeosciences*, **7**, 2795–849.

Ries, J.B., Cohen, A.L., and McCorkle, D.C. (2009). Marine calcifiers exhibit mixed responses to CO_2-induced ocean acidification. *Geology*, **37**, 1131–4.

Roberts, J.M., Wheeler, A.J., and Freiwald, A. (2006). Reefs of the deep: the biology and geology of cold-water coral ecosystems. *Science*, **312**, 543–7.

Rodolfo-Metalpa, R., Lombardi, C., Cocito, S., Hall-Spencer, J.M., and Gambi, M.C. (2010). Effects of ocean acidification and high temperatures on the bryozoan *Myriapora truncata* at natural CO_2 vents. *Marine Ecology*, **31**, 447–56.

Schneider, K. and Erez, J. (2006). The effect of carbonate chemistry on calcification and photosynthesis in the hermatypic coral *Acropora eurystoma*. *Limnology and Oceanography*, **51**, 1284–93.

Silverman, J., Lazar, B., and Erez, J. (2007). Effect of aragonite saturation, temperature, and nutrients on the community calcification rate of a coral reef. *Journal of Geophysical Research*, **112**, C05004, doi:10.1029/2006JC003770.

Silverman, J., Lazar, B., Cao, L., Caldeira, K., and Erez, J. (2009). Coral reefs may start dissolving when atmospheric CO_2 doubles. *Geophysical Research Letters*, **36**, L05606, doi:10.1029/2008GL036282.

Smith, A.D. and Roth, A.A. (1979). Effect of carbon dioxide concentration on calcification in the red coralline alga *Bossiella orbigniana*. *Marine Biology*, **52**, 217–25.

Smith, S.V. (1971). Budget of calcium carbonate, Southern California continental borderland. *Journal of Sedimentary Petrology*, **41**, 798–808.

Stanley, S.M., Ries, J.B., and Hardie, L.A. (2010). Increased production of calcite and slower growth for the major sediment-producing alga *Halimeda* as the Mg/Ca ratio of seawater is lowered to a 'calcite sea' level. *Journal of Sedimentary Research*, **80**, 6–16.

Steinacher, M., Joos, F., Frölicher, T.L., Plattner, G.-K., and Doney, S.C. (2009). Imminent ocean acidification projected with the NCAR global coupled carbon cycle-climate model. *Biogeosciences*, **6**, 515–33.

Stramma, L., Johnson, G.C., Sprintall, J., and Mohrholz, V. (2008). Expanding oxygen-minimum zones in the tropical oceans. *Science*, **320**, 655–8.

Tanzil, J.T.I., Brown, B. E., Tudhope, A.W., and Dunne, R.P. (2009). Decline in skeletal growth of the coral *Porites lutea* from the Andaman Sea, South Thailand between 1984 and 2005. *Coral Reefs*, **28**, 519–28.

Tittensor, D.P., Baco-Taylor, A.R., Brewin, P. *et al.* (2009). Predicting global habitat suitability for stony corals on seamounts. *Journal of Biogeography*, **36**, 1111–28.

Tribollet, A. (2008). The boring microflora in modern coral reef ecosystems: a review of its roles. In: M. Wisshak and L. Tapanila (eds), *Current developments in bioerosion*, pp. 67–94. Springer-Verlag, Berlin.

Tribollet, A., Godinot, C., Atkinson, M., and Langdon, C. (2009). Effects of elevated *p*CO_2 on dissolution of coral carbonates by microbial euendoliths. *Global Biogeochemical Cycles*, **23**, GB3008, doi:10.1029/2008GB003286.

Vecsei, A. and Berger, W.H. (2004). Increase of atmospheric CO_2 during deglaciation: constraints on the coral reef hypothesis from patterns of deposition. *Global Biogeochemical Cycles*, **18**, GB1035, doi:10.1029/2003GB002147.

Walter, L.M. and Morse, J.W. (1984). Reactive surface area of skeletal carbonate during dissolution: effect of grain size. *Journal of Sedimentary Petrology*, **54**, 1081–90.

Wollast, R. (1994). The relative importance of bioremineralization and dissolution of $CaCO_3$ in the global carbon cycle. In: F. Doumenge, D. Allemand, and A. Toulemont (eds), *Past and present biomineralization processes: considerations about the carbonate cycle*, pp. 13–34. Musée Océanographique, Monaco.

Wood, H.L., Spicer, J.I., and Widdicombe, S. (2008). Ocean acidification may increase calcification rates, but at a cost. *Proceedings of the Royal Society B: Biological Sciences*, **275**, 1767–73.

Yates, K.K. and Halley, R.B. (2006). CO_3^{2-} concentration and *p*CO_2 thresholds for calcification and dissolution on the Molokai reef flat, Hawaii. *Biogeosciences*, **3**, 357–69.

CHAPTER 8

Effects of ocean acidification on nektonic organisms

Hans-O. Pörtner, Magda Gutowska, Atsushi Ishimatsu, Magnus Lucassen, Frank Melzner, and Brad Seibel

8.1 Integrative concepts relevant in ocean acidification research

The average surface-ocean pH is reported to have declined by more than 0.1 units from the pre-industrial level (Orr *et al.* 2005), and is projected to decrease by another 0.14 to 0.35 units by the end of this century, due to anthropogenic CO_2 emissions (Caldeira and Wickett 2005; see also Chapters 3 and 14). These global-scale predictions deal with average surface-ocean values, but coastal regions are not well represented because of a lack of data, complexities of nearshore circulation processes, and spatially coarse model resolution (Fabry *et al.* 2008; Chapter 3). The carbonate chemistry of coastal waters and of deeper water layers can be substantially different from that in surface water of offshore regions. For instance, Frankignoulle *et al.* (1998) reported pCO_2 (note 1) levels ranging from 500 to 9400 µatm in estuarine embayments (inner estuaries) and up to 1330 µatm in river plumes at sea (outer estuaries) in Europe.[1] Zhai *et al.* (2005) reported pCO_2 values of > 4000 µatm in the Pearl River Estuary, which drains into the South China Sea. Similarly, oxygen minimum layers show elevated pCO_2 levels, associated with the degree of hypoxia (Millero 1996). These findings suggest that some coastal and mid-water animals, both pelagic and benthic, are regularly experiencing hypercap-

nic conditions (i.e. elevated pCO_2 levels), that reach beyond those projected in the offshore surface ocean. These organisms might, therefore, be pre-adapted to relatively high ambient pCO_2 levels. The anthropogenic signal will nonetheless be superimposed on the pre-existing natural variability.

These phenomena lead to the question of whether future changes in the ocean's carbonate chemistry pose a serious problem for marine organisms. Those with calcareous skeletons or shells, such as corals and some plankton, have been at the centre of scientific interest. However, elevated CO_2 levels may also have detrimental effects on the survival, growth, and physiology of marine animals more generally (Pörtner and Reipschläger 1996; Seibel and Fabry 2003; Fabry *et al.* 2008; Pörtner 2008; Melzner *et al.* 2009a). Global warming and expanding hypoxia (Stramma *et al.* 2008; Bograd *et al.* 2009) pose additional physiological challenges that may act synergistically with ocean acidification to impair various aspects of performance, including muscular exercise (Pörtner *et al.* 2005a,b; Pörtner and Farrell 2008; Rosa and Seibel 2008; Munday *et al.* 2009a; Pörtner 2010). The elevation in oxygen demand caused by warming is limited by the availability of ambient oxygen, which is reduced in warm compared with cold waters and even more so during environmental hypoxia. Additionally, CO_2 may impair oxygen transport by lowering blood pH, thereby

[1] For consistency within this volume, we use the lower case italic *p* as a symbol for partial pressure of CO_2 in seawater (pCO_2) and the capital italic *P* as in PCO_2 for body fluids. We thereby acknowledge that the use of symbols for partial pressure differs between disciplines. According to conventions in physics and physiology, the symbol of pressure is a capital italic *P* that also applies to the partial pressure of gases. In ocean chemistry a lower case italic *p* is used instead (1000 µatm = 101.3 Pa).

decreasing oxygen binding to transport proteins (e.g. haemoglobin or haemocyanin).

These synergies have been poorly explored to date. The concept of oxygen- and capacity-limited thermal tolerance (OCLTT) may provide a suitable matrix for integrating thermal effects with those of other climate-related factors such as CO_2 or hypoxia (Pörtner 2010). The concept explains why, where, and how the thermal window of perform-

ance in aquatic organisms is set and limited by the functional capacity of those tissues involved in oxygen uptake and distribution, i.e. mainly the cardiocirculatory system (cf. Pörtner 2006). As a result of capacity adjustments, the window of performance equals the window of aerobic metabolic scope (see Box 8.1). The term 'metabolic scope' refers to the amount of energy that can be allocated to activities beyond those required for basic

---

Box 8.1 Thermal windows shape and restrain energy availability

Schematic of the thermal window of a species or one of its life stages (after Pörtner and Farrell 2008), following the concept of oxygen- and capacity-limited thermal tolerance (OCLTT). The graph shows how the limits (vertical lines) of thermal tolerance shape the temperature-dependent performance curve and its optimum (T_{opt}) as reflected in the thermal window of aerobic scope determined in swimming trials of nektonic animals (Fry 1971). Aerobic scope is the difference between maximum and resting aerobic metabolic rates. In the warmth and cold, the first thermal limits experienced by the organism at pejus temperatures (T_{pej}, pejus means 'getting worse') result from the limited capacity of tissues overall and especially of cardiocirculation, and in addition, in invertebrates, ventilatory organs. Finally, the limitation in oxygen supply

leads to the transition to anaerobic metabolism beyond critical temperatures (T_{crit}) and before molecular denaturation sets in (at T_{den}). These limits can be shifted by the acclimation of functional capacity. Beyond pejus limits, passive tolerance to temperature extremes is supported by protective mechanisms including anaerobic metabolism, heat shock proteins and antioxidative defence. Hypoxia is expected to narrow the thermal window and lower performance optima (dashed lines, green arrows). CO_2 also causes such narrowing, but in the thermal optimum performance decrements may occur in sensitive species or life stages only. Energy allocation to life-sustaining functions will become constrained once aerobic scope is limited by temperature and its synergistic interaction with other environmental factors.

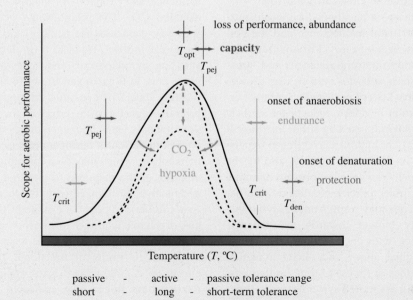

existence, and is therefore tightly coupled to the animal's fitness. In practice, metabolic scope is determined as the difference between energy dissipation (or, more conventionally, oxygen consumption) at maximum swimming activity (active metabolic rate) and at rest (standard or basal metabolic rate) (Fry 1971). By definition, the term metabolic scope refers to the sum of all energy-demanding biological activities, not exclusively to mechanical work. There is evidence of a significant positive relationship between metabolic scope, estimated from increments in activity costs, and other types of performance such as food intake (Mallekh and Lagardère 2002) and growth (Claireaux and Lefrançois 2007; Pörtner and Knust 2007), supporting the usefulness of the concept. Available data indicate that the characteristics of aerobic performance, the associated functional capacity of tissues, not least those supplying oxygen to mitochondria, shape the fitness of the organism and its role at the ecosystem level. The underlying OCLTT concept has been used successfully to explain the effects of climate-induced increases in temperature on species abundance and survival in the field (Pörtner and Knust 2007). The ecological relevance of the OCLTT concept in the context of ongoing warming supports its use as a matrix for studying the concomitant and synergistic effects of other factors (Pörtner 2010).

In animals, energy is partitioned between functions required for basal maintenance and the performance of those required for additional activities such as locomotion, immune defences and stress resistance, digestion, and biosynthesis (e.g. growth and reproduction). Loss of performance at the borders of the thermal envelope reflects the earliest level of thermal stress, caused by either insufficient functional capacity or hypoxemia (reduced oxygen partial pressure in the blood), or both, and by the resulting mismatch of oxygen supply and demand. In other words, under extreme conditions a limitation in aerobic scope may result from unfavourable shifts in energy budget towards maintenance. Oxygen deficiency limits one or more of the aerobic activities mentioned above and elicits transition from the sustenance of fitness to time-limited passive tolerance and associated systemic and cellular stress signals. These include hormonal responses or

oxidative stress, as well as the use of protective mechanisms such as heat shock proteins at thermal extremes (Anestis *et al.* 2007; Feidantsis *et al.* 2009; Kyprianou *et al.* 2010; Tomanek and Zuzow 2010; see Box 8.1). In general, limitations in the response of an organism to environmental factors first become effective at the highest levels of biological organization, the intact organism, which displays a higher sensitivity than any of the subordinate, cellular and molecular, functions (Pörtner 2002). Nevertheless, whole-organism limitations are ultimately based on the integration of molecular functions into functional and regulatory networks. Consequently, when studying the adaptation of organisms to a changing environment one needs to consider the function of individual molecules and their integration into higher organizational levels, up to the whole organism.

Evidence available for crustaceans and fishes indicates that the thermal window of aerobic performance is indeed affected by the synergistic impact of elevated CO_2 (Metzger *et al.* 2007; Munday *et al.* 2009a; Walther *et al.* 2009) or hypoxia (see Pörtner 2010). Recent findings by Findlay *et al.* (2010) can be interpreted in similar ways. Thermal acclimatization between seasons, or evolutionary adaptation to a climate regime, has implications for metabolic rate with consequences for capacity, performance, and probably for tolerance to hypoxia or elevated CO_2. The relationships between energy turnover, capacities for activity, and the width of thermal windows, leads to a fundamental understanding of adaptation and specialization to climate, and, in turn, of sensitivity to climate change (Pörtner 2006, 2010). Insufficient functional and respiratory capacity at the edges of the thermal window not only leads to oxygen deficiency but also to an accumulation of endogenous CO_2 produced by metabolism, which may then contribute to the exacerbation of the response to elevated ambient CO_2 and be involved in narrowing the thermal window. As a corollary, hypoxia or elevated CO_2 elicit strategies of passive tolerance to environmental extremes (e.g. metabolic depression) but lead the organism earlier to its limits of functional capacity. Such effects of climate-related stressors on functional relationships might also underpin any climate-induced changes in species interactions, and thus community responses in various ecosystems (Pörtner and Farrell 2008).

Temperature is the most pervasive environmental factor which affects all levels of biological organization (Box 8.1). The specific effects of ocean acidification are then also affected by temperature and exacerbated at thermal extremes. An overarching hypothesis is that disturbances in acid–base status play a key role in mediating the specific physiological effects of elevated CO_2 tensions (Pörtner 2008; Melzner et al. 2009a). The mechanisms of acid–base regulation strive to compensate for such disturbances, and thereby the organism escapes the detrimental effects of ambient stressors on physiological functions. Some of these mechanisms differ between animals breathing air and those breathing water. In general, all water-breathing animals require much higher ventilation volumes than air breathers for access to sufficient amounts of oxygen. As a result, CO_2 release is very efficient and diffusion gradients for CO_2 between body fluids and ambient media are about 30-fold lower than in air breathers. Acid–base regulation in air breathers may involve large respiratory adjustments in body fluid PCO_2. This process plays only a minor role in water breathers, where acid–base regulation largely occurs through branchial ion transport processes (Heisler 1986; Claiborne et al. 2002). The initial compensation of acid–base disturbances occurs through the pre-existing trans-epithelial ion transport mechanisms. On longer timescales further regulation may involve the dynamic modulation of the content and functional properties of ion transport proteins, including the mRNA and protein expression of specific isoforms and their post-translational modification.

The capacity for acid–base regulation will determine the extent of compensation of acid–base status and the potential for shifts in crucial physiological functions or their balance. Extracellular acid–base status is generally more susceptible to environmentally induced changes, whereas intracellular pH is much more stable and thus better protected. Extracellular rather than intracellular pH has been demonstrated to cause shifts in physiological functioning at the molecular (within membranes), cellular, tissue, and systemic levels (e.g. Reipschläger and Pörtner 1996; Pörtner et al. 1998, 2000).

Accordingly, extracellular pH (pH_e) is thought to play a crucial and integrating regulatory role in shaping the short- or long-term whole-organism response to environmental stressors such as elevated CO_2 (Pörtner 2008; Fig. 8.1). By lowering extracellular pH, ocean acidification may act on performance at various levels, such as somatic (and shell or skeleton) growth, reproduction, and behaviour. This includes an impact on calcification and, thereby, the structural stability, shape, or size of the (exo- and endo-) skeleton. The compensatory accumulation of bicarbonate can ameliorate this impact (see below). The ecological success of marine organisms facing ocean acidification may thus largely depend on the degree to which they can maintain pH homeostasis and their capacities to regulate acid–base status. The associated energetic costs are likely to comprise a significant fraction of an organism's energy budget, and are thus related to the rate of energy metabolism. Low-performance, especially hypometabolic, marine invertebrates, which are characterized by a low capacity to compensate for disturbances in extracellular ion and acid–base status, are probably more sensitive than high-performance species with a high capacity for ion and acid–base regulation (Seibel and Walsh 2001).

Other specific effects of CO_2 may occur, either independently or linked to acid–base status in ways that are still unknown. For example, CO_2-induced acidosis leads to the accumulation of adenosine, which acts as an inhibitory neurotransmitter in nervous tissue and can cause metabolic depression under extreme hypoxia or hypercapnia (Reipschläger et al. 1997). The metabolism of other amino acid neurotransmitters is probably also influenced, via the sensitivity of carboxylation and decarboxylation reactions to changes in cellular or mitochondrial CO_2 or bicarbonate levels (e.g. Hardewig et al. 1994; Mühlenbruch 2004; Stark 2008; Fig. 8.1). Functional responses to these changes have not been explored to date. At the level of the organism, olfactory disturbances may occur and lead to atypical behaviours such as those reported for tropical fishes (Munday et al. 2009b).

The acid–base balance and associated physiological functions of organisms respond acutely to environmental change. On longer timescales, however, organisms may acclimate by changing

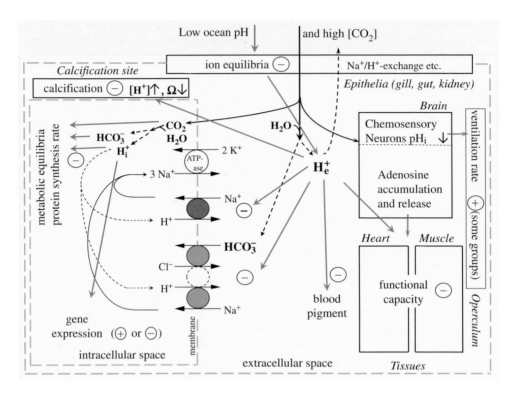

Figure 8.1 Role of extracellular pH (or proton activity, H_e^+) in modulating and coordinating the rates and capacity of various physiological functions (after Pörtner 2008). Changes in individual functions (neuronal functions affected by adenosine accumulation in the central nervous system, muscular excitability, ventilatory rates, metabolic equilibria, protein synthesis rates, calcification, ion exchange) integrate into changes in performance and fitness of the whole organism. Changes in extracellular pH influence the rate of pH regulation through Na^+-dependent proton exchange, and thus the rate of use of ATP -dependent Na^+/K^+-ATPase. Efficient maintenance of extracellular pH by proton-equivalent ion exchange across epithelia (for example, gill, gut, and kidney epithelia in marine teleosts) minimizes such disturbing influences.

gene expression patterns and the concentrations of functional proteins, and thereby the capacities of molecular functions. For example, it is well established that seasonal temperature fluctuations can cause acclimatization responses in animals which involve changes in the gene expression of key functional proteins. The capacity for such responses is probably greater in ectotherms from temperate climate zones than in polar or tropical climates. In contrast, acclimatization responses under changing CO_2 levels have scarcely been investigated and may also vary between organisms from different climate zones. Studies of the effects of hypercapnia on the transcriptome and proteome, with a focus on ion and acid–base regulation, have identified long-term compensatory responses in sensitive versus vulnerable species (Hofmann and Todgham 2010).

In the following sections we examine to what extent this general picture applies to nektonic organisms, particularly fishes and cephalopods. By definition, nektonic organisms comprise those that swim actively and freely and are generally independent of water currents. This implies that their metabolic and exercise capacities are higher than those of sessile organisms (Seibel and Drazen 2007). Active marine taxa that have high metabolic rates and strong ion regulatory abilities are considered to be the most tolerant to future changes in seawater carbonate chemistry (Seibel and Walsh 2003; Pörtner 2008; Melzner *et al.* 2009a). This poses a challenge to identify small but relevant levels of sensitivity to ocean acidification, which may become visible only with concomitant changes in other environmental factors, such as temperature. The physiological principles involved are those that have also been operative on evolutionary timescales, including mass extinction events (Pörtner *et al.* 2004, 2005a; Knoll *et al.* 2007). Many nektonic organisms are valued by society for their beauty and impressive performances as well as for being a relevant food source to humans. The well-being and conservation of these charismatic species under scenarios of global climate change are thus of concern to society.

8.2 Effects of ocean acidification on fishes

8.2.1 Acid–base regulation

Among fishes, freshwater teleosts have been more intensely studied than marine fishes with respect to effects of hypercapnia and the mechanisms and scope of acid–base regulation (Heisler 1986; Claiborne *et al.* 2002; Evans *et al.* 2005; Marshall and Grosell 2006). The information available for marine fishes suggests that they are capable of maintaining blood pH at control levels even when seawater pCO_2 rises above 5000 µatm (Toews *et al.* 1983; Larsen *et al.* 1997; Hayashi *et al.* 2004; Michaelidis *et al.* 2007). Exposure to elevated seawater pCO_2 results in elevated internal PCO_2 and requires the net excretion of acid to compensate for the respiratory acidosis in body fluids. Similar to cephalopods (see below) or crustaceans (Wheatly and Henry 1992), the gills are the primary sites of acid–base regulatory processes in fishes (Perry and Gilmour 2006). The capacity of acid–base regulation depends on the organism's lifestyle and is linked to metabolic and thus exercise capacity (Melzner *et al.* 2009a). Branchial acid–base regulation is complemented by ion exchange through the kidney and gut. Marine teleosts drink ambient seawater to avoid dehydration. They absorb water from the gut to replace the water lost to the hyperosmotic environment. The resulting enrichment of Ca^{2+} in the gut fluid leads to the precipitation of calcium carbonates. The bicarbonate required for this process is secreted by the intestine, thereby leading to an acidification of the blood plasma (Cooper *et al.* 2010). The excretion of calcium carbonate ($CaCO_3$) via the gut of marine teleosts is substantial and accounts for between 3 and 15% of the $CaCO_3$ produced in the ocean. It thus represents a significant, previ-

ously overlooked component of the global inorganic carbon cycle (Wilson *et al.* 2009).

The aerobic scope of fishes can be exploited by increasing oxygen uptake via the gills, oxygen supply to tissues through the circulatory system, or oxygen delivery across tissue capillary beds, or a combination of these three processes. Farrell *et al.* (2009) pointed out that increased cardiac output and enhanced arterio-venous difference in oxygen content both play a crucial role in meeting increased tissue oxygen demand. The high efficiency of acid–base regulation in fishes supports the maintenance of metabolic scope under elevated CO_2 tensions. Even when the fish is exposed to 10 000 µatm or higher, extracellular pH will be nearly restored to pre-hypercapnic levels within a few days (Heisler 1986), thereby supporting blood oxygen saturation under hypercapnia. In addition, fish have the ability to finely regulate the intracellular pH of red blood cells through the release of catecholamines, triggered by decreasing blood oxygen levels in the early phase of hypercapnia (Perry *et al.* 1989). Increasing oxygen transport capacity through the release of red blood cells stored in the spleen does occur during exercise (Nikinmaa 2006), but not usually during hypercapnic exposure at rest (Ishimatsu *et al.* 1992; Gallaugher and Farrell 1998).

Efficient acid–base regulation under hypercapnia causes an accumulation of bicarbonate in body fluids, to higher levels in the plasma than in the intracellular space. In teleosts, this is clearly paralleled by an equimolar decrease in plasma Cl^-. Branchial cells that are active in acid secretion contain electroneutral Na^+/H^+ exchangers (NHE) or V-type H^+-ATPases, coupled energetically to apical Na^+-channels (EnaC). The ion gradients driving the stoichiometric exchange of Na^+ (uptake) for H^+ as well as of Cl^- (uptake) for HCO_3^- (Wood 1991; Perry *et al.* 2003) are established, directly or indirectly, by the Na^+/K^+-ATPase. Na^+/K^+-ATPase activity uses a large fraction of the cellular and epithelial energy budget and is tightly regulated. It has therefore been used as a marker of the overall capacity for ion and acid–base regulation.

Regulation of Na^+/K^+-ATPase activity by gene expression under hypercapnia has been poorly explored to date and with variable results. In developing Atlantic salmon, mRNA levels of branchial

Na^+/K^+-ATPase were reduced after 4 days at a pCO_2 of 20 000 µatm, while enzyme activity remained constant (Seidelin *et al.* 2001). In the gills of Japanese flounder, activity increased during exposure to 10 000 and 50 000 µatm (unpublished data, cited in Ishimatsu *et al.* 2005). Recent studies in two marine teleosts found an increase in the capacity of branchial Na^+/K^+-ATPase during acclimation to hypercapnia at 10 000 µatm (Deigweiher *et al.* 2008; Melzner *et al.* 2009b). In the common eelpout, *Zoarces viviparus*, the specific contents of Na^+/K^+-ATPase mRNA and functional protein as well as its maximum activity rose rapidly with the onset of hypercapnia, and increased up to twofold above control levels during 6 weeks of exposure to a pCO_2 of 10 000 µatm (Deigweiher *et al.* 2008). Long-term acclimation (4 to 12 months) of Atlantic cod (*Gadus morhua*) also led to increased Na^+/K^+-ATPase activities and protein concentrations at a pCO_2 of 6000 µatm (Melzner *et al.* 2009b). Changes in enzyme activity were small at 3000 µatm, indicating that capacities under control conditions were sufficient to cope with the additional ion-regulatory effort required under moderate hypercapnia. It remains to be explored to what extent this statement not only holds close to the thermal optimum of the species but also at extreme temperatures.

Teleosts with high regulatory capacities may also draw from a functional reserve that can be activated upon demand, e.g. by protein modification through phosphorylation (see Ramnanan and Storey 2006). This mechanism may be important during the early regulatory phase and may partly explain any discrepancy between changing mRNA, protein, and functional levels. Long-term acclimation may then respond to the requirement for higher functional rates of ion and acid–base regulation at higher concentrations of the respective transporter.

The electrochemical gradient provided by Na^+/K^+-ATPase is exploited by gradient-dependent transporters and channels such as NHE, Na^+/HCO_3^- cotransporter (NBC), and Cl^-/HCO_3^- exchanger (AE). NBC and AE are both solute carrier 4 bicarbonate transporters, which were first characterized in gills of freshwater fishes (Wilson *et al.* 2000; Hirata *et al.* 2003; Perry *et al.* 2003). Recent findings indicate that they also play an important role during hypercapnic acclimation of marine teleosts (Deigweiher

et al. 2008). Similar to transcriptional changes of AE1 and two NHE1 isoforms, mRNA levels of NBC decreased transiently during early acclimation to acute elevations in CO_2. However, levels were enriched about threefold at the end of the 6-week exposure period, whereas AE1 and NHE1s were restored to control levels (Deigweiher *et al.* 2008). In freshwater fish, the level of NBC1 mRNA increased immediately after external acidification in *Tribolodon hakonensis* (Hirata *et al.* 2003) and under environmental hypercapnia in *Oncorhynchus mykiss* (Perry *et al.* 2003). During long-term acclimation, higher levels of the NBC1 transporter may thus support high levels of extracellular HCO_3^-, which is instrumental during the compensation of acidosis. The finding of constant Na^+ levels over time suggests that the enhanced activity of NBC1 might be counteracted by mechanism(s) that extrude Na^+.

The difference in timing between the responses of NBC1 in freshwater and seawater may relate to the more 'relaxed' situation for acid–base regulation in seawater, which is characterized by high levels of bicarbonate and counter-ions that support ion and acid–base regulation. The putative freshwater origin of teleost fishes (Fyhn *et al.* 1999) and the existence of extant anadromous and katadromous teleost species corroborate the conclusion that fishes are characterized by highly flexible ion and acid–base regulation systems that are well adapted to handle the largely different ionic compositions of their respective ambient media.

The pathways contributing to acid–base regulation may be more complicated than discussed here owing to the large number of transporters and channels present in gill epithelia, their heterogeneous localization and orientation in apical and basolateral membranes, and also due to the presence of different cell types within gill epithelia (see, e.g., the discussion in Perry *et al.* 2003). Extensive transcriptomic studies may be useful in the identification and characterization of the transporters and channels involved. Building on current mechanistic knowledge of marine teleost ion and pH regulation (Fig. 8.2) Deigweiher *et al.* (2008) carried out a comprehensive analysis of transcriptomic changes in gill ion exchange mechanisms under hypercapnia. They found an apical Na^+/H^+-exchanger isoform 2 (NHE2), two V-type H^+-ATPase (HA) isoforms, as well as Cl⁻ channels in a cDNA library from the gills of *Z. viviparus* selected for genes upregulated after 24 h of hypercapnia. Furthermore, three differentially expressed ESTs (expressed sequence tags) of carbonic anhydrase were identified (Deigweiher 2009). Carbonic anhydrase is a highly conserved enzyme that catalyzes the reversible hydration of CO_2, and thus

enhances the rate at which acid–base equilibria are reached. The apparent adjustment of carbonic anhydrase through individually expressed isoforms emphasizes its importance in facilitating ion exchange. The diverse responses of primary and secondary transporters led to an extension of the proposed model of ion transport regulation (Fig. 8.2).

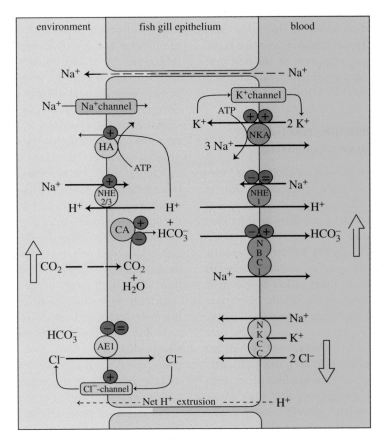

Figure 8.2   Working model for ion transport regulation in marine teleosts under hypercapnia (modified after Deigweiher *et al.* 2008, based on Claiborne *et al.* 2002; Evans *et al.* 2005). The model summarizes results from different mRNA and protein studies in common eelpout (Deigweiher *et al.* 2008; Deigweiher 2009). Red circles indicate the response of gene expression during short-term exposure (24 to 96 h) while green circles indicate the long-term (about 6 weeks) response to environmental hypercapnia. After diffusive entry CO_2 is hydrated by carbonic anhydrase (CA) producing HCO_3^- and H^+. Acute pH compensation is achieved by non-bicarbonate buffering together with net H^+ extrusion to the water supported by the transitional down-regulation of the basolateral Na^+/H^+ exchanger (NHE1) and Na^+/HCO_3^- cotransporter (NBC1) and the up-regulation of apical NHE2. Na^+/K^+-ATPase (NKA) expression is up-regulated. A delayed down-regulation of apical Cl^-/HCO_3^- anion exchanger (AE1) supports the maintenance of higher bicarbonate levels in the cell and plasma when extracellular pH is already restored. During long-term compensation, the net accumulation of extracellular HCO_3^- occurs via an increase in the abundance of basolateral NBC1 and the maintenance of AE1 and NHE1 at control levels. Net decrease of Cl^- in blood may be mediated by a basolateral $Na^+/K^+/2Cl^-$ co-transporter (NKCC) and an apical Cl^- channel which is up-regulated during short-term exposure. NHE1 is operating at control levels under long-term steady-state conditions. Net proton extrusion is possibly achieved by apical NHE2/3. The driving force for the new steady state is provided by elevated Na^+/K^+-ATPase density in the basolateral membrane. Excess Na^+ can diffuse via leaky tight junctions into the surrounding water. The mRNA levels of the respective transporters are depicted as up-regulated (+), down-regulated (–) or unchanged (=). Abbreviations of enzymes: HA, H^+-ATPase; NHE1/2/3, Na^+/H^+ exchanger isoforms; CA, carbonic anhydrase; AE1, Cl^-/HCO_3^- exchanger; NKA, Na^+/K^+-ATPase; NBC1, Na^+/HCO_3^- co-transporter; NKCC, $Na^+/K^+/2Cl^-$ co-transporter. Hollow arrows indicate changes in concentrations.

Over time, a biphasic response was found at the level of the transcriptome of common eelpout. An initial down-regulation in the message of gradient-dependent transporters at stable capacities of Na^+/K^+-ATPase was followed by an up-regulation of Na^+/K^+-ATPase message and activity in the gills during long-term acclimation. During long-term CO_2 exposure (weeks) the gills may respond to enhanced rates of ion regulation by increasing transcription and translation levels. In line with these observations, hypercapnia caused an increase in the energy demand of ion regulation in isolated perfused gills from two species of Antarctic notothenioid fishes (Deigweiher et al. 2010).

8.2.2   Functional capacities, sensitivities, and energy budget in various climates

Enhanced rates of ion regulation elicited by long-term CO_2 exposure resemble observations during cold adaptation. Eurythermal marine and freshwater teleost species including common eelpout display cold-compensated Na^+/K^+-ATPase capacity (reviewed by Pörtner et al. 1998, 2005b). Conversely, reduced capacity of this carrier appears to be an important element of acclimatization to elevated temperatures depending on the season or climate zone. The cost of ion exchange contributes significantly to the standard metabolic rate, such that thermal acclimatization probably contributes to sustained energy efficiency. In fact, the high costs of eurythermal cold adaptation are likely to reduce the energy available for growth and thus cause lower temperature-specific growth rates in sub-Arctic and Arctic populations of Atlantic cod (Pörtner et al. 2008). Under future ocean scenarios, an elevated energy demand for ion regulation in response to elevated pCO_2 levels may reduce the rates and capacities of other functions, as a trade-off in the animal's energy budget. Such constraints may affect fitness even in high-performance species which display a higher capacity for ion and acid–base regulation than the more sluggish species (Melzner et al. 2009a).

A recent study examined the relationships between ion regulation capacity, metabolic rate, and swimming performance in Atlantic cod under long-term hypercapnia (Melzner et al. 2009b). Routine metabolic rates and aerobic scope remained unchanged in North Sea cod (Gadus morhua; NSC) incubated for 4 months at a pCO_2 of 3000 µatm. Gill Na^+/K^+-ATPase activity, as a general indicator of ion regulatory capacity, did not differ from control values, suggesting that existing ion regulatory capacities were sufficient to maintain acid–base homeostasis, and also that the cost of pH compensation is low enough not to significantly affect maximum swimming performance. However, cod from the Barents Sea (north-eastern Arctic Cod; NEAC) responded by enhancing the gene expression of Na^+/K^+-ATPase after 12 months of exposure to a pCO_2 of 6000 µatm while standard and active metabolic rates, critical swimming speeds, and aerobic scope remained at control levels. However, effects on other components of the energy budget, like growth, cannot be excluded. A reduction in growth due to elevated energy demand for ion and acid–base regulation would be consistent with earlier observations in cold-exposed NSC and NEAC. In the cold, the expression of genes coding for aerobic enzymes was stronger in NEAC (Lucassen et al. 2006), reflecting enhanced mitochondrial ATP synthesis capacity and proton leakage in this population. The high cost of proton leakage might contribute to lower growth of NEAC (see above). Long-term elevated Na^+/K^+-ATPase capacities in cod and eelpout under elevated CO_2 levels (Deigweiher et al. 2008; Melzner et al. 2009b) are thus indicative of a persistent shift in the energy budget, possibly at the expense of other factors, such as growth.

The increased gene expression of Na^+/K^+-ATPase does not necessarily indicate a response of acid–base regulation. A high capacity of Na^+/K^+-ATPase represents an enhanced driving force for those mechanisms of acid–base regulation which exploit the ion gradient set up by Na^+/K^+-ATPase. Considering their high capacity for pH homeostasis, it is unlikely that adult teleosts will experience an acidosis in their body fluids in response to increasing seawater pCO_2. However, they may still experience potentially elevated costs of acid–base regulation and an associated shift in their energy budget (Deigweiher et al. 2008, 2010). Direct evidence for a significant energetic cost of acid–base regulation, as part of the energy budget allocated to ion regulation, has only recently been provided for marine organisms. Specifically, inhibition of trans-membrane proton equivalent ion transport and a reduction in oxygen demand was caused by a reduction in extracellular pH in a sipunculid (*Sipunculus nudus*), in Mediterranean mussels (*Mytilus edulis*) (Pörtner *et al.* 1998, 2000; Michaelidis *et al.* 2005), and in isolated cell preparations of two Antarctic fishes, *Pachycara brachycephalum* and *Lepidonotothen kempi*, under a pCO_2 of 10 000 μatm (Langenbuch and Pörtner 2003). These findings demonstrate that the mechanism of cellular metabolic inhibition through extracellular acidosis also exists in marine fishes. However, marine teleosts do not normally allow extracellular pH to fall permanently during long-term hypercapnia, such that oxygen uptake remains unchanged at the whole-organism level. The stimulation of acid–base regulation may even cause a stimulation of whole–organism energy turnover. Fishes indeed maintain, or transiently increase, oxygen uptake when exposed to elevated pCO_2 at rest (Ishimatsu *et al.* 2008). However, CO_2 may also affect the mode of energy metabolism. From changes in metabolic enzyme activities in the skeletal and heart muscles, Michaelidis *et al.* (2007) concluded that a shift from aerobic to anaerobic metabolism occurred in the marine fish *Sparus auratus* exposed to a pCO_2 of 5000 μatm for 10 days. This may indicate a loss in aerobic scope, which may compromise processes that depend on aerobic scope, such as growth. As a corollary, it must be emphasized that a clear comprehensive picture does not yet exist. At the CO_2 levels expected during ocean acidification scenarios and lower than 10 000 ppmv, metabolic depression may not occur even in the invertebrates, but metabolic stimulation may occur instead due to the stimulation of ion and acid–base regulation.

While the stoichiometric energy cost of some individual carriers has been quantified, the total cost of acid–base regulation in a cell or whole organism has not yet been estimated. It depends upon the rates of proton-equivalent ion exchange at cellular and whole-organism levels and on the fractional contribution of various carriers to net ion flux. A high cost of acid–base regulation may influence the energy budget and reduce the aerobic scope available for other functions. However, available studies

have not found a significant influence of pCO_2 below 10 000 μatm on somatic growth rates in marine teleosts, as shown by Fivelstad *et al.* (1998) in post smolts of *Salmo salar* studied for 6 weeks, by Foss *et al.* (2003) in juveniles of *Anarhichas minor* studied for 10 weeks, or by Foss *et al.* (2006) in juveniles of *Gadus morhua* studied for 9 weeks as well as by Munday *et al.* (2009c) in juveniles of tropical marine fishes. In all of these studies seawater pH remained unbuffered, no sodium bicarbonate salt was added, and all were carried out in full-strength seawater. Thus seawater carbonate chemistry resembled extreme ocean acidification. The aforementioned studies were, however, performed in dense monoculture conditions with *ad libitum* feeding. The effects of long-term exposure to elevated CO_2 on ecologically relevant parameters, such as individual fitness, competition, and predation influences have not been experimentally addressed yet.

The extended duration of many growth experiments (30–275 days; Ishimatsu *et al.* 2005) may overcome the pitfalls associated with acute studies and allow for acclimatory responses, which compensate for CO_2-induced shifts in energy budget. Such compensation was observed in a study that recorded growth rates of juvenile spotted wolfish (*Anarhichas minor*) on a weekly basis at a pCO_2 of 10 000 μatm. Despite a 20% decrease in growth during the first 3 weeks of exposure, specific growth rates returned to control values for the remainder of the 10-week-long trials (Foss *et al.* 2003). Differing acclimation patterns between short- and long-term exposures have also been documented in expression levels of acid–base relevant ion transporters in gills of the eelpout *Z. viviparus* over a time course of 6 weeks at a pCO_2 of 10 000 μatm (Deigweiher *et al.* 2008). To further elucidate interspecies variability and acclimation plasticity in teleosts, studies that examine baseline transporter expression profiles and distinguish between acute and long-term responses to ocean acidification are necessary.

Recent studies on acutely exposed tropical cardinal fishes from coral reefs have documented a surprisingly high sensitivity to ocean acidification. Aerobic scope and critical swimming speed decreased significantly in both *Ostorhinchus doederleini* and *Ostorhinchus cyanosoma* after 1 week of acclimation to 1000 μatm (Munday *et al.* 2009a).

Aerobic performance decreased by 37 and 47%, respectively. This finding contrasts with that in Atlantic cod (*G. morhua*) where there was no significant change in swimming capacity after 12 months of exposure to both 3000 and 6000 µatm (Melzner *et al.* 2009b). This difference may indicate a higher sensitivity of warm-water fishes but may also relate to the different durations of the respective acclimation periods (12 months versus 1 week). It remains to be investigated whether long-term acclimation to elevated seawater pCO_2 would alleviate the response found in the two tropical fish species. It is also important to consider that the cod study was conducted well within the thermal range of the species; this may be one reason why exposure to elevated CO_2 did not affect the resting and active oxygen consumption rates. Unchanged aerobic scope at intermediate temperatures coincides with the generalized conclusion that fish growth is unaffected until pCO_2 becomes much higher than 15 000 µatm (Ishimatsu *et al.* 2008). Overall, most studies indicate that adult temperate zone teleosts, if in the midst of their thermal range, will not be sensitive to increases of seawater pCO_2 of about 1000 µatm above normocapnic control values.

Nearly all experiments examining the sensitivity of marine teleosts to ocean acidification have been conducted with temperate fishes. Therefore, the finding of a surprisingly high sensitivity to ocean acidification in the two species of tropical cardinal fishes emphasizes the need to conduct studies in a broad range of habitats and climate zones. The decrease in aerobic scope of the coral reef fishes occurred especially at very low or high temperatures. Also, the acclimation temperature was very different (~5°C for cod and 29 to 32°C for the tropical cardinal fishes). These differences probably coincide with different capacities for acid–base regulation and cardiovascular or ventilatory systems. At 5°C, metabolic rates are low and the seawater oxygen concentration is more than 30% higher than in tropical waters. In light of the principles of the concept of OCLTT, tropical fishes live closer to the edge of oxygen limitation than temperate fishes. This may render tropical fishes more sensitive to environmental change in general, and even more so to the combined effects of temperature extremes and ocean acidification. In fact, the data of

Munday *et al.* (2009a) illustrate that sensitivity to elevated CO_2 is strongly enhanced at the low and high ends of the thermal window, similar to earlier findings in crustaceans (Walther *et al.* 2009), thereby confirming the projections by Pörtner and Farrell (2008) for the effect of elevated pCO_2 on marine fauna. It remains to be explored to what extent the combined effects of temperature and elevated CO_2 can be compensated for by acclimatization. It is conceivable that at the limits of thermal acclimatization capacity, the capacity to acclimatize to elevated pCO_2 levels is also reduced, and vice versa.

8.3 Effects of ocean acidification on cephalopods

Among nektonic animals the oceanic squids are the most active cephalopods, competing with active pelagic fishes such as tuna and marlin. These muscular squid display a very high oxygen demand (Seibel 2007; Seibel and Drazen 2007) and are hypothesized to live near the limit to oxygen availability. This high demand reflects high activity levels, as dictated by the pelagic environment, as well as the low energy efficiency that characterizes the squid's jet propulsion relative to other forms of locomotion (O'Dor and Webber 1986). The metabolic capacity of these muscular squids is surprising if one considers the inherent constraints on their metabolism and oxygen transport system. The oxygen-carrying capacity of squid blood is low relative to similarly active fishes due to viscosity-related constraints associated with an extracellular respiratory protein. Although squid are rather eurythermal due to their mode and rate of metabolism, they are considered vulnerable to the combined impact of climate-related variables (Pörtner and Reipschläger 1996; Pörtner and Zielinski 1998; Pörtner 2002; Rosa and Seibel 2008). However, information on the sensitivity to ocean acidification is much more limited in cephalopod molluscs than in teleost fishes.

Oxygen transport in the blood of cephalopods occurs via the extracellular pigment haemocyanin, which is highly responsive to pH changes (Pörtner 1990; Bridges 1994), a property that facilitates the delivery of oxygen to demanding tissues. In ommastrephid squids, such pH sensitivity is maximized and fine-tuned to support the full oxygen loading

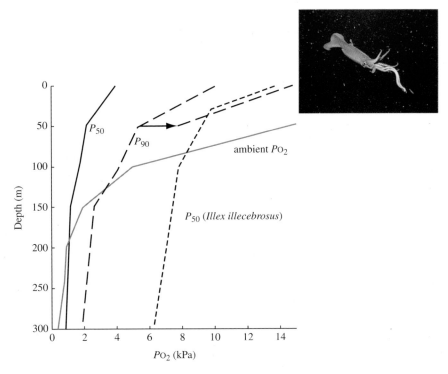

Figure 8.3 Relationship between blood-oxygen saturation and depth in the Humboldt squid, *Dosidicus gigas* (B. Seibel, unpublished data). The ambient oxygen content of water in the Gulf of California decreases as a function of depth, resulting in reduced oxygen saturation of the blood. The PO_2 resulting in 50 and 90% saturation in *D. gigas* (P_{50} and P_{90}) decreases with depth (i.e. the affinity of haemocyanin for oxygen increases) due to reduced temperature at depth. For comparison, the P_{50} for another ommastrephid squid, *Illex illecebrosus*, is shown from the more oxygenated Atlantic ocean (dashed line; data from Pörtner 1990). The lower P_{50} of *D. gigas* reflects an adaptation to the pronounced oxygen minimum zone in the eastern Pacific. At depths above ~200 m, the blood of *D. gigas* is at least 50% saturated. At depths shallower than ~130 m, its blood is at least 90% saturated. In contrast, *I. illecebrosus* would have to stay at depths shallower than 100 m to achieve 50% saturation if it lived in the eastern Pacific. The critical oxygen partial pressure (P_{crit}) occurs near 160 m for *D. gigas*. These numbers will change significantly with exposure to elevated CO_2 levels and uncompensated changes in blood pH as illustrated in shallow water for *D. gigas* (see arrow). A Pco_2 of 2000 µatm results in a 0.15 pH unit decrease in arterial pH assuming a buffering capacity similar to *I. illecebrosus* (Pörtner 1990). The Bohr coefficient of the *D. gigas* blood ($\Delta \log P_{50}/\Delta pH = -1.1$) results in a decrease in oxygen affinity (P_{90} at Pco_2 = 2000 µatm, dashed line). This effect may narrow the width of habitable water layers.

and then unloading of the pigment on each cycle through the body, leaving no venous oxygen reserve. Maximum oxygen transport in the blood, supplemented by oxygen provision via the skin of the working mantle musculature, reflects the optimized capacity and maximum use of the oxygen supply machinery in some swimming squids (Pörtner 1990, 1994). Consequently, muscular squids are thought to live chronically on the edge of oxygen limitation (Pörtner 2002) and are not well poised to adapt to future environmental changes that influence oxygen supply and demand (Fig. 8.3). Maintenance of extracellular pH is particularly important, and any lowering of pH in arterial blood endangers the uptake

of oxygen from the water and its binding to haemocyanin. This may occur in hypoxic, CO_2-rich seawater as in oxygen minimum zones (OMZs).

Given such constraints, it is surprising that a large ommastrephid squid, the Humboldt squid *Dosidicus gigas*, is associated with the distribution of pronounced OMZs in tropical oceans. These develop where the water is thermally stratified and a high surface productivity supports formation of a high biomass of respiring organisms at depth. Nearly 8% of the world's oceans contain less than 20 µmol O_2 kg^{-1} (Paulmier and Ruiz-Pino 2009). Since oxygen consumption is accompanied by CO_2 production, OMZs also tend to have low pH and to be undersaturated with respect to $CaCO_3$ (Millero 1996; Fabry et al. 2008). In the eastern Tropical Pacific D. gigas migrates from warm shallow waters at night into a layer of cooler, but oxygen-depleted, hypercapnic water at a depth of about 300 m during the day (Gilly et al. 2006). The hypoxia they experience during these vertical excursions is sufficient to exclude many top vertebrate predators (Carey and Robison 1981; Prince and Goodyear 2006). In contrast, the Humboldt squid appears to thrive in OMZs. Over the last few years, D. gigas has greatly extended its tropical/subtropical range as far north as Canada and Alaska, a range expansion correlated with the expanding OMZ and with the decline of some commercial fish stocks (Zeidberg and Robison 2007).

Some authors have proposed using naturally acidified water, such as that found near hydrothermal vents or in OMZs, as a natural laboratory for studying the mechanisms that organisms may use to adjust to ocean acidification (Fabry et al. 2008; Tunnicliffe et al. 2009). However, Rosa and Seibel (2008, 2011) demonstrated that the effects of elevated pCO_2 on activity and metabolism in the OMZ are confounded by, and negligible in comparison to, the effects of the hypoxia. Metabolism and activity are suppressed at temperatures and oxygen levels consistent with the daytime habitat depth of D. gigas. While in some organisms metabolic suppression (Hand 1991) is acutely triggered by high CO_2 tensions (10 000 µatm) and the associated low level of extracellular pH (Reipschläger and Pörtner 1996), a pCO_2 of 1000 µatm had no effect on routine or resting metabolism or on activity levels of D. gigas at 10°C (Rosa and Seibel 2008). However, after a 12-h acclimation period and a 24-h exposure period to a pCO_2 of 1000 µatm, active metabolism decreased at higher activity levels and higher temperatures (e.g. 25°C), from 70 to 48 µmol O_2 g^{-1} h^{-1} on average, a reduction by 31%. Although the OMZ environment is characterized by both hypoxia and high CO_2 levels these results indicate that for the squid in its cold, hypoxic daytime habitat at depth the effect of oxygen availability on metabolic rate overwhelms the more subtle CO_2 effect. However, at warmer temperatures and higher activity levels in the squid's night-time habitat, CO_2 has an important influence on the metabolic rate of D. gigas. This pattern matches the projection developed above

that CO_2 effects become stronger at the edges of the thermal window.

To date, the response of acid–base regulation to elevated pCO_2 has been examined in a single nekto-benthic cephalopod species, the cuttlefish, Sepia officinalis. This species exhibits active compensation of blood pH during exposure to a pCO_2 of 6000 µatm. Unlike teleosts, pH compensation is only partial (Gutowska et al. 2010a). Despite a decrease in extracellular pH by 0.2 units (Fig. 8.4), S. officinalis maintained control standard metabolic rates during acute exposure, and control growth rates over a 6-week period (Fig. 8.5). The oxygen saturation of the arterial blood was not significantly modified by a pH decrease of 0.2 units. The tolerance of the cuttlefish to hypercapnia may thus partly be attributed to its high capacity to maintain blood pH, which is higher than in other invertebrates, and the comparatively low oxygen demand and insensitivity of its haemocyanin to acidosis.

To increase our understanding of the sensitivity of cephalopods to ocean acidification, the mechanisms of acid–base regulation and ion transport in their gills need to be characterized at the same level of detail as in fish (see Section 8.2). In S. officinalis, the transporter Na^+/K^+-ATPase has been localized in the basolateral membranes of the gills and displays activity levels similar to those in teleosts (Schipp et al. 1979; Hu et al. 2010). A recent study found a transient elevation (15%) in maximum activity of Na^+/K^+-ATPase during the first 10 days of exposure to a pCO_2 of 3000 µatm in juvenile S. officinalis (M. Y. Hu et al. pers. comm.). Activity returned to control values after 6 weeks of exposure. Furthermore, there were no significant CO_2-related expression responses in transcripts of 29 proteins from various gene ontology classes in the gills (e.g. ion transporters, metabolic enzymes, stress-response proteins, and transcription factors). The electrogenic Na^+/HCO_3^- cotransporter NBC1 may play an important role in the hypercapnic response of cephalopods, similar to its role in fishes. The presence and electrogenic nature of NBCs have recently been demonstrated in the giant fibre lobe of squid (Loligo pealei; Piermarini et al. 2007). Whether this or other isoforms contribute to the hypercapnic response in the branchial tissues of cephalopods remains to be established. Similar to teleosts, it can

(A)

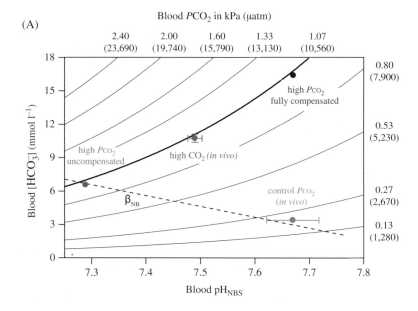

(B)

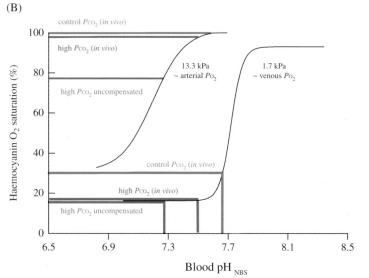

Figure 8.4 Regulation of blood pH in the cephalopod *Sepia officinalis*. (A) pH–bicarbonate diagram, displaying blood bicarbonate (HCO_3^-) concentration, pH, and PCO_2 isobars (modified from Gutowska *et al.* 2010a). Under control conditions blood PCO_2 in cephalopods is typically 0.2 to 0.3 kPa (about 2000 to 3000 µatm) higher than ambient in order to maintain a diffusive flux of respiratory CO_2 out of the organism. When seawater pCO_2 increases to ~0.6 kPa (about 5900 µatm), blood PCO_2 increases to ~1 kPa (about 9900 µatm, thick black isobar). Such an increase to a high blood PCO_2 would lead to a blood pH of ~7.3 if no active compensation occurred, such that blood pH follows the buffering capacity of the blood proteins (dashed buffering line). However, *S. officinalis* actively modifies the carbonate system of its blood to stabilize pH at a higher value *in vivo* (~7.5) by means of HCO_3^- accumulation (equivalent to net proton excretion, see Melzner *et al.* 2009a for further discussion). For full pH compensation, *S. officinalis* would have to increase blood HCO_3^- to more than 15 mmol l^{-1} (top right). Blood pH regulation has large implications for the function of the extracellular respiratory pigment haemocyanin. (B) *In vitro* blood haemocyanin oxygen-binding curves for *S. officinalis* at two different oxygen partial pressures that correspond to arterial and venous values recorded *in vivo* (modified from Johansen *et al.* 1982; Zielinski *et al.* 2001; Gutowska *et al.* 2010a). Lines indicate saturation of the pigment under arterial and venous conditions under control and high PCO_2 conditions (uncompensated and partially compensated as *in vivo*). On the arterial side, it appears that the partial pH compensation observed *in vivo* is crucial for maintaining full oxygen saturation of haemocyanin in the gills. Uncompensated blood pH would lead to <80% oxygen saturation under arterial conditions. As cephalopods are chronically oxygen limited (e.g. O'Dor and Webber 1991; Hochachka 1994; Pörtner 1994; Finke *et al.* 1996), such a decrease in arterial haemocyanin saturation could significantly decrease fitness.

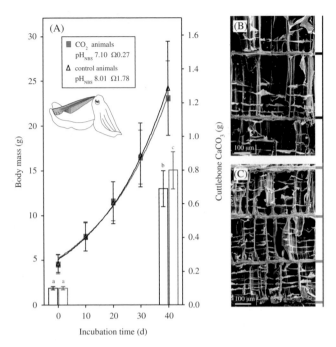

Figure 8.5 (A) Growth (left *y*-axis) and CaCO₃ gain (bars, right *y*-axis) in the cuttlefish, *Sepia officinalis*, during exposure to control (black, 400 µatm) and elevated pCO₂ (6000 µatm; grey) (after Gutowska *et al.* 2008). Means for cuttlebone CaCO₃ not sharing the same letter are significantly different (*n* = 20 in each group). Scanning electron micrographs show cuttlebone sections calcified in cuttlefish kept under control conditions (B) and under a pCO₂ of 6000 µatm (C). Increased CaCO₃ accretion during exposure to elevated seawater pCO₂ is also evident in the microstructure of the cuttlebones (data are means ± SD).

be concluded that standard ion regulatory capacities in the cuttlefish are sufficient to support acid–base regulation during exposure to a pCO₂ of 3000 µatm, and that the cost of regulation does not significantly affect energy allocation to somatic growth, at least in the centre of the thermal window. It needs to be emphasized that cuttlefish have lower metabolic rates than the active squids discussed above, and may thus not be as sensitive to high CO₂.

During exposure to ocean acidification, changes in the acid–base parameters of extracellular compartments have the potential to influence not only metabolism but also calcification. Organisms that are unable to compensate for chronic acidosis in their extracellular space could experience reduced calcification rates due to decreased CaCO₃ saturation conditions at the site of calcification. In contrast, CaCO₃ saturation will increase at elevated pCO₂ in taxa that exhibit a strong regulatory response. With rising pCO₂ in body fluids, extracellular HCO₃⁻ and CO₃²⁻ will significantly accumulate once control extracellular pH values have been

successfully stabilized. Interestingly, increased calcification has been measured in both teleosts and cephalopods during exposure to elevated seawater pCO₂. Otolith mass (calculated from changes in surface area) was found to increase by approximately 12% and 25% in 7-day-old *Atractoscion nobilis* reared under pCO₂ values of 1000 and 2500 µatm, respectively (Checkley *et al.* 2009). Similar hypercalcification of an aragonitic structure has also been found in *S. officinalis*. Cuttlebones of juvenile cuttlefish maintained for 6 weeks at 6000 µatm accreted 22–55% more CaCO₃ (depending on final body mass) than those maintained under control conditions. At elevated pCO₂ the cuttlebone lamellar spacing decreased from 384 ± 26 µm to 195 ± 38 µm and pillar thickness increased from 2.6 ± 0.6 µm to 4.9 ± 2.2 µm (means ± SD; Gutowska *et al.* 2010b; Fig. 8.5). Further studies are required to examine the influence of hypercalcification at elevated pCO₂ levels on the functional control of otoliths for orientation, and on the function of the cuttlebone as a buoyancy regulation device.

8.4   Conclusions and perspectives

Surface seawater will continue to warm and its pH will continue to decline during the remainder of this century. Higher temperatures will stimulate metabolism and require enhanced performance from thermally constrained oxygen transport systems in cephalopods and fishes. In cephalopods in particular, oxygen supply capacity will be impaired by ocean acidification. Thus warming and acidification may cause ventilatory and circulatory stresses that restrict aerobic scope and impair swimming activity. Together these variables may reduce an animal's ability to respond to external stimuli, leaving it more vulnerable to its main predators. Ocean warming and acidification will increasingly be accompanied by expanding hypoxic zones (Stramma et al. 2008; Seibel 2011). If the OMZs continue to expand vertically, the Humboldt squid, D. gigas, for example, will have to retreat to even shallower waters at night in order to reduce any accumulated oxygen debt and to hunt actively. Thus, acidification and warming of the surface ocean may create a ceiling that will preclude these squid from entering shallow waters, while the expanding hypoxic zone will increase the depth below which they cannot penetrate during their night-time recovery from hypoxia. The synergistic effect of these three climate-related variables may be to vertically compress the habitable depth range of the species.

The first level of sensitivity of animals to extreme temperatures, hypoxia, or ocean acidification involves a limitation in the functional capacity of tissues, including those responsible for oxygen supply to cells and their mitochondria. The sustainable performance or fitness of an animal thus becomes constrained once its functional capacity is limited and/or it cannot exploit its metabolic capacity for aerobic energy production. Therefore, understanding the functional specialization of an animal to its environment, including the environmental variability of these factors (Pörtner 2006), involves an in-depth understanding of the mechanisms controlling oxygen supply and respiratory activity. Likewise, evolutionary adaptation to various climate zones, lifestyles, and associated exercise capacities as well as to variable natural CO_2 levels are likely to shape the sensitivity of marine fauna to ongoing ocean acidification, as illustrated by the contrasting results for tropical coral reef fishes and those from temperate regions.

In this context, the intuitively appealing concept of a 'respiration index' (RI) = $\log_{10}(pO_2/pCO_2)$ was recently proposed to identify the limits to metazoan life based upon perceived chemical constraints in aerobic metabolism (Brewer and Peltzer 2009). The authors suggested that with falling ambient oxygen and increasing CO_2 levels there is a limiting threshold RI value below which insufficient free energy change is produced from the oxidative metabolism of organic matter for the organism to be viable. Such limiting or lethal conditions are those where reactions would be forced into equilibrium by falling pO_2 and rising pCO_2. However, such limits are extreme and found far beyond the actual conditions limiting animal life, as mitochondria, the sites of oxidative metabolism, see lower than ambient oxygen and higher than ambient CO_2 concentrations. As a corollary, even if such limiting thermodynamic conditions can be defined for aerobic metabolism in a test tube or in a hypothetical unicellular organism with diffusion distances set to zero, the complexity of the metazoan body plan requires consideration. In animals, convective mechanisms (ventilation and circulation) are used to bridge the distance between ambient media and the mitochondria. Ventilation and circulation minimize diffusion gradients and compensate, within limits, for a decrease in ambient oxygen and an increase in CO_2 levels. The capacity of convective mechanisms is specific for a species and influenced by its adaptation to lifestyles and environmental parameters including those set by climate conditions. Consequently, the capacities to defend the intracellular concentrations of reactants (e.g. oxygen or CO_2) vary greatly between species. These concentrations are very different from those in the surrounding seawater. As a result of species-specific gas exchange capacities in large complex bodies, limiting RI values are species-specific values, which must be determined experimentally and are much higher than with diffusion distances set to zero as in the hypothetical organism above. In general, organisms are open systems that acquire free energy from their surroundings and operate approximately in a steady state. The concentrations of substrates and products in living systems (including those of oxygen and CO_2) are actively maintained at disequilibrium to ensure a constant Gibb's free energy change, and thus thermodynamic drive, for forward flux through the pathways of energy metabolism (cf. Seibel *et al.* 2009). These considerations preclude the use of RI as a simple indicator of limiting conditions in metazoan habitats.

Individual species have evolved unique tolerances to their habitats. Comparative physiology has defined critical gas tensions, which reflect species-specific adaptations and limitations in oxygen uptake and acid–base regulation that have evolved for each species within their specific habitat. However, such limits specific for one factor benefit from the integration of limits to other factors to take the full range of environmental stressors into account (Pörtner 2010). With respect to the specific interactions between hypoxia and hypercapnia, in addition to the evaluation of critical oxygen partial pressures (Pc_{O_2}) or CO_2 tensions (Pc_{CO_2}) as traditionally done by comparative and environmental physiologists, an experimentally determined, redefined RI, which integrates the whole-organism limits to both gases, may be useful as a proxy to qualify resistance to the synergistic effects of hypoxia and hypercapnia. Such a proxy would need to be tested with respect to whether the sensitivity threshold to hypoxia (Pc_{O_2}) is influenced by hypercapnia, and vice versa. Only on the basis of such empirical data would it be suitable to project species-specific critical levels of climate-induced hypoxia and ocean acidification or critical conditions of combined hypoxia and hypercapnia in OMZs. Any predictive use of such an indicator requires additional mechanistic understanding of physiological tolerance and its ecological implications.

As a general conclusion, data on the effects of hypercapnia are scarce in nektonic species as in other groups. This is especially true for the long-term effects on processes such as investments in reproduction, development, growth, foraging capacity, behaviour(s), and resistance to disease which are crucial in setting fitness levels and, as a result, competitiveness with other species. Multigeneration studies or studies of species populations in different climate zones (according to differences in ambient pCO_2, temperature, and associated CO_2 solubilities) are needed to address the evolutionary consequences of ocean acidification.

Increased mechanistic knowledge for specific groups of organisms according to climate and habitat characteristics can help in the development of long-term projections of ecosystem change. Identifying the physiological background of relative changes in performance and competitiveness is important for identifying winners and losers at ecosystem levels (Pörtner and Farrell 2008). For a comprehensive understanding of the effects of ocean acidification in the context of further environmental challenges, the concept of OCLTT provides a matrix suitable for the integration of various stressors (Pörtner 2010). The data currently available in marine invertebrates and fishes support the following framework: the relatively low sensitivity of nektonic, and thus active, organisms to ocean acidification is related to their high capacity for (extracellular) acid–base regulation. Efficient acid–base regulation results in an accumulation of base in the extracellular and intracellular fluids, which facilitates calcification from ions in 'internal' fluids. In 'good' acid–base regulators such internally controlled calcification then operates at a higher saturation of calcium carbonates (see Box 1.1, Chapter 1). However, elevated PCO_2 values in body fluids might at the same time reduce relevant performance capacities, for example by causing unfavourable shifts in cellular, tissue, and organismal energy budgets and depressing growth and reproduction, or by 'dampening' organism behaviour(s) through neurobiological effects (e.g. by triggering the accumulation of adenosine). Transient exposure to elevated pCO_2 levels combined with hypoxia as in the OMZ or in intertidal sediments may even be beneficial (Reipschläger *et al.* 1997), for example by supporting strategies for saving energy. A transient reduction in spontaneous activity may result, which, if it does not interfere with foraging success or other crucial behaviours, can support an increase in growth efficiency and at the same time a reduced oxygen demand, in similar ways as adaptation to progressively colder temperatures along a latitudinal cline enhances growth efficiency (Heilmayer *et al.* 2004). Depending on environmental scenarios, on the level of energy turnover and associated mode of life of exposed organisms, and on their specific physiological capacities for resistance, acclimatization, and adaptation, a diversity of responses appears conceivable. Disentangling such diversity and its ecosystem-level implications is a challenging task for future studies.

References

Anestis, A., Lazou, A., Pörtner, H.-O., and Michaelidis, B. (2007). Behavioural, metabolic and molecular stress indicators in the marine bivalve *Mytilus galloprovincialis* during long-term acclimation at increasing ambient temperature. *American Journal of Physiology*, **293**, R911–21.

Bograd, S.J., Schroeder, I., Sarkar, N., Qiu, X.M., Sydeman, W.J., and Schwing, F.B. (2009). Phenology of coastal upwelling in the California Current. *Geophysical Research Letters*, **36**, L01602, doi:10.1029/2008GL035933.

Brewer, P.G. and Peltzer, E.T. (2009). Limits to marine life. *Science*, **324**, 347–8.

Bridges, C.R. (1994). Bohr and Root effects in cephalopod haemocyanins—paradox or pressure in *Sepia officinalis*? In: H.-O. Pörtner, R.K. O'Dor, and D.L. MacMillan (eds), *Physiology of cephalopod molluscs: lifestyle and performance adaptations*, pp. 121–30. Gordon and Breach, New York.

Caldeira, K. and Wickett, M.E. (2005). Ocean model predictions of chemistry changes from carbon dioxide emissions to the atmosphere and ocean. *Journal of Geophysical Research*, **110**, C09S04, dpi:10.1029/2004JC002671.

Carey, F.G. and Robison, B.H. (1981). Daily patterns in the activities of swordfish, *Xiphias gladius*, observed by acoustic telemetry. *Fisheries Bulletin*, **79**, 277–92.

Checkley, D.M., Dickson, A.G., Takahashi, M., Radich, J.A., Eisenkolb, N., and Asch, R. (2009). Elevated CO_2 enhances otolith growth in young fish. *Science*, **324**, 1683.

Claiborne, J.B., Edwards, S.L., and Morrison-Shetlar, A.I. (2002). Acid-base regulation in fishes: cellular and molecular mechanisms. *Journal of Experimental Zoology*, **293**, 302–19.

Claireaux, G. and Lefrançois, C. (2007). Linking environmental variability and fish performance: integration through the concept of scope for activity. *Philosophical Transactions of the Royal Society B: Biological Sciences*, **362**, 2031–41.

Cooper, C.A., Whittamore, J.M., and Wilson, R.W. (2010). Ca^{2+}-driven intestinal HCO_3^- secretion and $CaCO_3$ precipitation in the European flounder *in vivo*: influences on acid-base regulation and blood gas transport. *American Journal of Physiology*, **298**, R870–6.

Deigweiher, K. (2009). *Impact of high CO_2 concentrations on marine life: molecular mechanisms and physiological adaptations of pH and ion regulation in marine fish*. PhD Thesis, Universität Bremen.

Deigweiher, K., Koschnick, N., Pörtner, H.-O., and Lucassen, M. (2008). Acclimation of ion regulatory capacities in gills of marine fish under environmental hypercapnia. *American Journal of Physiology*, **295**, R1660–R1670.

Deigweiher, K., Hirse, T., Bock, C., Lucassen, M., and Pörtner, H.-O. (2010). Hypercapnia induced shifts in gill energy budgets of Antarctic notothenioids. *Journal of Comparative Physiology B*, **180**, 347–59.

Evans, D.H., Piermarini, P.M., and Choe, K.P. (2005). The multifunctional fish gill: dominant site of gas exchange, osmoregulation, acid-base regulation, and excretion of nitrogenous waste. *Physiological Reviews*, **85**, 97–177.

Fabry, V.J., Seibel, B.A., Feely, R.A., and Orr, J.C. (2008). Impacts of ocean acidification on marine fauna and ecosystem processes. *ICES Journal of Marine Science*, **65**, 414–32.

Farrell, A.P., Elison, E.J., Sandblom, E., and Clark, T.D. (2009). Fish cardiorespiratory physiology in an era of climate change. *Canadian Journal of Zoology*, **87**, 835–51.

Feidantsis, K., Pörtner, H.-O., Lazou, A., Kostoglou, B., and Michaelidis, B. (2009). Metabolic and molecular stress responses of the gilthead seabream *Sparus aurata* during long term exposure to increasing temperatures. *Marine Biology*, **156**, 797–809.

Findlay, H., Burrows, M.T., Kendall, M.A., Spicer, J.I., and Widdicombe, S. (2010). Can ocean acidification affect population dynamics of the barnacle *Semibalanus balanoides* at its southern range edge? *Ecology*, **91**, 2931–40.

Finke, E., Pörtner, H.-O., Lee, P.G., and Webber, D.M. (1996). Squid (*Lolliguncula brevis*) life in shallow waters: oxygen limitation of metabolism and swimming performance. *Journal of Experimental Biology*, **199**, 911–21.

Fivelstad, S., Haavik, H., Lovik, G., and Olsen, A.B. (1998). Sublethal effects and safe levels of carbon dioxide of Atlantic salmon post-smolts (*Salmo salar*). *Aquaculture*, **160**, 305–16.

Foss, A., Rosnes, B.A., and Oiestad, V. (2003). Graded environmental hypercapnia in juvenile spotted wolfish (*Anarhichas minor* Olafsen): effects on growth, food conversion efficiency an nephrocalcinosis. *Aquaculture*, **220**, 607–17.

Foss, A., Kristensen, T., Atland, A. *et al.* (2006). Effects of water reuse and stocking density on water quality, blood physiology an growth rate of juvenile cod (*Gadus morhua*). *Aquaculture*, **256**, 255–63.

Frankignoulle, M., Abril, G., Borges, A., Bourge, I., Canon, C., Delille, B., Libert, E., and Théate, J.-M. (1998). Carbon dioxide emission from European estuaries. *Science*, **282**, 434–6.

Fry, F.E.J. (1971). The effect of environmental factors on animal activity. In: W.S. Hoar and D.J. Randall (eds), *Fish physiology*, Vol. 6, pp. 1–98. Academic Press, New York.

Fyhn, H.J., Finn, R.N., Reith, M., and Norberg, B. (1999). Yolk protein hydrolysis and oocyte free amino acids as key features in the adaptive evolution of teleost fishes to seawater. *Sarsia*, **84**, 451–6.

Gallaugher, P. and Farrell, A.P. (1998). Hematocrit and blood oxygen-carrying capacity. In: S.F. Perry and B. Tufts (eds), *Fish physiology*, Vol. 17, pp. 185–227. Academic Press, San Diego.

Gilly, W.F., Markaida, U., Baxter, C.H. *et al.* (2006). Vertical and horizontal migrations by the jumbo squid *Dosidicus gigas* revealed by electronic tagging. *Marine Ecology Progress Series*, **324**, 1–17.

Gutowska, M.A., Pörtner, H.-O., and Melzner, F. (2008). Growth and calcification in the cephalopod *Sepia officinalis* under elevated seawater *p*CO$_2$. *Marine Ecology Progress Series*, **373**, 303–9.

Gutowska, M.A., Melzner, F., Langenbuch, M., Bock, C., Claireux, G., and Pörtner, H.-O. (2010a). Acid-base regulatory ability of the cephalopod (*Sepia officinalis*) in response to environmental hypercapnia. *Journal of Comparative Physiology B*, **180**, 323–35.

Gutowska, M.A., Melzner, F., Pörtner, H.-O., and Meier, S. (2010b). Increased cuttlebone calcification during exposure to elevated seawater *p*CO$_2$ in the cephalopod *Sepia officinalis*. *Marine Biology*, **157**, 1653–63.

Hand, S.C. (1991) Metabolic dormancy in aquatic invertebrates. In: R. Gilles (ed.), *Advances in comparative and environmental physiology*, Vol. 8, pp. 1–50. Springer-Verlag, Heidelberg.

Hardewig, I., Pörtner, H.-O., and Grieshaber, M.K. (1994). Interactions of anaerobic propionate formation and acid-base status in *Arenicola marina*: an analysis of propionyl-CoA-carboxylase. *Physiological Zoology*, **67**, 892–909.

Hayashi, M., Kita, J., and Ishimatsu, A. (2004). Comparison of the acid-base responses to CO$_2$ and acidification in Japanese flounder (*Paralichthys olivaceus*). *Marine Pollution Bulletin*, **49**, 1062–5.

Heisler, N. (1986) Acid-base regulation in fishes. In: N. Heisler (ed.), *Acid–base regulation in animals*, pp. 309–56. Elsevier, Amsterdam.

Heilmayer, O., Brey, T., and Pörtner, H.-O. (2004). Growth efficiency and temperature in scallops: a comparative analysis of species adapted to different temperatures. *Functional Ecology*, **18**, 641–7.

Hirata, T., Kaneko, T., Ono, T. *et al.* (2003). Mechanism of acid adaptation of a fish living in a pH 3.5 lake. *American Journal of Physiology*, **284**, R1199–R1212.

Hochachka, P.W. (1994). Oxygen efficient design of cephalopod muscle metabolism. *Marine and Freshwater Behaviour and Physiology*, **25**, 61–7.

Hofmann, G.E. and Todgham, A.E. (2010). Living in the now: physiological mechanisms to tolerate a rapidly changing environment. *Annual Review of Physiology*, **72**, 127–45.

Hu, M.Y., Sucre, E., Charmantier-Daures, M., Charmantier, G., Lucassen, M., and Melzner, F. (2010). Localization of ion regulatory epithelia in embryos and hatchlings of two cephalopods. *Cell and Tissue Research*, **339**, 571–83.

Ishimatsu, A., Iwama, G.K., Bentley, T.B., and Heisler, N. (1992). Contribution of the secondary circulatory system to acid-base regulation during hypercapnia in rainbow trout (*Oncorhynchus mykiss*). *Journal of Experimental Biology*, **170**, 43–56.

Ishimatsu, A., Hayashi, M., Lee, K.S., Kikkawa, T., and Kita, J. (2005). Physiological effects on fishes in a high-CO$_2$ world. *Journal of Geophysical Research*, **110**, C09S09, doi:10.1029/2004JC002564.

Ishimatsu, A., Hayashi, M., and Kikkawa, T. (2008). Fishes in high-CO$_2$, acidified oceans. *Marine Ecology Progress Series*, **373**, 295–302.

Johansen, K., Brix, O., and Lykkeboe, G. (1982). Blood gas transport in the cephalopod, *Sepia officinalis*. *Journal of Experimental Biology*, **99**, 331–8.

Knoll, A.H., Bambach, R. K., Payne, J.L., Pruss, S., and Fischer, W.W. (2007). Paleophysiology and end-Permian mass extinction. *Earth Planetary Science Letters*, **256**, 295–313.

Kyprianou, T.-D., Pörtner, H.-O., Anestis, A., Kostoglou, B., and Michaelidis, B. (2010). Metabolic and molecular stress responses of gilthead seam bream *Sparus aurata* during exposure to low ambient temperature: an analysis of mechanisms underlying the winter syndrome. *Journal of Comparative Physiology B*, **180**, 1005–18.

Langenbuch, M. and Pörtner, H.-O. (2003). Energy budget of hepatocytes from Antarctic fish (*Pachycara brachycephalum* and *Lepidonotothen kempi*) as a function of ambient CO$_2$: pH-dependent limitations of cellular protein biosynthesis? *Journal of Experimental Biology*, **206**, 3895–903.

Larsen, B.K., Pörtner, H.-O., and Jensen, F.B. (1997). Extra- and intracellular acid-base balance and ionic regulation in cod (*Gadus morhua*) during combined and isolated exposures to hypercapnia and copper. *Marine Biology*, **128**, 337–46.

Lucassen, M., Koschnick, N., Eckerle, L.G., and Pörtner, H.-O. (2006) Mitochondrial mechanisms of cold adaptation in cod (*Gadus morhua*) populations from different climatic zones. *Journal of experimental Biology*, **209**, 2462–71.

Mallekh, R. and Lagardère, J.P. (2002). Effect of temperature and dissolved oxygen concentration on the metabolic rate of the turbot and the relationship between metabolic scope and feeding demand. *Journal of Fish Biology*, **60**, 1105–15.

Marshall, W.S. and Grosell, M. (2006). Ion transport, osmoregulation, and acid-base balance. In: D.H. Evans and J.B. Claiborne (eds), *The physiology of fishes*, 3rd edn, pp. 177–230. CRC Press, Boca Raton, FL.

Melzner, F., Gutowska, M.A., Langenbuch, M. *et al.* (2009a). Physiological basis for high CO_2 tolerance in marine ectothermic animals: pre-adaptation through lifestyle and ontogeny? *Biogeosciences*, **6**, 2313–31.

Melzner, F., Göbel, S., Langenbuch, M., Gutowska, M.A., Pörtner, H.-O., and Lucassen, M. (2009b). Swimming performance in Atlantic Cod (*Gadus morhua*) following long-term (4–12 months) acclimation to elevated seawater pCO_2. *Aquatic Toxicology*, **92**, 30–7.

Metzger, R., Sartoris, F.J., Langenbuch, M., and Pörtner, H.-O. (2007). Influence of elevated CO_2 concentrations on thermal tolerance of the edible crab *Cancer pagurus*. *Journal of Thermal Biology*, **32**, 144–51.

Michaelidis, B., Ouzounis, C., Paleras, A., and Pörtner, H.-O. (2005). Effects of long-term moderate hypercapnia on acid-base balance and growth rate in marine mussels *Mytilus galloprovincialis*. *Marine Ecology Progress Series*, **293**, 109–18.

Michaelidis, B., Spring, A., and Pörtner, H.-O. (2007). Effects of long-term acclimation to environmental hypercapnia on extracellular acid-base status and metabolic capacity in Mediterranean fish *Sparus aurata*. *Marine Biology*, **150**, 1417–29.

Millero, F.J. (1996). *Chemical oceanography*, 469 pp. CRC Press, Boca Raton, FL.

Mühlenbruch, K. (2004). *Hyperkapnie-induzierte Veränderungen des gewebespezifischen Aminosäurestoffwechsels beim marinen Invertebraten* Sipunculus nudus, 72 pp. Diplomarbeit, Universität Bremen und Alfred-Wegener-Institut, Bremerhaven.

Munday, P.L., Crawley, N.E., and Nilsson, G.E. (2009a). Interacting effects of elevated temperature and ocean acidification on the aerobic performance of coral reef fishes. *Marine Ecology Progress Series*, **388**, 235–42.

Munday, P.L., Dixson, D.L., Donelson, J.M. *et al.* (2009b). Ocean acidification impairs olfactory discrimination and homing ability of a marine fish. *Proceedings of the National Academy of Sciences USA*, **106**,1848–52.

Munday, P.L., Donelson, J.M., Dixson, D.L., and Endo, G.G.K. (2009c). Effects of ocean acidification on the early life history of a tropical marine fish. *Proceedings of the Royal Society B: Biological Sciences*, **276**, 3275–83.

Nikinmaa, M. (2006). Gas transport. In: D.H. Evans and J.B. Claiborne (eds), *The physiology of fishes*, pp. 153–74. CRC Press, Boca Raton, FL.

O'Dor, R. K. and Webber, D. M. (1986) The constraints on cephalopods: why squid aren't fish. *Canadian Journal of Zoology*, **64**, 1591–605.

O'Dor, R.K. and Webber, D.M. (1991). Invertebrate athletes: trade-offs between transport efficiency and power density in cephalopod evolution. *Journal of Experimental Biology*, **160**, 93–112.

Orr, J.C., Fabry, V.J., Aumont, O. *et al.* (2005). Anthropogenic ocean acidification over the twenty-first century and its impact on calcifying organisms. *Nature*, **437**, 681–6.

Paulmier, A. and Ruiz-Pino, D. (2009). Oxygen minimum zones (OMZs) in the modern ocean. *Progress in Oceanography*, **80**, 113–28.

Perry, S.F. and Gilmour, K.M. (2006). Acid-base balance and CO_2 excretion in fish: unanswered questions and emerging models. *Respiration Physiology and Neurobiology*, **154**, 199–215.

Perry, S.F., Kinkead, R., Gallaugher, P., and Randall, D.J. (1989). Evidence that hypoxemia promotes catecholamine release during hypercapnic acidosis in rainbow trout (*Salmo gairdneri*). *Respiration Physiology*, **77**, 351–64.

Perry, S.F., Furimsky, M., Bayaa, M. *et al.* (2003). Integrated responses of Na^+/HCO_3^- cotransporters and V-type H^+-ATPases in the fish gill and kidney during respiratory acidosis. *Biochimica Biophysica Acta*, **1618**, 175–84.

Piermarini, P.M., Choi, I., and Boron, W.F. (2007). Cloning and characterization of an electrogenic Na/HCO_3^- cotransporter from the squid giant fiber lobe. *American Journal of Physiology*, **292**, C2032–45.

Pörtner, H.-O. (1990). An analysis of the effects of pH on oxygen binding by squid (*Illex illecebrosus, Loligo pealei*) haemocyanin. *Journal of Experimental Biology*, **150**, 407–24.

Pörtner, H.-O. (1994). Coordination of metabolism acid-base regulation and haemocyanin function in cephalopods. In: H.O. Pörtner, R.K. O'Dor, and D.L. MacMillan (eds), *Physiology of cephalopod molluscs: lifestyle and performance adaptations*, pp. 131–48. Gordon and Breach, London.

Pörtner, H.-O. (2002). Environmental and functional limits to muscular exercise and body size in marine invertebrate athletes. *Comparative Biochemistry and Physiology A*, **133**, 303–21.

Pörtner H.-O. (2006). Climate dependent evolution of Antarctic ectotherms: an integrative analysis. *Deep Sea Research II*, **53**, 1071–104.

Pörtner, H.-O. (2008). Ecosystem effects of ocean acidification in times of ocean warming: a physiologist's view. *Marine Ecology Progress Series*, **373**, 203–17.

Pörtner, H.-O. (2010). Oxygen and capacity limitation of thermal tolerance: a matrix for integrating climate related stressors in marine ecosystems. *Journal of Experimental Biology*, **213**, 881–93.

Pörtner, H.-O. and Farrell, A.P. (2008). Physiology and climate change. *Science*, **322**, 690–2.

Pörtner, H.-O. and Knust R. (2007). Climate change affects marine fishes through the oxygen limitation of thermal tolerance. *Science*, **315**, 95–7.

Pörtner, H.-O. and Reipschläger, A. (1996). Ocean disposal of anthropogenic CO_2: physiological effects on tolerant and intolerant animals. In: B. Ormerod and M.V. Angel (eds), *Ocean storage of carbon dioxide, Workshop 2—Environmental impact*, pp. 57–81. IEA Greenhouse Gas R&D Programme, Cheltenham, UK.

Pörtner, H.-O. and Zielinski, S. (1998). Environmental constraints and the physiology of performance in squids. *South African Journal of Marine Science*, **20**, 207–21.

Pörtner, H.-O., Reipschläger, A., and Heisler, N. (1998). Metabolism and acid-base regulation in *Sipunculus nudus* as a function of ambient carbon dioxide. *Journal of Experimental Biology*, **201**, 43–55.

Pörtner, H.-O., Bock, C., and Reipschläger, A. (2000). Modulation of the cost of pH_i regulation during metabolic depression: a ^{31}P-NMR study in invertebrate (*Sipunculus nudus*) isolated muscle. *Journal of Experimental Biology*, **203**, 2417–28.

Pörtner, H.-O., Langenbuch, M., and Reipschläger, A. (2004). Biological impact of elevated ocean CO_2 concentrations: lessons from animal physiology and earth history. *Journal of Oceanography*, **60**, 705–18.

Pörtner, H.-O., Langenbuch, M., and Michaelidis, B. (2005a). Synergistic effects of temperature extremes, hypoxia, and increases in CO_2 on marine animals: from Earth history to global change. *Journal of Geophysical Research*, **110**, C09S10, doi: 10.1029/2004JC002561.

Pörtner, H.-O., Lucassen, M., and Storch, D. (2005b). Metabolic biochemistry: its role in thermal tolerance and in the capacities of physiological and ecological function. In: A.P. Farrell and J.F. Steffensen (eds), *The physiology of polar fishes. Fish physiology* Vol. 22, pp. 79–154. Academic Press, New York.

Pörtner, H.-O., Bock, C., Knust, R. *et al.* (2008). Cod and climate in a latitudinal cline: physiological analyses of climate effects in marine fishes. *Climate Research*, **37**, 253–70.

Prince, E.D. and Goodyear, C.P. (2006). Hypoxia-based habitat compression of tropical pelagic fishes. *Fisheries Oceanography*, **15**, 451–64.

Ramnanan, C.J. and Storey, K.B. (2006). Suppression of Na^+/K^+ ATPase activity during estivation in the land snail *Otala lactea*. *Journal of Experimental Biology*, **209**, 677–88.

Reipschläger, A. and Pörtner, H.-O. (1996). Metabolic depression during environmental stress: the role of extra- versus intracellular pH in *Sipunculus nudus*. *Journal of Experimental Biology*, **199**, 1801–7.

Reipschläger, A., Nilsson, G.E., and Pörtner, H.-O. (1997). A role for adenosine in metabolic depression in the marine invertebrate *Sipunculus nudus*. *American Journal of Physiology*, **272**, R350–R356.

Rosa, R. and Seibel, B.A. (2008). Synergistic effect of climate-related variables suggests future physiological impairment in a top oceanic predator. *Proceedings of the National Academy of Sciences USA*, **52**, 20776–80.

Rosa, R. and Seibel, B.A. (2010). Metabolic physiology of the Humboldt squid, *Dosidicus gigas*: implications for vertical migration in a pronounced oxygen minimum zone. *Progress in Oceanography*, **86**, 72–80.

Schipp, R., Mollenhauer, S., and Boletzky, S. (1979). Electron microscopical and histochemical studies of differentiation and function of the cephalopod gill (*Sepia officinalis* L.). *Zoomorphology*, **93**, 193–207.

Seibel, B.A. (2007). On the depth and scale of metabolic rate variation: scaling of oxygen consumption and enzymatic activity in the class Cephalopoda. *Journal of Experimental Biology*, **210**, 1–11.

Seibel, B.A. (2011). Critical oxygen levels and metabolic suppression in oceanic oxygen minimum zones. *Journal of Experimental Biology*, **214**, 326–336.

Seibel, B.A. and Drazen, J. C. (2007). The rates of metabolism in marine animals: environmental constraints, ecological demands and energetic opportunities. *Philosophical Transactions of the Royal Society B: Biological Sciences*, **362**, 2061–78.

Seibel, B.A. and Fabry, V.J. (2003). Marine biotic response to elevated carbon dioxide. *Advances in Applied Biodiversity Science*, **4**, 59–67.

Seibel, B.A. and Walsh, P.J. (2001). Potential impacts of CO_2 injection on deep-sea biota. *Science*, **294**, 319–20.

Seibel, B.A. and Walsh, P.J. (2003). Biological impacts of deep-sea carbon dioxide injection inferred from indices of physiological performance. *Journal of Experimental Biology*, **206**, 641–50.

Seibel B.A., Girguis P.R., and Childress J.J. (2009). Critique of the respiration index. *Science E-letter*, **13**. November 2009.

Seidelin, M., Brauner, C.J., Jensen, F.B., and Madsen, S.S. (2001). Vacuolar-type H^+-ATPase and Na^+, K^+-ATPase expression in gills of Atlantic salmon (*Salmo salar*) during isolated and combined exposure to hyperoxia and hypercapnia in fresh water. *Zoological Science*, **18**, 1199–205.

Stark, A. (2008). *CO_2-Effekte auf Metabolismus und Fitness von marinen Wirbellosen der Gezeitenzone*. Diplomarbeit. Universität Kiel und Alfred-Wegener-Institut Bremerhaven.

Stramma, L., Johnson, G.C., Sprintall, J., and Mohrholz, V. (2008). Expanding oxygen-minimum zones in the tropical oceans. *Science*, **320**, 655–8.

Toews, D.P., Holeton, G.F., and Heisler, N. (1983). Regulation of acid-base status during environmental hypercapnia in the marine teleost fish *Conger conger*. *Journal of Experimental Biology*, **107**, 9–20.

Tomanek L. and Zuzow M.J. (2010) The proteomic response of the mussel congeners *Mytilus galloprovincialis* and *M. trossulus* to acute heat stress: implications for thermal tolerance limits and metabolic costs of thermal stress. *Journal of Experimental Biology*, **213**, 3559–74.

Tunnicliffe, V., Davies, K.T.A., Butterfield, D.A., Embley, R.W., Rose, J.M., and Chadwick, W.W. (2009). Survival of mussels in extremely acidic waters on a submarine volcano. *Nature Geoscience*, **2**, 344–8.

Walther, K., Sartoris, F.J., Bock, C., and Pörtner, H.-O., (2009). Impact of anthropogenic ocean acidification on thermal tolerance of the spider crab *Hyas araneus*. *Biogeosciences*, **6**, 2207–15.

Wheatly, M.G. and Henry, R.P. (1992). Extracellular and intracellular acid–base regulation in crustaceans. *Journal of Experimental Zoology*, **263**,127–42.

Wilson, J.M., Laurent, P., Tufts, B.L. *et al*. (2000). NaCl uptake by the branchial epithelium in freshwater teleost fish: an immunological approach to ion-transport protein localization. *Journal of Experimental Biology*, **203**, 2279–96.

Wilson, R.W., Millero, F.J., Taylor, J.R. *et al*. (2009). Contribution of fish to the marine inorganic carbon cycle. *Science*, **323**, 359–62.

Wood, C.M. (1991). Branchial ion and acid-base transfer—environmental hyperoxia as a probe. *Physiological. Zoology*, **64**, 68–102.

Zeidberg, L.D. and Robison, B.H. (2007). Invasive range expansion by the Humboldt squid, *Dosidicus gigas*, in the eastern North Pacific. *Proceedings of the National Academy of Sciences USA*, **104**, 12948–50.

Zhai, W., Dai, M., Cai, W.J., Wang, Y., and Wang, Z. (2005). High partial pressure of CO_2 and its maintaining mechanism in a subtropical estuary: the Pearl River estuary, China. *Marine Chemistry*, **93**, 21–32.

Zielinski, S., Sartoris, F.J., and Pörtner, H.-O. (2001). Temperature effects on haemocyanin oxygen binding in an Antarctic cephalopod. *Biological Bulletin*, **200**, 67–76.

CHAPTER 9

Effects of ocean acidification on sediment fauna

Stephen Widdicombe, John I. Spicer, and Vassilis Kitidis

9.1 Introduction

The vast majority of the seafloor is covered not in rocky or biogenic reefs but in unconsolidated sediments and, consequently, the majority of marine biodiversity consists of invertebrates either residing in (infauna) or on (epifauna) sediments (Snelgrove 1999). The biodiversity within these sediments is a result of complex interactions between the underlying environmental conditions (e.g. depth, temperature, organic supply, and granulometry) and the biological interactions operating between organisms (e.g. predation and competition). Not only are sediments important depositories of biodiversity but they are also critical components in many key ecosystem functions. Nowhere is this more apparent than in shallow coastal seas and oceans which, despite covering less than 10% of the earth's surface, deliver up to 30% of marine production and 90% of marine fisheries (Gattuso *et al.* 1998). These areas are also the site for 80% of organic matter burial and 90% of sedimentary mineralization and nutrient–sediment biogeochemical processes. They also act as the sink for up to 90% of the suspended load in the world's rivers and the many associated contaminants this material contains (Gattuso *et al.* 1998). Human beings depend heavily on the goods and services provided, for free, by the marine realm (Hassan *et al.* 2005) and it is no coincidence that nearly 70% of all humans live within 60 km of the sea or that 75% of all cities with more than 10 million inhabitants are in the coastal zone (Small and Nicholls 2003; McGranahan *et al.* 2007) Given these facts, it is clear that any broad-scale environmental impact that affects the diversity, structure, and function of sediment ecosystems could have a consider-

able impact on human health and well-being. It is therefore essential that the impacts of ocean acidification on sediment fauna, and the ecosystem functions they support, are adequately considered. This chapter will first describe the geochemical environment within which sediment organisms live. It will then explore the role that sediment organisms play as ecosystem engineers and how they alter the environment in which they live and the overall biodiversity of sediment communities. It will identify how the impacts of ocean acidification could act to reduce the influence of these ecosystem engineers. Finally, the chapter will identify the physiological and behavioural mechanisms that currently allow infaunal animals to live in a 'high-CO_2' environment and discuss whether these mechanisms will make them less vulnerable to ocean acidification in the future.

9.2 Distribution of carbon dioxide (CO_2) and pH within sediments

The distribution of CO_2 and pH within marine sediments is largely controlled by microbially mediated redox reactions that are linked to the mineralization of organic matter as well as abiotic processes (e.g. mineral formation and dissolution). However, by mixing and ventilating the sediment, the activities of large infaunal organisms can affect the microbial community, mineral type, and the redox state of sediments through multiple feedback mechanisms. Consequently, sediment-dwelling organisms, both multicellular and microbial, are instrumental in setting the geochemical environment that surrounds them. However, this relationship between organisms and their environment acts in both directions, with organisms both driving and responding to the geochemical conditions within the sediment. Therefore, in order to understand the potential impact of ocean acidification on sediment fauna and the processes they support, it is useful to first consider the key processes that control the distribution of CO_2 and pH in sediments.

The amount of CO_2 and the subsequent value for pH can vary enormously between different sediments—to such an extent that when looking across a range of marine environments, including temperate and tropical coastal seas and deep-sea sediments, pH in the uppermost few centimetres of sediment ranges from 6.5 to 8.2 (Aller and Yingst 1978; Martens *et al.* 1978; Wenzhöfer *et al.* 2001; Hulth *et al.* 2002; Zhu *et al.* 2006b; Burdige *et al.* 2008). The reader should be aware that pH in these studies is measured on different pH scales which may differ by up to 0.17 units depending on temperature (over the range 10–30°C) and salinity (over the range 10–30). As temperature and salinity are not always specified in the literature, we refer to the relevant pH scales where examples of sediment pH values are given in this chapter. CO_2 partial pressure (pCO_2) across a range of coastal and deep-sea sediments was found to be between 0.4 and 16 000 μatm (Wenzhöfer *et al.* 2001; Zhu *et al.* 2006a). However, organic matter mineralization, when cou-

pled with relatively slow ventilation, typically leads to a build-up of CO_2 and reduced metals and nitrogen, with a reduction in sediment pH. Consequently, pH profiles down through the sediment (Fig. 9.1) typically follow a sharp decrease from the top sediment layer to a pH minimum situated just below the oxic zone (usually the top few millimetres or centimetres) and subsequently pH becomes invariable with depth (e.g. Burdige *et al.* 2008). Figure 9.2 shows that even under situations of extreme acidification in the overlying seawater, values for porewater pH_{NBS} become more similar to those values observed under control conditions as one moves deeper into the sediment. These depth profiles are driven by microbial mineralization of organic matter and involve a number of redox reactions which follow in the order of decreasing Gibbs free energy ($\Delta G°$). For each redox couple (e.g. $NO_3^--N_2$), the electron acceptor (in this case NO_3^-) must be depleted before the next most energetically favourable reaction becomes dominant. Thereby, oxygen (O_2) is used first during aerobic respiration, followed by nitrate (NO_3^-), manganese (Mn), iron (Fe), sulphate (SO_4^{2-}) and finally CO_2 or acetate during methanogenesis. The relevant idealized redox reactions are listed below in order of decreasing $\Delta G°$:

$$CH_2O + O_2 \rightarrow CO_2 + H_2O \text{ (aerobic respiration)}$$

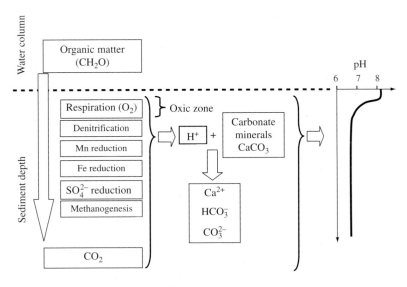

Figure 9.1 A summary of the diagenetic processes controlling pH distribution in sediments.

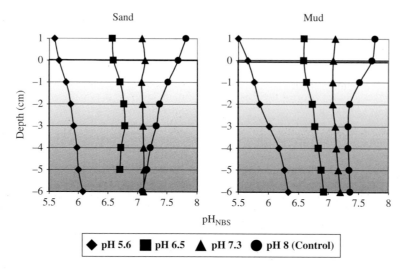

Figure 9.2 Impact of extreme seawater acidification (pH$_{NBS}$ of 5.6, 6.5, 7.3, and 8) on depth profiles in sand and mud (modified from Widdicombe et al. 2009).

$$CH_2O + NO_3^- + H^+ + e^- \rightarrow CO_2 + \tfrac{1}{2}N_{2(g)} + 2H_2O \text{ (denitrification)}$$

$$CH_2O + MnO_2 + 2e^- \rightarrow CO_2 + Mn^{2+} + H_2O \text{ (Mn reduction)}$$

$$CH_2O + Fe(OO)^- + e^- \rightarrow CO_2 + Fe^{2+} + H_2O \text{ (Fe reduction)}$$

$$CH_2O + SO_4^{2-} + 3H^+ \rightarrow CO_2 + HS^- + H_2O \text{ (sulphate reduction)}$$

$$4H_2 + CO_2 \rightarrow CH_4 + 2H_2O \text{ (methanogenesis)}$$

It is important to note that all of these processes, with the exception of methanogenesis, produce CO_2 and therefore lead to a reduction in pH.

In addition to CO_2 production, many microbial processes consume or produce ions which are also involved in acid–base equilibria and these ions will have an additional effect on sediment pH (Soetaert et al. 2007). For example, microbial nitrification ($NH_3 + 2O_2 \rightarrow HNO_3 + H_2O$) consumes ammonia ($NH_3$, a weak base: $NH_3 + H^+ \rightarrow NH_4^+$) and produces nitric acid (HNO_3) thereby reducing the pH (increasing [H^+]). With the exception of Mn and Fe reduction, the acid–base (non-CO_2) effect of these

microbial processes on pH (excluding CO_2 production), results in pH reduction from typical seawater values of 8.2 on the free scale (Soetaert et al. 2007). Soetaert et al. (2007) showed that the magnitude of the acid–base effect of various biogeochemical processes on pH (excluding CO_2 production) was of the order of 0.001 pH units per mole of substrate reduced or oxidized. The effect of CO_2 on pH is of the same order of magnitude evaluated using the CO2SYS software (Lewis and Wallace 1998). This observation suggests that the reduced pH commonly found in marine sediments is not due to CO_2 production alone and that acid–base reactions play a significant role in determining sediment pH. However, stoichiometric considerations dictate that during aerobic respiration of organic matter the molar production of CO_2 (which reduces pH) exceeds the molar production of NH_3 and PO_4^{3-} (which increase pH). Specifically, the production of CO_2 from respiration of organic matter with a molar C:N:P ratio of approximately 106:16:1 (Redfield et al. 1963), exceeds production of NH_3 and PO_4^{3-} by one and two orders of magnitude, respectively, on a molar basis. The aerobic respiration of organic matter therefore results in a net decrease of pH (Soetaert et al. 2007). Furthermore, adsorption of Mn, NH_3, and PO_4^{3-} onto minerals may remove substantial amounts of these solutes from porewaters (Berner

1977; Aller 1994) and thereby reduce their effect on sediment pH. The depth to which each electron acceptor, particularly O_2, penetrates sediments is critical in driving production of H^+. This penetration depth in turn depends on physical characteristics such as sediment porosity, sediment resuspension, and the supply and nature (fresh or degraded) of organic matter which control the rate of remineralization and thereby the rate with which each electron acceptor is successively depleted (Westrich and Berner 1984). In sediments deplete of large infauna, pH is laterally homogeneous (Zhu *et al.* 2006b), reflecting the fact that each of the processes outlined above dominates a discrete depth layer. However, studies of burrow structures of macrofauna have revealed that the chemical environment of burrows and burrow walls differs substantially from that of the surrounding sediments (Aller and Yingst 1978). Recent technological innovations in the determination of O_2, pH, and pCO_2 with optodes have highlighted the vertical, lateral, and temporal variability of these parameters (see below and review by Stockdale *et al.* 2009). Having introduced the main sources of H^+ in sediments, it is informative to consider relevant H^+ sinks which buffer pH changes.

Marine sediments have a high pH buffering capacity due to mineral dissolution–precipitation, adsorption–desorption reactions between solutes and minerals and acid–base reactions in porewaters. Field observations and modelling studies have shown that dissolution of calcium carbonate mineral (hereafter $CaCO_3$, including calcite, Mg-calcite, and aragonite) is a major sink for H^+ in marine sediments (Wenzhöfer *et al.* 2001; Jourabchi *et al.* 2005; Morse *et al.* 2006). However, as with the supply of CO_2 to sediments, there are also strong biological controls on pH buffering processes. Thereby, seasonal patterns in abundance of benthic foraminifera in coastal waters exert a strong influence on Ca^{2+} concentration and calcite saturation state in porewaters (Green *et al.* 1993). Macrofauna bioturbation plays a critical role in driving dissolution of $CaCO_3$ shells by maintaining the exchange of solutes, preventing the build-up of total alkalinity and thereby sustaining redox processes that lead to H^+ formation (Aller 1978, 1982; Green *et al.* 1993). Further examples of biological control of the pH buffering capacity of sediments can be found in the literature.

For example, model as well as experimental data suggest that seagrasses introduce O_2 directly into the sediments, increasing the mineralization of organic matter and thereby dissolution of $CaCO_3$ (Burdige *et al.* 2008). On the other hand, Burdige *et al.* (2008) showed that seagrass foliage decreased the bottom water flow, and thereby the advective uptake of O_2 by sediments. The net balance between these opposing effects depends on seagrass density, with carbonate dissolution from direct seagrass O_2 input becoming the dominant process at densities above 0.5 m^2 of leaf area per m^2 of seafloor (Burdige *et al.* 2008). Photosynthetic activity by microphytobenthic communities may also introduce O_2 directly into sediments following diel patterns (de Beer *et al.* 2005; Burdige *et al.* 2008). The result of carbonate mineral dissolution in marine sediments is an increase in total alkalinity (see Chapter 1) which in turn determines the concentrations of bicarbonate (HCO_3^-) and carbonate (CO_3^{2-}) in porewaters. It has been suggested that the increased bicarbonate due to the dissolution of carbonate minerals may re-precipitate as a different metastable mineral (e.g. aragonite dissolution followed by calcite precipitation; Hu and Burdige 2007). In addition to mineral dissolution–precipitation reactions, microbial redox processes (including those described above) affect the concentration of acid–base pairs in porewaters (Soetaert *et al.* 2007). However, the effect of microbial processes on sediment pH is partly cancelled out during re-oxidation reactions (e.g. Mn reduction increases pH, but Mn^{2+} re-oxidation decreases pH). Many of the microbially mediated redox reactions in sediments tend towards an equilibrium pH (>5.2) which is far lower than the typical seawater pH of 8.2 (free scale) (Soetaert *et al.* 2007). These equilibria contribute to pH buffering in marine sediments. Given the scale of the predicted changes in pH of the overlying water under ocean acidification, sediments are likely to buffer part of these changes through the dissolution of carbonate minerals. The dissolution of different phases of these minerals will depend on their respective mineral stability (see Chapter 7). For example, Morse *et al.* (2006) showed that high-Mg calcite is likely to respond earlier than aragonite in a progressive dissolution process. As some infaunal animals maintain their burrow pH at or near that of overlying

water, a reduction in pH in the latter will put additional constraints on the animals' ability to compensate. This is discussed further in Section 9.5.

9.3 The impact of macrofaunal activity on microbially driven geochemical processes

As stated above, the distribution of pH and $p\mathrm{CO_2}$
within the sediment is largely determined by microbial and abiotic processes which in turn depend on
the supply of various substrates (e.g. organic matter, $\mathrm{NO_3^-}$, etc.). The supply of these substances is
highly variable both spatially and temporally, and
this variability is often associated with the activity
of sediment fauna. One activity that has a large
effect on pH is the building and ventilation of permanent or semi-permanent burrows. These burrows increase the surface area between the reduced
sediment and the overlying water, create additional
habitats for important microbial groups, and provide a mechanism for the active transport of organic
matter and solutes into and out of the sediment. All
of these effects enhance the degradation of organic
matter and turnover and potentially lead to a reduction in sediment pH. For example, Aller and Yingst
(1978) showed that redox cycling in the burrow
walls of the polychaete *Amphitrite ornata* was substantially enhanced compared with surrounding
sediments. The timescale over which macrofauna
affect the chemical environment of marine sediments is relatively fast. A study following the *in situ*
transport and biological incorporation of ^{13}C-la-
belled organic material in sediments showed that
this process is very rapid; the tracer was incorporated into macrofauna and bacteria to a depth of 10
cm within 3 days (Witte *et al.* 2003).

In addition to the effects mentioned above, burrow ventilation can also drive pH down by actively
enhancing the supply of $\mathrm{O_2}$ to anoxic sediments. For
example, burrow ventilation by the polychaete
Arenicola marina was estimated to account for up to
25% of the $\mathrm{O_2}$ supplied to intertidal sediments in the
southern North Sea (de Beer *et al.* 2005). Burrow
ventilation can also be temporally variable with diel
patterns in $\mathrm{O_2}$ penetration depth and $\mathrm{O_2}$ uptake by
sediments having been attributed to the activity of
the polychaete *Nereis diversicolor* (Wenzhöfer and

Glud 2004). Such temporal variability in dissolved
constituents, particularly $\mathrm{O_2}$, gives rise to the development of microniches where sharp gradients in
pH and $\mathrm{CO_2}$ may be found. Hulth *et al.* (2002)
reported a sharp pH gradient of 1.5 units (NBS
scale) over a distance of 1.1 mm across a ribbon
worm (phylum Nemertea) burrow. In a similar
study, Zhu *et al.* (2006b) showed the development of
pH minima around the walls of actively ventilated
burrows of the polychaete *Nereis succinea* with a pH
in the burrow similar to that of the overlying water.
The authors termed this feature a growing 'low pH
radial halo' while the pH in uninhabited burrows
reverted to the surrounding porewater values
within 2 days (Zhu *et al.* 2006b). Burrow ventilation
by the polychaete presumably enhanced $\mathrm{H^+}$
production in this region of the sediment by introducing more energetically favourable electron
acceptors for organic matter mineralization. In
addition to ventilation, faunal activity may transport organic matter into subsurface sediments leading to enhanced $\mathrm{H^+}$ production. For example, a
small (5 mm × 15 mm), low-pH hot spot in a coastal
sediment sample was attributed to the mineralization of faecal aggregates in an abandoned *Nemertea*
spp. burrow (Hulth *et al.* 2002). Similarly, Zhu *et al.*
(2006b) showed a rapid reduction of pH associated
with the decay of a dead *Nereis* spp. polychaete
($\mathrm{pH_T}$ minimum ~5.9). This low-pH feature was
indistinguishable from the surrounding sediments
within 5 days ($\mathrm{pH_T}$ ~ 6.6).

Two-dimensional measurements using optodes
have now revealed that whilst the burrows may
reduce the pH within the surrounding sediments,
the pH of the water in the burrow itself is often
maintained above that of the surrounding sediment (Fig. 9.3). However, burrow-water pH also
differs substantially between organisms. For example, $\mathrm{pH_T}$ in active burrows of *Nereis succinea* is
maintained at the same level as overlying water
(~8.2), while burrows of the polychaete *Nephthys
incisa* were maintained at lower $\mathrm{pH_T}$ (~7.4; Zhu
et al. 2006b). Whilst Hulth *et al.* (2002) showed that
active nemertean burrows are maintained at an
even lower $\mathrm{pH_{NBS}}$ of 6.5, Zhu *et al.* (2006a) showed
that the $p\mathrm{CO_2}$ in active *N. incisa* burrows was elevated with respect to overlying water, but lower
than in the surrounding sediments. They also

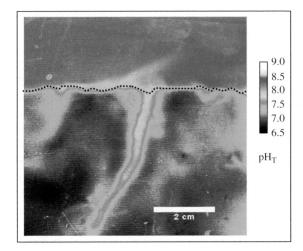

Figure 9.3   pH planar optode image showing the pH$_T$ distribution around a burrow of the polychaete *Hediste diversicolor*. The dotted line represents the sediment surface. Image by courtesy of Morten Larsen and Ronnie N. Glud.

observed a CO_2 plume rising from a burrow into the overlying water, thereby demonstrating clearly the effect of fauna on solute exchange between the sediment and overlying water. Active ventilation removes potentially toxic solutes (e.g. CO_2, NH_3). On the other hand, mucus used to cement the burrow walls may be inhibiting the diffusion of such solutes (Hannides *et al.* 2005) thereby allowing macrofauna to conserve energy which would otherwise be spent ventilating the burrow. Thus, different strategies of dealing with potentially toxic solutes emerge: burrow ventilation, insulation with cementing mucus, physiological adaptation, etc. Each one of these strategies carries an energetic cost and is discussed further below.

9.4 Sediment fauna as 'ecosystem engineers'

It is clear from the processes described in the previous section that the activities of infaunal organisms such as building and irrigating burrows as well as transporting and mixing sediment (collectively termed 'bioturbation') can have a significant effect on the sediment environment. Levinton (1995) reviewed the role of deposit-feeding marine invertebrates in modifying their sediment habitat and concluded that bioturbation is the main driving force behind the

transport of organic matter and the modulation of chemical reactions within sediments. In addition, by modification of the chemistry and physical structure of the sediment, together with the disturbance and displacement of other species, bioturbation may regulate the species composition and diversity of entire sediment assemblages (see review by Widdicombe and Austen 2005). As such, large bioturbating infaunal organisms must be considered as ecosystem engineers as defined by Lawton (1994):

Ecosystem engineers are organisms that directly or indirectly modulate the availability of resources (other than themselves) to other species, by causing physical state changes in biotic or abiotic materials. In so doing they modify, maintain and/or create habitats.

The importance of ecosystem engineering in controlling the biogeochemical and ecological processes occurring within marine sediments has highlighted the need to describe and quantify bioturbation in greater detail. In doing so this research has highlighted two key issues.

Firstly, the impact of large burrow-building species on biogeochemical cycles is not simply the result of extending the sediment–water interface across which nutrients may pass. It is becoming clear that the burrow is a specialized environment that often contributes more to nutrient cycling than would be expected solely based on the additional area of sediment–water interface provided. As seen above, and in contrast to surface sediments, the burrow environment is physically but not chemically stable (Kristensen *et al.* 1985 and references therein) and contains strong chemical gradients which exert significant control over biogeochemical processes. It has also been shown that the burrow environment supports highly diverse microbial communities that are different and more active than their sediment–surface counterparts (e.g. Laverock *et al.* 2010). This has most readily been seen in the case of nitrogen cycling where studies have shown that the presence of burrow builders increase the rates of nitrification and denitrification above the rates that would be expected purely from an increase in exchange surface area (e.g. Kristensen *et al.* 1985; Pelegrí *et al.* 1994; Howe *et al.* 2004; Webb and Eyre

2004). For example, the nitrification potential of *Nereis virens* burrow walls is estimated to be 1.7 to 4.1 times greater than that of surface sediment (Kristensen *et al.* 1985), whilst Pelegrí *et al.* (1994) showed that the burrows of the amphipod *Corophium volutator* stimulated denitrification fuelled from nitrification threefold and denitrification fuelled by nitrate from the overlying water fivefold. These processes can have a significant impact on the exchange of dissolved nutrients across the sediment–water interface (e.g. Webb and Eyre 2004; Widdicombe and Needham 2007).

Whilst it is important to think of burrows as a different environment from the sediment surface, it should also be remembered that the nature of the burrow environment is primarily controlled by the activity of the burrow builder. This leads to the second issue, which is that sediment communities are taxonomically and functionally diverse. This is supported by molecular studies of the bacterial communities inhabiting the burrow walls which also demonstrated that there are large differences between the communities inhabiting the burrows of different species. Whilst the bacterial communities inhabiting the burrows of some macrofauna more closely resemble those in the surrounding subsurface sediment (Lucas *et al.* 2003; Papaspyrou *et al.* 2005), others are more similar to the communities found at the sediment surface (Steward *et al.* 1996; Bertics and Ziebis 2009). Recent methodological advances have allowed the incorporation of individual species function into community-level estimates of bioturbation (e.g. Solan *et al.* 2004; Teal *et al.* 2009). Unfortunately, many studies and models still generalize the effects of 'deposit feeders', despite the fact that this group encompasses a wide range of species with different feeding behaviours to extract the organic material from the sediment. These differences, together with those in body size, life-history characteristics, and mobility, make it impossible to consider bioturbation as a single uniform process.

Bioturbation is now recognized as a globally important process (Teal *et al.* 2008) and changes in its precise nature and intensity will have enormous implications for the biodiversity and function of coastal ecosystems. Therefore, in the light of ocean acidification and warming, it is essential that we consider the potential impacts of elevated CO_2 on

the health and activity of fauna that burrow in, irrigate, and mix the sediment. This consideration should start, but certainly not end, with an understanding of the direct impacts on an organism's physiology (see Chapter 8). From there the challenge is to predict the biological and ecological responses that will ultimately shape the structure and function of bioturbating communities. At the population level, these responses (summarized in Fig. 9.4) include bioturbator size (growth), abundance (reproductive success), activity (respiration, feeding), and supply of nutrients (metabolic activity). From this point, additional ecological processes, such as competition for space and resources, further define the structure of bioturbating communities and the overall function of sediment ecosystems.

9.5 Assessing the potential impacts of ocean acidification on infaunal organisms

It is evident that the differing activities of infaunal organisms can create huge variability between the structure and function of the burrows in which they live. These differences create a variety of different burrow environments and these differences may therefore also be reflected in the physiological or behavioural mechanisms used by different species to cope with life in low-pH sediments. For example, Zhu *et al.* (2006b) found that values of extracellular fluid (i.e. haemolymph) pH_T differed substantially between two species of polychaete worm, *Nereis succinea* and *Nephthys incise* (~8.0 and ~6.8 respectively), and that these values also reflected pH_T conditions within the burrow (~8.2 and ~7.4 respectively). It may be concluded that since infaunal organisms live in an environment that is often high in CO_2, they will be inherently more immune to ocean acidification than organisms that live on the sediment surface (epifauna). However, before such a conclusion is reached, there should be an examination of the physiological mechanisms used by infaunal species to cope with a high-CO_2 environment.

There is growing evidence that marine organisms do not respond uniformly when exposed to CO_2-acidified seawater (Fabry *et al.* 2008; Widdicombe and Spicer 2008; Melzner *et al.* 2009; Ries *et al.* 2009; Hendriks *et al.* 2010; see also Chapters 6 and 7). The question is, given the current state of physiological

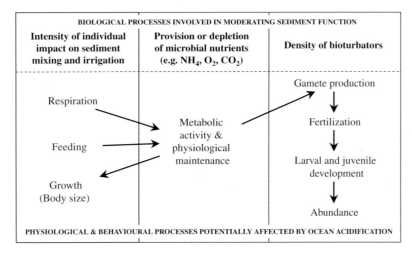

Figure 9.4   The relevance of CO_2-induced changes in organism physiology and behaviour to the biological control of sediment function.

knowledge, can we inform any generalizations about the response of sediment assemblages to elevated CO_2 and reduced pH? Studies of the effects of ocean acidification on both present-day and fossilized organisms suggest that there may be specific characteristics or traits that could indicate whether animals are tolerant of elevated CO_2 (hypercapnia) or low pH. For example, Bambach *et al.* (2002) suggested that organisms will be least affected by hypercapnia if they: (1) are 'buffered' against a range of chemically related physiological stresses, (2) have high rates of metabolism, (3) have well-developed gills and circulatory systems, and (4) possess skeletons limited in mass or made of materials other than $CaCO_3$ (see also Chapter 4). Melzner *et al.* (2009) compiled physiological data from taxonomic levels higher than species. He highlighted that compensation for extracellular acid–base disturbance together with high (specific) metabolic rates (and high levels of mobility/activity), may characterize CO_2-resistant taxa. But how robust are such generalizations? To find out, this chapter reviews current understanding and assumptions about large infauna before testing the currently limited knowledge against some of the key features identified above. In doing so, an attempt will be made to identify differences in the responses of infaunal and epifaunal species to hypercapnia or low pH.

A reasonable body of literature exists on the general biology of large infauna, but most focuses on adaptations intended to help the organism maintain contact with the surface via specialized structures or by inhabiting permanent or semi-permanent burrows, the deepest of which are actively ventilated (Eltringham 1971; Meadows and Meadows 1991; Little 2000). However, two seemingly opposing assumptions underpin much of the literature. Firstly, it is often assumed that these adaptations reduce the possibility of exposure to low O_2 tension (hypoxia) and high CO_2 tension (hypercapnia). While this is partially true, measurements made at different depths within sediments, or in burrow water, indicate that environmental hypoxia and hypercapnia are still issues for burrowing animals. The second common assumption is that large infaunal organisms are physiologically better equipped to tolerate and respond to hypoxia and hypercapnia than their epifaunal or open-water counterparts (e.g. Royal Society 2005; Knoll *et al.* 2007). While there is reasonable evidence that physiological adaptation to hypoxia is found in a wide range of infauna, including annelids, crustaceans, bivalves, and teleost fish (Atkinson and Taylor 1991; Mill 1997), empirical evidence of the degree of adaptation within physiological responses to hypercapnia is scarce. The responses of the oxygen exchange and transport systems to hypoxia were thought to greatly exceed any requirement for CO_2 excretion (Cameron 1986, 1989; Sundin *et al.* 2007). What is striking when collating data on the effects of ocean acidification on benthic species is that our informa-tion is heavily restricted to very few species distrib-uted amongst a few taxonomic groups (crustaceans, molluscs, echinoderms, fish, and to a lesser extent annelid and sipunculid worms). The list is reduced even further when the focus is restricted to infaunal species. Although our understanding has been slightly advanced by the review of the effect of hypercapnia on burrowing and non-burrowing ani-mals by Widdicombe and Spicer (2008) there is still nowhere near a rigorous phylogenetically based assessment of burrower versus non-burrower responses to hypercapnia. Interestingly, while it is now very difficult to publish a comparative study without taking phylogeny into account, the extent to which phylogenetic relationships are important for the comparison of physiological features is still debatable. On the one hand it seems that some physiological traits exhibit little phylogenetic con-straint (particularly when comparisons are made at the high taxonomic levels), such that species can justifiably be treated as independent data points for analysis. However, there is seldom sufficient a pri-ori knowledge to be able to reject a role for phylog-eny, and thereby avoid explicitly accounting for it in analyses (Spicer and Gaston 1999). With those uncertainties in mind it is still valuable to assess which of the available literature is at least consistent with some of the features thought to be characteris-tic of hypercapnia-tolerant animals.

9.5.1   High extracellular buffering capacity

Extracellular buffering of protons can be provided by bicarbonate or non-bicarbonate protein buffers and a good ion-regulatory capacity (Seibel and Walsh 2003; see Chapter 8). While knowledge of extracellular acid–base regulation in response to hypercapnia is growing for epifaunal crabs, mol-luscs, and echinoderms (Widdicombe and Spicer 2008, Melzner et al. 2009) the same is not true for infaunal species. Two notable exceptions are the sipunculid worm *Sipunculus nudis* and the lug-worm *Arenicola marina* where there appears to be little extracellular acid–base regulation (Melzner et al. 2009). In spangatoid burrowing urchins in the longer term (in excess of 2 months) any initial res-piratory acidosis is compensated for to an extent

that has not yet been reported in epifaunal urchins (J. I. Spicer et al. unpubl.). Searching the literature for extracellular protein concentration values as an indication of regulatory ability shows that while there are dramatic differences both between and within phyla (echinoderms have very low lev-els, crabs have intermediate levels, and fish have quite high levels), there are no consistent pat-terns for burrowing and non-burrowing phyla. Spangatoid echinoids have marginally lower con-centrations of extracellular proteins than epifaunal forms such as *Psammechinus miliaris* (Miles et al. 2007). The same is also true for bivalves. Similarly, deep-burrowing thalassinidean shrimps and infaunal crabs do not have dramatically higher haemolymph protein concentrations than epifau-nal crustaceans (Taylor et al. 2000). In fact, some-times the burrowing species actually show lower protein levels than their non-burrowing counter-parts (Atkinson and Taylor 2005).

9.5.2   Higher rates of metabolism

In the sipunculid worm *Sipunculus nudus*, hyper-capnia was shown to elicit metabolic depression, via acidification of body fluids (Reipschläger et al. 1997; Pörtner et al. 1998, 2000), and this response was accompanied by a reduction in protein turno-ver (Langenbuch et al. 2006). However, this is a rela-tively short-term response and is elicited only at very high (e.g. 10 000 μatm) CO_2 levels. Although comparisons of metabolic rates between different studies are notoriously difficult, there is some evi-dence that hypometabolism may be a feature of burrowing forms. For example, the nereid polycha-ete *Nereis virens* consumes less O_2 when confined in tubes than when living freely in water (Hyman 1932) and the deep-burrowing *Lumbrineris zonata* has a lower oxygen consumption than the shallow-burrowing *Nephtys hombergii* (Banse et al. 1971). In echinoderms, where two studies using exactly the same experimental technique are available, the infaunal brittlestar *Amphiura filiformis* shows a hypercapnia-induced (pH_{NBS} = 7.7) reduction in oxygen uptake whereas the epifaunal *Ophiura ophiura* does not (Wood et al. 2008, 2010). The former also generally exhibits a greater rate of oxygen uptake than the latter. In this case, it is the infaunal species which has the greatest oxygen uptake, in line with our original prediction. It is concluded therefore that support for the 'increased metabolism' hypothesis as a way of differentiating infauna from epifauna is currently equivocal.

9.5.3 Greater tolerance to hypercapnia in terms of survival and behaviour

As mentioned above, tolerance to hypercapnia is assumed to be a feature of infaunal organisms. Certainly the spatangoid heart urchin *Echinocardium cordatum* shows no mortality in sediments exposed to acidified water (7 weeks at $pH_{NBS} = 7.5$; Dashfield *et al.* 2008). Widdicombe *et al.* (2009) noted the persistence of capitellid worms in mesocosms exposed to extreme hypercapnia ($pH_{NBS} < 6$) for over 1 month. The same was true for *Nereis virens* in previous acidification experiments (Batten and Bamber 1996; Widdicombe and Needham 2007) with no observed deleterious effects observed at pH_{NBS} as low as 6.5. However, these few observations of apparently CO_2-tolerant burrowing species should not be extrapolated to the entire infaunal community. Using data extracted from Ries *et al.* (2009) it can be shown that there is no pattern in CO_2-related mortality when comparing infaunal (*Mya arenaria* and *Mercenaria mercenaria*) with epifaunal (*Mytilus edulis* and *Argopecten irradians*) bivalves. Admittedly, such a comparison was not the aim of their study but the data still call for some explanation. It is interesting, in this connection, that Knoll *et al.* (2007; see also Chapter 4) supported their hypothesis of a link between the end-Permian extinction and hypercapnia with geological evidence that the extinction of epifaunal bivalves was double that of infaunal species. The underlying assumption is that infaunal bivalves were already used to higher CO_2 in their burrows than their epifaunal counterparts.

The number of behavioural studies is limited, but it seems that hypercapnia does not affect the behaviour of some infaunal organisms. For example, there was no effect of seawater acidification (5 weeks, $pH_{NBS} = 7.21–7.30$) on the burrowing activity or the mortality of *N. virens*, a polychaete worm which creates a semi-permanent U-shaped burrow (Widdicombe and Needham 2007). Only at the extremely low pH ($pH_{NBS} < 6.5$) used in an earlier

study by Batten and Bamber (1996) was there disruption of the burrowing behaviour. Similarly, bioturbation by the brittlestar *Amphiura filiformis* is not affected on exposure to a pH_{NBS} of 7.7 or 7.3 for 40 days (Wood *et al.* 2009). Finally, for the polychaetes *Hyalinoecia tubicola* and *Diopatra cuprea*, which inhabit tubes in soft sediments, exposure to hypercapnia ($pH_{NBS} = 7.5$) causes no significant differences in the frequency or duration of respiratory irrigation (Dales *et al.* 1970).

9.5.4 Reduction in reliance on calcification for skeleton or shell construction

There is little doubt that some infaunal soft-bodied species can show a high level of tolerance to hypercapnia. Studies by Batten and Bamber (1996) and Widdicombe and Needham (2007), both on the burrowing polychaete worm *N. virens*, suggest that an infaunal organism can tolerate seawater with a pH_{NBS} down to 6.5. Similarly, the nemertean *Procephalothrix*, which inhabits coarse sand or lives beneath stones in the intertidal zone, is tolerant of pH_{NBS} in the range 5.0 to 9.2, at least over short (96 h) time periods (Yanfang and Shichun 2006). Are there infaunal species or groups with closely related epifaunal representatives where there is a reduction in reliance on calcification for producing skeletal or defensive structures? There does not appear to be any such reduction in bivalves, although it could be argued that the tests of infaunal echinoids are considerably thinner and less substantial than epifaunal forms. A recent study (Ries *et al.* 2009) has compared rates of net calcification in a large number of animal groups including infaunal and epifaunal forms, one of the first studies of this kind. The results are equivocal and no definitive pattern between infaunal and epifaunal species can be discerned. The response of the rate of calcification under elevated CO_2 concentrations of the infaunal clams *Mya arenaria* and *Mercenaria mercinaria* is different. The former exhibits a negative rate, the latter a threshold response. Of the two epifaunal species, *Argopecten irradians* shows a negative rate and *Mytilus edulis* shows a neutral response (see Table 1 in Ries *et al.* 2009). Finally, in the infaunal brittlestar *Amphiura filiformis*, arm regeneration increases at low pH whereas no such effect is detectable (at comparable temperature) in the epifaunal *Ophiura ophiura* (Wood *et al.* 2008, 2010). In both cases net calcification was maintained at low pH_{NBS} (undersaturated Ω_a) but extensive muscle wastage in low-pH treatments was found only in *A. filiformis* (after 40 days).

9.6   Summarizing the vulnerability of infaunal organisms to ocean acidification

It is clear that pH in sediments is frequently lower than the minimum pH of surface waters that is predicted to occur under projected ocean acidification scenarios. In addition, sediments have a high pH buffering capacity (Leclercq *et al.* 2002; Andersson *et al.* 2003). These observations could suggest that ocean acidification will have a lesser impact on infaunal sediment communities than on communities that live on the sediment surface or in the pelagic zone. Indeed, at first glance the literature review above would seem to suggest that during short-term exposure at least, infaunal organisms may cope better with hypercapnic conditions than epifaunal ones. However, there are a number of important caveats.

Firstly, formal comparisons of the responses of infaunal species across different phyla are fraught with difficulties, for example the comparability of methods and holding conditions as well as the possible phylogenetic dependence of physiological responses and functions. Secondly, there may be as much differential sensitivity across infaunal species, even amongst those that possess substantial $CaCO_3$ exo- or meso-skeletons, than between infaunal and epifaunal taxa. Thirdly, the majority of the experimental evidence used in this chapter has been generated in short-term exposure experiments which only address a limited fraction of an organism's entire and often complex life history. It is generally accepted that early life-history stages of invertebrates are more sensitive to CO_2 than juveniles or adults (Dupont and Thorndyke 2009; but see Kroeker *et al.* 2010). Since many infaunal species rely on a planktonic larval stage, the most severe effects of ocean acidification on benthic species could be on these early life stages. In this respect, differences between infaunal and epifaunal species may be far less clear cut. For example, adult forms of the epifaunal oyster *Crassostrea virginica* survived much better to elevated CO_2 levels than either the epifaunal scallop *Argopecten irradians* or the infaunal *Mercenaria mercinaria* (Talmage and Gobler 2009). However, exposure to CO_2-induced acidification (650 μatm) resulted in delayed metamorphosis and a smaller body size in the planktonic stages of these three bivalve species. Fourthly, we still know very little about the physiological adaptations that enable one species to adopt a fossorial existence and another species to remain epifaunal. Finally, one of the major assumptions running through much of what has been presented above is that animals which already experience hypercapnic conditions and demonstrate physiological adaptation to such conditions will be the least affected by ocean acidification (e.g. Raven *et al.* 2005). However, this assumption is dependent on such animals having much wider tolerance of hypercapnia than they currently require. But what if these animals are already working at the limits of their resistance to, and capacity to cope with, hypercapnia? Then the assumption is turned on its head, and it is those species that, comparatively speaking, are currently most tolerant that may be under greatest threat from additional ocean acidification. An analogous situation exists in the effect of temperature increase on intertidal animals. Stillman (2003) showed that, counter-intuitively, it was the porcelain crabs which inhabit the upper shore that were most at threat from the increase in temperature. Unlike those of the lower shore, the high-shore species were already living at, or close to, the upper limits of their thermal tolerance, and so were most vulnerable to warming.

It is clear, as argued by Garland and Adolph (1994), that care is needed when making deductions from the 'two species' study approach that one is continually forced to adopt in reviews such as this one, where pairs of closely related species or genera, one fossorial the other not, are compared. Given that targeted comparative studies are logistically complex it is worth considering possible alternative approaches that might help identify hypercapnic tolerance in and between infaunal groups. For example, it is interesting that in the three main environments where seawater can become markedly hypercapnic, namely intertidal and subtidal burrows, estuarine waters, and tide-pools, hypercapnia is regularly associated with hypoxia. Burnett (1997) comments that the negative correlation between oxygen and pH in estuaries is striking while also noting that pH is not affected directly by oxygen. Animals consume oxygen, and correspondingly produce a given amount of CO_2 (the exact ratio predicated on the energy substrate used to fuel metabolism), thereby reducing pH in surrounding waters. When O_2 and CO_2 are measured in water sampled at different depths in a burrow, the negative correlation between these two gases as highlighted by Burnett (1997) is evident. We know more about responses to O_2 than to CO_2. Therefore it is reasonable to assume (until we have more CO_2 studies) that the extent to which organisms are physiologically adapted to hypoxia may act as a surrogate for physiological tolerance to hypercapnia. For example, deposit-feeding thalassinidean shrimps (such as *Callianassa*) show better adaptation to hypoxia in terms of their respiratory physiology (namely ventilation and haemolymph O_2 binding characteristics) than their filter-feeding relatives (such as *Upogebia*; Atkinson and Taylor 2005).

9.7 Conclusions

It was initially assumed that, as sediment environments experience high levels of CO_2, infaunal organisms may be more tolerant to ocean acidification than those that live on the sediment surface or in the overlying water column. This assumption was based on the fact that infaunal organisms appear to have developed a number of behavioural and physiological strategies in order to cope with a fossorial lifestyle. However, many of these strategies have been primarily developed to avoid or withstand periods of hypoxia or anoxia rather than hypercapnia. Consequently, before we can assume with any great confidence that infaunal organisms are less vulnerable to elevated CO_2 than their counterparts living in other habitats, in the same way that they are more tolerant to hypoxia, more comparative data from a wide range of taxa and life strategies are needed. This is particularly the case for the many infaunal species that rely on a planktonic early life stage, as the vulnerability of infaunal adults to elevated CO_2

is likely to be different from that of their pelagic larvae or settling juveniles. For example, Green *et al.* (2009) showed that early settlement mortality in bivalves can be significantly affected by a reduction in the aragonite saturation state at the sediment surface. Another consideration is that many infaunal organisms may actually be living at the very limit of their CO_2 tolerance, and consequently even a small change in CO_2 levels could be detrimental to the long-term survival and functioning of key bioturbating species. All of which means that there remains a pressing need for carefully targeted comparative studies that will help identify the behaviour and physiological traits that define an organism's vulnerability to ocean acidification.

9.8 Acknowledgements

This work is a contribution to the 'European Project on Ocean Acidification' (EPOCA) which received funding from the European Community's Seventh Framework Programme (FP7/2007–2013) under grant agreement no. 211384. S.W. and V.K. acknowledge support from the NERC funded project Oceans 2025. The authors would also like to thank Andreas Andersson and Brad Seibel for their constructive comments on an initial draft of this chapter.

References

Aller, R.C. (1978). Experimental studies of changes produced by deposit feeders on pore water, sediment, and overlying water chemistry. *American Journal of Science*, **278**, 1185–234.

Aller, R.C. (1982). Carbonate dissolution in nearshore terrigenous muds—the role of physical and biological reworking. *Journal of Geology*, **90**, 79–95.

Aller, R.C. (1994). The sedimentary Mn cycle in Long-Island Sound - its role as intermediate oxidant and the influence of bioturbation, O_2, and C(Org) flux on diagenetic reaction balances. *Journal of Marine Research*, **52**, 259–95.

Aller, R.C. and Yingst, J.Y. (1978). Biogeochemistry of tube-dwellings - study of sedentary polychaete *Amphitrite ornata* (Leidy). *Journal of Marine Research*, **36**, 201–54.

Andersson, A.J., Mackenzie, F.T., and Ver, L.M. (2003). Solution of shallow-water carbonates: an insignificant buffer against rising atmospheric CO_2. *Geology*, **31**, 513–16.

Atkinson, R.J.A. and Taylor, A.C. (1991). Burrows and burrowing behavior of fish. *Symposium of the Zoological Society, London*, **63**, 133–55.

Atkinson, R.J.A. and Taylor, A.C. (2005). Aspects of the physiology, biology and ecology of thalassinidean shrimps in relation to their burrow environment. *Oceanography and Marine Biology*, **43**, 173–210.

Bambach, R.K., Knoll, A.H., and Sepkoski, J.J. Jr (2002). Anatomical and ecological constraints on Phanerozoic animal diversity in the marine realm. *Proceedings of the National Academy of Sciences USA*, **99**, 6854–9.

Banse, K., Nichols, F.N., and May, D.R. (1971). Oxygen consumption by the seabed. III. On the role of the macrofauna at three stations. *Vie et Milieu Supplément*, **22**, 31–52.

Batten, S.D. and Bamber, R.N. (1996). The effects of acidified seawater on the polychaete *Nereis virens* Sars, 1835. *Marine Pollution Bulletin*, **32**, 283–7.

Berner, R.A. (1977). Stoichiometric models for nutrient regeneration in anoxic sediments. *Limnology and Oceanography*, **22**, 781–6.

Bertics, V.J. and Ziebis, W. (2009). Biodiversity of benthic microbial communities in bioturbated coastal sediments is controlled by geochemical microniches. *ISME Journal*, **3**, 1269–85.

Burdige, D.J., Zimmerman, R.C., and Hu, X.P. (2008). Rates of carbonate dissolution in permeable sediments estimated from pore-water profiles: the role of sea grasses. *Limnology and Oceanography*, **53**, 549–65.

Burnett, L.E. (1997). The challenges of living in hypoxic and hypercapnic aquatic environments. *American Zoologist*, **37**, 633–40.

Cameron, J.N. (1986). Acid-base equilibria in invertebrates. In: N. Heisler (ed.), *Acid–base regulation in animals*, pp. 357–94. Elsevier, New York.

Cameron, J.N. (1989). Acid-base homeostasis: past and present perspectives. *Physiological Zoology*, **62**, 845–65.

de Beer, D., Wenzhöfer, F., Ferdelman, T.G. *et al.* (2005). Transport and mineralization rates in North Sea sandy intertidal sediments, Sylt-Romo Basin, Wadden Sea. *Limnology and Oceanography*, **50**, 113–27.

Dales, R.P., Mangum, C.P., and Tichy, J.C. (1970). Effects of changes in oxygen and carbon dioxide concentrations on ventilation rhythms in onuphid polychaetes. *Journal of the Marine Biological Association of the United Kingdom*, **50**, 365–80.

Dashfield, S.L., Somerfield, P.J., Widdicombe, S., Austen, M.C., and Nimmo, M. (2008). Impacts of ocean acidification and burrowing urchins on within-sediment pH profiles and subtidal nematode communities. *Journal of Experimental Marine Biology and Ecology*, **365**, 46–52.

Dupont, S. and Thorndyke, M.C. (2009). Impact of CO_2-driven ocean acidification on invertebrates early life history – what we know, what we need to know and what we can do. *Biogeosciences Discussions*, **6**, 3109–31.

Eltringham, S.K. (1971). *Life in mud and sand*, 218 pp. English Universities Press, London.

Fabry, V.J., Seibel, B.A., Feely, R.A., and Orr, J.C. (2008). Impacts of ocean acidification on marine fauna and ecosystem processes. *Journal of Marine Science*, **65**, 414–32.

Garland, T. and Adolph, S.C. (1994). Why not to do 2-species comparative studies – limitations on inferring adaption. *Physiological Zoology*, **67**, 797–828.

Gattuso, J.-P., Frankignoulle, M., and Wollast, R. (1998). Carbon and carbonate metabolism in coastal aquatic ecosystems. *Annual Review of Ecological Systems*, **29**, 405–34.

Green, M.A., Aller, R.C., and Aller, J.Y. (1993). Carbonate dissolution and temporal abundances of foraminifera in Long-Island Sound sediments. *Limnology and Oceanography*, **38**, 331–45.

Green, M.A., Waldbusser, G.G., Reilly, S.L., Emerson, K., and O'Donnell, S. (2009). Death by dissolution: sediment saturation state as a mortality factor for juvenile bivalves. *Limnology and Oceanography*, **54**, 1037–47.

Hannides, A.K., Dunn, S.M., and Aller, R.C. (2005). Diffusion of organic and inorganic solutes through macrofaunal mucus secretions and tube linings in marine sediments. *Journal of Marine Research*, **63**, 957–81.

Hassan, R., Scholes, R., and Ash, M. (eds) (2005). *Ecosystems and human well-being: current status and trends, Volume 1*, 921 pp. Island Press, Washington, DC.

Hendriks, I.E., Duarte, C.M., and Álvarez, M. (2010). Vulnerability of marine biodiversity to ocean acidification: a meta-analysis. *Estuarine, Coastal and Shelf Science*, **86**, 157–64.

Howe, R.L., Rees, A.P., and Widdicombe, S. (2004). The impact of two species of bioturbating shrimp (*Callianassa subterranea* and *Upogebia deltaura*) on sediment denitrification. *Journal of the Marine Biological Association of the United Kingdom*, **84**, 629–32.

Hu, X.P. and Burdige, D.J. (2007). Enriched stable carbon isotopes in the pore waters of carbonate sediments dominated by seagrasses: evidence for coupled carbonate dissolution and reprecipitation. *Geochimica et Cosmochimica Acta*, **71**, 129–44.

Hulth, S., Aller, R.C., Engstrom, P., and Selander, E. (2002). A pH plate fluorosensor (optode) for early diagenetic studies of marine sediments. *Limnology and Oceanography*, **47**, 212–20.

Hyman, L.H. (1932). Relation of oxygen tension to oxygen consumption in *Nereis virens*. *Journal of Experimental Biology*, **61**, 209–21.

Jourabchi, P., Van Cappellen, P., and Regnier, P. (2005). Quantitative interpretation of pH distributions in aquatic sediments: a reaction-transport modeling approach. *American Journal of Science*, **305**, 919–56.

Knoll, A.H., Bambach, R.K., Payne, J.L., Pruss, S., and Fischer, W.W. (2007). Paleophysiology and end-Permian mass extinction. *Earth and Planetary Science Letters*, **256**, 295–313.

Kroeker, K.J., Kordas, R.L., Crim, R.N., and Singh, G.G. (2010). Meta-analysis reveals negative yet variable effects of ocean acidification on marine organisms. *Ecology Letters*, **13**, 1419–34.

Kristensen, E., Jensen, M.H., and Andersen, T.K. (1985). The impact of polychaete (*Nereis virens* Sars) burrows on nitrification and nitrate reduction in estuarine sediments. *Journal of Experimental Marine Biology and Ecology*, **85**, 75–91.

Langenbuch, M., Bock, C., Leibfritz, D., and Pörtner, H.O. (2006). Effects of environmental hypercapnia on animal physiology: a ^{13}C NMR study of protein synthesis rates in the marine invertebrate *Sipunculus nudus*. *Comparative Biochemistry and Physiology A*, **144**, 479–84.

Laverock, B., Smith, C.J., Tait, K., Osborn, A.M., Widdicombe, S., and Gilbert, J.A. (2010). Changes to the microbial community structure in the burrows of two species of bioturbating shrimp. *ISME Journal*, **4**, 1531–44.

Lawton, J.H. (1994). What do species do in ecosystems? *Oikos*, **71**, 367–74.

Leclercq, N., Gattuso, J.-P., and Jaubert, J. (2002). Primary production, respiration, and calcification of a coral reef mesocosm under increased CO_2 partial pressure. *Limnology and Oceanography*, **47**, 558–64.

Lewis, E. and Wallace, D.W.R. (1998). *Program developed for CO_2 system calculations*. ORNL/CDIAC-105. Carbon Dioxide Information Analysis Center, Oak Ridge National Laboratory, US Department of Energy, Oak Ridge, TN.

Levinton, J.S. (1995). *Marine biology. Function, biodiversity, ecology*, 420 pp Oxford University Press, New York.

Little, C. (2000). *The biology of soft shores and estuaries*, 264 pp. Oxford University Press, Oxford.

Łucas, F.S., Bertru, G., and Höfle, M.G. (2003). Characterization of free-living and attached bacteria in 530 sediments colonized by *Hediste diversicolor*. *Aquatic Microbial Ecology*, **32**, 165–74.

McGranahan, G., Balk, D., and Anderson, B. (2007). The rising tide: assessing the risk of climate change and human settlement in low elevation coastal zones. *Environment and Urbanization*, **19**, 17–37.

Martens, C.S., Berner, R.A., and Rosenfeld, J.K. (1978). Interstitial water chemistry of anoxic Long-Island Sound sediments. 2. Nutrient regeneration and phosphate removal. *Limnology and Oceanography*, **23**, 605–17.

Meadows, P.S. and Meadows, A. (eds) (1991). *The environmental impact of burrowing animals and animal burrows: the proceedings of a symposium held at the Zoological Society of London on3rd and 4th May 1990*, 349 pp. Clarendon Press, Oxford.

Melzner, F., Gutowska, M.A., Langenbuch, M. *et al.* (2009). Physiological basis for high CO_2 tolerance in marine ectothermic animals: pre-adaptation through lifestyle and ontogeny? *Biogeosciences*, **6**, 2313–31.

Miles, H., Widdicombe, S., Spicer, J.I., and Hall-Spencer, J. (2007). Effects of anthropogenic seawater acidification on acid-base balance in the sea urchin *Psammechinus miliaris*. *Marine Pollution Bulletin*, **54**, 89–96.

Mill, P.J. (1997). Invertebrate respiratory systems. In: W.H. Dantzler (ed.) *Handbook of physiology*, pp. 1009–96. Oxford University Press, Oxford.

Morse, J.W., Andersson, A.J., and Mackenzie, F.T. (2006). Initial responses of carbonate-rich shelf sediments to rising atmospheric pCO_2 and 'ocean acidification': role of high Mg-calcites. *Geochimica et Cosmochimica Acta*, **70**, 5814–30.

Papaspyrou, S., Gregersen, T., Cox, R.P., Thessalou-Legaki, M., and Kristensen, E. (2005). Sediment properties and bacterial community in burrows of the ghost shrimp *Pestarella tyrrhena* (Decapoda: Thalassinidea). *Aquatic Microbial Ecology*, **38**, 181–90.

Pelegrí, S.P., Nielsen, L.P., and Blackburn, T.H. (1994). Denitrification in estuarine sediment stimulated by the irrigation activity of the amphipod *Corophium volutator*. *Marine Ecology Progress Series*, **105**, 285–90.

Pörtner, H.O., Reipschläger, A., and Heisler, N. (1998). Acid-base regulation, metabolism and energetics in *Sipunculus nudus* as a function of ambient carbon dioxide level. *Journal of Experimental Biology*, **201**, 43–55.

Pörtner, H.O., Bock, C., and Reipschläger, A. (2000). Modulation of the cost of pH$_i$ regulation during metabolic depression: a P-NMR study in invertebrate (*Sipunculus nudus*) isolated muscle. *Journal of Experimental Biology*, **203**, 2417–28.

Royal Society (2005). *Ocean acidification due to increasing atmospheric carbon dioxide*. The Royal Society, London.

Raven, J., Caldeira, K., Elderfield, H. *et al.* (2005). *Ocean acidification due to increasing atmospheric carbon dioxide*. Royal Society Policy Document 12/05. The Clyvedon Press Ltd, Cardiff, UK.

Redfield, A.C., Ketchum, B.H., and Richards, F.A. (1963). The influence of organisms on the composition of seawater. In: N.M. Hill (ed.), *The sea*, pp. 26–77. Interscience, New York.

Reipschläger, A., Nilsson, G.E., and Pörtner, H.O. (1997). A role for adenosine in metabolic depression in the marine invertebrate *Sipunculus nudus*. *American Journal of Physiology*, **272**, 350–6.

Ries, J.B., Cohen, A.L., and McCorkle, D.C. (2009). Marine calcifiers exhibit mixed responses to CO_2-induced ocean acidification. *Geology*, **37**, 1131–4.

Seibel, B.A. and Walsh, P.J. (2003). Biological impacts of deep-sea carbon dioxide injection inferred from indices of physiological performance. *Journal of Experimental Biology*, **206**, 641–50.

Small, C. and Nicholls, R.J. (2003). A global analysis of human settlement in coastal zones. *Journal of Coastal Research*, **19**, 584–99.

Snelgrove, P.V.R. (1999). Getting to the bottom of marine biodiversity: sedimentary habitats. *BioScience*, **49**, 129–38.

Soetaert, K., Hofmann, A.F., Middelburg, J.J., Meysman, F.J.R., and Greenwood, J. (2007). The effect of biogeochemical processes on pH. *Marine Chemistry*, **105**, 30–51.

Solan, M., Cardinale, B.J., Downing, A.L., Engelhardt, K.A.M., Ruesink, J.L., and Srivastava, D.S. (2004). Extinction and ecosystem function in the marine benthos. *Science*, **306**, 1177–80.

Spicer, J.I. and Gaston, K.J. (1999). *Physiological diversity and its ecological implications*, 241 pp. Blackwell Science, Oxford.

Steward, C.C., Nold, S.C., Ringelberg, D.B., White, D.C., and Lovell, C.R. (1996). Microbial biomass and community structures in the burrows of bromophenol producing and non-producing marine worms and surrounding sediments. *Marine Ecology Progress Series*, **133**, 149–65.

Stillman, J.H. (2003). Acclimation capacity underlies susceptibility to climate change. *Science*, **301**, 65.

Stockdale, A., Davison, W., and Zhang, H. (2009). Microscale biogeochemical heterogeneity in sediments: a review of available technology and observed evidence. *Earth Science Reviews*, **92**, 81–97.

Sundin, L., Burleson, M.L., Sanchez, A.R. *et al.* (2007). Respiratory chemoreceptor function in vertebrates – comparative and evolutionary aspects. *Integrated Comparative Biology*, **47**, 592–600.

Talmage, S.C. and Gobler, C.J. (2009). The effects of elevated carbon dioxide concentrations on the metamorphosis, size, and survival of larval hard clams (*Mercenaria mercenaria*), bay scallops (*Argopecten irradians*), and Eastern oysters (*Crassostrea virginica*). *Limnology and Oceanography*, **54**, 2072–80.

Taylor, A.C., Astall C.M., and Atkinson R.J.A. (2000). A comparative study of the oxygen transporting properties of the haemocyanin of five species of thalassinidean mud-shrimps. *Journal of Experimental Marine Biology and Ecology*, **244**, 265–83.

Teal, L.R., Bulling, M.T., Parker, E.R., and Solan, M. (2008). Global patterns of bioturbation intensity and mixed depth of marine soft sediment. *Aquatic Biology*, **2**, 207–18.

Teal, L.R., Parker, R., Fones, G., and Solan, M. (2009). Simultaneous determination of *in situ* vertical transitions of color, pore-water metals, and visualization of infaunal activity in marine sediments. *Limnology and Oceanography*, **54**, 1801–10.

Webb, A.P. and Eyre, B.D. (2004). Effect of natural populations of burrowing thalassinidean shrimp on sediment irrigation, benthic metabolism, nutrient fluxes and denitrification. *Marine Ecology Progress Series*, **268**, 205–20.

Wenzhöfer, F., Adler, M., Kohls, O. *et al.* (2001). Calcite dissolution driven by benthic mineralization in the deepsea: In situ measurements of Ca^{2+}, pH, pCO_2 and O_2. *Geochimica et Cosmochimica Acta*, **65**, 2677–90.

Wenzhöfer, F. and Glud, R.N. (2004). Small-scale spatial and temporal variability in coastal benthic O_2 dynamics: effects of fauna activity. *Limnology and Oceanography*, **49**, 1471–81.

Westrich, J.T. and Berner, R.A. (1984). The role of sedimentary organic-matter in bacterial sulfate reduction - the G model tested. *Limnology and Oceanography*, **29**, 236–49.

Widdicombe, S. and Austen, M.C. (2005). Setting diversity and community structure in subtidal sediments: the importance of biological disturbance. In: J. Kostka, R. Haese, and E. Kristensen (eds), *Interactions between macro- and microorganisms in marine sediments*, pp. 217–31. Coastal and Estuarine Studies 60. American Geophysical Union, New York.

Widdicombe, S. and Needham, H.R. (2007). Impact of CO_2-induced seawater acidification on the burrowing activity of *Nereis virens* and sediment nutrient flux. *Marine Ecology Progress Series*, **341**, 111–22.

Widdicombe, S. and Spicer, J.I. (2008). Predicting the impact of ocean acidification on benthic biodiversity: what can physiology tell us? *Journal of Experimental Marine Biology and Ecology*, **366**, 187–97.

Widdicombe, S., Dashfield, S.L., McNeill, C.L. *et al.* (2009). Effects of CO_2 induced seawater acidification on infaunal diversity and sediment nutrient fluxes. *Marine Ecology Progress Series*, **379**, 59–75.

Witte, U., Aberle, N., Sand, M., and Wenzhöfer, F. (2003). Rapid response of a deep-sea benthic community to POM enrichment: an *in situ* experimental study. *Marine Ecology Progress Series*, **251**, 27–36.

Wood, H.L., Spicer, J.I., and Widdicombe, S. (2008). Ocean acidification may increase calcification rates, but at a cost. *Proceedings of the Royal Society B: Biological Sciences*, **275**, 1767–73.

Wood, H.L., Widdicombe, S., and Spicer, J.I. (2009). The influence of hypercapnia and macrofauna on sediment nutrient exchange? *Biogeosciences*, **6**, 2015–24.

Wood, H.L., Spicer, J.I., Lowe, D.M., and Widdicombe, S. (2010). Interaction of ocean acidification and temperature: the high cost of survival in the brittlestar *Ophiura ophiura*. *Marine Biology*, **157**, 2001–13.

Yanfang, Z. and Shichun, S. (2006). Effects of salinity, temperature and pH on the survival of the nemertean *Procephalothrix stimulus* Iwata, 1952. *Journal of Experimental Marine Biology and Ecology*, **328**, 168–76.

Zhu, Q.Z., Aller, R.C., and Fan, Y.Z. (2006a). A new ratiometric, planar fluorosensor for measuring high resolution, two-dimensional pCO_2 distributions in marine sediments. *Marine Chemistry*, **101**, 40–53.

Zhu, Q.Z., Aller, R.C., and Fan, Y.Z. (2006b). Two-dimensional pH distributions and dynamics in bioturbated marine sediments. *Geochimica et Cosmochimica Acta*, **70**, 4933–49.

CHAPTER 10

Effects of ocean acidification on marine biodiversity and ecosystem function

James P. Barry, Stephen Widdicombe, and Jason M. Hall-Spencer

10.1 Introduction

The biodiversity of the oceans, including the striking variation in life forms from microbes to whales and ranging from surface waters to hadal trenches, forms a dynamic biological framework enabling the flow of energy that shapes and sustains marine ecosystems. Society relies upon the biodiversity and function of marine systems for a wide range of services as basic as producing the seafood we consume or as essential as generating much of the oxygen we breathe. Perhaps most obvious is the global seafood harvest totalling over 100 Mt yr^{-1} (82 and 20 Mt in 2008 for capture and aquaculture, respectively; FAO 2009) from fishing effort that expands more broadly and deeper each year as fishery stocks are depleted (Pauly *et al.* 2003). Less apparent ecosystem services linked closely to biodiversity and ecosystem function are waste processing and improved water quality, elemental cycling, shoreline protection, recreational opportunities, and aesthetic or educational experiences (Cooley *et al.* 2009).

There is growing concern that ocean acidification caused by fossil fuel emissions, in concert with the effects of other human activities, will cause significant changes in the biodiversity and function of marine ecosystems, with important consequences for resources and services that are important to society. Will the effects of ocean acidification on ecosystems be similar to those arising from other environmental perturbations observed during human or earth history? Although changes in biodiversity and ecosystem function due to ocean acidification have not yet been widely observed, their onset may

be difficult to detect amidst the variability associated with other human and non-human factors, and the greatest impacts are expected to occur as acidification intensifies through this century.

In theory, large and rapid environmental changes are expected to decrease the stability and productivity of ecosystems due to a reduction in biodiversity caused by the loss of sensitive species that play important roles in energy flow (i.e. food web function) or other processes (e.g. ecosystem engineers; Cardinale *et al.* 2006). In practice, however, most research concerning the biological effects of ocean acidification has focused on aspects of the performance and survival of individual species during short-term studies, assuming that a change in individual performance will influence ecosystem function. By their nature, controlled experimental studies are limited in both space and time, and thus may not capture important processes (e.g. acclimatization and adaptation, multispecies biological interactions, chronic low-level impacts) that can ultimately play large roles in the response of marine systems to ocean acidification. This 'scaling up' from individual- to ecosystem-level effects is the most challenging goal for research on the potential effects of ocean acidification.

To plan for the future, society needs information to understand how ocean acidification and other environmental changes will affect fisheries, aquaculture, and other services deriving from the efficient function of marine ecosystems. The influx of fossil fuel CO_2 into the upper ocean from the atmosphere is altering the chemistry of ocean waters at a faster rate and greater magnitude than is thought to have occurred on earth for at least a million years and perhaps as much as 40 Myr (Pelejero *et al.* 2010; see Chapter 2). In this chapter, the influence of this important change in ocean chemistry on the biodiversity and function of marine ecosystems is considered, from basic physiological responses of individual organisms and species, to the potential changes in various ocean environments.

10.2   Biodiversity and ecosystem function

The term biodiversity is broadly defined, and used to characterize aspects of the biological complexity of natural systems. Often cast simply as the number of species in a region (i.e. *species richness*), biodiversity has a far larger scope that spans the variation within and among systems and organisms over multiple scales and levels of genetic, organismal, ecological, or ecosystem diversity. Measures of biodiversity attempt to estimate the richness and evenness of biological characteristics at different levels, such as *species richness* and *species diversity* (the number and evenness of species in a region), *taxonomic diversity* (not just species richness, but diversity at higher taxonomic levels), *genetic diversity* (genetic variation in a population or species), *habitat* or *ecosystem diversity* (range of habitats or ecosys-

tems in a region), or *functional diversity* (the number of functional roles performed by the species present) (see Fig. 10.1; Petchey and Gaston 2006). Biodiversity is a dynamic feature of natural systems, reflecting the continual evolutionary response of species in a region to selection across a broad range of environmental and ecological pressures. Biodiversity can expand and contract as species diversity or other elements of biological diversity are created, maintained, or lost to extinction.

The function of ecosystems is wholly dependent on biodiversity that allows energy to flow through trophic webs and biological networks. Moreover, the stability and resilience of ecosystem functions, from nutrient cycling and energy flow, to the population dynamics of species, are thought to be sensitive to the loss of biodiversity caused by perturbations of both human and non-human origin. This concept has long been considered theoretically, with more diverse and trophically complex systems expected to be more stable and resistant to perturbations. *Species complementarity* (different species have similar ecological roles) and *species redundancy* (different species perform the same function) are thought to provide 'insurance' for ecosystem function in diverse systems by promoting functional diversity and maintaining energy flow among trophic levels (i.e. ecosystem function)

Figure 10.1   Species diversity does not necessarily represent functional diversity. Note that both groups have equal species richness and diversity (a single individual from each of eight molluscan species), but the group on the right has greater taxonomic distinctness and functional diversity. Photo R. M. Warwick.

upon a reduction or loss of species in response to environmental variation or other factors (Loreau *et al.* 2001).

Field studies in terrestrial and marine ecosystems over the past two decades have provided both support (e.g. Steneck *et al.* 2002) and moderate controversy concerning the diversity–stability concept (Loreau *et al.* 2001; Cardinale *et al.* 2006). The controversy relates, in part, to the disproportionate role of key species in many systems studied, without which stability is reduced. This issue may be important in several marine ecosystems where key prey species (e.g. pteropods, krill, anchovies, and squid in coastal systems) or taxa that play an important structural role (e.g. habitat-forming corals) can be critical resources for other taxa. Thus, reduced biodiversity due to the loss of prey taxa or habitat-forming species may not have large effects on energy flow or ecosystem function and stability, so long as key species that maintain the functional diversity of the system are relatively unaffected (Tilman *et al.* 1997). Although this debate continues, a recent examination of experimental studies of nearshore marine communities (Duffy 2009) and long-term studies of large marine ecosystems (Worm *et al.* 2006) provides fairly strong support for the role of biodiversity in ecosystem function and stability. For many marine systems, however, the current understanding of the natural history and functional roles of most species is poor; thus it will be challenging to predict how ocean acidification may affect ecosystem function.

10.3 Acclimatization and adaptation

Organisms and species faced with ocean acidification or other environmental changes have four options—migration, acclimatization (i.e. tolerance), adaptation, or extinction. Migration, by individuals or by a population successively through generations, may be possible in some cases, but the global nature of ocean acidification coupled with range limitations imposed by other environmental parameters (e.g. temperature) may limit this option. For example, meridional gradients in aragonite saturation (see Fig. 14.5 in Chapter 14) may allow the ranges of some mid-latitude taxa to shift toward more saturated waters of the tropics as acidification intensifies, unless other factors (e.g. tempera-

ture) are intolerable. Range shifts along other gradients in pH or carbonate saturation (e.g. depth-related or horizontally in coastal regions) may also be possible.

Whether marine organisms, from microbes to long-lived megafauna, will be able to acclimatize or adapt to future ocean acidification is an important, but unresolved, question. Acclimatization is the process by which individuals adjust to environmental changes (i.e. physiological adjustment). This may result in a change in energy costs (positive or negative) associated with living. Adaptation is the adjustment of species to environmental change between generations, through natural selection of individuals tolerant of new conditions. Tolerance or acclimatization by at least some individuals in the population allows for adaptation, assuming that the tolerant traits are heritable and sufficient time is available for selection to increase the frequency of tolerant genotypes through multiple generations. If individuals in most populations are able to acclimatize to reduced ocean pH or carbonate saturation through the adjustment of physiological homeostasis, changes in ocean biodiversity could be mild—perhaps only a contraction in genetic diversity within species with minor effects on ecosystem functions.

Although it is expected that there will be 'winners' and 'losers' in response to ocean acidification, many vulnerable species (such as corals; see Kleypas and Yates 2009) may suffer from reduced performance and survival, and have limited scope for adaptation due to the expected pace and magnitude of ocean acidification in the future. Taxa with short generation times and immense population sizes such as phytoplankton and microbes (e.g. with one to three generations per day) have perhaps the greatest capacity to adapt, given that upwards of 35 000 to 100 000 generations are possible over the next 100 years as ocean acidification intensifies. Collins and Bell (2006), however, found little evidence of adaptation to high CO_2 levels in a pond alga (*Chlamydomonas*) after 1000 generations. In contrast, the scope for adaptation by species with long generation times (e.g. 10 to 30 yr for some fishes) and relatively small population sizes is expected to be limited when selection for tolerant genotypes is constrained to just a few generations.

10.4   Effects of environmental change

Environmental variation over space or time can have positive and negative effects on biodiversity and ecosystem function related to the rate, magnitude, duration, and spatial scale of environmental change (Knoll *et al.* 2007). Habitats with greater spatial heterogeneity provide variable environmental conditions that typically support higher biodiversity than relatively homogeneous habitats. Temporal environmental variation also plays a key role in regulating local diversity. Some level of change in environmental factors (e.g. physical disturbance or variability in temperature, oxygen, or other parameters) or biological factors (e.g. variation in the abundance of predators or competitors) can lead to enhanced local diversity (Connell 1978). Moreover, environmental variability can promote genetic diversity by selecting for a broad range of genotypes to match environmental patterns, or allowing for adaptive radiation as species emerge to fill new ecological space. Species, populations, or genotypes originate through evolutionary divergence to exploit novel habitats, or through specialization in response to environmental variation (temperature, habitat complexity, oxygen concentration, light, etc.) and biological interactions (trophic, competitive, or mutualistic).

Coral reef ecosystems, typified by high topographic complexity that promotes species-packing, have been shown to have been cradles of diversification throughout the Phanerozoic, with high rates of species origination that are often exported to offshore and deeper regions (Kiessling *et al.* 2010; see Chapter 4). For example, habitat complexity generated by highly branched scleractinian corals such as *Acropora* spp. provide habitat for a remarkably diverse array of fishes and invertebrates. Such branching corals are keystone functional groups, and are threatened by ocean acidification and other environmental changes (Bellwood *et al.* 2004). Reduced coral growth and weaker carbonate cementation will increase the probability of damage to most structurally complex corals during storms, probably leading to reef flattening and reduced reef biodiversity (Hofmann *et al.* 2010). Similar impacts on scleractinian corals, and perhaps other structure-forming groups in deep-sea systems (Guinotte *et al.*

2006), may also influence biodiversity and ecosystem function due to the loss of critical habitat for various coral-associated taxa.

Evidence for shifts in biodiversity during periods of environmental change is common in the fossil record. Mayhew *et al.* (2008) report significant positive correlations between variation in global temperature and the rates of origination and extinction of families and genera through the Phanerozoic. Correlations were higher for originations, but not for extinctions, when diversity lagged behind temperature by 10 Myr, suggesting that diversification occurs mainly after a period of extinction driven by global warming. For marine genera, the effect of variation in CO_2 levels on extinction rates was stronger than that of temperature, suggesting that ocean carbon levels have influenced marine species diversity throughout the Phanerozoic. Five major mass extinction events that caused the loss of 75% or more of all species (Jablonski and Chaloner 1994; see also Chapter 4) are the most striking features of the fossil record. For many (if not all) of these events, rapid change in environmental parameters, such as temperature, oxygen levels, and ocean pH, coupled with ecological factors, appear to have played a large role in the high extinction rates (Knoll *et al.* 2007; Chapter 4). Marine life recovered following each extinction event, but required millions of years, potentially due to slow rates of evolutionary diversification or persistently unfavourable environmental conditions, or both (Knoll *et al.* 2007).

Is it likely that ocean acidification will reduce the biodiversity of marine ecosystems and drive significant shifts in their function? The response of marine ecosystems will be linked to the rate and magnitude of changes in ocean chemistry in relation to the potential rates of acclimatization, adaptation, and evolution of marine organisms, from microbes to vertebrates. The ongoing large and rapid changes in ocean pH and carbonate saturation are expected to drive environmental changes unseen in the recent evolutionary history of marine organisms, posing an evolutionary challenge to acclimatize or adapt. At a minimum, the genetic diversity of various marine taxa is likely to change. It remains unknown whether ocean acidification will drive species to extinction, but it is possible, based on the growing literature concerning the sensitivity and performance of marine organisms under future, high-CO_2 conditions (e.g. Fabry *et al.* 2008; Widdicombe and Spicer 2008; Doney *et al.* 2009).

Extinction is not required for ocean acidification to affect the function of marine ecosystems, since changes in the relative abundance and activities of species can affect biological interactions (e.g. food web function) and the way taxa are in the ocean pH a 005; Hofm s of ocean ac luc-tivity of in marine foo em function are h cts within the ecosy ol-luscs sensitive to re n-ditions (Feely *et al.* 2 meau *et al.* 2009) may be less abundant under future ocean conditions, leading to reduced availability for their current predators, which must then find alternative prey. Likewise, species that may depend on pteropods in other ways (e.g. *Pteropagurus* sp., a hermit crab that depends upon pteropod shells as habitat; McLaughlin and Rahayu 2008), may suffer from reduced pteropod abundance. Simultaneously, pteropod prey (e.g. copepods, diatoms) and potential competitors (e.g. krill, salps) are likely to experience reduced predation and competition, respectively, and thus, may increase in abundance. Consequently, a significant reshuffling of the structure of marine communities and ecosystem function in response to ocean acidification is possible if there are marked shifts in the abundance of 'losers' and 'winners', particularly if key species are affected.

10.5 The effects of ocean acidification on organisms

Several physiological processes, such as photosynthesis, calcification, acid–base homeostasis, respiration and gas exchange, and metabolic rate, can be influenced by changes in ocean carbonate chemistry (Gattuso *et al.* 1999; Seibel and Walsh 2003; Melzner *et al.* 2009b; Chapter 8). High ocean carbon levels are expected to affect primary producers in different ways, perhaps leading to a shift in the structure of phytoplankton populations (Hall-Spencer *et al.* 2008; Doney *et al.* 2009; see Chapters 6 and 7). Taxa that are currently carbon-limited (e.g. some cyanobacteria) may be among the 'winners' in a high-CO_2 ocean. For other autotrophs (e.g. coccolithophores), photosynthesis and growth, as well as calcification, may be affected, with complex responses among species. A suite of experiments on marine phytoplankton have shown that the responses of coccolithophorids (calcitic phytoplankton) to elevated CO_2 levels vary, but generally exhibit reduced rates of calcification (Ridgwell *et al.* 2009; Hendriks *et al.* 2010). Reduced calcification has been measured in a variety of taxa, particularly corals and molluscs (Michaelidis *et al.* 2005; Gazeau *et al.* 2007; Kuffner *et al.* 2008; Doney *et al.* 2009), and is the most widely observed and consistent effect of ocean acidification (Hendriks *et al.* 2010; Kroeker *et al.* 2010). Exposure times have typically been short for most calcification studies, and may often be too short to detect acclimatization, which has been shown to require about 6 weeks for a marine fish (Deigweiher *et al.* 2008). In contrast, coccolithophores may acclimatize to high CO_2 levels within hours (Barcelos e Ramos *et al.* 2010). Increased rates of calcification in low-pH waters have been observed for a few taxa (e.g. crustaceans, Ries *et al.* 2009), but it appears that, at least for some species, higher calcification in low-pH waters may require energetic trade-offs that reduce overall performance (Wood *et al.* 2008).

Even for taxa tolerant of low-pH waters, the physiological 'cost of living' is expected to change the energy required for basic biological functions (Pörtner *et al.* 2000). Immersion in high-CO_2 waters can disrupt the acid–base status of many marine animals, leading to reduced respiratory efficiency, reduced enzyme activity, and metabolic depression, with potentially large effects on overall metabolic performance (Seibel and Walsh 2003; see Chapter 8). Assuming a constant total energy budget, a change in the 'cost of living' is expected to result in a reallocation of energy for growth and reproduction (Fig. 10.2). For taxa affected by ocean acidification, individual physiological stress can lead to reduced growth, size, reproductive output, and survival. On a population level, impaired individual performance and survival have consequences for populations and species that may include reduced abundance, productivity, and resilience to disturbance, as well as increased likelihood of extinction. For taxa benefiting either directly or indirectly from high CO_2 levels, the opposite may be true. It is also important to consider the cumulative effects of environmental stressors on the demography and productivity of populations. Effects on different life stages can sum to significant impacts on population success. For example, during periods of low sea-surface temperature (<13.1°C), exposure to low-pH waters reduces the survival of early life stage barnacles along the coast of the south-west United Kingdom by 25%, potentially leading to reduced local population abundance (Findlay *et al.* 2010).

Sensitivity to ocean acidification is expected to be coupled primarily to fundamental physiological adaptations linked closely to phylogeny. Marine organisms with a natural capacity for gas exchange (i.e. organisms with well-developed respiratory and circulatory systems, as well as respiratory proteins allowing high O_2 and CO_2 fluxes) that support high metabolic rates and high aerobic scope (e.g. fishes, decapod crustaceans, and cephalopods) are pre-adapted for many of the stresses related to ocean acidification (Melzner *et al.* 2009b; see Chapter 8). This is due in part to the overlapping physiological challenges posed by metabolic CO_2 generation during intense aerobic activity (e.g. coping with internal acid–base disruption) and the effects of ocean acidification. Many taxa in habitats with variable or low pH (e.g. vesicomyid clams, vestimentiferan tubeworms, mussels in vent or seep environments) also have adaptations that allow them to thrive in naturally hypoxic and low-pH waters (Goffredi and Barry 2002; Tunnicliffe *et al.* 2009). Mobile crustaceans and fishes may benefit somewhat in a high-CO_2 ocean, based on their generally higher rates of growth and calcification in low-pH waters (Ries *et al.* 2009; Kroeker *et al.* 2010). However, even taxa with the capacity to cope with activity-related hypercapnia can experience impaired physiological performance in high-CO_2 waters. Rosa and Seibel (2008) found that activity levels in jumbo squid declined by 45% under a 0.3 unit reduction in pH. In contrast, cod exposed to a large pH perturbation (–1 pH unit) for several months displayed no evidence of impaired maximal swimming speed (Melzner *et al.* 2009a).

Taxa with weaker control over internal fluid chemistry may be at greater risk from ocean acidification. For example, echinoderms, brachiopods, and lower invertebrates (e.g. sponges, cnidarians, and ctenophores) lack respiratory organs and exchange gases with seawater by molecular diffusion across various body tissues. Although physiological tolerance to ocean acidification has not been examined closely in most of these groups (other than rates of calcification, see below), their postulated weak control of internal fluid chemistry (e.g. sea urchins, Miles *et al.* 2007) is expected to increase their sensitivity to changing ocean chemistry. Echinoderms appear less tolerant of low-pH waters than many groups, as indicated by their conspicuous absence from habitats with naturally high CO_2 levels such as hydrothermal vents (Grassle 1986) and low-pH areas near shallow CO_2 vents off Italy (Hall-Spencer *et al.* 2008). Notably, various taxa with limited physiological capabilities (many cnidarians and sponges) appear to tolerate low or variable pH, due to their occurrence in low-pH habitats such as hydrothermal vents and other natural CO_2 venting sites. Moreover, generalities based on short-term studies of organism physiology or survival, as are most common in the literature, may differ from the eventual long-term consequences of ocean acidification.

For calcifying taxa, the type of carbonate minerals formed can influence their vulnerability to ocean

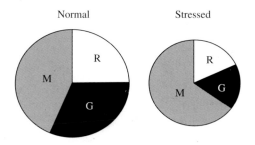

Figure 10.2 Hypothetical energy budget for normal and stressed organisms. Under normal conditions, the energetic cost of maintenance (M) is a significant portion of the total energy budget. If ocean acidification or other environmental changes are stressful, maintenance costs (e.g. ion regulation) can increase, leaving less energy available for growth (G) or reproduction (R). In addition, if metabolic depression is induced by ocean acidification, the total energy budget may decrease, hence the smaller pie size for the energy budget of the stressed organism.

acidification. Carbonate skeletal structures of marine taxa vary considerably both in terms of how much calcium carbonate ($CaCO_3$) is included, from nearly 100% $CaCO_3$ to mixtures of chitin and $CaCO_3$ (e.g. many crustacean shells). The form of $CaCO_3$ also varies, with most taxa precipitating aragonite or calcite, the latter which may include some percentage of magnesium. Of these, high-Mg calcite is the most soluble in seawater, and thus most susceptible to dissolution by ocean acidification, followed by aragonite and calcite (see Box 1.1 in Chapter 1).

Although calcification rates by organisms are generally impaired under low-pH conditions, there is considerable variation among the responses of major taxonomic groups (Hendriks *et al.* 2010; Kroeker *et al.* 2010). Scleractinian corals (aragonite) exhibit the largest reduction in calcification and most consistent response to low-pH waters. Coccolithophores (calcite) and molluscs (mostly calcitic) had somewhat weaker, variable, and non-significant changes in calcification. Individual studies have reported generally reduced rates of calcification for bivalve and gastropod molluscs under high-CO_2 conditions (Gazeau *et al.* 2007; Doney *et al.* 2009; Ries *et al.* 2009; Hendriks *et al.* 2010). In contrast, echinoderms (calcite) are highly variable in response, mainly due to the great variability in degree of calcification within the phylum (e.g. Wood *et al.* 2008). Crustaceans (chitin, calcite, amorphous carbonate) are the single group showing a significant increase in calcification rate under high-CO_2 conditions (Kroeker *et al.* 2010).

Reef-building corals in particular, due to their aragonitic skeletons, are perceived to be at high risk from ocean acidification, based on the projected future reduction in aragonite saturation throughout the world's oceans (Kleypas *et al.* 1999). These projections are consistent with the existing global distribution of deep-sea aragonitic corals, which are most abundant in the Atlantic and relatively rare in habitats with low aragonite saturation, such as the Pacific Basin (Guinotte *et al.* 2006; Manzello 2010). Surprisingly, some corals may survive acidic conditions without carbonate skeletons. Two Mediterranean species (*Oculina patagonica* and *Madracis pharencis*) survived a 12-month exposure to acidic (pH_T = 7.3–7.6) waters but lost their carbonate skeletons, which dissolved in the corrosive

waters (Fine and Tchernov 2007). Upon immersion in ambient pH waters (pH_T = 8.3) the corals recalcified. This type of recovery is unlikely for many other coral taxa with modes of life linked strongly to their structural framework.

Calcified shells and skeletons can play important roles for organisms coping with environmental variability. In some cases, more robust calcification may increase the survival (and presumably the fitness) of organisms, which can affect the biodiversity and function of marine communities. For example, the intertidal snail *Littorina littorea* thickens its shell after exposure to chemical cues produced by its main predator, the green shore crab, *Carcinus maenas*, in effect increasing its defence against predation (Bibby *et al.* 2007). Such shell thickening does not occur under high-CO_2 conditions, presumably due to the increased energetic cost of calcification at lower-pH, less saturated conditions, thereby increasing their risk of predation. Very few studies have examined the effects of ocean acidification on behavioural responses that mediate interactions between interacting species and populations.

10.6   Habitats

The risk of changes in the biodiversity and function of marine ecosystems due to ocean acidification is likely to vary considerably among habitats. Seawater carbonate chemistry is affected by temperature and biological processes, leading to significant patterns in carbonate chemistry across zonal and meridional gradients (Kleypas *et al.* 1999; Feely *et al.* 2004), with depth, and in relation to biological productivity. Colder high-latitude regions have naturally lower saturation states for aragonite and calcite due to the higher solubility of CO_2 at low temperatures, and will be the first surface waters to be persistently undersaturated with respect to aragonite (Orr *et al.* 2005; see Chapter 2), with potentially significant effects on marine calcifiers. The influence of ocean acidification on sediment ecosystems is considered in Chapter 9.

10.6.1   Upper ocean

Changes in environmental conditions through this century, including ocean acidification and warming, could lead to a restructuring of phytoplankton assemblages with consequences that would reverberate throughout marine communities. For example, the composition of phytoplankton assemblages in the north-east Atlantic changed as waters warmed during the latter half of the last century, with a poleward expansion of warm-temperate plankton and recession of colder forms (Hays *et al.* 2005). In the English Channel, coccolithophores and dinoflagellates have become more abundant and dominant over the past ~20 yr, while diatoms and *Phaeocystis* have decreased (Widdicombe *et al.* 2010). The expansion of dinoflagellates may be expected based on the emergence of these groups during warmer, high-CO_2 periods in earth's history (Beardall and Raven 2004). Cyanobacteria, which thrived under high CO_2 levels earlier during earth's history, are also expected to benefit from ocean acidification. Photosynthetic rates of two dominant oceanic cyanobacterial genera (*Synechococcus* and *Trichodesmium*) were observed to increase markedly under expected future climate conditions, so long as nutrient levels were sufficiently high, while other closely related taxa (*Prochlorococcous* and *Nodularia*) showed little change or even lower rates (Fu *et al.* 2007). An increase in the diversity, abundance, or productivity of cyanobacteria could increase rates of global nitrogen fixation (Hutchins *et al.* 2009), which would be likely to drive responses in primary production by other groups. High CO_2 levels have also been shown to cause enhanced production of dimethyl sulphide by natural phytoplankton assemblages, which could promote climate homeostasis by stimulating cloud formation (see Chapter 11).

Elevated pCO_2 can also affect phytoplankton communities and the entire water column community through changes in elemental uptake or calcification by major groups. Increased C:N and C:P ratios have been measured in mixed phytoplankton assemblages in response to high CO_2 levels (Riebesell *et al.* 2007; see Chapter 6), potentially changing their nutritional value to consumers and leading to changes in growth and reproduction of zooplankton. Reduced rates of calcification for coccolithophores and foraminifera, two major planktonic calcifying groups, may also affect the flux of organic debris to deeper waters, due to changes in

ballasting of organic aggregates (Fabry *et al.* 2008; Ridgwell *et al.* 2009).

Holoplankton can also be affected by changing ocean chemistry, perhaps especially in weakly saturated waters of high-latitude systems. Euphausiids and thecosomatous pteropods, two key planktonic groups thought to be critical linkages in global food webs, may be affected differently by ocean acidification. Pteropods (e.g. *Limacina* sp. and *Clio* sp.) are important planktonic calcifiers in open-ocean food webs and represent major prey taxa for higher predators (e.g. many salmon species), and thus are a key link in open-ocean food webs and energy flow (Fabry *et al.* 2009). Immersion of shelled pteropods in high-CO_2, low-CO_3^{2-} waters is known to weaken their aragonitic shells (Orr *et al.* 2005) and is expected to reduce their survival and productivity (Comeau *et al.* 2009, 2010). To date, only minor decadal-scale changes have been observed globally in these groups. In the California Current ecosystem, pteropod abundance has not declined, and may have increased over the past 50 yr (Ohman *et al.* 2009). In the Southern Ocean, where aragonite undersaturation is predicted to begin as early at 2030 (McNeil and Matear 2008), ocean acidification may already be affecting pteropod populations. Roberts *et al.* (2008) reported that the shell weights of pteropods collected in sediment traps deployed in sub-Antarctic waters (47°S) have decreased over the past decade. This decrease was not correlated with chlorophyll abundance or temperature, but was consistent with changes in aragonite saturation, and thus the potential influence of ocean acidification cannot be rejected. The loss of *Limacina* or other key pteropod taxa in undersaturated waters due to reduced calcification (e.g. Comeau *et al.* 2009, 2010) could have significant implications for their predators and energy flow through open-ocean food webs.

Euphausiids, copepods, and other planktonic crustaceans are dominant elements of food webs worldwide, and are often suitable alternative prey for many oceanic pteropod predators (Cooley *et al.* 2009). In contrast to pteropods, krill and other crustaceans may not be strongly affected by ocean acidification, and some taxa may even benefit, but few studies have examined this topic. Kroeker *et al.* (2010) report that the literature available to date indicates significantly higher calcification rates and marginally higher growth for crustaceans in low-pH waters, though survival was somewhat reduced. Kurihara (2008) found lower hatching success for krill exposed to low-pH waters and mixed effects on other crustacean taxa. It remains questionable how krill populations will be affected by ocean acidification, and how the effects, if any, will influence marine biodiversity and food web function.

Sparingly few studies are available to assess the effects of ocean acidification on gelatinous taxa. Jellyfish outbreaks have been reported more commonly over the past decades, with several factors (warming, overfishing, habitat modification, eutrophication, species introductions) being implicated (Richardson *et al.* 2009). Several gelatinous groups are important elements of open-ocean food webs from the tropics to the poles, including larvaceans, chaetognaths, salps, and siphonophores. It remains unclear how these taxa will respond to ocean acidification.

Meroplankton, taxa that live only part of their lives (often early life-history phases) in open waters, are expected to be particularly sensitive to ocean acidification. Recent reviews have shown generally negative or mixed results concerning the effects of ocean acidification on early life stages (Dupont *et al.* 2010; Kroeker *et al.* 2010). Kurihara (2008) reports generally negative effects on eggs, larvae, and other early phases for a variety of marine calcifiers. The vulnerability of early life-history stages can have large effects on population survival and demography, even though adults are somewhat unaffected. For some taxa, the development and survival of early life stages are impaired, and in others delayed, but the larvae develop fully in low-pH waters, albeit more slowly than in control treatments (Dupont *et al.* 2010). Slow development can put early life-history phases at prolonged risk to predators. Although the literature remains sparse concerning the impacts of ocean acidification or climate-related environmental change on the survival and development of meroplankton in general, changing ocean conditions could drive important changes in the population dynamics of various species, with indirect effects throughout marine food webs.

10.6.2 Deep-sea ecosystems

Deep-sea ecosystems may experience some of the most profound changes in biodiversity and ecosystem function in response to ocean acidification. Dramatic shoaling of the aragonite and calcite saturation boundaries (see Chapter 2) will cause very large shifts in habitat quality for deep-sea calcifiers. Aragonite and calcite undersaturation of deep-sea waters is likely to restrict deep-sea aragonitic corals (and perhaps many calcitic forms) from much of their existing bathymetric ranges (Tittensor *et al.* 2010). As the saturation states for aragonite (Ω_a) and calcite (Ω_c) drop, it becomes energetically more costly to precipitate $CaCO_3$ (Cohen and Holcomb 2009), and where Ω drops below 1, exposed $CaCO_3$ is subject to dissolution (Hall-Spencer *et al.* 2008; Manzello *et al.* 2008). Recent surveys of the global distributions of aragonitic scleractinian corals indicate that few taxa are currently found below the saturation depth for aragonite (Guinotte *et al.* 2006). *Lophelia* sp., a common aragonitic deep-sea coral, has been shown to calcify 30 to 56% more slowly in waters with pH perturbations 0.15 to 0.3 units lower than ambient (Maier *et al.* 2009). Although calcification proceeds even when Ω_a drops below 1, continued reductions in $CaCO_3$ saturation appear very likely to have an effect on deep-sea corals in the future. Many other deep-sea corals (e.g. gorgonians) precipitate less soluble calcite, but could be affected as the calcite saturation depth rises with increasing ocean CO_2.

Changes in the biodiversity of deep-sea corals are likely to affect the function of deep-sea ecosystems. Deep-sea coral communities are often considered to be hot spots for biodiversity, with high species diversity of structure-forming corals (often dominated by octocorals) as well as many other taxa associated with the heterogeneous habitat structure (Roberts *et al.* 2006). Such communities are common on many seamounts, which number upwards of 50 000 worldwide. Impacts on deep-sea corals could also require long periods for recovery, even in suitable habitats, considering the slow growth rates and high longevity of many species, with ages reaching from decades to centuries (Roberts *et al.* 2006) or longer (Roark *et al.* 2009).

Abyssal sedimentary habitats are not immune to the potential effects of ocean acidification. Echinoderms, including a diverse assemblage of ophiuroids, echinoids, and holothurians, commonly form a dominant guild of abyssal benthic invertebrates, along with decapod crustaceans and fishes. The weakly calcified tests of deep-sea urchins suggest that calcification is either unimportant as a protection against predators, or is energetically costly, or both. Some taxa, such as *Tromikosoma* sp. in the North Pacific, have little or no carbonate in their test, which is proteinaceous. As anthropogenic CO_2 penetrates to the abyss in the future, seawater will become corrosive to aragonite and calcite, presumably making it even more difficult for many echinoids and other carbonate-bearing taxa to form their skeletons. The absence of echinoderms from areas in the Okinawa Trough exposed continuously to high-CO_2 vent fluids (A. Boetius, pers. comm.), suggests that ocean acidification could act selectively against this often dominant abyssal phylum. Weaker calcification under more acidic conditions could affect the survival of a variety of taxa. Mussels (*Bathymodiolus brevior*) inhabiting low-pH hydrothermal vent systems in the western Pacific survive and grow, but have poorly calcified shells, making them more vulnerable to predation by decapod crabs (*Paralomus* sp.) than are conspecifics with thickly calcified shells that inhabit less corrosive sites (Tunnicliffe *et al.* 2009).

The strong link between communities at the surface and in the deep sea suggests that changes in biodiversity in the upper ocean due to ocean acidification could initiate shifts in biodiversity and ecosystem function in the deep sea. Changes in the export of organic debris from surface waters due to ocean acidification, perhaps in combination with other environmental changes, could affect bathypelagic, abyssal, and benthic ecosystems in the deep sea. Recycling of organic material in the upper water column may increase due to increased dissolution of coccolithophores and foraminifera, leading to a reduction in carbonate ballast within organic aggregates and reduced export of organic carbon to deep waters. These potential effects of ocean acidification on the rate of carbonate rain and the biological pump are not yet well understood (see Chapter 6). For the food-limited deep sea, however, changes

in sinking organic flux, in addition to altered pH and carbonate saturation, may drive important changes in ecosystem function.

10.6.3 Coastal ecosystems

Coastal ecosystems, including coastal upwelling zones, coral reefs, mangroves, kelp forests, seagrass beds, estuaries, and other nearshore systems, are by far the most important ecosystems that humans depend upon for finfish and shellfish fisheries and aquaculture, as well as recreation, and thus are critically important with respect to future impacts from ocean acidification and other environmental changes (Cooley *et al.* 2009). Coastal systems span a wide range of physical and oceanographic regimes from high to low latitudes, upwelling systems to western boundary currents, and both benthic and pelagic assemblages. The seawater chemistry and biological processes in these disparate environments vary greatly, and thus their sensitivity to ocean acidification is also expected to vary. Anthropogenic changes in oceanographic and ecological processes in coastal systems related to fossil fuel emissions and other human activities (e.g. coastal nutrient loading) are not fully understood (e.g. Feely *et al.* 2010), but will probably also differ among ecosystems. Considering the diversity of coastal ecosystems, it is beyond the scope of this chapter to provide a comprehensive treatment of their vulnerability to ocean acidification. Instead, we touch on several features of ocean acidification in coastal systems, using upwelling systems and coral reefs as examples.

10.6.3.1 Upwelling zones
Upwelling zones off the western US coast, along most eastern boundary currents, and several other regions worldwide typically have a wider range in oxygen, pH, and other carbonate system parameters than most open-ocean systems, due largely to the boom and bust productivity of surface waters and remineralization of organic material at depth. Surface pH and carbonate saturation vary through these cycles, and waters in the oxygen minimum zone (OMZ) several hundred metres below the surface can be suboxic and corrosive to $CaCO_3$. Estuaries can also have quite strong gradients in

oxygen, pH, and carbonate saturation (Miller *et al.* 2009). Along the Californian coast (and eastern Pacific margin) the concentration of oxygen can approach anoxia (less than 5% saturation, ~10 µmol kg^{-1}) in the core of the OMZ near a depth of 700 m. Wind-driven coastal upwelling and other processes can transport hypoxic, low-pH waters from the upper OMZ toward the surface, creating conditions of undersaturation with respect to $CaCO_3$ for surface organisms and coastal benthic communities (Feely *et al.* 2008, 2010). The highly variable conditions found in coastal upwelling zones act as a physiological filter, allowing species to thrive only if all of their life stages can tolerate the variable local conditions. Natural environmental variability has been shown to have large effects on the structure and function of the California Current ecosystem (e.g. Chavez *et al.* 2003), which may mask or interact with the effects of ocean acidification or other anthropogenic changes. Increasing acidification of coastal environments through this century due to elevated CO_2 emissions or other factors is likely to pose new physiological challenges for coastal taxa.

The response of coastal species to future ocean chemistry has been examined in various recent studies indicating mixed, but generally negative, responses (Kroeker *et al.* 2010). As noted above, some photosynthetic taxa may benefit from ocean acidification (e.g. some seagrasses; Hall-Spencer *et al.* 2008; Hendriks *et al.* 2010). However, even though they thrive in highly variable environments, coastal heterotrophs have generally responded to low-pH or low-carbonate-saturation waters with impaired performance (e.g. acid–base balance, calcification, growth, or survival), including adults (e.g. barnacles, Findlay *et al.* 2010; bivalves, Gazeau *et al.* 2007; ophiuroids, Wood *et al.* 2008; urchins, Miles *et al.* 2007) and early life-history phases (barnacles, Findlay *et al.* 2010; echinoderms, Dupont *et al.* 2010; molluscs, Kurihara 2008). Not all coastal fauna respond strongly or negatively to ocean acidification. Adults of some higher taxa have shown little sensitivity to acidification (e.g. Dungeness crab, Pane and Barry 2007; cod, Melzner *et al.* 2009a), and Ries *et al.* (2009) observed mixed responses to ocean acidification for juveniles of various taxa, including enhanced calcification and growth for a lobster and a shrimp.

Changes in the biodiversity and function of coastal ecosystems due to ocean acidification are likely to be linked to the vulnerability of key intermediate prey taxa, with broader effects driven indirectly through trophic dependences. Top predators in these systems, including fishes, cephalopods, birds, and mammals, are expected to have greater physiological capacities (at least as adults) to cope with elevated CO_2 levels, than many of their prey. Large, active cephalopods, especially those inhabiting suboxic habitats, may experience respiratory problems with future acidification (e.g. Rosa and Seibel 2008), but most vertebrates, especially air-breathing birds and mammals, are not expected to be affected directly by ocean acidification. However, the indirect effects of ocean acidification for these higher predators could be substantial. Impacts on highly sensitive taxa (e.g. thecosomatous pteropods) could cascade through food webs, as has been observed in relation to other environmental changes (warming). For example, the breeding success of planktivorous and piscivorous seabirds along the central California coast varies among warm and cold periods, but is linked directly to the availability of key prey taxa rather than specific physical parameters (Sydeman *et al.* 2001). It remains unknown whether the potentially negative effects of ocean acidification on some groups (e.g. pteropods) will be balanced by positive responses in other groups with similar functional roles (e.g. krill). In coastal systems with generally reduced functional redundancy among species (Micheli and Halpern 2005), the likelihood of indirect food web effects due to ocean acidification and other environmental changes may be high.

Benthic communities in coastal systems and seabed habitats throughout the world's oceans are centres of biodiversity where much of the species richness of the oceans is found. These communities include a wide range of organisms that play key roles as ecosystem engineers, providing habitat for other taxa (e.g. corals, kelp, seagrasses, burrowing taxa, and oyster beds), and play important roles in elemental fluxes of carbon and nitrogen through bioturbation and other effects of burrowing (Widdicombe and Spicer 2008). Benthic systems in coastal zones may be more vulnerable to ocean acidification than some more offshore sites because they are likely to contain waters that are corrosive to $CaCO_3$ (at least along coastal areas with strong oxygen minimum zones) earlier than many surface communities (e.g. Feely *et al.* 2008), with potentially important impacts for calcification and survival. Other chapters in this volume give a more thorough discussion of benthic (Chapter 7) and sedimentary (Chapter 9) habitats.

10.6.3.2 Coral reefs

Coral reefs are perhaps the ecosystems understood the best in terms of potential impacts of ocean acidification for marine biodiversity. Precipitation of $CaCO_3$ by corals and calcifying algae forms the physical structure and much of the habitat complexity of coral reefs upon which additional biodiversity develops. Tropical reef systems have also been shown to be 'cradles of evolution', where more species have originated during earth's history than any other region (Kiessling *et al.* 2010). Likewise, tropical reefs have been shown to disappear from the fossil record during several mass extinctions (Veron 2008), indicating that they are also vulnerable to environmental change (see Chapter 4).

Ocean acidification, coupled with ocean warming, can pose important risks for the biodiversity and function of coral reef systems in several ways, ranging from basic changes in the biomineralization and erosion of the physical foundation of reefs, to less understood changes in biological interactions among species. Numerous studies have documented a reduction in calcification by corals and coralline algae in response to ocean acidification (Doney *et al.* 2009). De'ath *et al.* (2009) documented a recent decline in calcification rates on the Great Barrier Reef attributable to warming, ocean acidification, or both. In addition to reduced calcification rates, the strength of cementation may also be reduced in waters with a lower pH, promoting higher rates of physical and bio-erosion (Manzello *et al.* 2008). Processes other than calcification are also affected by ocean acidification. Competition between corals and other taxa, particularly non-calcifying macroalgae, may also be mediated by reduced rates of calcification or other physiological processes linked to ocean acidification, which are likely to reduce the ability of corals to compete for space (Kuffner *et al.* 2008). Reproduction and recruit-

ment of corals can be affected by low-pH waters (Cohen and Holcomb 2009). Reduced calcification coupled with poor cementation can promote erosion and reef flattening, which severely reduces the structural heterogeneity of reefs and lowers its potential to support biodiversity (Alvarez-Filip *et al.* 2009). Reefs experiencing a loss of structural complexity also experience a loss or change in fish assemblages, lower densities of commercially important species, and lower rates of larval fish recruitment (Feary *et al.* 2007). Weaker reef cementation also increases the potential for reef damage as storm frequency and intensity increases with continued global warming, leading to further reef degradation. Eventually, erosion of poorly cemented (and non-accreting) forereef habitats will open lagoon and backreef areas to erosion and provide less effective shoreline protection for coastal communities.

The effect of ocean acidification on the function of entire communities has been addressed only rarely. In a mesocosm-based study of a temperate rocky shore community in the UK, Hale *et al.* (2011) observed a reduction in the species diversity and evenness of the faunal assemblage with reduced pH, with a general ranking of vulnerability from echinoderms (most sensitive), to molluscs, crustaceans, and polychaetes (least sensitive). Nematode abundance increased as pH was reduced, most likely due to a release from biological disturbance (predation and competition) rather than a direct benefit from ocean acidification. The effects of elevated temperature differed among pH treatments, highlighting the difficulty of extrapolating from studies involving single species or single environmental factors, to predict the effects of future environmental changes on natural communities.

Natural CO_2 venting sites offer perhaps the strongest evidence of shifts in the biodiversity of entire communities in response to high ocean CO_2 levels. Benthic communities along a persistent gradient in pH and carbonate saturation near CO_2 vents in intertidal to shallow subtidal depths adjacent to the island of Ischia off the southern coast of Italy show dramatic faunal and floral patterns that are apparently a result of differential tolerance to ocean acidity (Hall-Spencer *et al.* 2008). Continuous seafloor venting of nearly pure CO_2 alters seawater

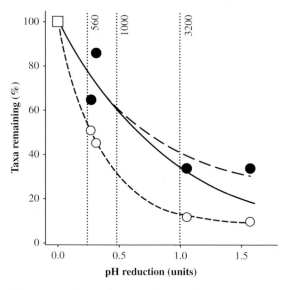

Figure 10.3 Change in diversity as a function of pH reduction for organisms living near the Ischia CO_2 vents. The biodiversity remaining (per cent of taxa that occur in areas with no pH reduction, open square) is shown for calcifying taxa (51 taxa total, white circles) and non-calcifying taxa (71 taxa total, black circles). Atmospheric pCO_2 levels (ppmv CO_2) that would be required to cause pH changes in ocean surface waters equivalent to those observed at three locations along the pH gradient at Ischia are indicated by dotted vertical lines. For calcifiers (short dashed curve), non-calcifiers (long dashed curve), and all taxa combined (solid curve, data not shown), exponential regressions explained 99%, 90%, and 88% of the variance, respectively. Fitted regressions indicate a loss of biodiversity of ~40% for non-calcifiers and all taxa, and ~70% for calcifiers, for a pH reduction corresponding to the atmospheric pCO_2 level expected by 2100. Data from Hall-Spencer *et al.* (2008).

chemistry at this site. CO_2 mixes as it is advected downstream to produce a spatial gradient in ocean pH from normal (pH_T ~8.2) to acidic (pH_T ~6.6). Hall-Spencer *et al.* (2008) documented patterns of species richness at sites along the pH gradient created by the venting CO_2, finding a general decrease in biodiversity with pH for all floral and faunal taxa examined. Over all taxa, biodiversity declined exponentially with pH ($R^2 = 0.88$; $P < 0.001$), with a slope indicating a 65% loss in taxa richness for a 1 unit decrease in pH (Fig. 10.3). Taxa richness for calcifiers declined more rapidly with pH than non-calcifying taxa with a ~60% loss of species under a ~0.5 unit pH reduction. Of 12 calcifying algal taxa, 10 were lost with just a 0.27 unit pH reduction, and none were present in the lower-pH sites. Calcifying animals were also less diverse in lower-pH areas,

but 62% of 39 taxa were still present in the mildly acidic sites (reduction of 0.27 pH units). In contrast, the pattern of diversity loss with pH was similar for non-calcifying fauna and algae. Although this system and other natural CO_2 venting sites do not completely mimic the scale and effects of global ocean acidification, they are strong indicators of its potential effects on biodiversity and ecosystem function, as well as natural laboratories that complement other efforts to understand the effects of a changing ocean.

10.7 Implications of biodiversity loss

Although studies available to date do not provide a clear picture of future changes in ecosystem function due to ocean acidification, much evidence suggests that ocean acidification could have increasingly profound effects in several marine ecosystems (coral reefs, deep-sea systems, high-latitude systems), particularly in combination with other anthropogenic environmental changes. Some reshuffling of dominance in phytoplankton communities appears likely, but it is still not known if the long-term effects of higher ocean CO_2 levels will cause a decrease or increase in primary production (see Chapter 6). If key species in intermediate trophic levels (e.g. thecosomatous pteropods) are affected either positively or negatively by ocean acidification, food webs may be destabilized to some extent, perhaps altering the path and efficiency of energy transfer to upper trophic levels. If biodiversity is reduced within food webs, it is expected that the productivity and predictability of fisheries will be reduced (Worm *et al.* 2006). It is also possible, however, that a simpler food web structure and potentially higher primary production will enable greater trophic transfer from the base to the top of food webs, thereby increasing fisheries yields.

Changes in biodiversity could have important effects on ecosystem services for society, with the greatest impacts being on island nations which rely heavily on seafood harvests and have less opportunity for agriculture (Cooley *et al.* 2009). Expected declines in coral reefs will affect coastal fisheries as well as tourism, an economic base for many tropical island nations. Reduced calcification, growth, and survival of calcifying organisms, especially molluscs, could compromise aquaculture efforts around the world, reducing production as demand for seafood rises with population. In addition, negative impacts on larvae could place a greater reliance on cultured larvae rather than natural seeding. For example, recent failures of oyster reproduction for both natural and cultured larvae in the Pacific Northwest appear to be linked strongly to low-pH waters (Feely *et al.* 2010). Chapter 13 gives a broader discussion of the societal impacts of ocean acidification.

Will the effects of ocean acidification on ecosystem function be comparable to those of other anthropogenic environmental changes? Overfishing has had very large effects on the distribution and abundance of marine fish communities across the globe (Pauly *et al.* 2003) and, at least at present, is exerting greater influence on the function of marine ecosystems than ocean acidification. In terrestrial systems, the effects of climate change are expected to be second only to land-use changes in the projected rapid decline in terrestrial biodiversity by 2100, with somewhat smaller effects of elevated CO_2 levels (Sala *et al.* 2000). Anthropogenic activities including fossil fuel emissions, pollutants, and nutrient additions to coastal and open-ocean waters are all expected to increase through much of this century and drive increasingly negative impacts on the function of ocean ecosystems (Doney 2010).

Ecologists are now developing innovative methods to move forward from short-term, single-species studies that have provided important information on species' responses to ocean acidification, to experimental approaches that capture longer-term, ecosystem-level effects and provide predictions of future ecosystem responses useful for resource managers and other stakeholders (Fig. 10.4). The design of experiments that integrate the effects of multiple environmental factors (e.g. warming, ocean acidification, eutrophication) on population and ecosystem performance is challenging, but is essential to gain an understanding of the real-world effects of ocean acidification. Multispecies responses to ocean acidification, in combination with other stressors, may be considerably different from results extrapolated from single-species, single-factor studies. For example, experimental approaches that examine linkages between physiological responses to ocean acidification and long-term organism performance and activity, particularly in the context of multispecies interactions would move beyond simpler mechanistic studies of physiological performance. A greater understanding of the potential for adaptation by individual species and entire communities are also important elements of future research programmes. Longer-term experiments, preferably including multiple factors and encompassing multispecies com-

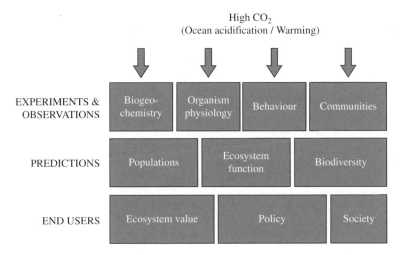

Figure 10.4 Integration of ocean acidification research activities. Experimental studies concerning the effects of ocean acidification and warming, or other environmental changes, on individuals to communities will provide input for understanding and modelling the effects on population processes and ecosystem function. Ultimately, integrated research and outreach will provide information for various stakeholders and society.

munities, will be essential for examining the effects of chronic ocean acidification. One important goal of such studies will be to understand if and how ocean acidification and other anthropogenic environmental changes will ultimately affect the function of marine communities. Will ocean acidification drive communities past 'tipping points' with dramatic shifts in ecosystem function and ecosystem services, as has been proposed for coral reef systems (Hoegh-Guldberg *et al.* 2007)? New experimental approaches to evaluate the response of organisms to ocean acidification and other concurrent environmental changes and human influences, including (1) the potential for acclimatization and adaptation by key taxa, (2) scaling up from individuals to populations, multispecies assemblages, and communities over long timescales and in various ecosystems, and (3) effects on energy flow through marine food webs and the eventual consequences for top predators, including marine fisheries, will provide society with new tools and understanding concerning how the resources and services society depends upon may be affected in a changing ocean.

10.8 Conclusion

Ocean acidification represents a large and very rapid change in the chemistry of the ocean, with the potential to affect the biodiversity and function of a variety of marine ecosystems. Although broad effects of anthropogenic ocean acidification have not yet been observed, they may be difficult to detect amid the influence of other factors, and are expected to emerge and increase through this century. Ocean acidification is considered a threat mainly due to the expected reduction in calcification in various species as the pH of ocean waters decreases. Reduced calcification, growth, and survival by corals in coral reef habitats and deep-sea coral-dominated communities are expected in response to ocean acidification, with ecosystem-wide impacts due to the indirect effects on coral-associated taxa. Shifts in phytoplankton communities and intermediate prey groups due to differential responses to ocean acidification may drive changes in energy flow through marine food webs, ultimately influencing the productivity and stability of marine fisheries. Ocean acidification is currently considered to have lesser impacts on the

function of marine ecosystems than other anthropogenic activities, but its influence in various communities is expected to increase as ocean acidification intensifies through this century. In some ecosystems, it is possible that ocean acidification, along with other anthropogenic environmental changes, may alter ecosystem function by reducing biodiversity and either approaching or potentially crossing ecological tipping points, with unpredictable effects on ecosystem function and services for society.

10.9 Acknowledgements

J.P.B. is grateful for support from the David and Lucille Packard Foundation, and to F. Melzner and W. Fischer who provided important and very helpful comments on the manuscript.

References

Alvarez-Filip, L., Dulvy, N.K., Gill, J.A., Côté, I.M., and Watkinson, A.R. (2009). Flattening of Caribbean coral reefs: region-wide declines in architectural complexity. *Proceedings of the Royal Society B: Biological Sciences*, **276**, 3019–25.

Barcelos e Ramos, J., Müller, M.N., and Riebesell, U. (2010). Short-term response of the coccolithophore *Emiliania huxleyi* to abrupt changes in seawater carbon dioxide concentrations. *Biogeosciences*, **6**, 177–86.

Beardall, J. and Raven, J.A. (2004). The potential effects of global climate change on microalgal photosynthesis, growth and ecology. *Phycologia*, **43**, 26–40.

Bellwood, D.R., Hughes, T.P., Folke, C., and Nyström, M. (2004). Confronting the coral reef crisis. *Nature*, **429**, 827–33.

Bibby, R., Cleall-Harding, P., Rundle, S., Widdicombe, S., and Spicer, J. (2007). Ocean acidification disrupts induced defences in the intertidal gastropod *Littorina littorea*. *Biology Letters*, **3**, 699–701.

Cardinale, B.J., Srivastava, D.S., Duffy, J.E. *et al.* (2006). Effects of biodiversity on the functioning of trophic groups and ecosystems. *Nature*, **443**, 989–92.

Chavez, F.P., Ryan, J., Lluch-Cota, S.E., and Niquen, M. (2003). From anchovies to sardines and back: multidecadal change in the Pacific Ocean. *Science*, **299**, 217–21.

Cohen, A.L. and Holcomb, M. (2009). Why corals care about ocean acidification: uncovering the mechanism. *Oceanography*, **22**, 117–27.

Collins, S. and Bell, G. (2006). Evolution of natural algal populations at elevated CO_2. *Ecology Letters*, **9**, 129–35.

Comeau, S., Gorsky, G., Jeffree, R., Teyssié, J.-L., and Gattuso, J.-P. (2009). Impact of ocean acidification on a key arctic pelagic mollusc (*Limacina helicina*). *Biogeosciences*, **6**, 1877–82.

Comeau, S., Jeffree, R., Teyssié, J.-L., and Gattuso, J.-P. (2010). Response of the arctic pteropod *Limacina helicina* to projected future environmental conditions. *PLoS One*, **5**, e11362.

Connell, J.H. (1978). Diversity in tropical rain forests and coral reefs. *Science*, **199**, 1302–10.

Cooley, S.R., Kite-Powell, H.L., and Doney, S.C. (2009). Ocean acidification's potential to alter global marine ecosystems services. *Oceanography*, **22**, 172–81.

De'ath, G., Lough, J.M., and Fabricius, K.E. (2009). Declining coral calcification on the Great Barrier Reef. *Science*, **323**, 116–19.

Doney, S.C. (2010). The growing human footprint on coastal and open-ocean biogeochemistry. *Science*, **328**, 1512–16.

Doney, S.C., Fabry, V.J., Feely, R.A., and Kleypas, J.A. (2009). Ocean acidification: the other CO_2 problem. *Annual Review of Marine Science*, **1**, 169–92.

Deigweiher, K., Koschnick, N., Pörtner, H.-O., and Lucassen, M. (2008). Acclimation of ion regulatory capacities in gills of marine fish under environmental hypercapnia. *American Journal of Physiology – Regulatory, Integrative and Comparative Physiology*, **295**, R1660–R1670.

Duffy, J.E. (2009). Why biodiversity is important to the functioning of real-world ecosystems. *Frontiers in Ecology and the Environment*, **7**, 437–44.

Dupont, S., Ortega-Martinez, O., and Thorndike, M. (2010). Impact of near-future ocean acidification on echinoderms. *Ecotoxicology*, **19**, 449–62.

Fabry, V.J., Seibel, B.A., Feely, R.A., and Orr, J.C. (2008). Impacts of ocean acidification on marine fauna and ecosystem processes. *ICES Journal of Marine Science*, **65**, 414–32.

Fabry, V.J., McClintock, J.B., Mathis, J.T., and Grebmeier, J.M. (2009). Ocean acidification at high latitudes: the bellwether. *Oceanography*, **22**, 160–71.

FAO (2009). *The state of world fisheries and aquaculture (2008)*, 178 pp. Food and Agriculture Organization of the United Nations (FAO), Rome, Italy.

Feary, D.A., Almany, G.R., McCormick, M.I., and Jones, G.P. (2007). Habitat choice, recruitment and the response of coral reef fishes to coral degradation. *Oecologia*, **153**, 727–37.

Feely, R.A., Sabine, C.L., Lee, K. *et al.* (2004). Impact of anthropogenic CO_2 on the $CaCO_3$ system in the oceans. *Science*, **305**, 362–6.

Feely, R.A., Sabine, C.L., Hernandez-Ayon, J., Ianson, D. and Hales, B. (2008). Evidence for upwelling of corrosive 'acidified' water onto the continental shelf. *Science*, **320**, 1490–2.

Feely, R.A., Alin, S.R., Newton, J. *et al.* (2010). The combined effects of ocean acidification, mixing and respiration on pH and carbonate saturation in an urbanized estuary. *Estuarine, Coastal, and Shelf Science*, **88**, 442–9.

Findlay, H.S., Burrows, M.T., Kendall, M.A., Spicer, J.I., and Widdicombe, S. (2010). Can ocean acidification affect population dynamics of the barnacle *Semibalanus balanoides* at its southern range edge? *Ecology*, **9**, 2931–40.

Fine, M. and Tchernov, D. (2007). Scleractinian coral species survive and recover from decalcification. *Science*, **315**, 1811.

Fu, F-X., Warner, M.E., Zhang, Y., Feng, Y., and Hutchins, D.A. (2007). Effects of increased temperature and CO_2 on photosynthesis, growth and elemental ratios in marine *Synechococcus* and *Prochlorococcus* (cyanobacteria). *Journal of Phycology*, **43**, 485–96.

Gattuso, J.-P., Allemand, D., and Frankignoulle, M. (1999). Photosynthesis and calcification at cellular, organismal and community levels in coral reefs: a review on interactions and control by carbonate chemistry. *American Zoologist*, **39**, 160–83.

Gazeau, F., Quiblier, C., Jansen, J.M., Gattuso, J.-P., Middelberg, J., and Heip, C.H.R. (2007). Impact of elevated CO_2 on shellfish calcification. *Geophysical Research Letters*, **34**, L07603, doi:10.1029/2006GL028554.

Goffredi, S.K. and Barry, J.P. (2002). Species-specific variation in sulfide physiology between closely related vesicomyid clams. *Marine Ecology Progress Series*, **225**, 227–38.

Grassle, J.F. (1986). The ecology of deep-sea hydrothermal vent communities. *Advances in Marine Biology*, **23**, 301–62.

Guinotte, J.M., Orr, J., Cairns, S., Freiwald, A., Morgan, L., and George, R. (2006). Will human-induced changes in seawater chemistry alter the distribution of deep-sea scleractinian corals? *Frontiers in Ecology and the Environment*, **4**, 141–6.

Hale, R., Calosi, P., McNeill, L., Mieszkowska, N., and Widdicombe, S. (in press). Predicted levels of future ocean acidification and temperature rise could alter community structure and biodiversity in marine benthic communities. *Oikos*, **120**, 661–674.

Hall-Spencer, J.M., Rodolfo-Metalpa, R., Martin, S. *et al.* (2008). Volcanic carbon dioxide vents show ecosystem effects of ocean acidification. *Nature*, **454**, 96–9.

Hays, G.C., Richardson, A.J., and Robinson, C. (2005). Climate change and marine plankton. *Trends in Ecology and Evolution*, **20**, 337–44.

Hendriks, I.E., Duarte, C.M., and Alvarez, M. (2010). Vulnerability of marine biodiversity to ocean acidification: a meta-analysis. *Estuarine, Coastal and Shelf Science*, **86**, 157–64.

Hoegh-Guldberg, O., Mumby, P.J., Hooten, A.J. *et al.* (2007). Coral reefs under rapid climate change and ocean acidification. *Science*, **318**, 1737–42.

Hofmann, G.E., Barry, J.P., Edmunds, P.J. *et al.* (2010). The effect of ocean acidification on calcifying organisms in marine ecosystems: an organism to ecosystem perspective. *Annual Review of Ecology, Evolution and Systematics*, **41**, 127–47.

Hutchins, D.A., Mulholland, M.R., and Fu, F. (2009). Nutrient cycles and marine microbes in a CO_2-enriched ocean. *Oceanography*, **22**, 128–45.

Jablonski, D. and Chaloner, W.G. (1994). Extinctions in the fossil record. *Philosophical Transactions of the Royal Society B: Biological Sciences*, **344**, 11–17.

Kiessling, W., Simpson, C., and Foote, M. (2010). Reefs as cradles of evolution and sources of biodiversity in the Phanerozoic. *Science*, **327**, 196–8.

Kleypas, J.A., Buddemeier, R.W., Archer, D., Gattuso, J.-P., Langdon, C., and Opdyke, B.N. (1999). Geochemical consequences of increased atmospheric carbon dioxide on coral reefs. *Science*, **284**, 118–20.

Kleypas, J.A. and Yates, K.K. (2009). Coral reefs and ocean acidification. *Oceanography*, **22**, 108–17.

Knoll, A.H., Bambach, R.K., Payne, J.L., Pruss, S., and Fischer, W.W. (2007). Paleophysiology and the end-Permian mass extinction. *Earth Planetary Science Letters*, **256**, 295–313.

Kroeker, K.J., Kordas, R.L., Crim, R.N., and Singh, G.G. (2010). Meta-analysis reveals negative yet variable effects of ocean acidification on marine organisms. *Ecology Letters*, **13**, 1419–34.

Kuffner, I.B., Andersson, A.J., Jokiel, P.L., Rodgers, K.S., and Mackenzie, F.T. (2008). Decreased abundance of crustose coralline algae due to ocean acidification. *Nature Geoscience*, **1**, 114–17.

Kurihara, H. (2008). Effects of CO_2-driven ocean acidification on the early developmental stages of invertebrates. *Marine Ecology Progress Series*, **373**, 275–84.

Loreau, M., Naeem, S., Inchausti, P. *et al.* (2001). Biodiversity and ecosystem functioning: current knowledge and future challenges. *Science*, **294**, 804–8.

McLaughlin, P.A. and Rahayu, D.L. (2008). *Pteropagurus* and *Catapagurus* (Decapoda, Anomura, Paguidae): resource sharing or 'any port in a storm'? *Zoosystema*, **30**, 899–916.

McNeil, B.I. and Matear, R.J. (2008). Southern ocean acidification: a tipping point at 450-ppm atmospheric CO_2. *Proceedings of the National Academy of Sciences USA*, **105**, 18860–4.

Maier, C., Hegeman, J., Weinbauer, M.G., and Gattuso, J.-P. (2009). Calcification of the cold-water coral *Lophelia pertusa* under ambient and reduced pH. *Biogeosciences*, **6**, 1671–80.

Manzello, D.P. (2010). Ocean acidification hot spots: spatiotemporal dynamics of the seawater CO_2 system of eastern Pacific coral reefs. *Limnology and Oceanography*, **55**, 239–48.

Manzello, D.P., Kleypas, J.A., Budd, D.A., Eakin, C.M., Glynn, P.W., and Langdon, C. (2008). Poorly cemented coral reefs of the eastern tropical Pacific: possible insights in reef development in a high-CO_2 world. *Proceedings of the National Academy of Sciences USA*, **105**, 10450–5.

Mayhew, P.J., Jenkins, G.B., and Benton, T.G. (2008). A long-term association between global temperature and biodiversity, origination and extinction in the fossil record. *Proceedings of the Royal Society, B, Biological Sciences*, **275**, 47–53.

Melzner, F., Göbel, S., Langenbuch, M., Gutowska, M., Pörtner, H.-O., and Lucassen, M. (2009a). Swimming performance in Atlantic Cod (*Gadus morhua*) following long-term (4–12 months) acclimation to elevated seawater Pco_2. *Aquatic Toxicology*, **92**, 30–7.

Melzner, F., Gutowska, M.A., and Langenbuch, M., (2009b). Physiological basis for high CO_2 tolerance in marine ectothermic animals: pre-adaptation through lifestyle and ontogeny? *Biogeosciences*, **6**, 2313–31.

Michaelidis, B., Ouzounis, C., Paleras, A., and Pörtner, H.O. (2005). Effects of long-term moderate hypercapnia on acid-base balance and growth rate in marine mussels *Mytilus galloprovincialis*. *Marine Ecology Progress Series*, **293**, 109–18.

Micheli, F. and Halpern, B.S. (2005). Low functional redundancy in coastal marine assemblages. *Ecology Letters*, **8**, 391–400.

Miles, H., Widdicombe, S., Spicer, J.I., and Hall-Spencer, J. (2007). Effects of anthropogenic seawater acidification on acid-base balance in the sea urchin *Psammechinus miliaris*. *Marine Pollution Bulletin*, **54**, 89–96.

Miller, A.W., Reynolds, A.C., Sobrino, C., and Riedel, G.F. (2009). Shellfish face uncertain future in high CO_2 world: influence of acidification on oyster larvae calcification and growth in estuaries. *PLoS One*, **4**, e5661.

Ohman, M.D., Lavaniegos, B.E., and Townsend, A.W. (2009). Multi-decadal variation in calcareous holozooplankton in the California Current system: thecosome pteropods, heteropods and foraminifera. *Geophysical Research Letters*, **36**, L18608, doi:10.1029/2009GL033901.

Orr, J.C., Fabry, V.J., Aumont, O. *et al.* (2005). Anthropogenic ocean acidification over the twenty-first century and its impact on calcifying organisms. *Nature*, **437**, 681–6.

Pane, E.F. and Barry, J.P. (2007). Inefficient acid-base regulation in the deep-sea decapod crab (*Chionoecetes tanneri*) during short-term hypercapnia. *Marine Ecology Progress Series*, **334**, 1–9.

Pauly, D., Alder, J., Bennett, E., Christensen, V., Tyedmers, P., and Watson, R. (2003). The future of fisheries. *Science*, **302**, 1359–61.

Pelejero, C., Calvo, E., and Hoegh-Guldberg, O. (2010). Paleo-perspectives on ocean acidification. *Trends in Ecology and Evolution*, **25**, 332–44.

Petchey, O.L. and Gaston, K.J. (2006). Functional diversity: back to basics and looking forward. *Ecology Letters,* **9**, 741–58.

Pörtner, H-O., Bock, C., and Reipschläger, A. (2000). Modulation of the cost of pHi regulation during metabolic depression: a ^{31}P-NMR study in invertebrates (*Sipunculus nudus*) isolated muscle. *Journal of Experimental Biology*, **203**, 2417–28.

Richardson, A.J., Bakun, A., Hays, G.C., and Gibbons, M.J. (2009). The jellyfish joyride: causes, consequences and management responses to a more gelatinous future. *Trends in Ecology and Evolution*, **24**, 213–322.

Ridgwell, A., Schmidt, D.N., Turley, C *et al*. (2009). From laboratory manipulations to earth system models: predicting pelagic calcification and its consequences. *Biogeosciences*, **6**, 2611–23.

Riebesell, U., Schulz, K.G., Bellerby, R.G.J. *et al*. (2007). Enhanced biological carbon consumption in a high CO_2 ocean. *Nature*, **450**, 545–8.

Ries, J.B., Cohen, A.L., and McCorkle, D.C. (2009). Marine calcifiers exhibit mixed responses to CO_2-induced ocean acidification. *Geology*, **37**, 1131–4.

Roark, E.B., Guilderson, T.P., Dunbar, R.B., Fallon, S.J., and Mucciarone, D.A. (2009). Extreme longevity in proteinaceous deep-sea corals. *Proceedings of the National Academy of Sciences USA*, **106**, 5204–8.

Roberts, D., Howard, W.R., Moy, A.D. *et al*. (2008). Interannual variability of pteropod shell weights in the high-CO_2 Southern Ocean. *Biogeosciences Discussions*, **5**, 4453–80.

Roberts, J.M., Wheeler, A.J., and Freiwald, A. (2006). Reefs of the deep: the biology and geology of cold-water coral ecosystems. *Science*, **312**, 543–7.

Rosa, R. and Seibel, B.A. (2008). Synergistic effects of climate-related variables suggest future physiological impairment in a top oceanic predator. *Proceedings of the National Academy of Sciences USA*, **105**, 20776–80.

Sala, O.E., Chapin, F.S. III, Armesto, J.J. *et al*. (2000). Global biodiversity scenarios for the year 2100. *Science*, **287**, 1770–4.

Seibel, B.A. and Walsh, P.J. (2003). Biological impacts of deep-sea carbon dioxide injection inferred from indices of physiological performance. *Journal of Experimental Biology*, **206**, 641–50.

Steneck, R.S., Graham, M.H., Bourque, B.J. *et al*. (2002). Kelp forest ecosystems: biodiversity, stability, resilience and future. *Environmental Conservation*, **29**, 436–59.

Sydeman, W.J., Hester, M.M., Thayer, J.A., Gress, F., Martin, P., and Buffa, J. (2001). Climate change, reproductive performance and diet composition of marine birds in the southern California Current system, 1969–1997. *Progress in Oceanography*, **49**, 309–29.

Tilman, D., Knops, J., Wedin, D., Reich, P., Ritchie, M., and Siemann, E. (1997). The influence of functional diversity and composition on ecosystem processes. *Science*, **277**, 1300–2.

Tittensor, D.P., Baco, A.R., Hall-Spencer, J.M., Orr, J.C., and Rogers, A.D. (2010). Seamounts as refugia from ocean acidification for cold-waters stony corals. *Marine Ecology*, **31**(Suppl. 1), 212–25.

Tunnicliffe, V., Davies, K.T.A., Butterfield, D.A., Embley, R.W., Rose, J.M., and Chadwick, W.W. Jr (2009). Survival of mussels in extremely acidic waters on a submarine volcano. *Nature Geoscience*, **2**, 344–8.

Veron, J.E.N. (2008). Mass extinctions and ocean acidification: biological constraints on geological dilemmas. *Coral Reefs*, **27**, 459–72.

Widdicombe, C.E., Eloire, D., Harbour, D., Harris, R.P., and Somerfield, P.J. (2010). Long-term phytoplankton community dynamics in the western English Channel. *Journal of Plankton Research*, **32**, 643–55.

Widdicombe, S. and Spicer, J.I. (2008). Predicting the impact of ocean acidification on benthic biodiversity: what can animal physiology tell us? *Journal of Experimental Marine Biology and Ecology*, **366**, 187–97.

Wood, H.L., Spicer, J.I., and Widdicombe, S. (2008). Ocean acidification may increase calcification rates, but at a cost. *Proceedings of the Royal Society B: Biological Sciences*, **275**, 1767–73.

Worm, B., Barbier, E.B., Beaumont, N. *et al*. (2006). Impacts of biodiversity loss on ocean ecosystem services. *Science*, **314**, 787–90.

CHAPTER 11

Effects of ocean acidification on the marine source of atmospherically active trace gases

Frances Hopkins, Philip Nightingale, and Peter Liss

11.1 Introduction

A wide range of trace gases, including dimethyl sulphide (DMS) and organohalogens, are formed in the surface oceans via biological and/or photochemical processes. Consequently, these gases become supersaturated in seawater relative to the overlying marine air, leading to a net flux to the atmosphere. Upon entering the atmosphere, they are subject to rapid oxidation or radical attack to produce highly reactive radical species which are involved in a number of important atmospheric and climatic processes. Organohalogens can affect the oxidizing capacity of the atmosphere by interacting with ozone, with implications for air quality, stratospheric ozone levels, and global radiative forcing. DMS and iodine-containing organohalogens (iodocarbons) can both contribute to direct and indirect impacts of aerosols on climate through the production of new particles and cloud condensation nuclei (CCN) in the clean marine atmosphere. Therefore, marine trace gases are considered a vital component of the earth's climate system, and changes in the net production rate and subsequent sea-to-air flux could have an impact on globally important processes. In recent years, attention has turned to the impact that future ocean acidification may have on the production of such gases, with the greatest focus on DMS and organohalogens. In this chapter, the current state-of-the-art in this growing area of research is outlined.

11.1.1 DMS

The oceans are a major source of sulphur (S), an element essential to all life, and marine emissions of the gas DMS (chemical formula $(CH_3)_2S$) represent a key pathway in the global biogeochemical sulphur cycle. The surface oceans are supersaturated with DMS relative to the atmosphere, resulting in a one-way flux from sea to air (Lovelock *et al.* 1972; Watson and Liss 1998).

DMS is a breakdown product of the biogenically produced dimethyl sulphoniopropionate (DMSP):

$$(CH_3)_2S^+CH_2CH_2COO^- \rightarrow \\ (CH_3)_2S + CH_2CHCOOH \qquad (11.1) \\ \text{(acrylic acid)}$$

Single-celled marine phytoplankton are the chief producers of DMSP, and this reaction is catalysed intra- and extracellularly by the enzyme DMSP-lyase (Malin *et al.* 1992; Liss *et al.* 1997). The capacity of phytoplankton to produce DMSP varies between species, with prymnesiophytes considered to be the most prolific (Malin *et al.* 1992; Liss *et al.* 1997; Watson and Liss 1998). The production of DMSP is thought to fulfil a number of roles within the algal cell, including osmoregulation (Vairavamurthy *et al.* 1985), cryoprotection (Malin *et al.* 1992; Liss *et al.* 1997), and protection against grazing (Wolfe *et al.* 1997) and oxidative stress (Sunda *et al.* 2002). Production is often associated with senescence, grazing, and viral lysis, when cells begin to break down enabling DMSP to

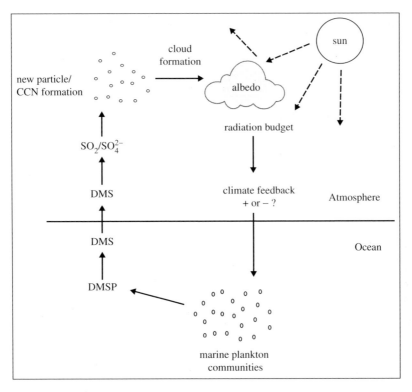

Figure 11.1   The CLAW hypothesis: DMS–CCN–climate interactions. The mechanism suggested by Charlson *et al.* (1987) by which marine DMS emissions from marine plankton communities influence climate via impacts on cloud properties.

come into contact with DMSP-lyase (Malin *et al.* 1992, 1998). Due to the biogenic nature of DMS production, the highest sea-surface DMS concentrations are associated with areas of high biological activity, particularly upwelling and frontal regions and coastal and nearshore regions. Large seasonal variations in the occurrence of DMS are seen in temperate and boreal regions where productivity is at a maximum in the spring months and very low in the winter (Kettle and Andreae 2000).

DMS has received a great deal of attention in recent decades due to its influence on atmospheric chemistry and climate. Upon entering the atmosphere, it is rapidly oxidized by free radicals including OH, BrO, Cl, and NO_3. This results in the production of methane sulphonic acid (MSA) and/or non-sea salt sulphate (nss-SO_4^{2-}) via several intermediate reactions (von Glasow and Crutzen 2004). The formation of H_2SO_4 makes a significant contribution to the atmosphere's natural acidity levels. Furthermore, this marine-derived aerosol can have direct and indirect aerosol effects, both scattering and absorbing incoming solar radiation, and contributing to the production of CCN, thereby modifying the planetary radiation budget (Charlson *et al.* 1987). This connection led to the CLAW hypothesis (named after the four authors of the study; Charlson *et al.* 1987), according to which the production of DMS by phytoplankton may contribute to the regulation of earth's climate (Fig. 11.1). This generated a large amount of research, which is ongoing to this day, and remains both highly topical and controversial.

11.1.2   Organohalogens

The oceans constitute the greatest source of volatile halogenated organic compounds (organohalogens) to the atmosphere, a result of the vast oceanic reservoir of halogens (chlorine, Cl; iodine, I; bromine, Br; fluorine, F; Manley 2002). Organohalogens are formed in seawater through biological and photochemical processes, resulting in supersaturations in

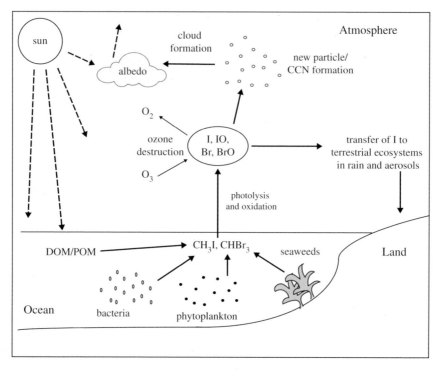

Figure 11.2   Pathways of organohalogen formation in the surface ocean, and the atmospheric and climatic consequences of the sea-to-air flux. DOM, dissolved organic matter; POM, particulate organic matter.

the surface oceans that lead to a net flux of these gases to the atmosphere (see Fig. 11.2).

The production of organohalogens has been confirmed in a wide range of marine organisms, including phytoplankton (e.g. Tokarczyk and Moore 1994; Moore *et al.* 1996; Manley and Cuesta 1997), macroalgae (Manley and Dastoor 1987; Nightingale *et al.* 1995), and bacteria (Amachi *et al.* 2001). As a result, high-productivity regions of the oceans, such as upwellings and coastal areas, are considered to be strong source regions of volatile organohalogens. Biological production by macroalgae and phytoplankton occurs via two mechanisms: (1) methylation of organic halogens (common haloform reaction) to produce monohalogenated compounds (CH_3X, X = Br, Cl, I; Urhahn and Ballschmiter 1998) or (2) halogenation of organic precursors, catalysed by haloperoxidases (BrPO and IPO) produced in algal cells (Nightingale *et al.* 1995; Manley 2002). Production by macroalgae appears to be a response to, or by-product of, photo-oxidative or mechanical stress (Nightingale *et al.* 1995; Manley 2002). In phy-

toplankton, production of organohalogens may serve as a method of cellular halide excretion, protect against photo-oxidative stress, chemically deter grazers, or simply represent a by-product of normal metabolism (Manley 2002). Recent work has also implicated marine bacteria as potential globally important producers of organohalogens, with a wide range of bacteria displaying methylating capabilities (Amachi *et al.* 2001).

Photochemical production of organohalogens involves a reaction between photochemically produced methyl radicals and halogen atoms (Moore and Zafiriou 1994; Richter and Wallace 2004). As the methyl radicals are likely to be derived from a biological source, an indirect biogenic control on the process exists (Richter and Wallace 2004). Dihalogenated compounds can also be formed through the photochemical degradation of other organohalogens. For example, CH_2I_2 is photolysed to produce CH_2ClI, with a yield of 25 to 30% (Martino *et al.* 2005). A further, recently described, source of organohalogens in seawater involves ozone-mediated reactions between dissolved organic matter (DOM) and hypoiodous acid/molecular iodine (Martino *et al.* 2009).

The production of organohalogens in surface seawater is balanced by a number of loss processes, including hydrolysis and nucleophilic attack (Moelwyn-Hughes 1938; Elliott and Rowland 1993), photochemical loss (Jones and Carpenter 2005; Martino *et al.* 2005), and uptake by bacteria (Goodwin *et al.* 1997, 1998; King and Saltzman 1997). The remainder of the volatile organohalogen pool undergoes exchange to the atmosphere. Here it is subject to photolysis and oxidation to produce highly reactive halogen radicals and aerosols, which are involved in a number of atmospheric and climatic processes (Fig. 11.2). Similarly to DMS-derived aerosols, iodine oxide (IO) radicals are aerosols which can indirectly influence climate through involvement with CCN and cloud formation, and thereby impact on the reflection of solar radiation (Andreae and Crutzen 1997). Furthermore, organohalogen-derived free radicals (I, IO, Br, and BrO) act as effective catalytic ozone-depleting species (Chameides and Davis 1980; Solomon *et al.* 1994; Davis *et al.* 1996), with a potentially significant influence on the atmosphere's oxidative capacity, global radiative forcing, and air quality. Finally, the sea-to-air flux of iodine represents a crucial step in its biogeochemical cycling—an element essential to the health of terrestrial organisms, including humans (Fuge and Johnson 1986).

11.2 Effects of ocean acidification on DMS production and its impact on climate

11.2.1 Summary of experimental evidence

To date, DMS and DMSP concentrations have been measured during three CO_2 perturbation experiments, all performed at the large-scale mesocosm facilities at the Marine Biological Field Station, Raunefjorden, Norway (60.3°N, 5.2°E): 2003 (Pelagic Ecosystem Enrichment Experiment; PeECE II; Avgoustidi 2007), 2005 (PeECE III; Wingenter *et al.* 2007; Vogt *et al.* 2008), and 2006 (UK NERC Microbial Metagenomics experiment; Hopkins *et al.* 2010). The data are summarized in Fig. 11.3 and Table 11.1.

The concept and design of each of the experiments were very similar. Polyethylene enclosures were suspended in the fjord from a moored floating raft (volume in 2003 and 2005: 20–25m³; 11m³ in 2006). Each enclosure was filled with nutrient-poor unfiltered fjord water, and the tops of the mesocosms were covered with tetrafluoroethylene films in order to form a tent covering more than 90% of the surface area of the mesocosm. Seawater pCO_2 and pH were initially set to target values by aerating the water with CO_2/air mixtures. Nutrients were added in order to stimulate phytoplankton blooms, and the bloom was monitored over the course of 20–23 days. In 2003, three treatments were used: 180 µatm (glacial), 370 µatm (present), and 700 µatm, and the 2005 experiment again compared three pCO_2 levels: 375 µatm (present), 750 µatm (future), and 1150 µatm (far future). In 2006, two CO_2 treatments were used: 380 µatm (present) and 750 µatm (future). In order to encourage diatom blooms, the 2003 experiment received an addition of silicate, as well as nitrate and phosphate, on day 0 and again on day 7, whereas the 2005 and 2006 experiments received only nitrate and phosphate on day 0. The concentration of chlorophyll *a* varied between experiments; relatively low concentrations were experienced in 2003 with a maximum of 4.6 mg m^{-3}, whilst a maximum of 15.0 mg m^{-3} was recorded in 2005. There were no significant differences in chlorophyll *a* between these treatments in either experiment. By contrast, chlorophyll *a* concentrations during the 2006 study showed a significant 40% decrease under high CO_2 during the bloom phase, with maximum concentrations of 6 mg m^{-3} and 10.3 mg m^{-3} under high CO_2 and present-day CO_2, respectively.

The response of the plankton communities to the CO_2 perturbation varied between experiments. As coccolithophores are considered to be prolific producers of DMSP it is pertinent to firstly consider their response to high-CO_2 conditions. The abundance of the coccolithophore *Emiliania huxleyi* showed some variation between experiments. In the 2003 experiment a strong coccolithophore bloom occurred, with maximum abundances of 56×10^6 cells ml^{-1}. There was a small reduction in *E. huxleyi* numbers under high CO_2, but the differences between treatments were not significant (Engel *et al.* 2008). *Emiliania huxleyi* was notably less prolific in the 2005 experiment, with maximum abundances of

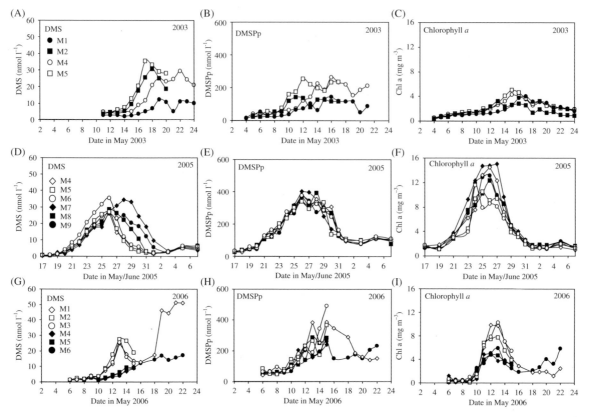

Figure 11.3 Concentrations of DMS, DMSPp (particulate DMSP) and chlorophyll *a* during nutrient-induced blooms of phytoplankton in mesocosm enclosures in Raunefjord (Bergen, Norway) in May 2003, May/June 2005, and May 2006. Data from Avgoustidi (2007) (A, B, C), Vogt *et al.* (2008) (D, E, F), and Hopkins *et al.* (2010) (G, H, I). Nominal CO_2 treatments at the beginning of the experiments were as follows (M = mesocosm): in 2003, ~750 µatm in M1 and M2, ~380 µatm in M4 and M5; in 2005 ~350 µatm in M4, M5, and M6, ~700 µatm in M7, M8, and M9; in 2006 ~750 µatm in M1, M2, and M3, ~300 µatm in M4, M5, and M6.

only 5.5×10^6 cells ml^{-1} and with no significant differences between CO_2 treatments (Riebesell *et al.* 2008; Vogt *et al.* 2008). In 2006, *E. huxleyi* was even less abundant, with maximum numbers of only 3×10^3 cells ml^{-1}, but large and significant reductions in abundance under high CO_2 were observed (Hopkins *et al.* 2010).

Changes in other components of the phytoplankton community also need to be considered. During the 2003 experiment, a general community shift was observed in response to high CO_2, with a significant change in the structure of the autotrophic community. The fraction of the community comprising pico- and nanoplankton appeared to be particularly sensitive to higher CO_2 concentrations, with potential impacts on the microbial food web and quality

of the DOM (Engel *et al.* 2008). By contrast, no significant differences were observed during the 2005 experiment in either phytoplankton community composition or bacterial abundance and diversity (Allgaier *et al.* 2008; Paulino *et al.* 2008). In 2006, most components of the nano- and picoplankton (coccolithophores, large picoeukaryotes, cryptophytes, and *Synechococcus*) were significantly lower under high CO_2. The overall plankton biomass was also greatly diminished, with a 28% decrease in total biomass under high CO_2, which comprised an 80% reduction in diatom biomass, a 56% reduction in autotrophic dinoflagellate biomass, and a 35% decrease in ciliates (Hopkins *et al.* 2010).

These differences in biological characteristics and community structure, as well as the resilience or

Table 11.1 Summary of the ocean acidification mesocosm experiments performed at the Bergen Large-scale Facility, Raunefjord, Norway in 2003, 2005, and 2006, including date and duration of experiment, initial nutrient concentrations, total mean and maximum (Max.) mean chlorophyll (Chl) a, DMS, and particulate DMSP (DMSPp) concentrations under high (~700–750 μatm) and present-day pCO_2 (~380 μatm). Total mean = the mean of all data from all mesocosms of each CO_2 treatment. Max. (mean) = the mean of the maximum values measured in all mesocosms of each CO_2 treatment.

Study	Date	Duration (days)	Time of nutrient addition: initial conc. (μmol l⁻¹)		Chl a (mg m⁻³) High pCO_2	Chl a (mg m⁻³) Present pCO_2	DMS (nmol l⁻¹) High pCO_2	DMS (nmol l⁻¹) Present pCO_2	DMSPp (nmol l⁻¹) High pCO_2	DMSPp (nmol l⁻¹) Present pCO_2
Avgoustidi (2007)	May 2003	20	Days 0 and 7: 0.5 (PO_4), 9 (NO_3), 12 ($Si(OH)_4$)	Total mean	1.7	2.3	9.7	16.4	78.6	136.0
				Max. (mean)	4.0	4.6	21.5	32.3	143.8	258.7
Vogt et al. (2008)	May/June 2005	22	Day 0: 0.7 (PO_4), 15 (NO_3)	Total mean	13.5	4.3	11.7	9.3	175.0	172.7
				Max. (mean)	5.1	11.0	27.4	29.5	366.0	370.0
Hopkins et al. (2010)	May 2006	23	Day 0: 0.8 (PO_4), 15 (NO_3)	Total mean	2.5	3.2	5.7	14.1	139.0	175.6
				Max. (mean)	6.0	10.3	11.8	34.7	262.2	414.3

sensitivity of various components of the community to the CO_2 perturbation, may have produced some of the variations in response of DMS and DMSP concentrations that were observed between the three experiments (Fig. 11.3). The 2003 and 2006 experiments gave similar results, with significantly lower DMS and DMSP under high CO_2. Avgoustidi (2007) reported a 59% decrease in DMS concentrations in the high-CO_2 mesocosm M1 relative to the present-day CO_2 mesocosm M4 (see Fig. 11.3). Similarly, Hopkins et al. (2010) observed a highly significant 46% decrease in mean DMS concentrations under high CO_2, a result that was strongly replicated in all three high-CO_2 mesocosms (see Fig. 11.3). By contrast, during the 2005 experiment small increases in production of DMS under high CO_2 were reported, although the differences were not significant (Vogt et al. 2008). However, the temporal development of DMS did display small but statistically significant differences between CO_2 treatments. Vogt et al. (2008) concluded that as there were no significant differences in species composition or succession between treatments, the small observed differences in DMS concentrations were most likely due to differences in bacterial or viral activity. The observed significant differences in plankton communities seen during the 2003 and 2006 experiments are likely to explain some of the observed differences in DMS/P.

During the 2006 study, the DMSP concentration was less significantly affected by high CO_2 than the DMS concentration, suggesting that the impact was greatest on the conversion of DMSP to DMS rather than on the initial production of DMSP. This implicates an effect on secondary factors, such as grazing, viral lysis, and bacterial metabolism of DMSP. In addition, the significant decreases in DMS can also be directly attributable to significant changes in ecosystem composition. For all of the studies, a direct impact on DMS concentrations may be the result of bacterial consumption of DMS (Kiene and Bates 1990), a process that may have been either enhanced (2003 and 2006) or diminished (2005) under high CO_2.

One study has investigated the combined impacts of rising ocean acidity and increasing temperatures on DMSP production (Lee et al. 2009). During shipboard experiments, North Atlantic plankton com-

munities incubated in 'greenhouse' conditions (690 µatm CO_2, +4°C) exhibited a significant increase in the production of particulate DMSP compared with control conditions (390 ppmv CO_2, ambient temperature). However, the dissolved DMSP fraction decreased under high CO_2, a result of decreased grazing pressure.

Despite some conflicting results, assimilation of the currently available data suggests that a negative impact of ocean acidification on net DMS and/or DMSP production is likely. Avgoustidi (2007) performed in vitro experiments on natural seawater assemblages and monospecific cultures of E. huxleyi, and again observed a reduction in DMS concentrations under elevated CO_2. This supports the findings of the 2003 and 2006 mesocosm experiments (Avgoustidi 2007; Hopkins et al. 2010), and suggests that a decrease in net DMS and DMSP production is possible in a future high-CO_2 world. To further elucidate the impacts of ocean acidification on DMS production additional investigation is required, perhaps in the form of further mesocosm experiments or incubations of natural assemblages from various oceanic regions.

11.2.2 Atmospheric and climatic implications

Although clearly not applicable to the entire global ocean, mesocosm studies are arguably representative of highly productive regions such as high-latitude waters, coastal waters, and upwellings, with such areas expected to be rapidly, and in some cases dramatically, affected by ocean acidification (Orr et al. 2005; IPCC 2007; Feely et al. 2008; Steinacher et al. 2009; see also Chapter 3). Therefore, it is pertinent to assess the impact that changes in seawater concentrations of DMS may have on the climate-regulating properties of DMS and its atmospheric oxidation products. It is important to note that, as yet, no modelling studies have considered the impact of ocean acidification on marine DMS emissions (see Chapter 12).

By influencing both the reflection and absorption of solar radiation through a variety of complex radiative and microphysical processes, atmospheric aerosols are able to directly exert a strong influence on the earth's radiative budget (Andreae and Crutzen 1997; Ramanathan et al. 2001). Aerosols also indirectly influence climate through involvement in CCN and cloud formation, further impacting on the reflection of solar radiation (Andreae and Crutzen 1997). Aerosols originate from both natural and anthropogenic sources. Anthropogenic sources include sulphate and carbonaceous materials produced during fossil fuel combustion and biomass burning. Such pollutants generally show localized distribution, more concentrated in the industrialized Northern Hemisphere (Ramanathan *et al.* 2001). Natural aerosols originate from both terrestrial and marine environments, most notably non-methane hydrocarbons such as terpenes that are emitted from forests and DMS from the oceans. It has been estimated that marine DMS contributes from 20 to 80% of the sulphates in air over the Northern Hemisphere, and more than 80% over most of the Southern Hemisphere (Chin and Jacob 1996).

When DMS is emitted to the atmosphere, it undergoes rapid oxidation in the marine boundary layer (MBL) via two major pathways (addition and/or abstraction), with their respective roles being dependent on environmental conditions such as atmospheric temperature, solar intensity, and cloud cover (von Glasow and Crutzen 2004; Vogt and Liss 2008). Through the addition pathway, DMS is initially oxidized to dimethyl sulphoxide (DMSO), followed by methane sulphinic acid (MSIA) and MSA. The growth of smaller particles is encouraged, as all intermediate and end products are taken up onto pre-existing particles. The abstraction pathway involves the transformation of DMS to SO_2, MSA, and H_2SO_4. Again, SO_2 and MSA are taken up by pre-existing particles, whilst formation of H_2SO_4 may lead to the production of new particles (von Glasow and Crutzen 2004; Vogt and Liss 2008). The extent of climate regulation by the oxidation products of DMS is dependent on a number of processes, but ultimately relies on there being an overall increase in the number concentration of CCN (particles ~0.05 μm in diameter) (Charlson *et al.* 1987; Andreae and Crutzen 1997; von Glasow and Crutzen 2004).

The CLAW hypothesis states that changes to oceanic DMS emissions would cause corresponding changes to atmospheric $[SO_4^{2-}]$ and hence to the number of particles that grow to the size of CCN (Charlson *et al.* 1987). Therefore a decrease in DMS production in the oceans as a result of ocean acidifi-

cation, as observed in the mesocosm studies discussed earlier, may ultimately lead to a reduction in CCN and marine stratus cloud albedo, and therefore produce a positive feedback on climate that would increase the warming that will occur as a result of anthropogenic greenhouse gases (GHGs).

The flux (*F*) of DMS to atmosphere can be described as follows:

$$F = A \times k \times \Delta c \tag{11.2}$$

where *A* is the total ocean surface area, *k* is the transfer velocity, and Δc is the concentration difference across the air–sea interface. Due to the highly supersaturated nature of the ocean relative to the atmosphere, Δc can be considered to be identical to the concentration of DMS in the surface oceans. Therefore, assuming no changes to other parameters, a significant decrease in seawater DMS concentrations, e.g. ~50% (Hopkins *et al.* 2010), as a result of ocean acidification, would be equivalent to a 50% decrease in Δc. This would lead to a proportional decrease of the flux of DMS to the atmosphere. Despite these assumptions, it is likely that the climate response would be substantial if such a change were seen over extensive ocean areas. For example, using a coupled ocean–atmosphere general circulation model, Gunson *et al.* (2006) predict a 1.6°C increase in surface air temperature in response to a halving of ocean DMS emissions.

11.2.3 Large uncertainties

However, despite such results, and for a variety of reasons discussed below, it is highly problematic to make quantitative predictions about the climatic implications of changes in marine DMS emissions.

11.2.3.1   Effects vary regionally

The effect of the number of CCN on albedo is more prominent at low particle numbers, resulting in a greater climatic effect in oceanic areas away from the influence of terrestrial air heavily laden with aerosols (Twomey 1991). As a demonstration of this, Twomey (1991) calculates that if equal quantities of sulphur were to enter the atmosphere in the two hemispheres, the impact on albedo would be 25 times more pronounced in the Southern than in the

Northern Hemisphere. A number of modelling studies on the effects of climate change (but not ocean acidification) on DMS production and aerosol formation have revealed large spatial heterogeneities in both DMS concentrations and fluxes, and the associated climatic impacts (Gabric *et al.* 1998, 2005; Bopp *et al.* 2003; Gunson *et al.* 2006). For example, Bopp *et al.* (2003) reported a mean decrease in global seawater DMS concentrations of ~1% in response to doubled atmospheric CO_2 concentrations, but the regional impacts were much more pronounced. The western Equatorial Pacific and the eastern Equatorial Atlantic saw a decrease of up to 50%, whilst the eastern Equatorial Pacific saw an increase of up to 50%. In addition, the subtropical/subantarctic convergence zone experienced a 20% enrichment, with smaller increases seen in the north Atlantic and Pacific. The global DMS flux showed similar regional variation, and resulted in heterogeneity in the climatic effects. Whilst the impact on radiative forcing was -1 W m^{-2} in the Southern Ocean, it was calculated to be $+0.5$ W m^{-2} in the tropics.

11.2.3.2 *Number of CCN derived from marine DMS*
Atmospheric aerosols come from a range of sources. Thus, it is difficult to distinguish the impact of DMS-derived aerosols on climate from that from other aerosols. It is also unknown how much the marine stratus cloud currently in the atmosphere is affected by them. Watson and Liss (1998) made attempts to calculate the current influence that DMS has on global albedo. By assuming that around half of the CCN in typical Southern Hemisphere marine air are due to DMS, they calculated that a doubling of the concentrations of CCN would lead to a 6 to 46% increase in marine stratus cloud albedo. If one-third of the globe is assumed to be covered by marine stratus clouds, this would result in a ~2% increase in global albedo, and a 3.8°C cooling of climate. However, the large projected range highlights the difficulties involved in quantifying the influence of marine DMS on aerosols and climate.

11.2.3.3 *Phytoplankton species differ in their ability to produce DMS/DMSP*
Haptophytes, in particular coccolithophorids, are considered to be the most prolific planktonic producers of DMSP/DMS, followed by *Phaeocystis*,

with lesser production by dinoflagellates and diatoms (Malin *et al.* 1992; Liss *et al.* 1997). Ecosystem shifts in response to global changes are likely to affect DMS production through changing species dominance. The differing ability of phytoplankton species to produce these compounds further complicates our capacity to make quantitative predictions about the impact of ocean acidification on DMS production and climate.

11.2.3.4 *Atmospheric effects of oxidized sulphur*
In terms of climate impacts, it is not the bulk quantity of DMS oxidation that is important, but the ability of the oxidation products to enhance the CCN number concentration (Charlson *et al.* 1987; Andreae and Crutzen 1997). However, the formation of CCN from the oxidation products of DMS (SO_2 and H_2SO_4) and the resulting climate sensitivity is complex and non-linear, and cannot be constrained using a simple, globally applicable model (Carslaw *et al.* 2009).

11.2.3.5 *Impact of other climate change effects on DMS–CCN–climate feedback*
The flux of DMS to the atmosphere is controlled by the concentration of DMS in surface seawater and the magnitude of the transfer velocity, both of which are influenced by climate variables. Seawater DMS concentrations are controlled by marine biological activity, which depends on solar irradiance, sea temperature, and ocean dynamics, whilst the transfer velocity is controlled by air and sea temperature and wind velocity, and related properties (e.g. waves and bubbles). Therefore, ocean acidification will not be the only process that has an impact on DMS production in the future oceans, and a range of other climatic changes will also have an effect. Increased solar irradiance and sea-surface temperature will ultimately result in lowered marine productivity in some regions, due to enhanced stratification and reduced upwelling of nutrient-rich water. Using an atmosphere–ocean general circulation model coupled to a marine biogeochemical model, Bopp *et al.* (2003) observed such an effect in the western Equatorial Pacific in response to doubled atmospheric CO_2, with a resultant lowering of particulate DMSP and DMS concentrations. By contrast, the modelling study by Gunson *et al.* (2006)

reported a promotion of phytoplankton growth and increased DMS production in response to increased temperatures and irradiance. Clearly, the response of the DMS system to changing global temperature is likely to be variable and difficult to predict or generalize, without taking into account the effects of ocean acidification which have not been included in any model study to date. Changes to wind velocity are also likely to have an impact. Again, in the modelling study performed by Bopp *et al.* (2003) a 19% increase in the DMS flux was observed from 30°S to 50°S, a quarter of which could be attributed to an increase in wind speed. In brief, three other climatic changes may influence DMS production. First, a reduction in Arctic sea ice cover due to increasing global temperatures. This would expose more seawater to sunlight, and result in an increase in phytoplankton productivity and DMS production (Gabric *et al.* 2005). Second, changes to atmospheric dust input to the oceans due to changing wind and land-use patterns. The iron fertilization effect would enhance productivity, and increase DMS production (Jickells *et al.* 2005). Third, changes to atmospheric convection patterns as a result of climatic shifts may affect the transport of DMS to the troposphere and subsequent CCN production (Shaw *et al.* 1998).

Thus, it is clear that the response of the DMS–CCN–climate system to global environmental changes is challenging to quantify, and despite more than 20 years of research since the CLAW hypothesis was first described, the sign of the feedback mechanisms that may be involved is still uncertain. At this stage, the extent of the impact and any accompanying climate feedbacks associated with changes in marine emissions of DMS as a result of ocean acidification are uncertain.

11.3 Impacts of ocean acidification on organohalogen production and atmospheric chemistry

11.3.1 Summary of experimental evidence

The response of iodocarbons to ocean acidification has been assessed in two mesocosm experiments (Wingenter *et al.* 2007; Hopkins *et al.* 2010). During the PeECE III experiment in 2005, Wingenter *et al.*

(2007) reported that time-integrated concentrations of chloroiodomethane (CH_2ClI) increased under high pCO_2, with a 46% increase under two-times present-day pCO_2 (~700 µatm) and a 131% increase under three-times present-day pCO_2 (~1050 µatm). Concentrations ranged from around 5 to 40 fmol l^{-1}, and maximum concentrations occurred 6 to 10 days after the peak of chlorophyll *a*. The authors suggest that differences in viral attack and phytoplankton lysis may have produced the observed difference in CH_2ClI concentrations between present-day and higher-pCO_2 treatments.

Following this, Hopkins *et al.* (2010) determined the concentrations of methyl iodide (CH_3I), ethyl iodide (C_2H_5I), diiodomethane (CH_2I_2), and chloroiodomethane (CH_2ClI) during an experiment at the Bergen mesocosm facility in 2006 (Fig. 11.4). These gases showed similar temporal trends, and maximum concentrations were observed over a 4-day period immediately after the peak of the bloom. The temporal development of these gases suggests a close association with biological activity. However, they do not appear to be directly related to phytoplankton growth, as maximum gas concentrations generally occurred after the maxima in chlorophyll *a*. Most importantly, the data strongly suggest that higher pCO_2 (lower pH) leads to a reduction in iodocarbon concentrations during the phytoplankton blooms. Over the course of the experiment, large, and in some cases significant, reductions in concentration of all iodocarbons were observed (CH_3I, –44%; C_2H_5I, –35%; CH_2I_2, –27%; CH_2ClI, –24%), and these differences were even more pronounced during the post-bloom phase of the experiment (CH_3I, –67%; C_2H_5I, –73%; CH_2I_2, –93%; CH_2ClI, –59%).

In this experiment, CH_3I and C_2H_5I concentrations within the mesocosm were representative of realistic open-ocean and coastal seawater (Abrahamsson *et al.* 2004; Chuck *et al.* 2005; Archer *et al.* 2007). C_2H_5I is considered to be a minor iodocarbon in seawater (Archer *et al.* 2007), and similarly it made up less than 1% of the total iodocarbon pool in the mesocosm. CH_2I_2 and CH_2ClI concentrations were somewhat elevated compared with most (but not all) oceanic measurements, and they dominated the iodocarbon pool, in common with a number of other field studies (Klick and Abrahamsson 1992; Abrahamsson *et al.* 2004; Archer *et al.* 2007).

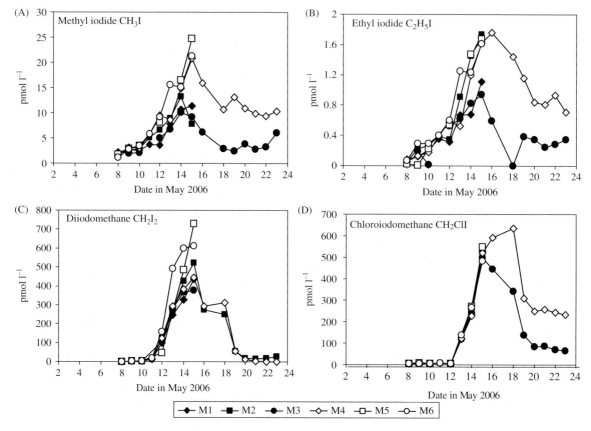

Figure 11.4.   Concentrations of methyl iodide (CH_3I) (A), ethyl iodide (C_2H_5I) (B), diiodomethane (CH_2I_2) (C) and chloroiodomethane (CH_2ClI) (D) during nutrient-induced blooms of phytoplankton in mesocosm enclosures in Raunefjord, Bergen, Norway, in May 2006. The nominal CO_2 treatments at the beginning of the experiment were as follows (M = mesocosm): ~750 µatm in M1, M2 and M3; ~300 µatm in M4, M5 and M6. Data from Hopkins *et al.* (2010).

CH_3I is often referred to as 'biogenic' as it is thought to be produced both directly by macroalgae, phytoplankton, and bacteria, and indirectly through photochemical reactions with organic matter (Moore and Zafiriou 1994; Archer *et al.* 2007). Less information is available on C_2H_5I; however, a number of studies, including Hopkins *et al.* (2010), have found significant correlations between C_2H_5I and CH_3I, suggesting similar production and removal mechanisms (see also Makela *et al.* 2002; Richter and Wallace 2004; Archer *et al.* 2007).

CH_2I_2 is considered to have a primarily biogenic source (Moore and Zafiriou 1994). It is subject to rapid photolysis in surface seawater (photolytic lifetime of around 12 min), with strong evidence that this reaction is an important source of CH_2ClI

(Martino *et al.* 2005). In the productive mesocosm environment described by Hopkins *et al.* (2010), the processes controlling net iodocarbon production were susceptible to lowered pH, and were likely to be a consequence of plankton community shifts in response to elevated $p CO_2$ conditions. The authors reported that small picoeukaryotes, cryptophytes, and *Synechococcus* were all significantly lower under high CO_2, and similarly to the iodocarbons, the differences between treatments were most pronounced during the post-bloom phase. If such organisms are involved in iodocarbon production or consumption, changes in their abundance may have had a direct impact on seawater concentrations during this experiment. Further evidence of a reduction in iodocarbon production in response to

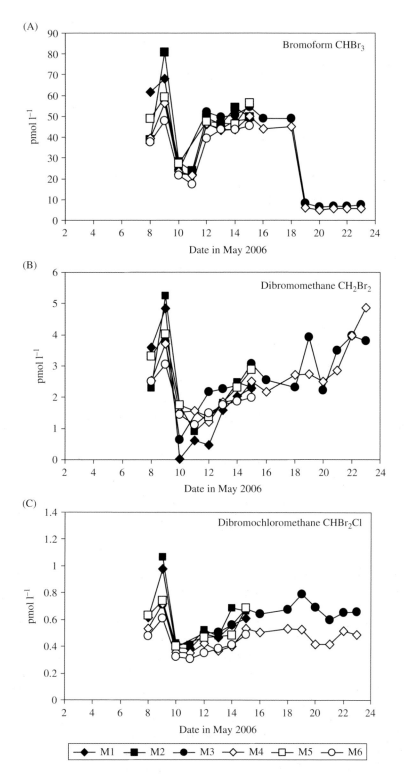

Figure 11.5 Concentrations of bromoform ($CHBr_3$) (A), dibromomethane (CH_2Br_2) (B), and dibromochloromethane ($CHBr_2Cl$) (C) during nutrient-induced blooms of phytoplankton in mesocosm enclosures in Raunefjord, Bergen, Norway, in May 2006. The nominal CO_2 treatments at the beginning of the experiment were as follows (M = mesocosm): ~750 µatm in M1, M2, and M3; ~300 µatm in M4, M5, and M6. Data from Hopkins *et al.* (2010).

ocean acidification was observed during laboratory incubations of natural plankton assemblages from UK coastal waters (Hopkins 2010).

Hopkins *et al.* (2010) also reported the response of a number of bromine-containing organohalogens (bromocarbons) to ocean acidification during the 2006 mesocosm experiment (Fig. 11.5). The temporal development of bromoform ($CHBr_3$), dibromomethane (CH_2Br_2), and dibromochloromethane ($CHBr_2Cl$) was substantially different from that of the iodocarbons, and these gases tended to show some increase in concentrations in response to high pCO_2. Mean concentrations were elevated under high pCO_2, particularly $CHBr_2Cl$ which was statistically higher during the whole experiment. However, as considerable differences in bromocarbon concentrations were apparent from the start of the experiment, and as the temporal development of bromocarbons did not reflect the growth and decline of the bloom, it is not clear whether these differences can be attributed to an effect of pCO_2.

11.3.2 Atmospheric and climatic implications

The oceans are naturally enriched in iodine, a result of volcanism earlier in the earth's history. Most marine iodine (>96%) is present in the thermodynamically stable form of iodate which is reduced to iodide (I^-) in surface-ocean waters by the activity of bacteria and phytoplankton (Elderfield and Truesdale 1980). Hence, in the euphotic zone the dissolved iodine pool may be dominated by iodide by up to 50% (Wong 1991). Iodide is subsequently taken up by seaweed and phytoplankton and can be released as volatile iodocarbons (CH_3I, CH_2I_2, etc.). These gases undergo sea–air exchange, and through photochemical reactions, iodine and its associated oxidized radicals (IO, OIO) are released to, or formed in, the atmosphere. As described above, decreased seawater concentrations of all measured iodocarbons were found during a mesocosm phytoplankton bloom experiment in 2006 (Hopkins *et al.* 2010), and this result is supported by observations from laboratory incubations of UK coastal plankton assemblages (Hopkins 2010). These findings suggest that ocean acidification can have an impact on the net production of these gases. Once released to the atmosphere, these compounds

play a number of important roles which may be affected by a decrease in their sea-to-air flux, and are discussed below.

11.3.2.1 *Oxidative capacity of the atmosphere*
Ozone (O_3) is a highly oxidizing gas that performs a number of important roles in the atmosphere. In the stratosphere (~25 km above the surface of the earth) it absorbs solar ultraviolet-B (UV-B) radiation, thus protecting the earth's living organisms from its harmful effects (Solomon 1999). By contrast, elevated concentrations of O_3 in the troposphere (the surface to 10 km) are not beneficial to life, adversely affecting plant, animal, and human health, and acting as a potent GHG (Ramaswamy *et al.* 2001). Therefore, a clear understanding of the processes that control O_3 levels in the lower atmosphere is vital for projections of future O_3 levels to be made. The production and use of anthropogenic chlorofluorocarbons (CFCs) has resulted in depletions to stratospheric O_3 allowing increased levels of UV-B to reach the earth's surface (Solomon 1999). Not only harmful to plant and animal life, an increase in the penetration of UV-B results in an enhancement in the chemical activity of the troposphere, with implications for a number of processes (Tang *et al.* 1998).

In polluted air, such as in the Northern Hemisphere, an increase in UV-B leads to an increase in tropospheric O_3 through the photo-oxidation of pollutants such as CH_4, NO_x, and volatile organic compounds (VOCs; Crutzen 1974; Tang *et al.* 1998). This process leads to an increase in hydroxyl (OH) radicals. These radicals exert an important control on the oxidative capacity of the atmosphere and are effective atmospheric cleansers, promoting the removal of GHGs and other pollutants (Tang *et al.* 1998). In clean, remote air, an increase in UV-B simply results in a decrease in tropospheric O_3 through photolysis in the presence of water vapour. This leads to an increase in OH radicals and an enhancement of the atmosphere's oxidative capacity (Tang *et al.* 1998).

Upon entering the atmosphere, marine iodocarbons undergo rapid photolysis to produce free radicals (I, IO, Br, BrO) which act as effective catalytic O_3-depleting species (Chameides and Davis 1980; Solomon *et al.* 1994; Davis *et al.* 1996). Due to weak iodine bonds, iodocarbons are very photochemically active, with an atmospheric lifetime of a few minutes to days (Solomon *et al.* 1994). This generally limits their O_3-depleting capacity to the troposphere, where a significant impact on O_3 levels is possible under certain conditions (Davis *et al.* 1996). Additionally, in regions that experience strong atmospheric convection, such as the tropics, iodocarbons can be rapidly transported to the upper troposphere and lower stratosphere, and can contribute to O_3 depletion at these levels (Solomon *et al.* 1994).

The regulation of the oxidative capacity of the atmosphere is clearly complex and is controlled by a number of processes, some of which have undergone significant anthropogenic perturbations. The impact that a decrease in marine emissions of volatile iodocarbons to the troposphere will have is therefore difficult to quantify, but may result in a decrease in tropospheric O_3 destruction. This would reduce the atmosphere's capacity to remove this potent GHG and air pollutant, enhancing global warming and contributing to negative impacts on human health and plant growth.

11.3.2.2 *New particle formation in the marine boundary layer*

The formation of new particles in the MBL from volatile iodocarbon precursors originating from marine macroalgae and kelp beds has been demonstrated by both observational and experimental studies (Makela *et al.* 2002; O'Dowd *et al.* 2002), suggesting that in coastal regions biogenic iodocarbons may exert a significant impact on local, and more speculatively global, radiative forcing.

Studies have shown that the oxidation of DMS in the MBL is connected to this process (O'Dowd *et al.* 2002). Oxidation of DMS represents the first step in the production of new particles, resulting in the production of small (~1 nm), thermodynamically stable clusters. In order to achieve an increase in particle number concentration, these stable clusters must rapidly grow to a size of about 3–4 nm to avoid colliding with larger pre-existing particles and being captured (Kulmala *et al.* 2000). The role, in this second process, of condensable iodine vapours (CIVs) produced by the photolysis of CH_2I_2 in the presence of O_3 has been confirmed by observational and modelling studies (O'Dowd *et al.* 2002; Pechtl *et al.* 2007). The influence of Br and Cl oxides on DMS chemistry has also received some attention. In a modelling study, von Glasow *et al.* (2002) found that when atmospheric halogen chemistry was included, DMS oxidization in the MBL increased by about 63%. Therefore, the potential climate impact of both DMS- and halocarbon-derived new particles/CCN is closely related. Future modelling studies on the impacts of ocean acidification on marine biogeochemistry and climate feedbacks need to consider the synergistic impacts of changes in the net production of these gases. A combined decrease in both DMS and iodocarbons would result in a decrease in two of the steps involved in new particle formation in the MBL and lead to an overall positive feedback to global warming.

The work of O'Dowd *et al.* (2002) was based on studies of new particle bursts at Mace Head, Eire, over dense beds of kelp. Therefore, it is most applicable to coastal regions with significant seaweed cover, and it becomes problematic to directly extrapolate such observations to the open oceans, resulting in uncertainty in the significance of any climatic impact. The situation in the open ocean is less clear, as particle burst events are less frequent and lower in intensity than their coastal analogues. Consequently, O'Dowd *et al.* (2002) expanded their observational and experimental work by simulating the process using a marine aerosol model. Further modelling work by Pechtl *et al.* (2007) confirmed the importance of iodine oxides in both primary particle formation and secondary growth of particles in the clean marine atmosphere. These simulations suggest that concentrations of iodocarbons over the open ocean may be high enough to influence marine particle production. Thus, pelagic open-ocean production of iodocarbons may exert a significant influence on climate through the production of new particles and CCN. In order to fully understand and quantify the role of phytoplankton, and achieve an understanding of the possible global climatic impacts, further knowledge of production and consumption of halocarbons and DMS by phytoplankton and bacteria in surface seawater, as well as the process of particle formation in the atmosphere, is required.

Thus, changes in the source of marine iodocarbon species in the atmosphere may significantly influence CCN concentration (O'Dowd et al. 2002), with implications for global radiative forcing and climate. During the 2006 mesocosm experiment, the time-integrated mean concentration of CH_3I was 44% lower in the high-CO_2 mesocosms (pCO_2 ~750 µatm) than in the control mesocosms (pCO_2 ~280 µatm), and 35% and 27% lower for C_2H_5I and CH_2I_2, respectively (Hopkins et al. 2010). The current flux of iodine from the oceans to the atmosphere, using a globally averaged marine atmosphere surface-mixed-layer height of 300 cm, is 1.4×10^3 atoms cm^{-3} s^{-1} (O'Dowd et al. 2002). With a mean 42% reduction in iodocarbons in the future high-CO_2 world (the mean decrease in iodocarbon concentrations observed during mesocosm experiments) and assuming no changes to other parameters (e.g. sea-surface temperature, mixed-layer depth, wind speed), the flux would decrease to 8.3×10^2 atoms cm^{-3} s^{-1}. Such a decrease in net input of particles into the aerosol population would result in a comparable percentage decrease in CCN (O'Dowd et al. 2002). As around 10% of new particles survive to CCN sizes (O'Dowd et al. 2002), this would correspond to an approximate 4.2% decrease in available CCN in the clean marine atmosphere. It is difficult to quantify the impact that this decrease would have on radiative forcing. The previous discussion on the problems of attempting to quantify the impact of DMS-derived aerosols is similarly applicable to iodine-derived aerosols. Additionally, understanding is lacking in a number of areas, ranging from the initial production and consumption of organohalogens by phytoplankton and bacteria in seawater, to the influence of I-derived particles on the present-day climate. Further research is required to enable quantification of the climatic impacts of a decrease in the production of marine biogenic iodocarbons as a result of ocean acidification.

11.4 Conclusions and future research needs

In this chapter, an assessment has been made of the currently available information on the effects of ocean acidification on the production of atmospherically important marine trace gases. Firstly, it is

important to acknowledge that the data are of a limited and somewhat contradictory nature; so far, only a small number of studies have been performed, and of those studies, there is little consistency in the observed responses of trace gases to ocean acidification. Furthermore, as almost all of these studies have been carried out at the mesocosm facilities in Bergen in early summer during a phytoplankton bloom, the data are very specific to one region, season, and situation, with very little diversity in experimental conditions. For these reasons, and also due to the complex atmospheric chemistry processes described in Sections 11.2.2 and 11.3.2, it is problematic to make global extrapolations from the available data. There is also a lack of understanding of the underlying mechanisms responsible for the observed responses, with little information on how and why ocean acidification may affect trace gas concentrations. In addition, only DMS and organohalogens have been considered, a small representation of a large number of important gases produced in the surface oceans. Therefore, in order to advance our understanding, there are a number of areas that warrant further research:

11.4.1 Marine trace gases

The studies reported so far have only assessed the response of DMS/P and organohalogens to ocean acidification. The oceans are a vital source (and in some cases, sink) of other atmospherically and climatically important trace gases. Alkyl nitrates ($RONO_2$), including methyl nitrate ($MeONO_2$) and ethyl nitrate ($EtONO_2$), are a major source of oxidized nitrogen to the remote marine atmosphere. They form a significant part of the 'odd nitrogen' (NO_y) reservoir and play an important role in regulating tropospheric ozone (Neu et al. 2008). Oxygenated volatile organic compounds (OVOCs) are a group of trace gases that include ketones (e.g. acetone), aldehydes (e.g. acetaldehyde), and alcohols (e.g. methanol, propanol). The occurrence of these compounds in the troposphere can strongly influence the oxidative capacity and ozone-forming potential of the atmosphere through the production of HO_x (OH and HO_2) free radicals (Singh et al. 2001). Non-methane hydrocarbons such as isoprene (2-methyl-1,3-butadiene) are thought to be produced by phytoplankton in the surface oceans (Broadgate *et al.* 1997). Upon emission to the atmosphere, isoprene is oxidized to produce secondary aerosols that contribute to particle growth and cloud droplet number, affect the chemical composition of CCN, and potentially exert a significant impact on climate feedback mechanisms (Meskhidze and Nenes 2006). The oceans are also a source of a number of long-lived, radiatively active and oxidizing gases, including methane (CH_4) and nitrous oxide (N_2O). CH_4, produced by methanogenic bacteria in anoxic sediments, contributes around 15% to current greenhouse forcing (IPCC 2007). N_2O, a by-product of microbial denitrification and nitrification, has 300 times the radiative forcing capacity of CO_2 and participates in catalytic ozone removal (Law and Owens 1990). Clearly, changes in the source rate and sea-to-air flux of any of the above trace gases could have significant implications for atmospheric chemistry and climatic processes. The role that ocean acidification may play in such changes now requires further investigation.

11.4.2 Mesocosm experiments

Mesocosm experiments are currently the best available tool for assessing the impacts of ocean acidification on pelagic ecosystems and their associated processes, in large volumes of water (compared to laboratory studies) and under quasi-natural meteorological and oceanic conditions (Riebesell *et al.* 2008). The results of mesocosm studies are most relevant in terms of highly productive regions (high-latitude waters, coastal waters, bloom events, and upwelling regions). Such regions are not only expected to experience the greatest changes as a result of anthropogenic ocean acidification and other global climatic changes (Orr *et al.* 2005; IPCC 2007; Feely *et al.* 2008; Steinacher *et al.* 2009; see also Chapter 3), but also represent important source regions of a number of climatically important trace gases (Class and Ballschmiter 1988; Carpenter and Liss 2000; Quack and Wallace 2003; Quack *et al.* 2004, 2007; Chuck *et al.* 2005). For these reasons, the continued use of mesocosm experiments is vital for furthering our understanding of the impacts of ocean acidification on the pelagic community, and the associated trace gas production.

11.4.3 'Natural analogue' sites

Regions that experience naturally lowered seawater pH may lend themselves as useful sites for studying the long-term effects of ocean acidification on a range of marine organisms and processes. Volcanically acidified sites, such as sites around the island of Ischia, Italy, have received some attention because the high-CO_2 conditions have prevailed for relatively long time periods (hundreds to thousands of years), allowing long-term adaptation of marine communities to the perturbed conditions (Hall-Spencer *et al.* 2008; see Chapter 10). The Ischia site has served as a good natural analogue for studying the effects of ocean acidification on the benthic communities that inhabit the area (Hall-Spencer *et al.* 2008). However, the site is considered less ideal for studies of the effects of ocean acidification on planktonic communities and their associated processes due to rapidly fluctuating seawater pH over small temporal and spatial scales, and rapid overturning of seawater into and out of the site (Hopkins 2010). Therefore other natural analogue sites need to be identified, ideally in more open-ocean situations. Oceanic upwelling regions that experience prolonged periods of undersaturation of $CaCO_3$ and low seawater pH may have some potential in this kind of research (Feely *et al.* 2008; see Box 1.1 in Chapter 1).

11.4.4 Regions sensitive to ocean acidification

A number of regions of the oceans have been identified as being particularly sensitive to future ocean acidification. High-latitude polar seas are particularly vulnerable due to low seawater temperatures and increased solubility of CO_2. Such regions may experience undersaturation with respect to aragonite as soon as 2023 (Steinacher *et al.* 2009; see Chapter 3). Similarly, upwelling regions are naturally acidified by the influx of CO_2-rich water from depth, although anthropogenic CO_2 is now increasing the extent of such acidification (Feely *et al.* 2008). Temperate coastal regions are expected to be substantially impacted by ocean acidification, which is likely to be enhanced by the deposition of sulphur and nitrogen from human activities (Doney *et al.* 2007). Such regions also experience high phytoplankton productivity, and act as important source regions of a range of climatically active halocarbons (Class and Ballschmiter 1988; Carpenter and Liss 2000; Quack and Wallace 2003; Quack *et al.* 2004, 2007; Chuck *et al.* 2005). Therefore, future research efforts should be focused on such regions, and involve: (1) long-term *in situ* monitoring strategies to detect changes and make distinctions between natural variability and anthropogenic impacts and (2) bioassay experimental work, such as on-deck incubations of water from a range of oceanic locations, to assess the response of complex ecosystems and populations to elevated pCO_2.

11.4.5 Furthering a mechanistic understanding

A response of net iodocarbon and DMS production to ocean acidification has been observed during the work reviewed here. Although there is some understanding of the processes involved for DMS, information is lacking on the biological mechanisms that result in net production of organohalogens and on their cycling in seawater. The photochemistry (Moore and Zafiriou 1994; Richter and Wallace 2004; Jones and Carpenter 2005; Martino *et al.* 2005), nucleophilic substitution, and hydrolysis of organohalogens (Elliott and Rowland 1993; Jeffers and Wolfe 1996) have been investigated. In addition, the use of stable isotope tracer techniques (e.g. production of $H^{14}CO_3$ from $^{14}CHBr_3$) has provided new insight on the biological loss rates of brominated methanes in both freshwater and seawater, and bacteria are greatly involved in these loss processes (Goodwin *et al.* 1997; King and Saltzman 1997; Goodwin *et al.* 1998; Tokarczyk *et al.* 2001). Furthermore, bacteria are probably involved in the cycling of iodocarbons in seawater (Amachi *et al.* 2001). As the iodocarbons and DMS/P display a strong response to ocean acidification (Avgoustidi 2007; Hopkins *et al.* 2010), it is important to further our understanding of the processes controlling the net concentrations of these gases in seawater.

^{14}C- or ^{13}C-labelled compounds may be employed to derive loss and production rates of trace gases in seawater, focusing both on whole plankton communities and on the bacterial fraction, and employing molecular techniques to gain detailed information on the bacterial strains that may be involved. Such approaches could be applied to ocean acidification experiments, in mesocosms or smaller-scale bioassay incubations, to assess the impacts of the perturbation on the processes controlling the net concentrations of trace gases in seawater.

11.4.6 Modelling studies

The ability to scale up from relatively small-scale perturbation experiments and field observations to regional and global scales using model simulations is critical to ocean acidification research, enabling researchers to assess the temporal and spatial changes that may occur over the coming decades (see Chapter 12). As DMS and organohalogens are considered to exert a considerable influence on climatic processes, the inclusion of the effects of ocean acidification on these compounds in global ocean–atmosphere modelling studies is vital to furthering our understanding of how the entire earth system may respond to future climate change and ocean acidification.

11.5 Acknowledgements

We thank Jean-Pierre Gattuso, Lina Hansson, Marion Gehlen, and Laurent Bopp for reviewing this chapter. Their helpful comments greatly improved the manuscript. We acknowledge financial support from the Natural Environment Research Council (NERC) NER/S/A/2005/13686, the Leverhulme Trust F/00 204/AC, and Oceans 2025 (PML's NERC-funded core program). This work is a contribution to the 'European Project on Ocean Acidification' (EPOCA) which received funding from the European Community's Seventh Framework Programme (FP7/2007–2013) under grant agreement no 211384.

References

Abrahamsson, K., Lorén, A., Wulff, A., and Wängberg, S.-Å. (2004). Air-sea exchange of halocarbons: the influence of diurnal and regional variations and distribution of pigments. *Deep Sea Research Part II: Topical Studies in Oceanography*, **51**, 2789–05.

Allgaier, M., Riebesell, U., Vogt, M., Thyrhaug, R., and Grossart, H.P. (2008). Coupling of heterotrophic bacteria to phytoplankton bloom development at different pCO$_2$ levels: a mesocosm study. *Biogeosciences*, **5**, 1007–22.

Amachi, S., Kamagata, Y., Kanagawa, T., and Muramatsu, Y. (2001). Bacteria mediate methylation of iodine in marine and terrestrial environments. *Applied and Environmental Microbiology*, **67**, 2718–22.

Andreae, M.O., and Crutzen, P.J. (1997). Atmospheric aerosols: biogeochemical sources and role in atmospheric chemistry. *Science*, **276**, 1052–58.

Archer, S.D., Goldson, L.E., Liddicoat, M.I., Cummings, D.G., and Nightingale, P.D. (2007). Marked seasonality in the concentrations and sea-to-air flux of volatile iodocarbon compounds in the western English Channel. *Journal of Geophysical Research*, **112**, C08009, doi:10.1029/2006JC003963.

Avgoustidi, V. (2007). *Dimethyl sulphide production in a high CO$_2$ world*. PhD Thesis, University of East Anglia, Norwich.

Bopp, L., Aumont, O., Belviso, S., and Monfray, P. (2003). Potential impact of climate change on marine dimethyl sulphide emissions. *Tellus B*, **55**, 11–22.

Broadgate, W.J., Liss, P.S., and Penkett, S.A. (1997). Seasonal emissions of isoprene and other reactive hydrocarbon gases from the ocean. *Geophysical Research Letters*, **24**, 2675–78.

Carpenter, L.J. and Liss, P.S. (2000). On temperate sources of bromoform and other reactive organic bromine gases. *Journal of Geophysical Research*, **105**, 20539–47.

Carslaw, K.S., Boucher, O., Spracklen, D.V. *et al.* (2009). Atmospheric aerosols in the earth system: a review of interactions and feedbacks. *Atmospheric Chemistry and Physics Discussions*, **9**, 11087–183.

Chameides, W.L. and Davis, D.D. (1980). Iodine: its possible role in tropospheric photochemistry. *Journal of Geophysical Research–Atmospheres*, **85**, 7383–98.

Charlson, R.J., Lovelock, J.E., Andreae, M.O., and Warren, S.G. (1987). Oceanic phytoplankton, atmospheric sulphur, cloud albedo and climate, *Nature*, **326**, 655–61.

Chin, M. and Jacob, D.J. (1996). Anthropogenic and natural contributions to tropospheric sulphate: a global model analysis. *Journal of Geophysical Research*, **101**, 18691–99.

Chuck, A.L., Turner, S.M., and Liss, P.S. (2005). Oceanic distributions and air-sea fluxes of biogenic halocarbons in the open ocean. *Journal of Geophysical Research*, **110**, C10022, doi:10.1029/2004JC002741.

Class, T. and Ballschmiter, K. (1988). Chemistry of organic traces in air. VIII. Sources and distribution of bromo- and bromochloromethanes in marine air and surface water of the Atlantic Ocean. *Journal of Atmospheric Chemistry*, **6**, 35–46.

Crutzen, P.J. (1974). Photochemical reactions initiated by and influencing ozone in unpolluted tropospheric air. *Tellus*, **26**, 47–57.

Davis, D., Crawford, J., Liu, S. *et al.* (1996). Potential impact of iodine on tropospheric levels of ozone and other critical oxidants. *Journal of Geophysical Research*, **101**, 2135–47.

Doney, S. C., Mahowald, N., Limar, I. *et al.* (2007). Impact of anthropogenic atmospheric nitrogen and sulfur deposition on ocean acidification and the inorganic carbon system. *Proceedings of the National Academy of Sciences USA*, **104**, 14580–85.

Elderfield, H., and Truesdale, V.W. (1980). On the biophilic nature of iodine in seawater. *Earth and Planetary Science Letters*, **50**, 105–14.

Elliott, S. and Rowland, F.S. (1993). Nucleophilic substitution rates and solubilities for methyl halides in seawater. *Geophysical Research Letters*, **20**, 1043–46.

Engel, A., Schulz, K.G., Riebesell, U., Bellerby, R., Delille, B., and Schartau, M. (2008). Effects of CO$_2$ on particle size distribution and phytoplankton abundance during a mesocosm bloom experiment (PeECE II). *Biogeosciences*, **5**, 509–21.

Feely, R.A., Sabine, C.L., Hernandez-Ayon, J.M., Ianson, D., and Hales, B. (2008). Evidence for upwelling of corrosive 'acidified' water onto the continental shelf. *Science*, **320**, 1490–92.

Fuge, R. and Johnson, C.C. (1986). The geochemistry of iodine—a review. *Environmental Geochemistry and Health*, **8**, 31–54.

Gabric, A.J., Bo, Q.U., Patricia, M., and Anthony, H.C. (2005). The simulated response of dimethylsulfide production in the Arctic Ocean to global warming. *Tellus B*, **57**, 391–403.

Gabric, A.J., Whetton, P.H., Boers, R., and Ayers, G.P. (1998). The impact of simulated climate change on the air-sea flux of dimethylsulphide in the subantarctic Southern Ocean. *Tellus B*, **50**, 388–99.

von Glasow, R. and Crutzen, P.J. (2004). Model study of multiphase DMS oxidation with a focus on halogens, *Atmospheric Chemistry and Physics*, **4**, 589–608.

von Glasow, R., Sander, R., Bott, A., and Crutzen, P.J. (2002). Modelling halogen chemistry in the marine boundary layer 2. Interactions with sulfur and the cloud-covered MBL. *Journal of Geophysical Research*, **107**, 4323, doi: 10/1029JD000943.

Goodwin, K.D., Lidstrom, M.E., and Oremland, R.S. (1997). Marine bacterial degradation of brominated methanes. *Environmental Science and Technology*, **31**, 3188–92.

Goodwin, K.D., Schaefer, J.K., and Oremland, R.S. (1998). Bacterial oxidation of dibromomethane and methyl bromide in natural waters and enrichment cultures. *Applied and Environmental Microbiology*, **64**, 4629–36.

Gunson, J.R., Spall, S.A., Anderson, T.R., Jones, A., Totterdell, I.J., and Woodage, M. J. (2006). Climate sensitivity to ocean dimethylsulphide emissions. *Geophysical Research Letters*, **33**, L07701, doi:10.1029/2005GL024982.

Hall-Spencer, J.M., Rodolfo-Metalpa, R., Martin, S., *et al.* (2008). Volcanic carbon dioxide vents show ecosystem effects of ocean acidification. *Nature*, **454**, 96–99.

Hopkins, F.E. (2010). *Ocean acidification and marine biogenic trace gas production*, 364 pp. PhD Thesis, University of East Anglia, Norwich.

Hopkins, F.E., Turner, S.M., Nightingale, P.D., Steinke, M., and Liss P.S. (2010). Ocean acidification and marine biogenic trace gas production. *Proceedings of the National Academy of Sciences USA*, **107**, 760–65.

IPCC (2007). *Climate change 2007: Synthesis report. Contribution of Working Groups I, II and III to the Fourth Assessment Report of the Intergovernmental Panel on Climate Change*, 104 pp. Cambridge University Press, Cambridge.

Jeffers, P.M. and Wolfe, N.L. (1996). On the degradation of methyl bromide in sea water. *Geophysical Research Letters*, **23**, 1773–76.

Jickells, T.D., An, Z.S., Andersen, K.K. *et al.* (2005). Global iron connections between desert dust, ocean biogeochemistry, and climate. *Science*, **308**, 67–71.

Jones, C.E. and Carpenter, L.J. (2005). Solar photolysis of CH_2I_2, CH_2ICl, and CH_2IBr in water, saltwater and seawater. *Environmental Science and Technology*, **39**, 6130–37.

Kettle, J. and Andreae, M. (2000). Flux of dimethyl sulfide from the oceans: a comparison of updated data sets and flux models. *Journal of Geophysical Research*, **105**, 793–808.

Kiene, R.P. and Bates, T.S. (1990). Biological removal of dimethyl sulphide from seawater. *Nature*, **345**, 702–5.

King, D.B. and Saltzman, E.S. (1997). Removal of methyl bromide in coastal seawater: chemical and biological rates. *Journal of Geophysical Research*, **102**, 18715–21.

Klick, S. and Abrahamsson, K. (1992). Biogenic volatile iodated hydrocarbons in the ocean. *Journal of Geophysical Research*, **97**, 12683–87.

Kulmala, M., Pirjola, L. and Makela, J.M. (2000). Stable sulphate clusters as a source of new atmospheric particles. *Nature*, **404**, 66–69.

Law, C.S. and Owens, N.J.P. (1990). Significant flux of atmospheric nitrous oxide from the northwest Indian Ocean. *Nature*, **346**, 826–28.

Lee, P.A., Rudisill, J.R., Neeley, A.R. *et al.* (2009). Effects of increased pCO_2 and temperature on the North Atlantic spring bloom. III. Dimethylsulfoniopropionate. *Marine Ecology Progress Series*, **388**, 41–49.

Liss, P.S., Hatton, A.D., Malin, G., Nightingale, P.D., and Turner, S. (1997). Marine sulphur emissions. *Philosophical Transactions of the Royal Society B: Biological Sciences*, **352**, 159–69.

Lovelock, J.E., Maggs, R.J., and Rasmussen, R.A. (1972). Atmospheric dimethyl sulphide and the natural sulphur cycle. *Nature*, **237**, 452–53.

Makela, J.M., Hoffman, T., Holzke, C. *et al.* (2002). Biogenic iodine emissions and identification of end-products in coastal ultrafine particles during nucleation bursts. *Journal of Geophysical Research*, **107**, D198110, doi:8110/1029JD000580.

Malin, G., Turner, S.M., and Liss, P.S. (1992). Sulfur: the plankton/climate connection. *Journal of Phycology*, **28**, 590–97.

Malin, G., Wilson, W.H., Bratbak, G., Liss, P.S., and Mann, N.H. (1998). Elevated production of dimethyl sulphide resulting from viral infection of cultures of *Phaeocystis pouchetii*. *Limnology and Oceanography*, **43**, 1389–93.

Manley, S.L. (2002). Phytogenesis of halomethanes: a product of selection or a metabolic accident? *Biogeochemistry*, **60**, 163–80.

Manley, S.L. and de la Cuesta, J.L. (1997). Methyl iodide production from marine phytoplankton cultures. *Limnology and Oceanography*, **42**, 142–47.

Manley, S.L. and Dastoor, M.N. (1987). Methyl halide (CH_3X) production from the giant kelp, *Macrocystis*, and estimates of global CH_3X production by kelp. *Limnology and Oceanography*, **32**, 709–15.

Martino, M., Liss, P.S., and Plane, J.M.C. (2005). The photolysis of dihalomathanes in surface seawater. *Environmental Science and Technology*, **39**, 7097–101.

Martino, M., Mills, G.P., Woeltjen, J., and Liss, P.S. (2009). A new source of volatile organoiodine compounds in surface seawater. *Geophysical Research Letters*, **36**, L01609, doi:10.1029/2008GL036334.

Meskhidze, N. and Nenes, A. (2006). Phytoplankton and cloudiness in the Southern Ocean. *Science*, **314**, 1419–23.

Moelwyn-Hughes, E.A. (1938). The hydrolysis of the methyl halides. *Proceedings of the Royal Society of London A*, **164**, 295–306.

Moore, R.M. and Zafiriou, O.C. (1994). Photochemical production of methyl iodide in seawater. *Journal of Geophysical Research*, **99**, 16415–20.

Moore, R.M., Webb, M., and Tokarczyk, R. (1996). Bromoperoxidase and iodoperoxidase enzymes and production of halogenated methanes in marine diatom cultures. *Journal of Geophysical Research*, **101**, 20899–908.

Neu, J.L., Lawler, M.J., Prather, M.J., and Saltzman, E.S. (2008). Oceanic alkyl nitrates as a natural source of tropospheric ozone. *Geophysical Research Letters*, **35**, L13814, doi:10.1029/2008GL034189.

Nightingale, P.D., Malin, G., and Liss, P.S. (1995). Production of chloroform and other low-molecular weight halocarbons by some species of macroalgae. *Limnology and Oceanography*, **40**, 680–89.

O'Dowd, C.D., Jimenez, J.L., Bahreini, R. *et al.* (2002). Marine aerosol formation from biogenic iodine emissions. *Nature*, **417**, 632–36.

Orr, J.C., Fabry, V.J., Aumont, O. *et al.* (2005). Anthropogenic ocean acidification over the twenty-first century and its impact on calcifying organisms. *Nature*, **437**, 681–86.

Paulino, A.I., Egge, J.K., and Larsen A. (2008). Effects of increased atmospheric CO_2 on small and intermediate sized osmotrophs during a nutrient induced phytoplankton bloom. *Biogeosciences*, **5**, 739–48.

Pechtl, S., Lovejoy, E.R., Burkholder, J.B., and von Glasow, R. (2007). Modelling the possible role of iodine oxides in atmospheric new particle formation. *Atmospheric Chemistry and Physics*, **7**, 1381–93.

Quack, B. and Wallace, D.W.R. (2003). Air-sea flux of bromoform: controls, rates, and implications. *Global Biogeochemical Cycles*, **17**, 1023, doi:1010.1029/2002GB001890.

Quack, B., Atlas, E., Petrick, G., Stroud, V., Schauffler, S., and Wallace, D.W.R. (2004). Oceanic bromoform sources for the tropical atmosphere. *Geophysical Research Letters*, **31**, L23S05, doi:10.1029/2004GL020597.

Quack, B., Atlas, E., Petrick, G., and Wallace, D.W.R. (2007). Bromoform and dibromomethane above the Mauritanian upwelling: atmospheric distributions and oceanic emissions. *Journal of Geophysical Research*, **112**, D09312, doi:10.1029/2006JD007614.

Ramanathan, V., Crutzen, J., Kiehl, T., and Rosenfeld, D. (2001). Aerosols, climate, and the hydrological cycle. *Science*, **294**, 2119–24.

Ramaswamy, V., Boucher, O., Haigh, J. *et al.* (2001). Radiative forcing of climate change. In: J. T. Houghton *et al.* (eds) *Climate change 2001, the scientific basis: Contribution of Working Group I to the Third Assessment Report of the Intergovernmental Panel on Climate*, pp. 351–406. Cambridge University Press, Cambridge.

Richter, U. and Wallace, D.W.R. (2004). Production of methyl iodide in the tropical Atlantic Ocean. *Geophysical Research Letters*, **31**, L23S03, doi:10.1029/2004GL020779.

Riebesell, U., Bellerby, R.G.J., Grossart, H.-P., and Thingstad, F. (2008). Mesocosm CO_2 perturbation studies: from organism to community level. *Biogeosciences*, **5**, 1157–64.

Shaw, G.E., Benner, R.L., Cantrell, W., and Clarke, A.D. (1998). The regulation of climate: a sulfate particle feedback loop involving deep convection – an editorial essay. *Climatic Change*, **39**, 23–33.

Singh, H., Chen, Y., Staudt, A., *et al.* (2001). Evidence from the Pacific troposphere for large global sources of oxygenated organic compounds. *Nature*, **410**, 1078–81.

Solomon, S. (1999). Stratospheric ozone depletion: a review of concepts and history. *Reviews of Geophysics*, **37**, 275–316.

Solomon, S., Garcia, R.R., and Ravishankara, A.R. (1994). On the role of iodine in ozone depletion. *Journal of Geophysical Research*, **99**, 20491–99.

Steinacher, M., Joos, F., Frölicher, T.L., Plattner, G. K., and Doney, S.C. (2009). Imminent ocean acidification in the Arctic projected with the NCAR global coupled carbon cycle-climate model. *Biogeosciences*, **6**, 515–33.

Sunda, W., Kieber, D.J., Kiene, R.P., and Huntsman, S. (2002). An antioxidant function for DMSP and DMS in marine algae. *Nature*, **418**, 317–20.

Tang, X., Madronich, S., Wallington, T., and Calamari, D. (1998). Changes in tropospheric composition and air quality. *Journal of Photochemistry and Photobiology B: Biology*, **46**, 83–95.

Tokarczyk, R. and Moore, R.M. (1994). Production of volatile organohalogens by phytoplankton cultures. *Geophysical Research Letters*, **21**, 285–88.

Tokarczyk, R., Goodwin, K.D., and Saltzman, E.S. (2001). Methyl bromide loss rate constants in the North Pacific Ocean. *Geophysical Research Letters*, **28**, 4429–4432.

Twomey, S (1991). Aerosols, clouds and radiation. *Atmospheric Environment*, **25**, 2435–42.

Urhahn, T. and Ballschmiter, K. (1998). Chemistry of the biosynthesis of halogenated methanes: C1-organohalogens as pre-industrial chemical stressors in the environment? *Chemosphere*, **37**, 1017–32.

Vairavamurthy, A., Andreae, M.O., and Iverson, R.L. (1985). Biosynthesis of dimethylsulphide and dimethyl propiothetin by *Hymenomonas caterae* in relation to sulfur source and salinity variations. *Limnology and Oceanography*, **30**, 59–70.

Vogt, M. and Liss, P.S. (2008). Dimethylsulfide and Climate. In: C. le Quéré and E. Saltzman (eds), *Surface ocean–lower atmosphere processes*, pp. 197–232. American Geophysical Union, Washington, DC.

Vogt, M., Steinke, M., Turner, S. *et al.* (2008). Dynamics of dimethylsulphoniopropionate and dimethylsulphide under different CO_2 concentrations during a mesocosm experiment. *Biogeosciences*, **5**, 407–19.

Watson, A.J. and Liss, P.S. (1998). Marine biological controls on climate via the carbon and sulphur geochemical cycles. *Philosophical Transactions of the Royal Society B: Biological Sciences*, **353**, 41–51.

Wingenter, O.W., Haase, K.B, Zeigler, M. *et al.* (2007). Unexpected consequences of increasing CO_2 and ocean acidity on marine production of DMS and CH_2ClI: potential climate impacts. *Geophysical Research Letters*, **34**, L05710, doi:10.1029/2006GL028139.

Wolfe, G.V., Steinke, M., and Kirst, G.O. (1997). Grazing-activated chemical defence in a unicellular marine alga. *Nature*, **387**, 894–97.

Wong, G.T.F. (1991). The marine geochemistry of iodine. *Reviews in Aquatic Sciences*, **4**, 45–73.

CHAPTER 12

Biogeochemical consequences of ocean acidification and feedbacks to the earth system

Marion Gehlen, Nicolas Gruber, Reidun Gangstø, Laurent Bopp, and Andreas Oschlies

12.1 Introduction

By the year 2008, the ocean had taken up approximately 140 Gt carbon corresponding to about a third of the total anthropogenic CO_2 emitted to the atmosphere since the onset of industrialization (Khatiwala *et al.* 2009). As the weak acid CO_2 invades the ocean, it triggers changes in ocean carbonate chemistry and ocean pH (see Chapter 1). The pH of modern ocean surface waters is already 0.1 units lower than in pre-industrial times and a decrease by 0.4 units is projected by the year 2100 in response to a business-as-usual emission pathway (Caldeira and Wickett 2003). These changes in ocean carbonate chemistry are likely to affect major ocean biogeochemical cycles, either through direct pH effects or indirect impacts on the structure and functioning of marine ecosystems. This chapter addresses the potential biogeochemical consequences of ocean acidification and associated feedbacks to the earth system, with focus on the alteration of element fluxes at the scale of the global ocean. The view taken here is on how the different effects interact and ultimately alter the atmospheric concentration of radiatively active substances, i.e. primarily greenhouse gases such as CO_2 and nitrous oxide (N_2O).

Changes in carbonate chemistry have the potential for interacting with ocean biogeochemical cycles and creating feedbacks to climate in a myriad of ways (Box 12.1). In order to provide some structure to the discussion, direct and indirect feedbacks of ocean acidification on the earth system are distinguished (Table 12.1 and Fig. 12.1). Direct feedbacks are those which directly affect radiative forcing in the atmosphere by altering the air–sea flux of radiatively active substances. Indirect feedbacks are those that first alter a biogeochemical process in the ocean, and through this change then affect the air–sea flux and ultimately the radiative forcing in the atmosphere. For example, when ocean acidification alters the production and export of organic matter by the biological pump, then this is an indirect feedback. This is because a change in the biological pump alters radiative forcing in the atmosphere indirectly by first changing the near-surface concentrations of dissolved inorganic carbon and total alkalinity. These changes will in turn affect the air–sea flux of CO_2. The magnitude of the indirect feedbacks associated with biological pumps depends on at least three elements: (1) the magnitude of the impact that ocean acidification has on a particular aspect of the ocean's biological pumps, (2) how these changes in the biological pumps affect the ocean's carbonate chemistry, and (3) how these changes in carbonate chemistry affect the air–sea CO_2 flux. For example, the response of the air–sea CO_2 flux to a given change in the export of organic carbon from near-surface waters may vary by up to a factor of 10, depending on where and how this change occurs (e.g. Jin *et al.* 2008; Oschlies 2009).

By far the most important direct feedback is that associated with ocean acidification changing the buffer (Revelle) factor of the ocean. As discussed in detail in Chapter 3, the ocean's capacity to hold additional CO_2 from the atmosphere is inversely

---

Box 12.1 The concept of feedback and ocean acidification

A feedback is an interaction mechanism in which the result of an initial process drives changes in a second process that in turn influences the initial one. A positive feedback intensifies the original process, while a negative feedback reduces it. In the climate system, one of the main positive feedbacks is the tendency of warming to increase the quantity of water vapour in the atmosphere, and hence the greenhouse effect and earth's warming itself. Ocean acidification, a direct consequence of increasing levels of atmospheric CO_2, interacts with biogeochemical processes, alters air–sea

exchange of CO_2, and hence atmospheric CO_2. If ocean acidification leads to an increase (decrease) in atmospheric CO_2 then this represents a positive (negative) feedback. In this chapter, the notion of feedback is extended to include the effect of ocean acidification on climate not only through changes in atmospheric CO_2 (CO_2–acidification feedback), but also through changes in other atmospheric constituents (e.g. nitrous oxide and dimethyl sulphide) which do not represent, strictly speaking, feedbacks on CO_2.

---

proportional to the Revelle buffer factor, i.e. the larger this factor, the lower the capacity of the ocean to take up additional CO_2. Ocean acidification increases the Revelle buffer factor, causing a drastic decrease in the capacity of ocean water to take up CO_2 from the atmosphere, leading to a decrease in the rate of uptake and a transient accumulation of CO_2 in the atmosphere.

The majority of the indirect feedbacks are those affecting the ocean's biological pumps (both organic and carbonate). Two groups of indirect feedbacks can be identified: group 1, in which ocean acidification affects the biological pumps directly, and group 2, in which ocean acidification affects a particular ocean biogeochemical process, which in turn alters the biological pumps.

The stimulation of marine photosynthesis by increased levels of CO_2 (Rost et al. 2008) is an example of an indirect group 1 feedback. However, the extent of CO_2 fertilization depends on the physiological characteristics of individual phytoplankton groups. It is higher in organisms with an inefficient carbon acquisition pathway. A second indirect effect of the first group is the change in calcification (Fabry et al. 2008, but see also Iglesias-Rodriguez et al. 2008). An example of an indirect effect of the second group is the reported enhancement of dinitrogen (N_2) fixation by cyanobacteria at elevated pCO_2 concentrations (Hutchins et al. 2009). This process represents a major source of reactive nitrogen (N) to oligotrophic tropical and subtropical areas, and given the N-limited nature of these areas, has the potential to substantially increase primary production.

Over recent years, an increasing number of studies have addressed the effects of ocean acidification on isolated processes based on first-order chemical principles (e.g. trace metal speciation) or controlled process studies (e.g. N_2 fixation, calcification, photosynthesis). While these studies provided new and valuable insights into, for example, the vulnerability of specific processes in response to ocean acidification, they do not allow us to apprehend impacts at the scale of the marine biogeochemical cycle. Scaling up from the level of physiological processes to that of organisms and ecosystems is not straightforward and challenges global biogeochemical modelling efforts. It is further complicated by the fact that ocean acidification does not occur in isolation, but in synergy with ocean warming and related changes in the physical environment that might amplify or alleviate its impacts (e.g. Brewer and Peltzer 2009). All these changes taken together will alter the partitioning of climate-relevant gases between the ocean and the atmosphere.

This chapter presents a synthesis of our understanding of impacts of ocean acidification on marine biogeochemical cycles, including its interaction with climate change and feedbacks to the earth system. It starts with the discussion of the marine carbon cycle, an area for which experimental and modelling studies allow a first-order evaluation of impacts and feedbacks, moves to the nitrogen cycle and ends with atmospherically active trace gases. Impacts are discussed together with associated feedbacks and, when possible, taking into account climate change.

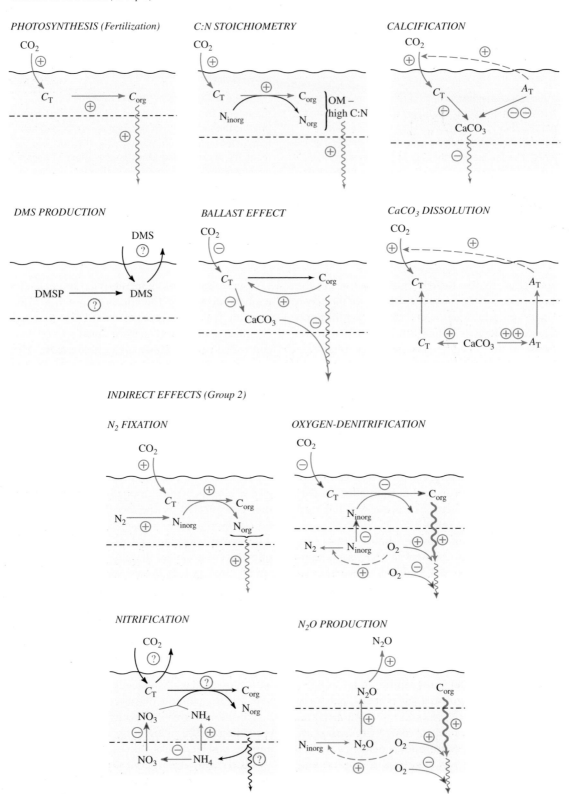

Figure 12.1 Summary of feedbacks between the ocean's biogeochemical cycles and radiative forcing induced by ocean acidification. Shown are the pathways that ultimately lead to a change in the atmospheric concentration of radiatively active components, namely CO_2, N_2O, and DMS. Refer to the text and Table 12.1 for a detailed discussion. Solid arrows indicate fluxes and/or transformations, while dashed arrows indicate that this quantity has an influence on the flux/transformation. The dot-dashed horizontal line depicts the boundary between the euphotic and the aphotic zones.

Table 12.1 Summary of effects and feedbacks. A positive (negative) feedback tends to increase (decrease) atmospheric CO_2 (see Box 12.1)

Process	Causality	Sign of feedback	Magnitude	Level of understanding
Direct feedbacks				
CO_2 buffer factor	Decrease of ocean uptake capacity	Positive	Large	High
Indirect group 1 feedbacks				
Photosynthesis	Enhancement of biological production and export by CO_2 fertilization	Negative	Medium	Medium
Export: stoichiometric ratio	Increase of C:N ratio, thereby enhancing marine productivity in terms of carbon	Negative	Small to medium	Low
Calcification	Decrease of marine calcification (BUT mixed experimental evidence)	Negative	Small to medium	Low to medium
$CaCO_3$ dissolution	Increase in dissolution of $CaCO_3$ in particles and sediments, thereby increasing ocean alkalinity	Negative	Small (short-term) — large on long timescales	Medium
Remineralization: ballast effect	Decrease of $CaCO_3$ production, leading to reduced org. matter export	Positive	Small to medium	Low
DMS production	Enhancement or reduction of DMS production	Unknown	Unknown	Low
Indirect group 2 feedbacks				
N_2 fixation	Enhancement of biological fixation of N_2, increasing N inventory, enhancing biological production	Negative	Medium	Medium
Oxygen denitrification	Reduction in penetration depth of organic matter, shallow remineralization creating higher O_2 demand in low-O_2 regions, causing expansion of these regions, enhancing denitrification, lowering N inventory, lower biological production	Positive	Medium	Low
Nitrification	Reduction in nitrification	Unknown	Small	Low
Nitrous oxide production	Decrease in oxygen concentrations, enhancing N_2O production	Positive	Medium	Low

12.2   The marine carbon cycle

The marine carbon cycle is an essential component of the earth system, participating in the long-term stabilization of atmospheric CO_2 levels. The oceans are by far the largest reservoir for carbon, with the exception of sedimentary rocks (Falkowski *et al.* 2000; see Chapter 2). They interact with the atmosphere on a variety of timescales ranging from hours (daily cycle of biological production) over seasons (mixed-layer dynamics), to several centuries (large-scale ocean circulation), and up to several thousands of years (interaction with marine sediments through the mechanism of $CaCO_3$ compensation). A crucial element of this interaction is the fact that the surface ocean is substantially depleted in dissolved inorganic carbon (C_T) relative to the deep ocean, requiring processes that transfer carbon from the near-surface to depth in order to maintain this downward gradient against homogenization by mixing and transport (Volk and Hoffert 1985). Three pathways have been identified as gradient makers ('pumps'): the solubility pump, the carbonate pump, and the soft tissue pump. The latter two are often collectively referred to as the biological pump. Ocean acidification, by modifying the carbonate chemistry of the surface ocean and environmental conditions for marine biota, interacts with all three pumps ultimately altering the air–sea balance of CO_2.

12.2.1   The solubility pump

The solubility pump refers to the physico-chemical process driving the uptake of CO_2 and its downward mixing and transport along with the large-scale ocean circulation. The solubility of CO_2 increases with decreasing temperature and is thus higher at high latitudes where deep-water formation takes place. The combined effects of solubility and deep-water formation result in a downward transport of CO_2-enriched water masses and thus higher C_T concentrations at depth (see Chapter 3).

The equilibrium reactions of the carbonate system are at the origin of the large uptake capacity of the ocean for CO_2. However, as for any buffer system, its capacity is not infinite. As discussed in Chapter 3, the Revelle factor, a measurement of the buffer capacity, increases (decreasing buffer capacity) with increasing atmospheric CO_2. As a result, the strength of the ocean sink for CO_2 is going to decrease in the future, a direct positive feedback of ocean acidification to atmospheric CO_2 levels and hence to the earth system. This positive feedback is very substantial, i.e. in a business-as-usual scenario it may be as large as 30% in the next 100 years (Sarmiento *et al.* 1995). Hydration of gaseous CO_2 and the equilibration between individual dissolved species of the carbonate system are dependent on temperature and salinity. Both will change in response to global climate change. Climate change aggravates the chemical effect of decreasing buffer capacity in two ways: (1) due to the inverse relationship between temperature and CO_2 solubility (Chapter 3) and (2) due to the increase in stratification and the anticipated slowdown of the surface-to-deep exchange of carbon (Sarmiento *et al.* 1998).

12.2.2 The carbonate pump

The carbonate pump is driven by the precipitation of $CaCO_3$ by marine organisms, the settling of carbonate particles across the water column, and their dissolution at depth in undersaturated waters and burial in sediments. Since the precipitation of $CaCO_3$ reduces the total alkalinity of the seawater more than it decreases its C_T, this process increases the concentration of dissolved CO_2 and thus the partial pressure of CO_2 (pCO_2) of near-surface waters. This can also be understood by recognizing that the removal of the carbonate ion (CO_3^{2-}) by the precipitation of $CaCO_3$ leads to a redistribution of the different species of the carbonate system in such a way

that the dominant C_T species at the pH of surface-ocean waters, i.e. bicarbonate (HCO_3^-), will dissociate in order to replenish the lost CO_3^{2-} but thereby generating dissolved CO_2 as well:

$$Ca^{2+} + CO_3^{2-} \rightarrow CaCO_3(s) \qquad (12.1a)$$

$$2HCO_3^- \rightarrow CO_3^{2-} + CO_2 + H_2O \qquad (12.1b)$$

or written as a summary equation:

$$Ca^{2+} + 2HCO_3^- \rightarrow CaCO_3(s) + CO_2 + H_2O. \qquad (12.1c)$$

Thus the carbonate pump tends to force CO_2 out of the ocean into the atmosphere, despite the fact that it leads to a depletion of C_T in the upper ocean.

The saturation state with respect to $CaCO_3$ decreases with depth, largely owing to the soft-tissue pump that acidifies the deep ocean as a result of the release of metabolic CO_2 during the remineralization of the organic matter transported downwards (Gruber and Sarmiento 2002). The saturation state has a direct impact on the formation and dissolution of carbonate structures. Throughout this discussion, the saturation state of seawater with respect to a $CaCO_3$ mineral (Ω) defined by Zeebe and Gattuso (see Box 1.1 in Chapter 1) is used. The stoichiometric solubility product increases with depth as a result of increasing pressure and decreasing temperature.

12.2.2.1 Calcium carbonate production

In the modern ocean, $CaCO_3$ formation is largely a biotic process. While inorganic precipitation and dissolution of $CaCO_3$ are a direct function of the saturation state (Morse *et al.* 2007), the mechanisms of calcification and their sensitivity to changes in carbonate chemistry are less well understood (see Chapters 6 and 7). The diversity of responses of calcifiers to a decrease in saturation state of seawater challenges global ocean biogeochemical models. These models represent $CaCO_3$ formation as a geochemical source/sink function of varying complexity with a limited number of studies including a dependency on carbonate chemistry. In its simplest expression, $CaCO_3$ production is implemented as a constant fraction of organic carbon production modulated by a dependency on saturation state

(Ridgwell *et al.* 2007). In Heinze (2004), pre-industrial $CaCO_3$ export production was computed as a constant function of the organic carbon export that is not driven by silicifying organisms. A third approach assigns $CaCO_3$ production to a specific phytoplankton functional type, the nanophytoplankton, which corresponds to the size class of coccolithophores (Gehlen *et al.* 2007). In this model $CaCO_3$ export production is a function of irradiance, nutrient availability, nanophytoplankton biomass, grazing by micro- and mesozooplankton, and saturation state with respect to calcite. These models consider the formation and dissolution of the less soluble $CaCO_3$ polymorph calcite. The different parameterizations of calcite production as a function of carbonate chemistry rely on a limited number of studies. Gangstø *et al.* (2008) extended the conceptual approach originally derived for calcite production by nanophytoplankton by Gehlen *et al.* (2007) to the production of aragonite by mesozooplankton. At the time of this model sensitivity study, no data were available to derive a parameterization of aragonite production by pteropods based on observations. Since the study by Gangstø *et al.* (2008) was published, experimental data on the calcification response of pteropods as a function of carbonate chemistry have become available. More recently, the models were extended further to include a parameterization specific for pteropods (Gangstø *et al.* 2011).

In all models, $CaCO_3$ production is linked to the production of particulate organic carbon (POC) through the rain ratio. The latter is defined as the ratio of $CaCO_3$ to POC flux and corresponds to the relevant quantity in terms of biogeochemical impacts and feedbacks to atmospheric CO_2 (e.g. Archer and Maier-Reimer 1994). On average, the rain ratio is about 0.09 (Jin *et al.* 2008). The POC production did not respond to changes in carbonate chemistry in any of these studies, despite some experimental evidence for an increase in POC production in response to increased levels of CO_2 (e.g. Zondervan *et al.* 2001).

12.2.2.2   Calcium carbonate dissolution

Dissolution of $CaCO_3$ is an abiotic process driven by thermodynamics, i.e. the degree of undersaturation. This implies that the dissolution of $CaCO_3$ is bound

to increase in response to decreasing saturation state. The resulting increase of total alkalinity favours CO_2 uptake, a negative indirect group 1 feedback.

The dissolution of $CaCO_3$ is usually described by a higher-order reaction rate law with respect to undersaturation:

$$R = k \times (1 - \Omega)^n \qquad (12.2)$$

where n is the reaction order and k is the dissolution rate parameter (time^{-1}).

Published estimates of n range from 1 to 4.5 (Keir 1980; Hales and Emerson 1997; Gehlen *et al.* 2005) based on laboratory studies and the evaluation of sediment porewater data. The higher-order rate law implies that dissolution rates are low at modest levels of undersaturation and increase following a power law as a function of increasing undersaturation. As a result, when normalized to a given value of Ω, the higher-order rate law translates to an initial lower sensitivity to decreases in saturation state compared with the linear rate expression.

The dissolution of sinking $CaCO_3$ particles is implemented into some global biogeochemical models as a first-order rate law, i.e. $n = 1$ (e.g. Heinze 2004; Gehlen *et al.* 2007; Gangstø *et al.* 2008). In contrast, Ridgwell *et al.* (2007) do not explicitly solve for $CaCO_3$ dissolution, but rather apply a constant depth-penetration profile to the $CaCO_3$ export flux, an approach used by most models that do not include a sensitivity of their biogeochemical processes to ocean acidification.

12.2.2.3   Interaction with carbonate sediments

The vast reservoir of mineral carbonates in the sediments provides the ultimate buffer against ocean acidification (see Chapter 2). If those carbonates were to dissolve readily in response to a decrease in the saturation state of the overlying waters, ocean acidification would not be a problem. The sediments would resupply the carbonate ions that are titrated away by the invading anthropogenic CO_2, thus keeping pH changes to a minimum. Unfortunately, the ocean's sediment pool will be reacting very slowly to the invasion of anthropogenic CO_2 and the associated ocean acidification. First, because it takes time for anthropogenic CO_2

to invade the ocean causing the CO_3^{2-} concentration to decrease and exposing increasing areas of marine carbonate sediments to undersaturated waters. Second, the dissolution rates are very small. As a result, it will take thousands of years for this compensation to occur (e.g. Archer 2005). Sundquist (1990) estimated that about 60% of the total buffering of an atmospheric CO_2 perturbation by ocean processes can be attributed to circulation with characteristic timescales of several centuries. The interaction with carbonate sediments will neutralize the remaining 40%. The majority of ocean biogeochemical models that are concerned about projecting the evolution of ocean acidification over the next centuries do not include the sediment pool. In strong contrast, if one wants to consider the long-term consequences of ocean acidification or model past geological events, it is absolutely critical to include this compartment as well (e.g. Ridgwell and Hargreaves 2007).

12.2.2.4   Future projections: impacts and feedbacks
Given that calcification increases surface-ocean pCO_2 (Eq. 12.1), a decrease in calcification in response to a decrease in the $CaCO_3$ saturation state would translate into an additional uptake of CO_2. The opposite would be the case if calcification were to increase. Thus, a decrease in calcification tends to act as a negative indirect group 1 feedback.

To illustrate the evolution of carbonate chemistry and its impact on the marine carbonate cycle in more detail, we use output from an ocean biogeochemistry model that was run following the standard scenario of the Coupled Model Intercomparison Project (CMIP; http://www-pcmdi.llnl.gov/projects/cmip/index.php). In this scenario, atmospheric pCO_2 increases at a rate of 1% yr^{-1} from 286 (referred to as 1× CO_2) to 1144 (4× CO_2) ppmv over a 140 year time period (Fig. 12.2A). Three sensitivity experiments were undertaken during this study (Gehlen *et al.* 2007). In experiment CAL01, both $CaCO_3$ production and dissolution responded to changes in carbonate chemistry, while in experiment CAL02, $CaCO_3$ production was kept constant at pre-industrial levels, but dissolution responded to ocean acidification. Finally, in experiment CAL03 production and dissolution of $CaCO_3$ were kept at pre-industrial levels.

The mean global saturation state of surface-ocean waters with respect to calcite (Ω_c) decreases from $\Omega_c > 5$ at year 0 to $\Omega_c = 2$ at the end of the acidification scenario (Fig. 12.2B). The model projects a decrease in $CaCO_3$ production of 27% (Fig. 12.2C). In experiment CAL01, $CaCO_3$ dissolution decreases by 16%, reflecting a reduction in $CaCO_3$ production and thus the availability of particles for dissolution (Fig. 12.2.D). When normalized to production, carbonate dissolution increases from 61% at 1× pCO_2 to 72% at 4× pCO_2. In experiment CAL02, water-column dissolution increases by 19% relative to the pre-industrial state. The reduction in $CaCO_3$ production (CAL01) drives an additional uptake of 5.9 Gt C relative to CAL03 over the course of the simulation (Fig. 12.2.E). This corresponds to a very modest decrease in atmospheric CO_2 of about 2.8 ppmv, i.e. equivalent to 2 years of current levels of growth of atmospheric CO_2. The increase in dissolution flux alone (CAL02) gives rise to an excess uptake of 1.2 Gt C. In this particular model study, the substantial decrease in $CaCO_3$ production combined with the increase in relative dissolution translates to an overall modest negative feedback to atmospheric CO_2.

How robust are these future projections of $CaCO_3$ production and dissolution across models? At the same atmospheric CO_2 level of about 1100 ppmv, Heinze (2004) predicts a global decrease in $CaCO_3$ production of approximately 38%, slightly larger than the 27% simulated by Gehlen *et al.* (2007). This difference may be the result of the former study reaching this pCO_2 level after 420 years compared with 140 years in the case of the latter one. The longer duration of the experiment allows for an amplification of ocean chemistry changes and contributes to the stronger decrease in calcification (see Chapter 14). Ridgwell *et al.* (2007) adjusted their parameterization to reproduce a dependency of $CaCO_3$ production on Ω_c similar to Gehlen *et al.* (2007). The corresponding calcification feedback ranged from 6.5 to 7.7 Gt C at 3× pCO_2, compared with 5.6 Gt C at 4× pCO_2 reported by Gehlen *et al.* (2007).

At first sight, one might thus conclude that the projected decrease in pelagic calcification and associated increase in atmospheric CO_2 uptake by the ocean converge to modest levels of less than 10 Gt C over the next century. However, these models rely largely on the same small experimental dataset for

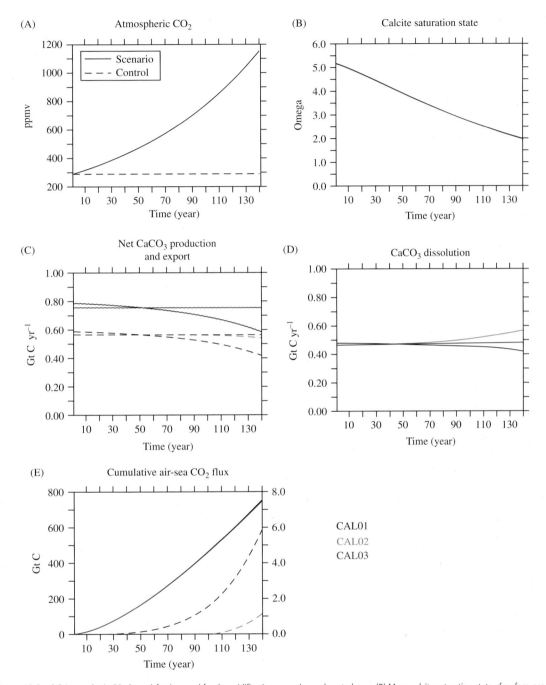

Figure 12.2 (A) Atmospheric CO_2 (ppmv) forcing used for the acidification scenarios and control runs. (B) Mean calcite saturation state of surface-ocean waters (0–100 m). (C) Global net $CaCO_3$ production (full line) and export at 100 m (stippled line). (D) Global $CaCO_3$ dissolution flux. (E) Cumulative air–sea flux of CO_2 on the left axis, and carbon uptake in excess to CAL03 on the right axis. CAL01, production and dissolution of $CaCO_3$ responds to changes in carbonate chemistry; CAL02, constant production, but dissolution responds to changes in carbonate chemistry; CAL03, constant production and dissolution (after Gehlen *et al.* 2007).

deriving the functional relationship between seawater chemistry and $CaCO_3$ production. Including a greater diversity of datasets results in a much larger spread of the calcification feedback estimates, ranging from 33 to 107 Gt C at 1400 ppmv for year 2300 (Ridgwell et al. 2007). New experimental data encompassing laboratory and mesocosm experiments, together with field observations, will allow us to derive process parameterizations that reflect a community response, rather than the upscaled sensitivity of a specific calcifying species (Ridgwell et al. 2009). Other effects, in particular warming and associated shifts in nutrient concentrations and irradiance, are likely to affect species distribution and primary production. It is not yet clear what impacts these concurrent changes will have on total pelagic calcification.

How will ocean acidification interact with climate change? Climate change translates into a net positive feedback on atmospheric CO_2 (Friedlingstein et al. 2006). The negative calcification feedback will thus counteract the impact of climate change on rising atmospheric CO_2 concentrations. In line with this consideration, Ridgwell et al. (2007) project an amplification of the calcification feedback by one-third in response to climate change by the year 3000.

Marine $CaCO_3$ production is a function of the biomass of calcifiers and their specific calcification rate. While ocean acidification will affect the latter, climate change is likely to drive major reorganizations of ecosystems, hence affecting the distribution and biomass of calcifying plankton. Isolating the direct effect of ocean chemistry on $CaCO_3$ production in future projections of the marine $CaCO_3$ cycle under climate change from the indirect effect of changes in POC production in response to temperature, irradiance, and nutrient availability is not straightforward. The point is well illustrated by Schmittner et al. (2008) who predict an approximate doubling of primary production and calcification by the year 4000 in response to global warming. While the discussion of future projections of primary and/or export production by coupled climate carbon cycle models is beyond the scope of this chapter, we note that some models converge in predicting their decrease (Steinacher et al. 2010). The picture gets even more complicated for models distinguishing

different phytoplankton functional types (PFT). Here, the overall decrease in primary production and export production might go along with a shift in the relative proportion of primary production related to specific PFTs. For example, Bopp et al. (2005) report a replacement of diatoms by nanophytoplankton (the calcifying PFT in the model) in mid latitudes in response to the projected increase in stratification and diminished nutrient supply. From the preceding, it becomes evident that future studies across a variety of models should focus on the synergy between climate change and ocean acidification to try to disentangle the direct effects of chemistry from the effects of changes in temperature, irradiance and nutrient availability.

12.2.3 The soft-tissue carbon pump

The soft-tissue pump (Volk and Hoffert 1985) starts in the surface ocean with the production of organic carbon during photosynthesis, followed by the gravitational settling of particles across the water column, their remineralization, and the incorporation of the remaining fraction to surface sediments. The soft-tissue pump affects the surface-ocean carbonate system and thus air–sea exchange through the uptake of C_T together with a small increase in total alkalinity, leading to a decrease in surface-ocean pCO_2 during photosynthesis. The C_T is replaced by CO_2 released by the respiration of organic C within the mixed layer (zero net effect on air–sea exchange of CO_2) or by mixing with deeper layers and is partly exchanged with the atmosphere. The time during which carbon bound into the particulate organic fraction is unavailable for exchange with the atmosphere is a function of remineralization depth and ranges from days for shallow remineralization in well-mixed waters to geological timescales for the fraction buried in marine sediments. A complete shutdown of the biological pump would yield an increase in atmospheric CO_2 of between 150 and 220 ppmv (e.g. Gruber and Sarmiento 2002).

12.2.3.1 Primary production

Photosynthesis appears to benefit from increased levels of CO_2 in some marine photosynthesizers (e.g. Rost et al. 2008). However, the extent of this

CO_2 fertilization depends on the physiological characteristics of individual phytoplankton groups, favouring in particular organisms with a comparable inefficient carbon acquisition pathway. Chapter 7 gives a detailed discussion of the effects of CO_2 fertilization on phytoplankton. Enhanced photosynthesis *per se* will not cause a change in the net air–sea balance of CO_2. This requires a net increase in net community production, i.e. the net balance between gross CO_2 fixation and respiration, as it is net community production that provides the organic carbon that can be exported to depth. Furthermore, marine primary production is limited by irradiance and nutrients, and the total biomass is kept in check by grazing by zooplankton. Thus, only if those limiting factors can be overcome by CO_2 fertilization, causing an additional export of organic carbon by the soft-tissue pump from the surface, will ocean acidification-induced changes in primary production cause earth system feedbacks.

12.2.3.2 Export

A powerful way for the soft-tissue pump to overcome the stringent control of nutrients on primary production and export is to alter the stoichiometric nutrient to carbon ratio of the organic matter produced and exported. Mesocosm experiments with natural plankton communities have indeed reported enhanced carbon drawdown under elevated CO_2 (Riebesell *et al.* 2007). In these experiments, the stoichiometry of the carbon-to-nitrogen drawdown increased from 6.0 at 350 µatm to 8.0 at about 1050 µatm. While the significance of this excess carbon drawdown is not fully established yet and the mechanisms not fully understood, a possible route is via enhanced carbon fixation by the phytoplankton at higher CO_2 levels, exudation of carbon-rich dissolved organic matter, and its subsequent export in form of aggregates (Arrigo 2007). To estimate the potential global impact of such a CO_2-sensitive stoichiometry of C:N drawdown and, possibly, export, Oschlies *et al.* (2008) extrapolated the mesocosm results to the global ocean by means of a simple ecosystem-circulation model. They found that enhanced C:N ratios could accomplish a negative feedback on atmospheric carbon levels for a business-as-usual scenario (SRES A2; see Chapter 15) amounting to 34 Gt C by the end of the century. This

is of similar magnitude to the negative feedbacks estimated by modelling studies for CO_2-sensitive calcification rates (Section 12.2.2.4). A dominant 'side' effect of the enhanced C:N ratios identified by Oschlies *et al.* (2008) is the enhanced oxygen consumption associated with the respiration of carbon-rich organic matter at depth. In their model, this leads to a 50% expansion of oceanic suboxic regions, with direct consequences for the amount of nitrogen loss by denitrification and anaerobic ammonium oxidation (anammox), and hence for the oceanic inventory of fixed nitrogen.

The negative feedback of enhanced inorganic carbon-to-nitrogen consumption depends on the export of the additional organic carbon produced out of the surface mixed layer. For a pelagic Arctic ecosystem, Thingstad *et al.* (2008) showed that this depends on the presence or absence of growth-limiting nutrients for both autotrophic and heterotrophic processes. Depending on the nutrient status, enhanced production of organic carbon can even lead to reduced phytoplankton biomass as a result of stimulated bacterial competition for nutrients. Ecological impacts can also be induced by changes in temperature, which may result in shifts among autotrophic and heterotrophic processes (Wohlers *et al.* 2009). Future studies are needed to examine the combined effects of elevated CO_2 and higher temperatures.

12.2.3.3 Remineralization

Most of the exported organic carbon is remineralized in the upper 1000 m, but about 10% escapes to the deep ocean, where it is remineralized or buried in sediments and sequestered from the atmosphere on geological timescales. Therefore, changes in the efficiency with which the organic carbon is transported to depth provide a powerful means to alter the overall efficiency of the soft-tissue pump and thereby alter the air–sea balance of carbon, giving rise to an indirect group 1 type feedback.

The analysis of deep fluxes (water depth > 1000 m) of particulate inorganic and organic carbon suggests a close association of both phases (Armstrong *et al.* 2002; Klaas and Archer 2002). While the exact mechanism behind this observation awaits further elucidation (Passow and De La Rocha 2006), Armstrong *et al.* (2002) proposed that $CaCO_3$ acts as the main carrier phase for POC to the deep ocean. Following their line of thought, $CaCO_3$ would provide POC with excess density (ballasting) thereby increasing its sinking speed. It is also hypothesized that the association between $CaCO_3$ and POC might protect the latter from bacterial degradation. If a control of POC fluxes by $CaCO_3$ is assumed, then a decrease in $CaCO_3$ production would imply less ballasting of POC fluxes, resulting in a decrease of its penetration depth. Particulate organic carbon would be remineralized at shallower depth and the overall efficiency of the biological pump would decrease resulting in a positive feedback to rising atmospheric CO_2.

Barker *et al.* (2003) were the first to address the combined calcification and ballast feedback. Their box model sensitivity study confirms that the ballast effect counteracts the negative feedback of reduced calcification and, depending on the penetration depth of particle fluxes, might overcome it completely. In line with these results, Heinze (2004) reported a positive feedback attributed to a decrease in ballasting of POC fluxes which counteracts the small excess uptake of CO_2 in response to a decrease in $CaCO_3$ production. Taking into account climate change does not modify the picture. By the year 3000, the combined effect of ballasting and reduced calcification yields a negative feedback to atmospheric CO_2 of 50 ppmv compared to 125 ppmv for the calcification feedback only (Hofmann and Schellnhuber 2009). This study projects a strong decrease in meridional overturning circulation in response to climate change, leading to a decrease in ventilation of intermediate water masses. A decrease in penetration depth of POC due to the ballast effect and remineralization of POC at shallower depths will increase the oxygen demand. Physical and biogeochemical processes combine to draw O_2 levels down and promote an extension of oxygen minimum zones. Oxygen minimum zones are sites of intense denitrification, a suboxic metabolic pathway yielding N_2O, a potent greenhouse gas. An increase of the ocean source of N_2O would correspond to a positive feedback on the earth's radiative balance. This example illustrates the potential for cascading effects of ocean acidification running across multiple biogeochemical processes and cycles.

12.3 The marine nitrogen cycle

Ocean acidification affects the marine nitrogen cycle in a myriad of ways. On the one hand, this is a consequence of many biologically mediated transformations of nitrogen-involving pH-dependent redox reactions (Fig. 12.3; Gruber 2008). On the other hand, many of these transformations are mediated by autotrophic organisms that require CO_2 for their growth, so that these organisms may become stimulated by the higher availability of dissolved CO_2 resulting from the uptake of anthropogenic CO_2 from the atmosphere (e.g. Rost *et al.* 2008). Given the intricate and tight connection of the marine nitrogen cycle with those of carbon, phosphorus, oxygen, and many other important biogeochemical elements, any alteration of the marine nitrogen cycle will invariably impact upon the cycles of these other elements, possibly leading to feedbacks to the earth system. While we have just begun to quantitatively understand the impact of ocean acidification on certain isolated processes of the marine nitrogen cycle (e.g. the recent review by Hutchins *et al.* 2009), such as N_2 fixation or nitrification, our knowledge of how these changes interact with each other and affect the other biogeochemical cycles is very poor. These interactions and their potential effects are addressed below, but the discussion and conclusions remain somewhat speculative.

12.3.1 Nitrogen fixation

Nitrogen fixation, the conversion of biologically unavailable N_2 into organic forms of nitrogen, plays a central role in the marine nitrogen cycle, as it resupplies a substantial fraction of the nitrogen that is lost from the biologically available fixed nitrogen pool by denitrification. This process is undertaken by photoautotrophic organisms, of which the cyanobacterium *Trichodesmium* is the best known and studied (Capone *et al.* 1997). With the advent of molecular and genetic tools (Jenkins and Zehr 2008), the number of species known to be able to fix N_2 is rapidly increasing (e.g. Zehr *et al.* 2001; Montoya *et al.* 2004), and the geographical areas where they have been reported to exist is expanding (Moisander *et al.* 2010).

Acidification experiments with *Trichodesmium* cultures have so far yielded a consistent positive

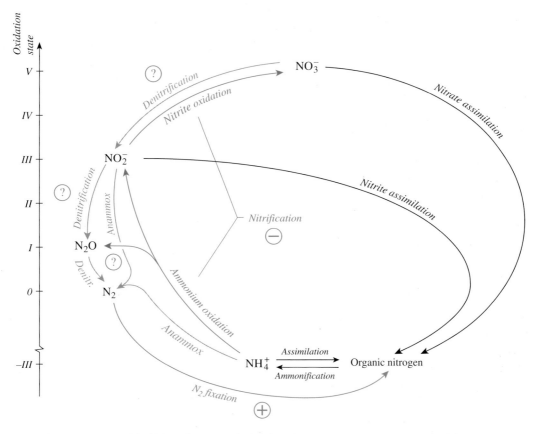

Figure 12.3   The marine nitrogen cycle highlighting the reactions that have been shown to be sensitive to changes in pH and the oceanic carbonate system. + and − indicate reactions stimulated or inhibited by ocean acidification. ?: effect of ocean acidification is unknown. anammox = anaerobic ammonium oxidation. Adapted from Gruber (2008).

response of this organism to elevated CO_2 and lowered pH (e.g. Barcelos e Ramos *et al.* 2007; Hutchins *et al.* 2007; Levitan *et al.* 2007; Kranz *et al.* 2009). As summarized by Hutchins *et al.* (2009), the mean enhancement of the rate of N_2 fixation at elevated CO_2 levels (750–1000 µatm) from the six reported experiments with isolated cultures amounts to about 40 ± 20%. The only experiment with natural populations of *Trichodesmium* collected from the Gulf of Mexico also yielded increased fixation rates, although of smaller amplitude, i.e. between 6 and 41% (Hutchins *et al.* 2009). The enhanced rate of N_2 fixation tends to be accompanied by enhanced rates of carbon fixation, so that the mean C:N ratio of these organisms remained relatively constant (Barcelos e Ramos *et al.* 2007). This suggests that the primary mechanism responsible for the enhanced

rate of N_2 fixation is the CO_2-induced increase in photosynthesis, providing additional energy for the fixation of N_2.

Experiments with non-*Trichodesmium* N_2 fixers are rare, in large part due to the fact that the majority of the identified species have not been cultured yet (Carpenter and Capone 2008). In the case of *Crocosphaera watsonii*, an enhancement was found that is very similar to that identified for *Trichodesmium* (Fu *et al.* 2008). In contrast, *Nodularia spumigena* exhibited almost no response to ocean acidification (Czerny *et al.* 2009). Thus, with the exception of one species, the majority of experiments point toward an enhancement of pelagic marine N_2 fixation in response to ocean acidification. However, given the fact that the majority of the experiments were performed with the same strain of *Trichodesmium*, it is premature to draw firm conclusions. If it is nevertheless assumed that all nitrogen fixers are stimulated by ocean acidification, the global rate of marine N_2 fixation may increase by more than 50 Tg N yr^{-1} by 2100. Relative to present-day rates of the order of 120 ± 50 Tg N yr^{-1} (Gruber 2008), this represents a substantial acceleration of the rate of input of fixed nitrogen into the marine fixed nitrogen pool.

In nitrogen-limited regions of the ocean, such an addition has the potential to substantially increase marine productivity and thereby increase the export of organic carbon from the surface to the interior. This tends to enhance the uptake of CO_2 from the atmosphere, i.e. this is a negative indirect group 2 type feedback (Table 12.1).

12.3.2 Nitrification

Nitrification, the combined processes of the oxidation of ammonium (NH_4^+) to nitrite (NO_2^-), and the oxidation of nitrite to nitrate (NO_3^-) is undertaken by two distinct classes of chemoautotrophic organisms which use the chemical energy released from these two processes as their source of energy: *Nitrosomonas* is responsible for the first oxidation step, i.e. the conversion from ammonium to nitrite, while *Nitrobacter* oxidizes nitrite to nitrate. Nitrification is inhibited by light, so it tends to be a dominant process only in the aphotic zone. However, a recent compilation by Yool *et al.* (2007) shows that a substantial amount of nitrification also occurs in the near-surface ocean.

Ocean acidification studies on marine nitrifiers are not as abundant as those that have been undertaken on marine N_2 fixers, but tend to show a negative response, i.e. reduced rates at lower pH and elevated CO_2 (Huesemann *et al.* 2002). This may be due to two effects (Hutchins *et al.* 2009). First, in the case of ammonium oxidation by *Nitrosomonas*, substrate limitation may be responsible. This is because these organisms prefer NH_3, which becomes substantially less abundant relative to NH_4^+ in a more acidified ocean (pK of NH_3/NH_4^+ is around 9.2). Second, both *Nitrosomonas* and *Nitrobacter* may be responding directly to the increased concentration of dissolved CO_2, which is their substrate for the synthesis of organic matter.

Fundamentally, as is the case for photoautotrophic organisms, one would expect a positive effect though, i.e. enhanced fixation of CO_2 due to a higher efficiency of the CO_2-fixing enzyme RubisCO (Rost *et al.* 2008).

Using results from the highly limited experiments, Hutchins *et al.* (2009) estimated a global-mean reduction of nitrification by about 10% for a business-as-usual scenario. This would result in a substantial shift of fixed nitrogen in the near-surface waters from the nitrate pool to the ammonium pool. However, the implications of this shift for earth system feedbacks are likely to be small because it does not directly alter the amount of fixed nitrogen in the ocean, keeping the impact on total production small. However, it is conceivable that this will cause alterations of the phytoplankton community structure, since not all phytoplankton can take up nitrate and ammonium equally well. For example, one could expect that species that are highly adapted to low nitrate and high ammonium concentrations, such as *Prochlorococcus* (Moore *et al.* 2002), would benefit at the expense of other phytoplankton that are more adapted to high nitrate and low ammonium concentrations (e.g. many diatoms).

A potentially larger effect may occur in shallow coastal sediments that are overlain by waters with relatively low nitrate concentrations. Here, the reduction of nitrate may cause a substrate-driven reduction in the rates of sedimentary denitrification (Blackford and Gilbert 2007). Given the fact that sedimentary denitrification is the largest sink for fixed nitrogen in the ocean (Gruber 2008), a reduction in this sink would increase the pool of fixed nitrogen in the ocean and hence probably increase biological productivity. However, there is little evidence that sedimentary denitrification is limited by nitrate on a global scale (Middelburg *et al.* 1996), so that this may remain a more local effect.

In summary, changes in nitrification induced by ocean acidification may lead to several indirect group 2 feedbacks. The direction of these feedbacks is unclear, but their magnitudes appear to be small. Given the limited extent of the studies undertaken so far, this conclusion remains tentative.

12.3.3 Denitrification

Denitrification is a dissimilatory process that occurs only at extremely low to non-existent oxygen concentrations, but at appreciable levels of nitrate. Under such circumstances, heterotrophic microorganisms can use nitrate rather than dissolved oxygen as a terminal electron acceptor, i.e. they 'breathe' nitrate instead of oxygen. The threshold for the onset of this process occurs at oxygen concentrations of around 5 to 10 μmol kg^{-1}, or at the boundary between hypoxia and suboxia. In the present-day ocean, this condition is only met in the water column at a few locations, namely the eastern tropical North and South Pacific, the Arabian Sea, and a few more localized coastal regions. In contrast, many sediments underlying productive regions are completely void of oxygen (anoxic) at a depth of a few centimetres, making them sites of intense denitrification. Globally, denitrification is two times higher in the sediments than in the water column (180 ± 50 Tg N yr^{-1} vs 65 ± 20 Tg N yr^{-1}), although there exists a considerable amount of debate about the exact magnitude of these processes (Gruber 2008).

Although so far unsupported by direct experiments, Hutchins *et al.* (2009) suggest that the heterotrophic microorganisms responsible for denitrification are not directly affected by ocean acidification. This is consistent with the fact that these organisms live and thrive in anoxic regions that naturally have a much lower pH than the rest of the ocean. However, ocean acidification-induced changes in the flux of organic matter that is entering such anoxic regions and changes in the extent and location of anoxic regions can lead to very substantial changes in marine denitrification, both in the water column and in the sediments. This may occur as a result of several mechanisms: first, a shallower remineralization of the organic matter sinking downward in response to reduced ballasting may lead to a higher oxygen demand in shallow waters at the expense of a smaller oxygen demand in deep waters (see Section 12.2.3.3). This may cause an expansion of anoxia, since it leads to the enhanced removal of oxygen in the upper thermocline where oxygen is already low in many locations (e.g. Keeling *et al.* 2010), while reducing the oxygen

demand further down the water column, where oxygen concentration tends to be higher. A second mechanism is the increase of the C:N ratio induced by ocean acidification, which enhances the downward flux of organic carbon, thereby increasing the oxygen demand throughout the water column (see Section 12.2.3.2).

Hofmann and Schellnhuber (2009) simulated the impact of the first mechanism in a global model for a business-as-usual scenario, and found a 75% reduction in the export of $CaCO_3$, a very substantial expansion of hypoxia (i.e. O_2 concentrations below 60 μmol kg^{-1}), and a moderate increase of anoxic/suboxic regions. Their model did not include denitrification, but the relatively small expansion of anoxic/suboxic waters would suggest that denitrification would not have increased substantially.

Oschlies *et al.* (2008) investigated the effect of changes in the C:N ratio on marine oxygen in a global model using a business-as-usual scenario, and found a 50% increase in the ocean volume with suboxic conditions by the end of this century, which caused an increase in water-column denitrification of about 60 Tg N yr^{-1}. If sustained, such a loss of fixed nitrogen would lead to a decrease in the marine nitrogen inventory of the order of 10% in 1000 years, causing a corresponding decrease in marine productivity.

In conclusion, while denitrification is probably not directly affected by ocean acidification, it plays an important role in the sequence of processes that may ultimately cause substantial changes in the earth system.

12.3.4 Nitrous oxide production

Nitrous oxide (N_2O) is produced in the ocean through at least two pathways (Fig.12.3; Gruber 2008). It is an intermediary product of denitrification, and under suboxic, but not completely anoxic, conditions its further reduction to N_2 tends to proceed less efficiently. N_2O is also produced during the oxidation of ammonium, and the fraction of the ammonium transformed into N_2O instead of nitrite also tends to increase with lower oxygen concentrations. Most of the N_2O produced in the ocean is emitted to the atmosphere, as only a small fraction is consumed in the anoxic regions of the ocean.

While there is no experimental evidence that ocean acidification will affect N_2O production directly, the decrease in oceanic oxygen induced by ocean acidification is bound to increase N_2O production substantially. The reduced ballast effect could lead to higher rates of nitrification in low-oxygen regions which would substantially enhance N_2O production associated with the nitrification pathway (Jin and Gruber 2003). The expansion of the anoxic regions caused by the altered C:N stoichiometry is likely to accelerate N_2O production by the denitrification pathway. To date, no simulation has been undertaken to quantify this effect, but it is conceivable that oceanic N_2O emissions could double in response to a doubling of the ocean's anoxic regions.

12.3.5   Interactive effects—the future marine nitrogen cycle

None of the above processes operate in isolation. Although the degree of coupling is intensively debated, nitrogen fixation and denitrification tend to be coupled (Deutsch *et al.* 2007; Gruber 2008). In addition, the marine oxygen content is likely to decrease substantially in response to global warming, irrespective of changes induced by ocean acidification. So what will be the response of the system as a whole? To date, no study has attempted to look at these nitrogen cycle-driven feedbacks holistically and in depth, so one can only provide a qualitative assessment. In addition, the answer is by nature speculative, since our understanding of how the different processes interact with each other is poor.

The marine nitrogen cycle will probably be accelerated in a high-CO_2 ocean, with a substantially elevated rate of marine nitrogen fixation and a higher rate of (water-column) denitrification. This will decrease the mean residence time of fixed nitrogen in the ocean. Given the enhanced rates of sources and sinks, it is not possible to conclude anything about the potential generation of imbalances which are required to cause net changes in the oceanic fixed nitrogen inventory and changes in the biological pump that could alter the air–sea CO_2 balance. It appears, however, that imbalances are not very likely and that ocean acidification-induced feedbacks to the earth system involving N_2 fixation

and denitrification will not become large at the global scale. On a regional level, and from the perspective of marine organisms that depend on sufficient oxygen levels to live, these changes will be relevant nevertheless—it is just that their impact on radiative forcing in the atmosphere will probably not be substantial. However, the level of confidence in this statement is very low. The story is different for N_2O because evidence is mounting that ocean acidification will increase its production and emission into the atmosphere.

12.4   The ocean as a source of atmospherically active trace gases

In addition to being a source or a sink of some major greenhouse gases (e.g. CO_2 and N_2O), the ocean is also a source of climatically active trace gases to the marine atmosphere (see Chapter 11 for detailed information). Among them, dimethyl sulphide (DMS) is a gaseous sulphur compound produced by marine biota in surface seawater. Once emitted to the atmosphere, it undergoes rapid oxidation to produce particles that can modify the optical properties of clouds, thereby influencing climate. Iodo- and bromocarbon gases are also produced in surface seawater and can be outgassed to the atmosphere. They represent a major source of halogens to the marine atmosphere where their oxidation can produce reactive radicals. These radicals play a role in the photochemical loss of tropospheric ozone (a major greenhouse gas), but also in the regulation of stratospheric ozone. In addition, they can contribute to particle formation and modify the optical properties of clouds, thereby affecting climate.

For more than 20 years now, it has been proposed that marine emissions of DMS are sensitive to climatic change and that the radiative budget of the earth is in turn sensitive to modifications of the marine DMS source (Charlson *et al.* 1987). Climate model simulations suggested that a 50% decrease in DMS emissions could result in a net increase in the mean surface temperature of 1.6°C (Gunson *et al.* 2006). Changes in the production and in the sea-to-air flux of DMS resulting from ocean acidification could thus have a significant impact on climate and hence form an indirect group 1 type feedback. A detailed presentation of the chemistry of DMS and the role of marine biota in controlling its cycle is given in Chapter 11. Impacts of ocean acidification on DMS emissions and implications for climate feedbacks are still largely unknown. Published studies report contrasting results. Potential effects have so far not been included in coupled climate–marine biogeochemistry models.

12.5   Conclusion and perspectives

This chapter has addressed biogeochemical impacts for which experimental evidence is available and which, with the exception of atmospheric trace gases, have been addressed in model studies. As a result, we have largely focused on the C and N cycles. It is likely that other major cycles also will be affected by ocean acidification either directly (Fe; Shi *et al.* 2010; Breitbarth *et al.* 2010) or indirectly through changes in elemental ratios of export production (P and Si; Hutchins *et al.* 2009 and references therein). At present, however, experimental evidence is too sparse, and often contradictory, to allow inferences of biogeochemical consequences across the whole spectrum of elements.

Not surprisingly, early research efforts on the impacts of ocean acidification focused on marine calcification. Because of its seemingly straightforward incorporation into biogeochemical models and the negative feedback to atmospheric CO_2 levels associated with its decrease, this particular impact was rapidly implemented in global ocean biogeochemical models. Ocean acidification research is a field of rapidly expanding knowledge, and recent evidence suggests that the response of marine calcifiers to changes in carbonate chemistry is more complex than originally expected. Within-species and between-species variability, as well as processes such as acclimatization and adaptation, challenge the approaches currently used in biogeochemical modelling. These mostly rely on the representation of a limited number of plankton groups, to which major biogeochemical functions are assigned, e.g. $CaCO_3$ production, biogenic silica formation and N_2 fixation. If environmental conditions (irradiance, temperature, nutrients) cross a critical threshold for a given plankton functional type, it disappears from the model world. There is no shift between species within a

given plankton functional type, nor potential for acclimation and adaptation. The increase in functional types in an effort to better capture ecosystem complexity is limited by the lack of available data to constrain model parameterizations. The evaluation of the impacts of ocean acidification and climate change might well require the development of a new generation of ecosystem models (Follows *et al.* 2007; Barton *et al.* 2010) or optimality-based adaptive models (Bruggeman and Kooijman 2007; Pahlow *et al.* 2008).

Elemental cycles are tightly coupled through reactions involved in organic matter synthesis and remineralization. One has just begun to evaluate how changes in one cycle spread to another as exemplified by studies linking changes in the export efficiency of the biological pump either mediated by $CaCO_3$ production (Hofmann and Schellnhuber 2009) or export stoichiometry (Oschlies *et al.* 2008) to ocean oxygen inventory and the nitrogen cycle. Clearly, the study of biogeochemical impacts of ocean acidification, its interaction with climate change, and feedbacks to the earth system are at an early stage.

12.6   Acknowledgements

We acknowledge financial support from grant GOCE-511176 (EU FP6 RTP project CARBOOCEAN) and grant 211384 (EU FP7 RTP project EPOCA) provided by the European Commission.

References

Archer, D. (2005). Fate of fossil fuel CO_2 in geologic time. *Journal of Geophysical Research*, **110**, C09S05, doi:10.1029/2004JC002625.

Archer, D. and Maier-Reimer, E. (1994). Effect of deep-sea sedimentary calcite preservation on atmospheric CO_2 concentration. *Nature*, **367**, 260–3.

Arrigo, K. (2007). Carbon cycle: marine manipulations. *Nature*, **450**, 491–2.

Armstrong, R.A., Lee, C., Hedges, J.I., Honjo, S., and Wakeham, S.G. (2002). A new, mechanistic model for organic carbon fluxes in the ocean based on the quantitative association of POC with ballast minerals. *Deep-Sea Research II*, **49**, 219–36.

Barcelos e Ramos, J., Biswas, H., Schulz, K.G., LaRoche, J., and Riebesell, U. (2007). Effect of rising atmospheric carbon dioxide on the marine nitrogen fixer *Trichodesmium*. *Global Biogeochemical Cycles*, **21**, GB2028, doi:10.1029/2006GB002898.

Barker, S., Higgins, J.A., and Elderfield, H. (2003). The future of the carbon cycle: review, calcification response, ballast and feedback on atmospheric CO_2. *Philosophical Transactions of the Royal Society A: Mathematical, Physical and Engineering Sciences*, **361**, 1977–1998.

Barton, A.D., Dutkiewicz, S., Flierl, G., Bragg, J., and Follows, M.J. (2010). Patterns of diversity in marine phytoplankton. *Science*, **327**, 1509–11.

Blackford, J.C. and Gilbert, F.J. (2007). pH variability and CO_2 induced acidification in the North Sea. *Journal of Marine Systems*, **64**, 229–41.

Bopp, L., Aumont, O., Cadule, P., Alvain, S., and Gehlen, M. (2005). Response of diatoms distribution to global warming and potential implications: a global model study. *Geophysical Research Letters*, **32**, L19606, doi:10.1029/2005GL023653.

Breitbarth, E., Bellerby, R.J., Neill, C.C. *et al.* (2010). Ocean acidification affects iron speciation during a coastal seawater mesocosm experiment. *Biogeosciences*, **7**, 1065–73.

Brewer, P.G. and Peltzer, E.T. (2009). Limits to marine life. *Science*, **324**, 347–8.

Bruggeman, J. and Kooijman, S.A.L.M. (2007). A biodiversity-inspired approach to aquatic ecosystem modeling. *Limnology and Oceanography*, **52**, 1533–44.

Caldeira, K. and Wickett, M.E. (2003). Anthropogenic carbon and ocean pH. *Nature*, **425**, 365.

Capone, D.G., Zehr, J.P., Paerl, H.W., Bergman, B., and Carpenter, E.J. (1997). *Trichodesmium*, a globally significant marine cyanobacterium. *Science*, **276**, 1221–9.

Carpenter, E.J. and Capone, D.G. (2008). Nitrogen fixation in the marine environment. In: D.G. Capone, D.A. Bronk, M.R. Mulholland, and E.J. Carpenter (eds), *Nitrogen in the marine environment*, pp. 141–98. Elsevier, Amsterdam.

Charlson, R.J., Lovelock, J.E., Andreae, M.O., and Warren, S.G. (1987). Oceanic phytoplankton, atmospheric sulfur, cloud albedo and climate. *Nature*, **326**, 655–61.

Czerny, J., Barcelos e Ramos, J., and Riebesell, U. (2009). Influence of elevated CO_2 concentrations on cell division and nitrogen fixation rates in the bloomforming cyanobacterium *Nodularia spumigena*. *Biogeosciences*, **6**, 1865–75.

Deutsch, C., Sarmiento, J.L., Sigman, D.M., Gruber, N., and Dunne, J.P. (2007). Spatial coupling of nitrogen inputs and losses in the ocean, *Nature*, **445**, 163–7.

Fabry, V.J., Seibel, B.A., Feely, R.A., and Orr, J.C. (2008). Impacts of ocean acidification on marine fauna and ecosystem processes. *ICES Journal of Marine Sciences*, **65**, 414–32.

Falkowski, P., Scholes, R.J., Boyle, E. *et al.* (2000). The global carbon cycle: a test of our knowledge of Earth as a system. *Science*, **290**, 291–6.

Follows, M.J., Dutkiewicz, S., Grant, S., and Chisholm, S.W. (2007). Emergent biogeography of microbial communities in a model ocean. *Science*, **315**, 1843–6.

Friedlingstein P., Cox, P., Betts, R. *et al.* (2006). Climate-carbon cycle feedback analysis: results from the C4MIP model intercomparison. *Journal of Climate*, **19**, 3337–53.

Fu, F-X., Mulholland, M.R., Garcia, N.S. *et al.* (2008). Interactions between changing pCO_2, N_2 fixation, and Fe limitation in the marine unicellular cyanobacterium *Crocosphaera*. *Limnology and Oceanography*, **53**, 2472–2484.

Gangstø, R., Gehlen, M., Schneider, B., Bopp, L., Aumont, O., and Joos, F. (2008). Modeling the marine aragonite cycle: changes under rising carbon dioxide and its role in shallow water $CaCO_3$ dissolution. *Biogeosciences*, **5**, 1057–72.

Gangstø R. Joos, F. and Gehlen, M. (2011). Sensitivity of pelagic calcification to ocean acidification, *Biogeosciences*, **8**, 433–58.

Gehlen, M., Bassinot, F., Chou, L., and McCorkle, D. (2005). Reassessing the dissolution of marine carbonates: II. Reaction kinetics. *Deep-Sea Research I*, **52**, 1461–76.

Gehlen, M., Gangstø, R., Schneider, B., Bopp, L., Aumont, O., and Ethé, C. (2007). The fate of pelagic $CaCO_3$ production in a high CO_2 ocean: a model study. *Biogeosciences*, **4**, 505–19.

Gruber, N. (2008). The marine nitrogen cycle: overview and challenges. In: D.G. Capone, D.A. Bronk, M.R. Mulholland, and E.J. Carpenter (eds), *Nitrogen in the marine environment*, pp. 1–50. Elsevier, Amsterdam.

Gruber, N. and Sarmiento, J.L. (2002). Large-scale biogeochemical/physical interactions in elemental cycles. In: A.R. Robinson, J.J. McCarthy, and B.J. Rothschild (eds), *The sea: biological–physical interactions in the oceans*, pp. 337–99. John Wiley and Sons, Inc., New York.

Gunson, J.R., Spall, S.A., Anderson, T.R., Jones, A., Totterdell, I.J., and Woodage, M.J. (2006). Climate sensitivity to ocean dimethylsulphide emissions. *Geophysical Research Letters*, **33**, L07701, doi:10.1029/2005GL024982.

Hales, B. and Emerson, S. (1997). Evidence in support of first order dissolution kinetics of calcite in seawater. *Earth and Planetary Science Letters*, **148**, 317–27.

Heinze, C. (2004). Simulating oceanic $CaCO_3$ export production in the greenhouse. *Geophysical Research Letters*, **31**, L16308, doi:10,1029/2004GL020613.

Hofmann, M. and Schellnhuber, H.-J. (2009). Oceanic acidification affects marine carbon pump and triggers extended marine oxygen holes. *Proceedings of the National Academy of Sciences USA*, **106**, 3017–22.

Huesemann, M.H., Skillman, A.D., and Crecelius, E.A. (2002). The inhibition of marine nitrification by ocean disposal of carbon dioxide. *Marine Pollution Bulletin*, **44**, 142–8.

Hutchins, D.A., Fu, F.-X., Zhang, Y. *et al.* (2007). CO_2 control of *Trichodesmium* N_2 fixation, photosynthesis, growth rates and elemental ratios: implications for past, present and future ocean biogeochemistry. *Limnology and Oceanography*, **52**, 1293–304.

Hutchins, D.A., Mulholland, M.R., and Fu, F. (2009). Nutrient cycles and marine microbes in a CO_2-enriched ocean. *Oceanography*, **22**, 128–45.

Iglesias-Rodriguez, M.D., Halloran, P.R., Rickaby, R.E.M. *et al.* (2008). Phytoplankton calcification in a high-CO_2 world. *Science*, **320**, 336–40.

Jenkins, B.D. and Zehr, J.P. (2008). Molecular approaches to the nitrogen cycle. In: D.G. Capone, D.A. Bronk, M.R. Mulholland, and E.J. Carpenter (eds), *Nitrogen in the marine environment*, pp. 1303–44. Elsevier, Amsterdam.

Jin, X. and Gruber, N. (2003). Offsetting the radiative benefit of ocean iron fertilization by enhancing N_2O emissions. *Geophysical Research Letters*, **30**, 2249, doi: 10.1029/2003GL018458.

Jin, X., Gruber, N., Frenzel, H., Doney, S.C., and McWilliams, J.C. (2008). The impact on atmospheric CO_2 of iron fertilization induced changes in the ocean's biological pump. *Biogeosciences*, **5**, 385–406.

Keeling, R.F., Körtzinger, A., and Gruber, N. (2010). Ocean deoxygenation in a warming world. *Annual Review of Marine Science*, **2**, 199–229.

Keir, R.S. (1980). The dissolution kinetics of biogenic carbonate in seawater. *Geochimica et Cosmochimica Acta*, **44**, 241–52.

Khatiwala, S., Primeau, F., and Hall, T. (2009). Reconstruction of the history of anthropogenic CO_2 concentrations in the ocean. *Nature*, **462**, 346–9.

Klaas, C. and Archer, D.E. (2002). Association of sinking organic matter with various types of mineral ballast in the deep sea: implications for the rain ratio. *Global Biogeochemical Cycles*, **16**, 1116, doi:10.1029/2001GB001765.

Kranz, S.A., Sültemeyer, D., Richter, K.-U., and Rost, B. (2009). Carbon acquisition by *Trichodesmium*: the effect of pCO_2 and diurnal changes. *Limnology and Oceanography*, **54**, 548–59.

Levitan, O., Rosenberg, G., Setlik, I. *et al.* (2007). Elevated CO_2 enhances nitrogen fixation and growth in the marine cyanobacterium *Trichodesmium*. *Global Change Biology*, **13**, 531–8.

Middelburg, J.J., Soetaert, K., Herman, P.M.J., and Heip, C.H.R. (1996). Denitrification in marine sediments: a model study. *Global Biogeochemical Cycles*, **10**, 661–73.

Moisander, P.H., Beinart, R.A., Hewson, I. *et al.* (2010). Unicellular cyanobacterial distributions broaden the oceanic N_2 fixation domain. *Science*, **327**, 1512–14.

Montoya, J.P., Holl, C.M., Zehr, J.P., Hansen, A., Villareal, T.A., and Capone, D.G. (2004). High rates of N_2-fixation by unicellular diazotrophs in the oligotrophic Pacific. *Nature*, **430**, 1027–32.

Morse, J.W., Arvidson, R.S., and Lüttge, A. (2007). Calcium carbonate formation and dissolution. *Chemical Reviews*, **107**, 342–81.

Moore, L.R., Post, A.F., Rocap, G., and Chisholm, S.W. (2002). Utilization of different nitrogen sources by the marine cyanobacteria *Prochlorococcus* and *Synechococcus*. *Limnology and Oceanography*, **47**, 989–96.

Oschlies, A. (2009). Impact of atmospheric and terrestrial CO_2 feedbacks on fertilization-induced marine carbon uptake. *Biogeosciences*, **6**, 1603–13.

Oschlies, A., Schulz, K.G., Riebesell, U., and Schmittner, A. (2008). Simulated 21st century's increase in oceanic suboxia by CO_2 enhanced biotic carbon export. *Global Biogeochemical Cycles*, **22**, GB4008, doi: 10.1029/2007GB003147.

Pahlow, M., Vezina, A.F., Casault, B. *et al.* (2008). Adaptive model of plankton dynamics for the North Atlantic. *Progress in Oceanography*, **76**, 151–91.

Passow, U. and De La Rocha, C.L. (2006). Accumulation of mineral ballast on organic aggregates. *Global Biogeochemical Cycles*, **20**, GB1013, doi:10.1029/2005GB002579.

Ridgwell, A. and Hargreaves, J.C. (2007). Regulation of atmospheric CO_2 by deep-sea sediments in an Earth System Model. *Global Biogeochemical Cycles*, **21**, GB2008, doi: 10.1029/2006GB002764.

Ridgwell, A., Zondervan, I., Hargreaves, J.C., Bijma, J., and Lenton, T.M. (2007). Assessing the potential long-term increase of oceanic fossil fuel CO_2 uptake due to CO_2-calcification feedback. *Biogeosciences*, **4**, 481–92.

Ridgwell, A., Schmidt, D.N., Turley, C. *et al.* (2009). From laboratory manipulations to Earth system models: scaling calcification impacts of ocean acidification. *Biogeosciences*, **6**, 2611–23.

Riebesell, U., Schulz, K.G., Bellerby, R.G.J. *et al.* (2007). Enhanced biological carbon consumption in a high CO_2 ocean. *Nature*, **450**, 545–8.

Rost, B., Zondervan, I., and Wolf-Gladrow, D. (2008). Sensitivity of phytoplankton to future changes in ocean carbonate chemistry: current knowledge, contradictions and research directions. *Marine Ecology Progress Series*, **373**, 227–37.

Sarmiento, J.L., Le Quéré, C., and Pacala, S.W. (1995). Limiting future atmospheric carbon dioxide. *Global Biogeochemical Cycles*, **9**, 121–37.

Sarmiento, J.L., Hughes, T.M.C., Stouffer, R.J., and Manabe, S. (1998). Simulated response of the ocean carbon cycle to anthropogenic climate warming. *Nature*, **393**, 245–9.

Schmittner, A., Oschlies, A., Matthews, H.D., and Galbraith, E.D. (2008). Future changes in climate, ocean circulation, ecosystems, and biogeochemical cycling simulated for a business-as-usual CO_2 emission scenario until year 4000 AD. *Global Biogeochemical Cycles*, **22**, GB1013, doi: 10.1029/2007GB002953.

Shi, D., Xu, Y., Hopkinson, B.M., and Morel, F.M.M. (2010). Effect of ocean acidification on iron availability to marine phytoplankton. *Science*, **327**, 676–9.

Steinacher, M., Joos, F., Frölicher, T.L. *et al.* (2010). Projected 21st century decrease in marine productivity: a multi-model analysis. *Biogeosciences*, **7**, 979–1005.

Sundquist, E.T. (1990). Influence of deep-sea benthic processes on atmospheric CO_2. *Philosophical Transactions of the Royal Society A: Mathematical, Physical and Engineering Sciences*, **331**, 155–65.

Thingstad, T.F., Bellerby, R.G.J., Bratbak, G. *et al.* (2008). Counterintuitive carbon-to-nutrient coupling in an Arctic pelagic ecosystem. *Nature*, **455**, 387–90.

Volk, T. and Hoffert, M.I. (1985). Ocean carbon pumps: analysis of relative strengths and efficiencies in ocean-driven atmospheric CO_2 changes. In: E.T. Sundquist and W.S. Broecker (eds), *The carbon cycle and atmospheric CO_2: natural variations archean to present*, pp. 99–110. American Geophysical Union, Washington, DC.

Wohlers, J., Engel, A., Zöllner, E. *et al.* (2009). Changes in biogenic carbon flow in response to sea surface warming. *Proceedings of the National Academy of Sciences USA*, **106**, 7067–72.

Yool, A., Martin, A.P., Fernández, C., and Clark, D.R. (2007). The significance of nitrification for oceanic new production. *Nature*, **447**, 999–1002.

Zehr, J.P., Waterbury, J.B., Turner, P.J. *et al.* (2001). Unicellular cyanobacteria fix N_2 in the subtropical North Pacific Ocean. *Nature*, **412**, 635–8.

Zondervan, I., Zeebe, R.E., Rost, B., and Riebesell, U. (2001). Decreasing marine biogenic calcification: a negative feedback on rising atmospheric pCO_2. *Global Biogeochemical Cycles*, **15**, 507–16.

CHAPTER 13

The ocean acidification challenges facing science and society

Carol Turley and Kelvin Boot

13.1 Introduction

Human development, inspiration, invention, and aspiration have resulted in a rapidly growing population, with each generation aspiring to greater wealth and well-being, so having greater needs than the previous generation (Fig. 13.1). Amongst the resulting negative impacts are over-exploitation of planetary resources and the build-up of gases in the atmosphere and oceans to the extent that they are changing earth's climate and ocean chemistry (IPCC 2007).

However, the history of humanity's relationship to the environment has shown that, if threatened, society can respond rapidly to environmental risks, introducing better practices, controls, regulations, and even global protocols, for example the reduction of city smog, the move from leaded to unleaded petrol, and reduction of chlorofluorocarbon (CFC) production to reduce loss of the ozone layer. Nearly all of these changes have led to direct and obvious positive gain to human health and well-being which has been a driving force in the production, agreement and implementation of the policies and laws that have brought them about. The spatial scale or 'ecological footprint' of these risks has increased with time, such that international agreements and protocols, like the Montreal Protocol for CFCs, have been increasingly necessary for reducing them. Along with the globalization of agriculture, business, industry, and financial markets and the expansion of the human population goes the globalization of risk to the environment. Climate change and ocean acidification are global issues with solutions that are only possible through global agreements and action.

Substantial proportions of nations' gross domestic product (GDP) were used to secure the banks and major industries in the economic crises that have swept the world in the last few years, far greater than the 1 to 2% per annum estimated to be required to mitigate climate change (Stern 2006). However, the response to the economic crisis does show that global society can react rapidly when it believes it is necessary. The question is, when do society and governments deem it necessary to act, and to act together? One issue may be time, the perceived immediacy of the crisis. While the economic crisis of 2007 to the present came upon us quickly and globally, climate change and ocean acidification have a stealth that does not perhaps attract public and government concern so immediately. Rather, they are perceived to be something to worry about in the far future, at least beyond the political cycles of governments, if at all.

Despite all the scientific evidence gathered and assimilated through collaboration across disciplinary and territorial borders, and all the political effort at the climate change negotiations, there is not yet a global treaty on emissions reductions. The scientific evidence is strong (IPCC 2007), the risks are high, and there is belief in many sectors that the need for action is increasingly urgent. The future of the planet as we know it and the lives of the millions who are supported by the earth's biosphere are greatly influenced by politics and economics. Ocean acidification and its potential risks are further factors supporting actions for CO_2 emissions reduction (see Chapter 14; Turley and Findlay 2009). However, politicians use multiple inputs in decision-making, scientific evidence being just one of them. The role of scientists is to build the strength of

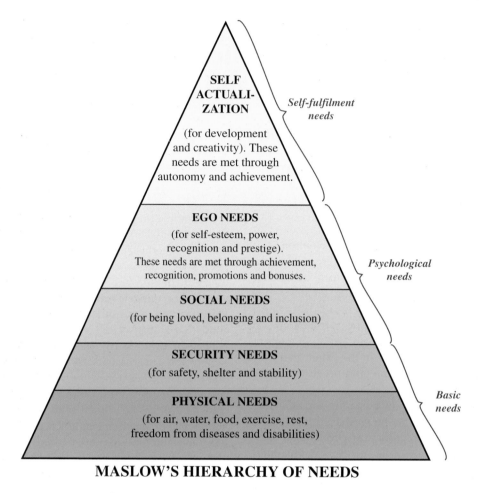

MASLOW'S HIERARCHY OF NEEDS

Figure 13.1 The American psychologist Abraham Maslow outlined a hierarchy of five levels of basic needs. Maslow believed that fulfilling these needs helped sustain human behaviour (amended from Maslow 1943).

the evidence based on good-quality science that is useful to the decision-making process. To understand what is relevant to policy, scientists must talk to policymakers. To understand the strength of the evidence and the uncertainties of the scientific process, policymakers must talk to scientists. And both must talk to the general public.

13.2 Why society should be concerned about ocean acidification

Ocean acidification is caused by uptake of anthropogenic CO_2 by the ocean (see Chapter 1); it is a global phenomenon and is happening now, it is measurable, and it will continue as more CO_2 is emitted (see Chapter 3). Already ocean acidity has increased by 30%, and by 2100, if we continue emitting CO_2 at the same rate, ocean acidity will have increased by 150% during this century (Orr et al. 2005; Chapter 1). Such a substantial alteration in basic ocean chemistry is likely to have wide implications for life in the ocean, especially for those food webs and ecosystems vulnerable to future levels of ocean acidification (detailed in Chapters 5–10). Tropical coral reef ecosystems, for example, may be irreparably damaged (Cao and Caldeira 2008; Hoegh-Guldberg et al. 2008; Chapter 7). These three-dimensional structures provide the habitat within which an estimated 1 million species live, and provide food, income, and coastal protection and well-being for around 500 million people throughout tropical coastal areas of the world (Hoegh-Guldberg *et al.* 2008).

Polar, subpolar, and deep-sea ecosystems are also at risk as ocean acidification will be most severe there (Orr *et al.* 2005; Guinotte *et al.* 2006; Turley *et al.* 2007, 2010a; Steinacher *et al.* 2009) and organisms playing important roles in those ecosystems are particularly vulnerable, for example the mollusc *Limacina helicina*, a key link in the polar and subpolar food chain, or the deep-water coral *Lophelia pertusa* (Maier *et al.* 2009), which is key in creating important deep-sea ecosystems.

Shallow productive seas are also vulnerable (e.g. Blackford and Gilbert 2007; Feely *et al.* 2008), with a growing list of organisms which may be affected by ocean acidification in a number of ways at some stage during their life history. As many of these provide food (Table 13.1) there may be a risk to food security (see Box 13.1).

The world's eyes were on Copenhagen in December 2009 where the 15th Conference of the Parties (COP15) to the UN Framework Convention on Climate Change (UNFCCC) met to negotiate future agreements for mitigating climate change. A binding agreement was not reached, rather 'The Copenhagen Accord', according to which the increase in global temperatures should be no greater than 2°C, was agreed by most parties although there was no firm resolution to reduce CO_2 emissions. The accord was effectively one of intent only and is not legally binding, parties to the accord merely agreed to 'take note' of it.

Post-Copenhagen, as the world moves towards the next round of negotiations, it seems sensible to take into account not only the impact of CO_2 emissions on the climate, but also its impacts on ocean chemistry and the value of the oceans.

13.3   Valuing the oceans

It is difficult and controversial to put a monetary value on ecosystem services because they are neither fully a part of commercial markets nor as easily quantifiable as economic services or manufacturing output. For this reason they do not currently carry

as much weight in policy decisions. However, in 1997, marine ecosystem services were estimated to have a total value of US$21 trillion per annum, representing 63% of the total value of earth's ecosystem services of US$33 trillion (Costanza *et al.* 1997). Whilst the accuracy of this figure has been hotly debated, there is no doubt that the marine environment accounts for a substantial proportion of the world's ecosystem services.

13.3.1   Provision of food and food products

The Food and Agriculture Organization (FAO) of the United Nations has monitored world fisheries since 1950. The term 'fish' includes finfish and other creatures of oceanic origin that are consumed by humans, including molluscs, crustaceans, echinoderms, and many other smaller animal groups. These invertebrates make up 15% of global marine catch.

Fish provide 16% of annual protein consumption for around 3 billion people worldwide (Fig. 13.2; FAO 2003), and global fish production in 2004 was valued at US$150 billion per annum (Kite-Powell 2009). Fish is the primary protein source for about 1 billion people. In low-income food-deficient countries fisheries can make up 22% of animal protein consumption, whilst in many coastal communities the percentage can be considerably higher (FAO 2003).

Food consumption per person varies globally (Fig. 13.2) and is related to need, availability, geography, and wealth. Seasonality plays a part too; when other food sources are unavailable, fish may be the only animal protein food available. The importance of small-scale fisheries in particular for food security is emphasized by the FAO (2005).

About 38 million people worldwide are employed in fisheries, 95% of whom live in developing countries where fish are often one of the cheapest protein sources available. Fish are not required simply for protein, but are also rich in essential fatty acids, micronutrients, and trace elements. In local subsistence communities, fish may also provide a small income stream enabling the import of other staples such as rice.

The demand for fish protein has increased with the growing human population, resulting in

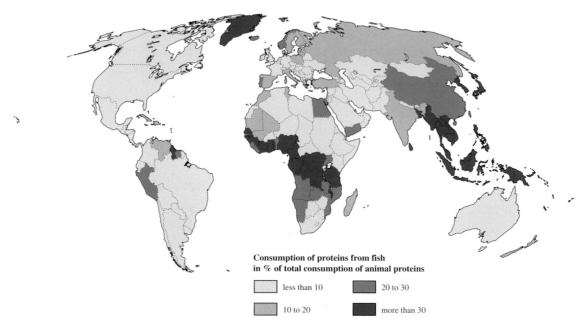

Figure 13.2 Consumption of proteins from fish as a percentage of total world consumption of animal proteins. From UNEP/GRID-Arendal Maps and Graphics Library. Retrieved 17 June 2010 from http://maps.grida.no/go/graphic/fish-protein-world-consumption.

increasing fishing pressure. World fisheries are recognized as being in decline, and despite equal or increased fishing effort, improved detection technology, and growing demand, fewer fish are being caught than previously (Pauly *et al.* 2005) and fish size is often smaller. With the global population projected to rise to 9 billion by 2040 there will be further demand for food production and provision of protein from the sea. World fisheries already face multiple challenges (Worm *et al.* 2006), but are now further subject to the twin stressors of climate change and ocean acidification. Tropical coral reefs, cold-water corals, and polar and upwelling ecosystems are all important in supporting fisheries. Recently they have all been highlighted as vulnerable to ocean acidification on a timescale of a decade to several decades (Turley *et al.* 2010a and references within).

Attempts at estimating the economic impact of ocean acidification for commercial fisheries have been made for the USA (Cooley and Doney 2009), the UK (Turley *et al.* 2010b), and the Mediterranean Sea (Hilmi *et al.* 2009). Commercial fisheries contributed $34 billion to the US gross national product in 2007, while in 2000 recreational marine fishing gen-

erated $12 billion of income in the USA and supported almost 350 000 jobs. Therefore loss of income or jobs from reduced catch or loss of habitat or fishing grounds as a result of ocean acidification would have significant economic impact. For example, loss to fisheries from ocean acidification by 2060 has been estimated to be 1.9 to 12.7% by value, depending on CO_2 emissions scenarios, based on integrated net present values (Cooley and Doney 2009), but other goods and services could be affected too (Cooley *et al.* 2009).

13.3.2 Carbon storage and regulation of gas and climate

The oceans are an enormous store of carbon, substantially greater than storage on land or in the atmosphere, and hence they play a key role in the global carbon cycle, especially the regulation of the amount of CO_2 in the atmosphere (Royal Society 2005). The oceans have already taken up around 28 to 34% of the CO_2 produced by humankind since the beginning of the Industrial Revolution (Sabine *et al.* 2004). This has reduced the extent of global warming but has come at the

Table 13.1 Summary of marine food-supplying organisms, their ecosystem role, contribution to the global food supply, significance as a food source, and reported potential vulnerability to ocean acidification

Group	Ecosystem role	Contribution to global seafood supply (% of weight)	Significance as food	Current understanding of vulnerability to ocean acidification	Relevant references
Molluscs: clams, scallops, mussels, oysters, pteropods, abalone, conchs	Mussels and oysters are important ecosystem engineers and an important food source for top predators. Pteropods are important food for salmon.	8% of global catch and increasing in aquaculture importance.	Valuable commercial fisheries and of local importance as a direct protein source in some island states. Locally important.	Some adults and juveniles (larval stages) have shown reduced calcification of shells, growth and health.	Beesley et al. (2008); Berge et al. (2006); Bibby et al. (2008); Comeau et al. (2009); Ellis et al. (2009); Gazeau et al. (2007); Green et al.(2009); Kurihara et al. (2007 and 2009); Marshall et al. (2008); Michaelidis et al. (2005); Orr et al. (2005); Parker et al. (2009); Ries et al. (2009); Talmage and Gobler (2009); Watson et al. (2009).
Echinoderms: sea urchins, sea cucumbers, starfish	Keystone species and food source for fish. Starfish can be important predators.	A small (0.1%) but valuable percentage.	Valuable commercial fisheries and locally important in some areas, increasingly regarded as a 'luxury' food item.	Few species studied, but all show vulnerability in adult and/or juvenile life stages, including egg fertilization and development.	Dashfield et al. (2008); Dupont et al. (2008); Gooding et al. (2009); Havenhand et al. (2008); Miles et al. (2007); Sheppard Brennand et al. (2010); Wood et al. (2008).
Crustaceans: shrimp, prawn, crabs, lobsters, copepods, and their relatives contributing to the zooplankton	Copepods play a central role in food webs. They feed on primary producers (the phytoplankton) so connecting them to larger parts of the food web that feed on the copepods which comprise a major part of the zooplankton. Zooplankton is a major direct food source for many fish and mammals.	7% of global consumption and rising. Wild catch and aquaculture species.	Valuable commercial fisheries for some species, especially shrimp, prawn, crab, and lobster. Significant fisheries.	Few species studied. Lobsters show vulnerability at a juvenile life stage and crab behaviour is affected, although adult shell formation is unaffected. Thermal tolerance of spider crab and edible crab reduced in high-CO_2 waters.	Arnold et al. (2009); Kurihara et al. (2004); Metzger et al. (2007); Pane and Barry (2007); Ries et al. (2009); Spicer et al. (2007); Walther et al. (2009).

Table 13.1 Continued

Group	Ecosystem role	Contribution to global seafood supply (% of weight)	Significance as food	Current understanding of vulnerability to ocean acidification	Relevant references
Finfish: small pelagic species (herrings, sardines, anchovies). Large pelagic species (tuna, bonitos, billfishes). Demersal species (flounders, halibut, cod, haddock). Miscellaneous coastal species.	Majority of finfish are carnivorous and hence close to or at the top of food webs or connect important trophic levels. Thus they play a major role in the balance of ecosystems.	Finfish as a group comprise the majority of the world's catch (over 80%). Small pelagics account for around 26%, large pelagics for 21%, demersals 15%, and miscellaneous coastal 6%.	Finfish are a significant proportion of human food consumption, fish oil manufacture, and fish meal provision. Dependence varies geographically with some communities having a high reliance, for food and income.	Direct effects on adult finfish unlikely other than possible orientation and behavioural changes due to variable otolith growth rates. Altered behaviour of larvae may affect replenishment of fish populations. Direct effects on reproduction, egg and larval development largely unknown. Indirect effects due to changes in prey and loss of habitats such as corals is likely.	Checkley et al. (2009); Cooley and Doney (2009); Fabry et al. (2008); Hoegh-Guldberg et al. (2008); Kleypas et al. (2006); Melzner et al. (2009); Munday et al. (2009, 2010); Silverman et al. (2009); see also Chapters 7–10.

Box 13.1 Ocean acidification and food security

'Food security exists when all people, at all times, have physical and economic access to sufficient, safe and nutritious food that meets their dietary needs and food preferences for an active and healthy life' (Rome Declaration on World Food Security and World Food Summit Plan of Action 1996).

Demand for fish and fishery products is projected to expand from 133 million tonnes in 2001 to 183 million tonnes by 2030 (FAO 1999). As world marine capture fisheries are projected to stagnate, there will be a mismatch between supply and demand which will need to be met from other sources. The most likely candidate for supplying the deficit is the burgeoning industry of aquaculture which is currently the fastest growing method of food production in the world with 7% growth per annum (by value), accounting for 47% of the world's food fish supply in 2006 and valued at US$78.8 billion (FAO 2008).

Aquaculture, at present and for the foreseeable future, relies heavily upon fish meal as a food stock—effectively using fish to feed fish. It is reported that for every kilogram of fish produced via aquaculture, up to 5 kg of capture fish are consumed by the process (Naylor *et al.* 2000). The species, caught in vast quantities, used for aquaculture feedstock fall into the category of 'forage fish'—important links in ocean food chains and key species for human consumption in some developing countries. So, critics argue, aquaculture is no 'silver bullet' or 'cure all' and is beset with its own challenges, amongst them environmental damage and inefficient use of dwindling resources.

It can be seen that the demand for marine protein and the available supply, including aquaculture, will be in a delicate balance. The entire fisheries system and the ecosystems associated with it are finely tuned yet highly pressured by fisheries and other stressors (Worm *et al.* 2006). Ocean acidification is yet another, although perhaps more insidious, pressure on these limited resources.

---

price of the substantial change to ocean chemistry under consideration here. This invaluable role of oceans in the earth's carbon cycle is performed by a series of biogeochemical processes regulated by marine organisms as well as the important physical processes of ocean mixing, tides, currents, and air–sea exchange. The ability of microscopic plants, phytoplankton, to fix CO_2 through photosynthesis and transfer a proportion of it to the deep sea via this 'biological pump' is a key part of the global carbon cycle (see Chapter 12). It is a surprise to many that oceans contribute around half of global primary production (Field *et al.* 1998), and while this is mostly due to phytoplankton, larger marine plants contribute as well. Any stressor reducing ocean productivity and the biological pump could have substantial feedback to climate (see Chapter 12). Marine microorganisms also play a less well known role in regulation of other climate changing gases such as dimethyl sulphide, methane, and nitrous oxide (see Chapter 11).

In contrast, and importantly, the larger marine plant communities such as mangroves, salt marshes, and seagrass and kelp beds store carbon, known as 'blue carbon' for longer periods (Duarte *et al.* 2005),

not unlike the 'green carbon' stores in forests, scrublands, savannas, grasslands, and tundra on land. These communities build up sediments rich in organic carbon similar to soils under rainforests (Nellemann *et al.* 2009). As on land, these marine ecosystems are degrading and shrinking, and this natural 'blue carbon' store is reducing. Climate change and ocean acidification are other pressures on these carbon-binding ecosystems to add to those of coastal development, aquaculture, and pollution. There are calls for a global 'blue carbon' fund for the protection and management of coastal and marine ecosystems, not unlike the Reducing Emissions from Deforestation and Degradation (REDD) initiative (Angelsen 2008) to preserve the 'green carbon' locked in rainforests, so that this marine carbon store can be preserved and replenished.

13.3.3 Nutrient regulation

Sediments play a crucial role in the cycling of carbon and nitrogen through the biochemical activities of microorganisms and the behaviour of larger organisms that live in and on the sediments. Their activities result in a release of nutrients from the sediment to the water column where they support primary productivity. Changes to the regeneration of nutrients supporting primary production could have worrying knock-on consequences for shelf sea productivity and food webs. Initial studies indicate that this process may be affected by ocean acidification (see Chapter 7).

13.3.4 Provision of leisure, recreation, and well-being

A significant component of leisure and recreation depends upon coastal marine biodiversity (e.g. bird watching, sea angling, rock pooling, snorkelling, and scuba-diving) which in turn supports employment and small businesses. If coastal biodiversity declines as a result of ocean acidification and other drivers (e.g. Feely *et al.* 2008; Hall-Spencer *et al.* 2008) the value of this sector will decrease, with a potential loss of revenue.

The enormous biodiversity supported by coral reefs underpins substantial tourist industries for many tropical countries, and often provides their main revenue. Countries with coral reefs attract millions of scuba-divers every year, yielding significant economic benefits to the host country. Globally, tourism is estimated to provide US$9.6 billion in annual net benefits (Cesar *et al.* 2003) and a multiple of this amount in tourism spending. Coral reef biodiversity also has a high research and conservation value, as well as a non-use value (the value of an ecosystem to humans, irrespective of whether it is used or not) estimated together at US$5.5 billion annually (Cesar *et al.* 2003). Loss of coral reefs and their biodiversity would have an impact on tourism in these areas (see Box 13.2).

13.3.5 Coastal protection and habitat provision

Coral reefs, mangroves, and salt marshes play an important role in shore protection and enhance local productivity and biodiversity. The protective function of reefs was valued at US$9 billion per annum by Cesar *et al.* (2003), so any loss of this role could

---

Box 13.2 The economic value of Hawaiian reefs: a case study

Tropical coral reefs have been a major focus for the evaluation of environmental goods and services provided to humankind, with an estimated annual value of US$30 billion provided to the world's economies by coral reefs (Cesar *et al.* 2003).

Table 13.2 shows the range of goods and services that can be used in assessing their value (Cesar *et al.* 2002), using the example of Hawaiian reefs. 'Use values' are benefits that arise from the actual use of the ecosystem, both directly and indirectly, such as fisheries, tourism, and beach-front property values. 'Non-use values' include an existence value, which reflects the value of an ecosystem to humans irrespective of whether it is used or not. The monetary value of the sum of the goods and services forms the total economic value (TEV).

Focusing on tourism, fisheries, coastal protection, amenity and property values, research and biodiversity, and cultural values the associated economic benefits of each of these goods and services was estimated for individual islands and for Hawaii as a whole (Cesar *et al.* 2002). For Hawaii as a whole the largest contribution (85%) to the yearly benefits of $364 million is the annual value added by recreation and tourism ($304 million). Second is the amenity/property value (40%), with benefits of $40 million per annum. The third most important benefit is biodiversity ($17 million), while fisheries income is smaller at $2.5 million.

Obviously, this is not the TEV of the Hawaiian reefs as some of the use and non-use values have not been included, but it does give an idea of the significance of reefs to the island society and economy. It also aids the decision-making process when considering benefits of active risk management of stressors on reef ecosystems. Such approaches can be applied to other marine environments to assess the value of national marine resources and the potential costs of the threat of ocean acidification (e.g. Cooley and Doney 2009; Turley *et al.* 2010b).

Table 13.2 Subdivision of the total economic value of coral reefs as used to estimate the economic value of Hawaiian reefs by Cesar *et al.* (2002). Direct-use values are the outputs and services that can be consumed directly, the functional benefits are those enjoyed indirectly, and the bequest, option, and existence values are those functions that value either the future use, expected new information, or are based on moral convictions.

Total economic value	Use values	Direct-use values	Extractive (capture fisheries, mariculture, aquarium trade, pharmaceutical). Non-extractive (tourism, recreation, research, education, aesthetic)
		Indirect-use values	Biological support to seabird, turtles, fisheries. Physical protection to other coastal ecosystems, coastline, navigation. Global life-support in terms of carbon storage.
	Non-use values	Bequest, option and existence values	Endangered and charismatic species. Threatened reef habitats, aesthetic reefscapes, and 'way of life' linked to traditional use.

have an impact on the people who depend on or enjoy these services, either through loss of low-lying land, habitat, and infrastructure or through a need for investment in shore protection. The *annual* economic damage of ocean-acidification-induced coral reef loss by 2100 has been estimated to be US$870 billion and US$500 billion for A1 and B2 SRES emission scenarios (IPCC 2001), representing 0.14 and 0.18%, respectively, of global GDP (Brander *et al.* 2009). Obviously, these aggregate global numbers do not reflect the distribution of the potential serious economic impact across different regions, especially those economically reliant on coral reefs.

13.4 The relevance of ocean acidification to individuals

The Deepwater Horizon oil spill disaster, in the highly productive and biologically diverse waters of the Gulf of Mexico in April 2010, was the largest offshore spill in US history and resulted in the loss of 11 lives and injury to 17. It impacted livelihoods through loss of goods and services provided by the Gulf, such that it became a major global environmental and political issue (http://www.whitehouse.gov/deepwater-bp-oil-spill/). Although the leak was staunched on 15 July, by September 2010 BP reported that its own expenditure on the oil spill had reached US$8 billion, including the cost of the spill response, containment, relief well-drilling, grants to the Gulf states, claims paid, and federal costs. Between 20 April and 25 June BP suffered a loss of US$105 billion to its total share value, result-

ing in a 54% loss to investors around the world. This disaster has tragically highlighted the importance of the more obvious marine goods and services to society, although the full range, such as non-use values, have not yet been included. As well as loss of revenue from tourism and fisheries, property prices fell and there were concerns about impacts on the mental and physical health of coastal residents. This tragedy clearly demonstrated that the loss of form and function of the marine environment affects individuals directly and indirectly on local to global scales.

With men and women in developed nations living longer, it is increasingly frequent that four generations of the same family are alive at one time (Fig. 13.3). This increased lifespan (on average 75 and 80 years for men and women, respectively) means that climate change and ocean acidification have the potential to affect many of those alive today, their children, and their children's children, and the younger of us may witness those effects on our relatives and descendants. From the perspective of an individual's needs and lifespan, the slow advance and invisible nature of ocean acidification creates the illusion that this very urgent problem can be tackled later.

However, ocean acidification is a global phenomenon, which if allowed to continue unabated will result in severe changes in carbonate chemistry occurring within the lifetime of children alive today and which will impact upon important ecosystems and the biodiversity they support such as the Arctic, Antarctic, and tropical waters (Fig. 13.3).

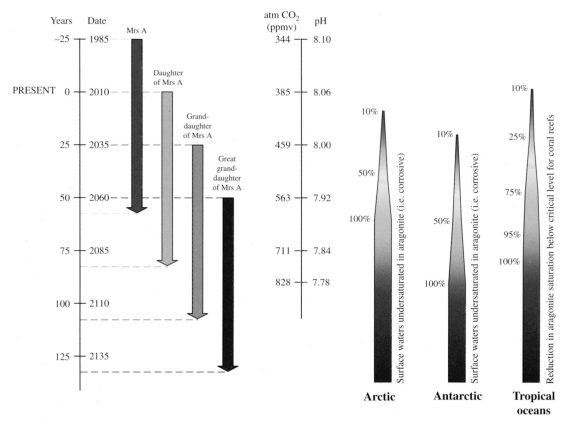

Figure 13.3   In western developed countries the lifespan for a woman averages 80 years. This means that it is possible for four generations of a family to be alive at any one time. If atmospheric CO_2 emissions continue unabated the uptake of this CO_2 by the oceans will result in decreasing ocean pH, and model projections indicate that the surface waters of the Arctic will become increasingly undersaturated in aragonite first, with the Antarctic following soon after (Orr *et al.* 2005; Steinacher *et al.* 2009; Turley *et al.* 2010a). The tropical oceans will not become undersaturated, but saturation will be reduced sufficiently to be critical to coral reef growth (Cao and Caldeira 2008; Hoegh-Guldberg *et al.* 2008). A mother and her daughter born today will both experience the onset and continued strengthening of ocean acidification. In her later years the daughter will experience a world where all the Arctic and much of the Antarctic will be corrosive to shelled organisms and nearly all tropical coral reefs will be in decline.

With the substantial investment that each individual and government makes to provide for present and future happiness, health, welfare, safety, shelter, and education, it seems remarkable that climate change and ocean acidification, which could affect all these benefits that we try to bestow on our children, are not considered as part of this care package. Unmitigated ocean acidification is a threat to many of the basic and psychological needs outlined by Maslow (1943; Fig. 13.1). How will we then face and respond to our children or their children or our great grandchildren? It may not just be the marine ecosystems that we will

have watched passively disappear or degrade; it may be the loss of fish protein for many nations that depend on it as their sole protein source (Table 13.1), as well as our basic 'social' and 'ego' needs. However, if we could reach for and engage our creativity and ability to achieve and focus on mitigation of climate change and ocean acidification, then those basic underlying needs, including those 'ego' and 'social' needs, could be met too. It might be then that we recognize that we are temporary caretakers of this planet, which is our life-support system and the life-support system of future generations.

13.5   Communicating ocean acidification to policy- and decision-makers

13.5.1   The Intergovernmental Panel on Climate Change

Policymakers need an objective source of information about the causes of climate change, its potential environmental and socio-economic consequences, and the adaptation and mitigation options to respond to it. Over the last 20 years, the Intergovernmental Panel on Climate Change (IPCC) has been the main source of collective evidence, scientific consensus, and assessment of the risks of climate change for policymaking (IPCC 2007).

The IPCC process was established by the World Meteorological Organization (WMO) and the United Nations Environment Programme (UNEP) (Fig 13.4). The IPCC is a scientific body. It reviews and assesses the most recent scientific, technical, and socio-economic information produced worldwide relevant to the understanding of climate change. It does not conduct any research nor does it monitor climate-related data or parameters. Around 4000 scientists from 130 countries contributed to the work of the IPCC 4th Assessment Report on Climate Change in 2007 on a voluntary basis. The enormous

challenge is the objective assessment of tens of thousands of scientific papers by three expert working groups on the 'Physical Science Basis' (Working Group I), 'Impacts, Adaptation and Vulnerability' (Working Group II), and 'Mitigation of Climate Change' (Working Group III). Following the three reports and a review process by experts and governments the information is further distilled in a Technical Annex and finally a Summary for Policymakers, the wording of which is agreed by governments. By endorsing the IPCC reports, governments acknowledge the authority of their scientific content. The work of the organization is therefore policy-relevant and yet policy-neutral, never policy-prescriptive. Such a distillation of information will almost certainly lead to generalizations and some loss of accuracy, but despite some errors, knowledge gaps, and uncertainty the conclusion that human-induced climate change is a threat to humankind remains (PBL 2010).

Ocean acidification was recognized by IPCC in the 4th Assessment Report as a risk caused by increasing CO_2 emissions (IPCC 2007). With evidence from ocean acidification research growing as new national and international programmes [for example the EU funded European Project on Ocean

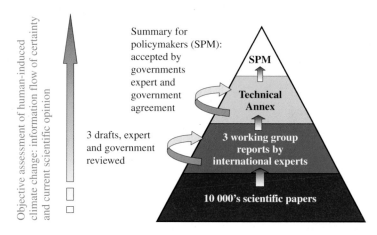

Figure 13.4   A simplified schematic of the Intergovernmental Panel on Climate Change (IPCC) process which was developed by the World Meteorological Organization and the United Nations Environment Programme for the assessment of human-induced climate change. Four key assessment reports on climate change have been produced so far, each informing the climate negotiating process. The IPCC 1st Assessment Report (1990) informed the Rio de Janeiro Summit in 1992, the IPCC 2nd Assessment Report (1995) informed the Kyoto Protocol in 1997, the IPCC 3rd Assessment Report (2001) informed the development of the United Nations Framework Convention on Climate Change and the Kyoto Protocol, and the IPCC 4th Assessment Report (2007) informed the Bali Roadmap and the 15th and 16th Conference of Parties (COP15 and COP16) climate negotiations in Copenhagen (2009) and Cancún (2010) in the progression towards a post-2012 deal on climate change.

Acidification (EPOCA), the German programme BIOACID and the UK Ocean Acidification Research Programme] are funded it is likely that this community will contribute a greater body of evidence in the 5th Assessment Report due after 2012.

The relatively recent and emerging understanding of the nature of ocean acidification and its potential consequences means that it has not achieved the 'pull-through' to policymakers that its nature might warrant. For example, the risk to marine ecosystems received little attention in the COP15 negotiations at Copenhagen. Scientists working on ocean acidification must therefore embrace the challenging task of communicating their science openly and understandably to policy- and decision-makers.

13.5.2 Communicating complex science to policymakers

One can circumvent the hierarchical method of dissemination that can occur in a complex organiza-

tional structure (depicted in the lower part of Fig. 13.5) by communicating directly and clearly to policymakers through expert groups and committees and the formation of a Reference User Group (RUG) of stakeholders. The pitfalls are numerous as one tries to communicate complex science to non-specialists (as indicated on the left of Fig. 13.5) and this should be avoided.

Communication is a two-way process which only works when the recipient understands what is being communicated. Confusion can occur when there is a lack of clarity in understanding between two groups or individuals. For instance, policymakers often use the word 'target' in relation to CO_2 emissions reductions; a 'target' is something to aim at, in the recognition that one might hit it or shoot below it or above it. However, a scientist might use the word 'threshold' to describe a point beyond which changes take place and so perhaps should be avoided. There could be instances when 'target' and 'threshold' are incorrectly taken to be the same

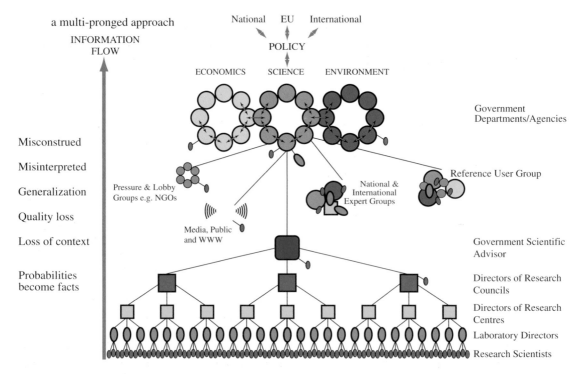

Figure 13.5 Schematic illustration of methods of information flow to policymakers (Turley 1999).

thing, so resulting in confusion; for CO_2 emissions the 'target' should be well below the 'threshold'.

13.5.3 Science as organized scepticism

Subsequent to the media focus following the 'hacked' or stolen emails from the Climatic Research Unit (CRU) at the University of East Anglia (coined 'Climategate' by the media) in December 2009, the science of climate change and the IPCC process of assessing climate change has been under extreme scrutiny by the media despite investigations finding no evidence of behaviour that might undermine the conclusions of the IPCC (e.g. House of Commons Science and Technology Committee 2010). Often the so-called 'climate sceptics' have been given greater exposure to put their case. It is interesting that the term 'sceptic' has, in a sense, been high-jacked from scientists by the 'climate change deniers'; indeed science could be defined as organized scepticism as opposed to denialism which is 'motivated by conviction rather than evidence…scepticism forms the intellectual cornerstone of scientific enquiry'(Kemp *at al.* 2010). It is this scepticism that scientists must retain or reclaim as it is essential for the scientific process. Additionally, the scientist sceptic has the ability to change his or her mind when presented with new evidence, that is, a scientist is sceptical only until adequate evidence is presented to convince him or her otherwise. Denialists, on the other hand, will rarely change their minds. It is therefore important to recognize denialism when confronted with it and accept that the normal academic response, to debate the evidence as a whole without deliberate distortions using principles of logic, will not necessarily be returned (Diethelm and McKee 2009). Undoubtedly the public perception of the scientific process and what it can and cannot offer society can be very different and needs clarifying.

13.5.4 Evidence, not advocacy

The ambiguity and uncertainty of science does not sit naturally with policymaking. Many politicians, the media, and the general public expect certainty, proof, facts, and answers from scientists. Scientists cannot usually speak in such certain terms as science is an open debate or an ongoing dialogue amongst peers who are sceptical by nature. So there is a misperception by society of the role of the scientist. Scientists would do well to recognize and perhaps address this issue. They should also recognize that scientific evidence and the expert advice that they provide are just two of the many links in the policymaking chain. Science supports sound policymaking when it informs risk management by describing the risks, how they can change through time, defining knowledge gaps that might be important in assessing the risks, and how the risks can change by action or by further research. At all costs scientists should steer clear of advocacy but rather focus on communicating the evidence in an understandable, impartial, and objective fashion.

13.5.5 Confidence and certainty

Assessing 'certainty' (Fig. 13.6) can be difficult for scientists giving evidence to policymakers. There are statistical methods for assessing the probabilities of data following specific trends; for instance, scientists frequently interrogate their data so that they may say that there is a probability that 99, 95, 90% etc. of their data will follow the expectation. Of course, this also means that the chances are that 1, 5, 10% of the data will not follow the trend. This behaviour of seeking trends in data is how our understanding of science grows and generalizations are made. However, assessing certainty for evidence-based policymaking is more qualitative; it tries to assess the level of agreement or consensus in datasets and model confidence, as well as the amount of evidence from theory, observations, or models. Because of the non-quantitative aspects, this may be divided into low, medium, and high confidence associated with the generalized statement. If we do this for ocean acidification, we immediately have a dilemma, as the confidence changes if we talk about what is already happening and whether we are referring to changes in chemistry, biology, biogeochemistry, or ecosystems. The Marine Climate Change Impacts Partnership (MCCIP) went through an interesting process of trying to assess confidence in what is happening in terms of ocean acidification now and what may happen in the future (Turley *et al.* 2010b).

The chemical changes (Fig. 13.6A, marked with X) are measurable and happening now; the chemistry is very well known with good agreement amongst datasets and experts, so certainty is high. However, there is little evidence for biological impacts (Fig. 13.6A marked with ●) of current ocean acidification (and therefore little agreement and low certainty) but that may be because there are no long-term databases that look at this, or it could be because current changes have had no impact. Thus,

as research in this area continues this assessment could rapidly change.

Assessment of what could happen in the future to ocean chemistry (Fig. 13.6B, marked with X) is based on good understanding of ocean carbonate chemistry and good agreement from circulation models and dependent on the rate of future CO_2 emissions to the atmosphere. Assessment of what could happen to biology in the future (Fig. 13.6B, marked with ●) is based on recent projections

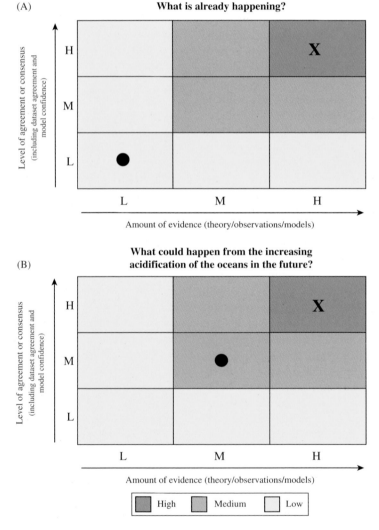

Figure 13.6 The Marine Climate Change Impacts Partnership (MCCIP) assessment of certainty based on level of agreement or consensus and the amount of evidence for (A) what is happening now in terms of ocean acidification and (B) what could happen in the future. Assessments for current and future chemical changes (X) are more certain than those for impacts now and in the future (●). See text for a detailed explanation (Turley *et al.* 2010b).

indicating vulnerability of the Arctic and coastal upwelling to carbonate undersaturation, further work highlighting the vulnerability of corals to decreasing $CaCO_3$ saturation and other experimental evidence increasingly revealing that juveniles, reproduction, physiology etc. of different organisms are vulnerable to ocean acidification, even if adults are not. There are papers with conflicting results which are being debated in the literature. The great majority of the growing evidence from field work, experiments, modelling, and the geological record indicates that the consequences for the future could be very serious, but there are still substantial knowledge gaps which together support the continued 'medium rating' (e.g. Kroeker *et al.* 2010). Even more uncertain are the biogeochemical impacts and impacts on food webs and ecosystems. This is due partly to a lack of research in the area but also due to increasing uncertainty with the complexity of marine systems. Thus, in terms of our understanding of the future impacts of ocean acidification there currently seems to be a gradient of certainty based on evidence from chemistry at the high-certainty end of the range to individual organisms with medium certainty, but as we move through populations, food webs, and ecosystems to biogeochemistry, the certainty is reduced. However, despite this uncertainty, expert opinion is generally that there is real concern that ecosystems (such as tropical coral reefs, cold-water coral communities, and polar ecosystems) or biogeochemical cycles could be at risk within this century if CO_2 emissions continue unabated (see Chapter 15).

13.5.6 Risk and applying the precautionary principle

The nature of a risk is important in assessing whether society should act and apply the precautionary principle to reduce it. In the case of ocean acidification, the hazard is caused by increasing CO_2 emissions in the future. This CO_2 is also causing another hazard to coral reefs through warming surface seawater which can cause coral bleaching. These two hazards may occur at the same time and therefore represent a greater risk than either on their own. Scientific evidence to date indicates that the only way to reduce these specific hazards on a global scale is to reduce CO_2 emissions to the atmosphere. However, there are other hazards to coral reefs such as coastal development, pollution, and overfishing that may amplify their vulnerability (IPCC 2007). Averting potential changes to major marine ecosystems such as coral reefs, that harbour great biodiversity, protect coastlines, and provide livelihoods and food for millions, may or may not be sufficient argument for mitigation and will depend on international politics and the proportionality (and cost) of proposed solutions. Policymakers will need to know the answers to a number of complex questions that have socio-economic relevance. If the non-CO_2-related hazards are removed, will that enable reefs to survive at higher CO_2 levels? What level of CO_2 will be safe for corals? If society invests in low-carbon energies and technology and reduces CO_2 emissions what will it gain for its effort? So, if CO_2 emissions peak at 550 ppmv then what proportion of the world's coral reefs will remain, what will their quality be, and which corals will survive? If CO_2 emissions are reduced even further, with a peak at 450 ppmv, what will be the further advantage in terms of the proportion, quality, and geographical distribution of surviving reefs? Are there 'tipping points' in ocean acidification which should be avoided? What are the costs and benefits or the risk:reward ratio of action to prevent and/or mitigate ocean acidification?

Such policy-relevant questions like these may not be in the mind of many field- or laboratory-based scientists, unless they have a close working relationship with those who influence policy. That is why there is a need for very close communication and understanding between scientists and policymakers.

13.6 Wider communication of ocean acidification

13.6.1 Communicating with stakeholders

In 2004, an ocean acidification RUG of key stakeholders was set up to run alongside the scientific research in the UK project 'Impact of CO_2 on Marine Organisms' (IMCO2, 2004–2008). In 2008, the RUG was further developed as part of the EU funded EPOCA programme. In 2010, the RUG further evolved to represent EPOCA, the new national UK Ocean Acidification Research Programme, and the German BIOACID programme.

This approach ensures a two-way exchange of knowledge between the scientists and stakeholders right from the start of the programme (Fig. 13.7)—and stakeholders gain knowledge that they can bring back to their organization. This approach has resulted in each stakeholder's organization becoming more aware of ocean acidification, often stimulating its own activities and in some instances collaboration in this area.

Scientists get direct feedback from a wide range of stakeholders on what the key policy questions and issues are and how they can make their science more credible, accessible, and useful to users. The RUG concept has become a standard for knowledge exchange between researchers and stakeholders, with feedback from the RUG already ensuring that scientists are more aware of the policy relevance of their work.

The RUG has a clear remit and terms of reference, advising on: the types of data and analyses and products; the format and nature of key messages arising from research; the dissemination procedures to ensure that the results from the research will be most useful to managers, policy advisers, decision-makers, and politicians, and are distributed to all potential end-users of the information; the transmission of key science developments into their own sector/parent organization.

The ocean acidification RUG has an international membership including individuals from both commerce and government, with interests spread across relevant climate policy, the environment, industry, and conservation. This knowledge exchange mechanism has been highly successful in promoting understanding between policymakers and other stakeholders.

The RUG produced two guides, 'Ocean acidification: the facts' and 'Ocean acidification: questions answered' (Ocean Acidification Reference User Group 2009 and 2010). These were written by stakeholders, the RUG members, as special introductory guides for policy advisers and decision-makers and the science content was checked by experts.

13.6.2 Communicating with the wider public through the press and media

Some public understanding of scientific issues is increasingly recognized as essential for broad acceptance of any political or economic actions that may be required to face and solve challenges, especially within the environmental sector. A growing trend is for 'communication' of science to be embraced within research funding, and for a 'public engagement plan' to be created at the outset. The press and media are key stakeholders in such engagement, but in the interests of being unbiased can be powerful supporters or significant detractors. Certainly the press and media are more scientifically aware than at any time in the past, sometimes having specialist science or environmental correspondents. Lessons learned from the long-running climate change 'debate' can be applied to

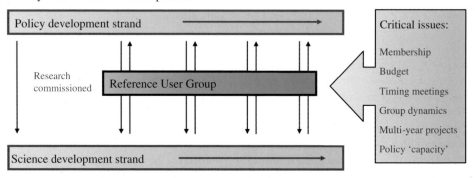

Figure 13.7 Showing the Reference User Group (RUG) approach to ensure better practice in linking policy and science. The interaction between the RUG stakeholders and the scientists continues throughout the programme so that science and policy develop together. Courtesy Dan Laffoley, IUCN.

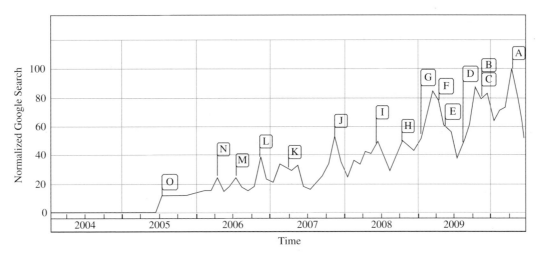

Figure 13.8   Google Insights search for 'ocean acidification' from 2004 to mid 2010 taken in July 2010. The data are displayed on a scale of 0 to 100, after normalization; each point on the graph has been divided by the highest point or 100. The flags indicate specific activities that occurred at the time: O, the launch of the Royal Society Working Group on ocean acidification (published in summer, Royal Society 2005); N, is when *Scientific American* and the *New Yorker* had articles on ocean acidification; M and L, the period of the Avoiding Dangerous Climate Change Symposium and book publication; K, the period of the NSF, NOAA, USGS report on the impacts of ocean acidification on coral reefs and other marine calcifiers (Kleypas *et al.* 2006); J, publication of the IPCC 4th Assessment Report on Climate Change (which featured ocean acidification for the first time); I, the period when key articles were published such as on upwelling of high-CO_2 water off the west coast of North America (Feely *et al.* 2008) and the vulnerability of calcifiers around natural CO_2 vents (Hall-Spencer *et al.* 2008); H, the Oceans in a High-CO_2 World Symposium II, Monaco; G, the Monaco declaration of concern; F, the attention received by the CBD legal proceedings against the EPA; E, the Interacademy Statement on ocean acidification; D, the threat of high-CO_2 waters to Alaskan fisheries; C, the launch of the film 'Acid test'; B, the speech by the UK Secretary of State for the Environment at Oceans Day and other ocean acidification activities during COP15. Activity remains high post-COP15 culminating in A, the publication of a paper on the behavioural response of fish to predators under high CO_2 (Munday *et al.* 2010). It should be noted that the increasing baseline interest in ocean acidification since 2004 is also in response to the increasing numbers of research publications over this period stimulating increasing 'take up' by stakeholders.

public communication of ocean acidification, its impacts, and potential consequences. Openness, honesty, and clarity of message are the key ingredients for building a strong relationship with these journalists, a relationship that must be founded on integrity and trust. Prior to and during the COP15 negotiations, the UK press and media, at least, became highly engaged with ocean acidification as a topic of interest to their audiences. The straightforward nature of the science, the pioneering stage the research has reached, and the potential impacts contain all of the ingredients for an interesting storyline. Ocean acidification scientists have shown through various publications (summarized in Fig.1.3 in Chapter 1) their desire to ensure that clear and accurate messages are provided to the press and media, demonstrating a willingness to engage with this key group of stakeholders and an understanding of the needs of journalists. A good example of this is the production of responses to

frequently asked questions posed by policymakers, the public, and climate change deniers by experts from around the world (Kleypas *et al.* 2010).

Google Insights for Search Index (Fig. 13.8) on 'ocean acidification' is an indication of the growing worldwide web attention that the topic has received since the publication of the Royal Society Report in 2005 and culminating in the ocean acidification activities on the lead up to, during, and after the COP15 negotiations in Copenhagen in December 2009.

13.6.3   Communication and the next generation

While global warming and climate change are well-known concepts among young people, ocean acidification is new to most of them. It is therefore important to share knowledge about ocean acidification with schools and the younger generation. For example, many resources developed for schools around the world are available under the Carboschools (www.carboschools.org) initiative and the EPOCA project (www.epoca-project.eu) which produce educational projects with a strong scientist–teacher–pupil interaction.

13.6.4   Children communicating the issue to each other

The 'viral' nature by which issues and trends are spread by young people has been exploited by advertisers, not least those concerned with music, clothing, and technology; word of mouth or peer recommendation is very powerful when spreading messages. Recognizing this phenomenon, the EPOCA project communicated with and inspired a group of students ranging in age from 11 to 15 years old to present the science and potential impacts of ocean acidification, in their language, to their peer group. They researched, wrote, filmed, and acted the parts for a short animated film—'The other CO_2 problem'. The film has been featured widely in the UK press and media, and currently enjoys a strong YouTube presence. Such has been its impact that many thousands of DVD copies have been requested and the film has been shown at prestigious international meetings and on television stations internationally, translated into a number of languages, and won awards for communicating science (view 'The other CO_2 problem' at www.epoca-project.eu/index.php/The-other-CO2-problem-animation.html).

13.7   The response of policymakers

The United Nations General Assembly is charged with undertaking an annual review and evaluation of the implementation of the United Nations Convention on the Law of the Sea and other developments relating to ocean affairs and legislation, including in relation to oceans and climate change. The United Nations holds a unique position for raising awareness and facilitating discussions among delegates and the international community during ocean-related and sustainable development-related events. As mentioned above, the IPCC has a special responsibility in this regard, and oceans and

ocean acidification will be included in the 5th Assessment Report in 2012.

Other opportunities to raise awareness and highlight options on ocean acidification include the United Nations Framework Convention for Climate Change, the Convention on Biological Diversity, the Intergovernmental Oceanographic Commission, the UNEP Governing Council, and the UN Commission on Sustainable Development. Indeed efforts on the issue of ocean acidification have been coordinated by the Intergovernmental Oceanographic Commission (IOC) of UNESCO and the UNEP. The Conference of the Parties of the Convention on Biological Diversity (CBD) has also considered the issue with respect to its impacts on marine biodiversity. The Commission on Sustainable Development will consider the issues of oceans and seas and small island developing states in 2014 and 2015 and will provide an opportunity to highlight ocean acidification and take policy decisions aimed at combating it.

The US Environmental Protection Agency (EPA) agreed to consider how states can address ocean acidification under the Clean Water Act on 15 March 2010, and solicited public comments on how to evaluate waters threatened by acidification and address the problem (Parks 2009). This was a response to a lawsuit brought under the Clean Water Act by the Center for Biological Diversity that challenged EPA's failure to recognize the impacts of acidification on coastal waters off the state of Washington.

13.8   Geoengineering and its relationship to ocean acidification

The propensity of humans to believe they can 'fix' the global warming problem by geoengineering the planet has led to a growing number of suggested 'solutions' recently assessed by the Royal Society (2009; see also Chapter 14). The techniques suggested so far fall into two categories: solar radiation management (SRM), reflecting the sun's energy away from earth, and carbon dioxide removal (CDR), direct removal of CO_2 from the atmosphere. SRM techniques do not take into account or resolve the issue of ocean acidification (Zeebe et al. 2008) while CDR, through removal of atmospheric CO_2, would reduce the CO_2 entering the ocean, but costs are thought to be high. Another option is CO_2 capture and storage (CCS), the extraction of CO_2 from the point of production (such as a power plant) and its long-term disposal in either the deep oceans or in suitable geological repositories thus removing it from contact with the atmosphere (IPCC 2005). Adding quicklime to the oceans may well remove CO_2 from the atmosphere and sequester it in the world's oceans (see www. Cquestrate.com, accessed July 2010), and may make the oceans more alkaline and directly combat ocean acidification. Nonetheless, the potential side-effects are unknown and the scale of the project enormous, with estimates of around 1.6 billion m^3 of limestone required to sequester each gigatonne (Gt) of carbon. Currently the ocean is absorbing about 2 Gt of anthropogenic carbon per year.

Another issue with geoengineering is that growing publicity may imply to society and policymakers that it provides ways to avoid the consequences of climate change, and that actions to reduce CO_2 emissions can be delayed. There may also be strong public opposition to attempts to engineer the planet on a scale substantially larger than that seen for a small-scale research project off Hawaii into ocean injection of CO_2 (E. Adams, pers. comm.), which was called off due to mounting public concern.

Efforts to mitigate the effects of ocean acidification could take two other forms: adaptation or remediation. Adaptation accepts that lower ocean pH will occur and have socio-economic impacts and so adjusts economic activity and resource management to take this into account. For instance, if marine conditions become unfavourable for commercially valuable fish species in their present habitat, fish farmers may choose to switch to controlled production environments in onshore tanks or ponds, or to take measures to control pH at a local scale in the marine environment around the fish farm.

Remediation, on the other hand, seeks to restore ocean pH to a level that would avoid negative effects. That is, it may be possible to influence local and regional ocean pH through geoengineering schemes over limited periods of time; but this has not been demonstrated and could carry with it costly ecological side-effects. For this reason research is necessary to provide scientific evidence to enable policymakers to select those solutions with the least risk of unintended environmental consequences that are the most ethical, economic (from the monetary and carbon perspective), and effective at doing the job. In this case, developing codes of practice and governance for future possible application of geoengineering techniques is an important focus for future policymaking. Independent international interdisciplinary teams are required to assess new geoengineering techniques as they emerge.

Clearly the most obvious and optimum policy option for mitigating ocean acidification is to reduce fossil fuel CO_2 emissions to the atmosphere through the UNFCCC climate change negotiations currently under way (see Chapter 14).

13.9   Conclusions

Ocean acidification is a recently recognized phenomenon, with the same cause as climate change—increased anthropogenic CO_2 emissions. Experiments, modelling, and an increasing number of *in situ* measurements point to the likelihood of significant and developing changes to ocean chemistry with the potential to affect natural marine systems, including chemical cycling and ecosystems. Such chemical changes have been shown to affect a wide range of marine organisms, especially, but not exclusively, those which utilize $CaCO_3$ to build protective and supporting structures, such as shells and reefs. The nature of the chemistry of CO_2 in seawater is such that some ocean areas, notably polar regions, are likely to be affected first. Coincidentally these high-risk areas are the sites of the greatest production of marine-based human food resources. With an increasing world population becoming more reliant upon marine resources for human consumption, and with existing pressures such as pollution, fisheries exploitation, and coastal development, any further stress to these resources through ocean acidification is of concern. If the future impact on marine systems is as some studies suggest, the ramifications for policymaking are wide ranging.

Communication regarding ocean acidification, the threats it may present, what these threats mean to the human population, and the timescale through which they may unfold requires careful management. Extravagant claims based on a new and still incomplete knowledge base will only prove counterproductive. A cautious approach, building upon growing credible scientific observation and experiments leading to reliable conclusions, might be seen as too defensive, especially when policymakers and stakeholders demand instant answers with definite conclusions and predictions. It is clear that open, well-informed responses, couched in terms that the various stakeholder groups can understand, based on knowledge rather than supposition, is paramount. Dialogue through initiatives such as the RUG concept has proved to be immediately useful, and is obviously an effective communication vehicle. Above all, scientists involved in ocean acidification research should 'stick to what they know best' and avoid being pressured into predicting the future, and advocacy regarding policymaking. Ocean acidification studies are still in their infancy, but the phenomenon is real and has the potential for widespread and significant effects that will have an impact on humans. Scientists, policymakers, and stakeholders must work together to understand the broader implications, and cooperate to overcome the threats that ocean acidification poses to human society.

13.10 Acknowledgements

This work is a contribution to the 'European Project on Ocean Acidification' (EPOCA) which received funding from the European Community's Seventh Framework Programme (FP7/2007–2013) under grant agreement no. 211384 and to the 'UK Ocean Acidification Research Programme' which is funded by NERC, Defra, and DECC (grant number ME5201). Our thanks to Dawn Ashby, Kelly-Marie Davidson, and Jessica Heard from Plymouth Marine Laboratory for help with drawing and researching the figures.

References

Angelsen, A. (2008). *Moving ahead with REDD: issues, options and implications*, 44 pp. CIFOR, Bogor, Indonesia.

Arnold, K.E., Findlay, H.S., Spicer, J.I., Daniels, C.L., and Boothroyd, D. (2009). Effects of hypercapnia-related acidification on the larval development of the European lobster, *Homarus gammarus* (L.). *Biogeosciences*, **6**, 1747–54.

Beesley, A., Lowe, D.M., Pascoe, C. K., and Widdicombe, S. (2008). Impact of CO_2 induced seawater acidification on the health of *Mytilus edulis*. *Climate Research*, **37**, 215–25.

Berge, J.A., Bjerkeng, B., Pettersen, O., Schaanning, M.T., and Øxnevad, S. (2006). Effects of increased sea water concentrations of CO_2 on growth of the bivalve *Mytilus edulis* L. *Chemosphere*, **62**, 681–7.

Bibby, R., Widdicombe, S., Parry, H., Spicer, J., and Pipe, R. (2008). Effects of ocean acidification on the immune response of the blue mussel *Mytilus edulis*. *Aquatic Biology*, **2**, 67–74.

Blackford, J.C. and Gilbert F.J. (2007). pH variability and CO_2 induced acidification in the North Sea. *Journal of Marine Systems*, **64**, 229–41.

Brander L.M., Rehdanz, K., Tol, R.S.J., and van Beukering, P.J.H. (2009). *The economic impact of ocean acidification on coral reefs*, 33 pp. The Economic and Social Research Institute Working Paper 282. Economic and Social Research Institute, Dublin.

Cao, L. and Caldeira, K. (2008). Atmospheric CO_2 stabilization and ocean acidification. *Geophysical Research Letters*, **35**, L19609, doi:10.1029/2008GL035072.

Cesar, H., van Beukering, P., Pintz, S., and Dierking, J. (2002). *Economic valuation of the coral reefs of Hawaii*, 144 pp. Cesar Environmental Economics Consulting (CEEC), Arnhem, The Netherlands.

Cesar, H., Burke, L., and Pet-Soede, L. (2003). *The economics of worldwide coral reef degradation*, 23 pp. Cesar Environmental Economics Consulting (CEEC), Arnhem, The Netherlands.

Checkley, D.M., Dickson, A.G., Takahashi, M., Radish, J.A., Eisenkolb, N., and Asch, R. (2009). Elevated CO_2 enhances otolith growth in young fish. *Science*, **324**, 1683.

Comeau, S., Gorsky, G., Jeffree, R., Teyssié, J.-L., and Gattuso, J.-P. (2009). *Limacina helicina* threatened by ocean acidification. *Biogeosciences*, **6**, 1877–82.

Cooley, S.R. and Doney, S.C. (2009). Anticipating ocean acidification's economic consequences for commercial fisheries. *Environmental Research Letters*, **4**, 1–8.

Cooley, S.R., Kite-Powell, H.L., and Doney, S.C. (2009). Ocean acidification's potential to alter global marine ecosystem services, *Oceanography*, **22**, 172–81.

Costanza, R., D'Arge, R., De Groot, R. *et al.* (1997). The value of the world's ecosystem services and natural capital. *Nature*, **387**, 253–60.

Dashfield, S.L., Somerfield, P.J., Widdicombe, S., Austen, M.C., and Nimmo, M. (2008). Impacts of ocean acidification and burrowing urchins within-sediment pH profiles and subtidal nematode communities. *Journal of Experimental Marine Biology and Ecology*, **365**, 46–52.

Diethelm, P. and McKee, M. (2009). Denialism: what is it and how should scientists respond? *The European Journal of Public Health*, **19**, 2–4.

Duarte, C.M., Middelburg, J., and Caraco, N. (2005). Major role of marine vegetation on the oceanic carbon cycle. *Biogeosciences*, **2**, 1–8.

Dupont, S., Havenhand, J., Thorndyke, W., Peck, L., and Thorndyke, M. (2008). CO_2-driven ocean acidification radically affects larval survival and development in the brittlestar *Ophiothrix fragilis*. *Marine Ecology Progress Series*, **373**, 285–94.

Ellis, R.P., Bersey, J., Rundle, S.D., Hall-Spencer, J.M., and Spicer J.I. (2009). Subtle but significant effects of CO_2 acidified seawater on embryos of the intertidal snail, *Littorina obtusata*. *Aquatic Biology*, **5**, 41–8.

Fabry, V.J., Seibel, B.A., Feely, R.A., and Orr, J.C. (2008). Impacts of ocean acidification on marine fauna and ecosystem processes. *Journal of Marine Science*, **65**, 414–32.

FAO (1999). *The state of the world fisheries and aquaculture 1998*. FAO, Rome. http://www.fao.org/docrep/w9900e/w9900e00.htm

FAO (2003). *Assessment of the world food security situation. 29th Session of the Committee on World Food Security, 12–16 May 2003* (http://www.fao.org/unfao/bodies/cfs/cfs29/CFS2003-e.htm).

FAO (2005). *FAO/Worldfish Center Workshop on Interdisciplinary Approaches to the Assessment of Small-Scale Fisheries, 20–22 September 2005*. FAO Fisheries Report 787. FAO, Rome.

FAO (2008). *The state of the world's fisheries and aquaculture*, 196 pp. FAO Fisheries and Aquaculture Department, Rome.

Feely, R.A., Sabine, C.L., Hernandez-Ayon, J.M., Lanson, D., and Hales, B. (2008). Evidence for upwelling of corrosive 'acidified' water onto the continental shelf. *Science*, **320**, 1490–2.

Field, C.B., Behrenfeld, M.J., Randerson, J.T., and Falkowski, P. (1998). Primary production of the biosphere: integrating terrestrial and oceanic components. *Science*, **281**, 237–40.

Gazeau, F., Quiblier, C., Jansen, J.M., Gattuso, J.-P., Middelburg, J.J., and Heip C.H.R. (2007). Impact of elevated CO_2 on shellfish calcification. *Geophysical Research Letters*, **34**, L07603, doi:10.1029/2006GL028554.

Gooding, R.A., Harley, C.D.G., and Tang, E. (2009). Elevated water temperature and carbon dioxide concentration increase the growth of a keystone echinoderm. *Proceedings of the National Academy of Sciences USA*, **106**, 9316–21.

Green, M.A., Waldbusser, G.G., Reilly, S.L., Emerson, K., and O'Donnell, S. (2009). Death by dissolution: sediment saturation state as a mortality factor for juvenile bivalves. *Limnology and Oceanography*, **54**, 1037–47.

Guinotte, J.M., Orr, J., Cairns, S., Freiwald, A., Morgan, L., and George, R. (2006). Will human induced changes in seawater chemistry alter the distribution of deep-sea scleractinian corals? *Frontiers in Ecology and Environment*, **4**, 141–6.

Hall-Spencer, J.M., Rodolfo-Metalpa R., Martin, S. *et al.* (2008). Volcanic carbon dioxide vents show ecosystem effects of ocean acidification. *Nature*, **454**, 96–9.

Havenhand, J.N., Buttler, F.-R., Thorndyke, M.C., and Williamson, J.E. (2008). Near-future levels of ocean acidification reduce fertilization success in a sea urchin. *Current Biology*, **18**, R651–R652.

Hilmi, N., Allemand, D., Jeffree, R.A., and Orr, J.C. (2009). Future economic impacts of ocean acidification on Mediterranean seafood: first assessment summary. In: E. Özhan (ed.), *Proceedings of the 9th International Conference on the Mediterranean Coastal Environment MEDCOAST09, Sochi, Russia*, Vol.1.

Hoegh-Guldberg, O., Mumby, P.J., Hooten, A.J. *et al.* (2008). Coral reefs under rapid climate change and ocean acidification. *Science*, **318**, 1737–42.

House of Commons Science and Technology Committee (2010). *The disclosure of climate data from the Climatic Research Unit at the University of East Anglia*, 61 pp. The Stationery Office, House of Commons, London.

IPCC (2001). *Climate change 2001: the scientific basis. Contribution of Working Group I to the Third Assessment Report of the Intergovernmental Panel on Climate Change* (ed. J.T. Houghton, Y. Ding, D.J. Griggs *et al.*). Cambridge University Press, Cambridge.

IPCC (2007). Climate Change 2007: The Physical Science Basis. Contribution of Working Group I to the Fourth Assessment Report of the Intergovernmental Panel on Climate Change (eds Solomon, S., D. Qin, M. Manning, Z. Chen, M. Marquis, K.B. Averyt, M. Tignor and H.L. Miller). Cambridge University Press, Cambridge.

IPCC (2005). IPCC Special Report on Carbon Dioxide Capture and Storage. Prepared by Working Group III of the Intergovernmental Panel on Climate Change (eds Metz, B., O. Davidson, H. C. de Coninck, M. Loos, and L. A. Meyer). Cambridge University Press, Cambridge.

Kemp, J, Milne, R. and Reay, D. (2010). Sceptics and deniers of climate change not to be confused. *Nature*, **464**, 673.

Kite-Powell, K.L. (2009). A global perspective on the economics of ocean acidification. *Journal of Marine Education*, **25**, 25–9.

Kleypas, J.A., Feely, R.A., Fabry, V.J., Langdon, C., Sabine C.L., and Robbins, L.L. (2006). *Impacts of ocean acidifica-*

tion on coral reefs and other marine calcifiers: a guide for future research, 88 pp. NSF, NOAA, and the US Geological Survey, St Petersburg, FL.

Kleypas, J.A., Feely, R., Gattuso, J.-P., and Turley, C. (2010). *Frequently asked questions about ocean acidification*, 16 pp. Ocean Carbon and Biogeochemistry Project, European Project on Ocean Acidification, and the United Kingdom Ocean Acidification Research Programme, Woods Hole, MA.

Kroeker, K.J., Kordas, R.L., Crim, R.N., and Singh, G.G. (2010). Meta-analysis reveals negative yet variable effects of ocean acidification on marine organisms. *Ecology Letters*, **13**, 1419–34.

Kurihara, H., Shimode, S., and Shirayama, Y. (2004). Effects of raised CO_2 concentration on the egg production rate and early development of two marine copepods (*Acartia steueri* and *Acartia erythraea*). *Marine Pollution Bulletin*, **49**, 721–7.

Kurihara, H., Kato, S., and Ishimatsu, A. (2007). Effects of increased seawater pCO_2 on early development of the oyster *Crassostrea gigas*. *Aquatic Biology*, **1**, 91–8.

Kurihara, H., Asai, T., Kato, S., and Ishimatsu, A. (2009). Effects of elevated pCO_2 on early development in the mussel *Mytilus galloprovincialis*. *Aquatic Biology*, **4**, 225–33.

Maier, C., Hegeman, J., Weinbauer, M.G., and Gattuso, J.-P. (2009). Calcification of the cold-water coral *Lophelia pertusa* under ambient and reduced pH. *Biogeosciences*, **6**, 1671–80.

Marshall, D.J., Santos, J.H., Leung, K.M.Y., and Chak, W.H. (2008). Correlations between gastropod shell dissolution and water chemical properties in a tropical estuary. *Marine Environmental Research*, **66**, 422–9.

Maslow, A.H. (1943). A theory of human motivation. *Psychological Review*, **50**, 370–96.

Melzner, F., Göbel, S., Langenbuch, M., Gutowska, M., Pörtner, H.-O., and Lucassen, M. (2009). Swimming performance in Atlantic cod (*Gadus morhua*) following long-term (4–12 months) acclimation to elevated seawater pCO_2. *Aquatic Toxicology*, **92**, 30–7.

Metzger, R., Sartoris, F.J., Langenbuch, M., and Pörtner, H.O. (2007). Influence of elevated CO_2 concentrations on thermal tolerance of the edible crab *Cancer pagurus*. *Journal of Thermal Biology*, **32**, 144–51.

Michaelidis, B., Ouzounis, C., Paleras, A., and Pörtner, H.O. (2005). Effects of long-term moderate hypercapnia on acid–base balance and growth rate in marine mussels *Mytilus galloprovincialis*. *Marine Ecology Progress Series*, **293**, 109–18.

Miles, H., Widdicombe, S., Spicer, J.I., and Hall-Spencer, J. (2007). Effects of anthropogenic seawater acidification on acid–base balance in the sea urchin *Psammechinus miliaris*. *Marine Pollution Bulletin*, **54**, 89–96.

Munday, P.L., Dixson, D.L., Donelson, J.M. *et al.* (2009). Ocean acidification impairs olfactory discrimination and homing ability of a marine fish. *Proceedings of the National Academy of Sciences USA*, **106**, 1848–52.

Munday, P.L., Dixson, D.L., McCormick, M.I., Meekan, M., Ferrari, M.C.O., and Chivers, D.P. (2010). Replenishment of fish populations is threatened by ocean acidification. *Proceedings of the National Academy of Sciences USA*, **107**, 12930–4.

Naylor, N.C., Goldburg, R.J., Primavera, J.H. *et al.* (2000). Effect of aquaculture on world fish supplies. *Nature*, **405**, 1017–24.

Nellemann, C., Corcoran, E., Duarte, C.M., *et al.* (eds) (2009). *Blue carbon. A rapid response assessment*, 78 pp. United Nations Environment Programme, GRID-Arendal, Norway.

Ocean Acidification Reference User Group (2009). *Ocean acidification: the facts. A special introductory guide for policy advisors and decision makers* (ed. D. d'A. Laffoley and J.M. Baxter). European Project on Ocean Acidification (EPOCA).

Ocean Acidification Reference User Group (2010). *Ocean acidification: questions answered: a special introductory guide for policy advisors and decision makers.* (ed. D. d'A. Laffoley and J.M. Baxter). European Project on Ocean Acidification (EPOCA).

Orr, J.C., Fabry, V.J., Aumont, O. *et al.* (2005). Anthropogenic ocean acidification over the twenty-first century and its impact on calcifying organisms. *Nature*, **437**, 681–6.

Pane, E.F. and Barry, J.P. (2007). Extracellular acid-base regulation during short-term hypercapnia is effective in a shallow water crab, but ineffective in a deep-sea crab. *Marine Ecology Progress Series*, **334**, 1–9.

Parker, L.M., Ross, P.M., and O'Connor, W.A. (2009). The effect of ocean acidification and temperature on the fertilization and embryonic development of the Sydney rock oyster *Saccostrea glomerata* (Gould 1850). *Global Change Biology*, **15**, 2123–36.

Parks, N. (2009). Is regulation on ocean acidification on the horizon? *Environmental Science and Technology*, **43**, 6118–19.

Pauly, D., Watson, R., and Alder, J. (2005). Global trends in world fisheries: impacts on marine ecosystems and food security. *Philosophical Transactions of the Royal Society B: Biological Sciences*, **360**, 5–12.

PBL (2010). *Assessing an IPCC assessment. An analysis of statements on projected regional impacts in the 2007 report.* Netherlands Environmental Assessment Agency (PBL).

Netherlands Environmental Assessment Agency, The Hague/Bilthoven.

Ries, J.B., Cohen, A.L., and McCorkle, D.C. (2009). Marine calcifiers exhibit mixed responses to CO_2-induced ocean acidification. *Geology*, **37**, 1131–4.

Rome Declaration on World Food Security and World Food Summit Plan of Action (1996). *World Food Summit, 13 November 1996*, pp. 1–43. Food and Agriculture Organization of the United Nations, Rome.

Royal Society (2005). *Ocean acidification due to increasing atmospheric carbon dioxide*, 68 pp. The Royal Society, London.

Royal Society (2009). *Geoengineering the climate: science, governance and uncertainty*, 98 pp. The Royal Society, London.

Sabine, C.L., Feely, R.A., Gruber, N. *et al.* (2004). The oceanic sink for anthropogenic CO_2. *Science*, **305**, 367–71.

Sheppard Brennand, H, Soars, N., Dworjanyn, S.A., Davis, A.R., and Byrne, M. (2010). Impact of ocean warming and ocean acidification on larval development and calcification in the sea urchin *Tripneustes gratilla*. *PLoS One*, **5**, e11372, doi:10.1371/journal.pone.0011372.

Silverman, J., Lazar, B., Cao, L., Caldeira, K., and Erez, J. (2009). Coral reefs may start dissolving when atmospheric CO_2 doubles. *Geophysical Research Letters*, **36**, L05606, doi:10.1029/2008GL036282.

Spicer, J.I., Raffo, A. and Widdicombe, S. (2007). Influence of CO_2-related seawater acidification on extracellular acid-base balance in the velvet swimming crab *Necora puber*. *Marine Biology*, **151**, 1117–25.

Steinacher, M., Joos, F., Frölicher, T.L., Plattner, G.-K., and Doney, S.C. (2009). Imminent ocean acidification in the Arctic projected with the NCAR global coupled carbon cycle-climate model. *Biogeosciences*, **6**, 515–33.

Stern, N. (2006). *Stern Review: the economics of climate change. Executive summary*. HM Treasury, London.

Talmage, S.C. and Gobler, C.J. (2009). The effects of elevated carbon dioxide concentrations on the metamorphosis, size, and survival of larval hard clams (*Mercenaria mercenaria*), bay scallops (*Argopecten irradians*), and Eastern oysters (*Crassostrea virginica*). *Limnology and Oceanography*, **54**, 2072–80.

Turley, C.M. (1999). Taking research to policy making. *Proceedings of the Third European Marine Science and Technology Conference, Lisbon. Research in Enclosed Seas*, vol. 6, pp. 13–19. EU Directorate – General Science Research and Development., Brussels.

Turley, C.M. and Findlay, H.S. (2009). Ocean acidification as an indicator for climate change. In: T. M. Letcher (ed.), *Climate and global change: observed impacts on planet earth*, pp. 367–90. Elsevier, Amsterdam.

Turley, C.M., Roberts, J.M., and Guinotte, J.M. (2007). Corals in deep-water: will the unseen hand of ocean acidification destroy cold-water ecosystems? *Coral Reefs*, **26**, 445–8.

Turley, C., Eby M., Ridgwell, A.J., *et al.* (2010a). The societal challenge of ocean acidification. *Marine Pollution Bulletin*, **60**, 787–92.

Turley, C., Brownlee, C., Findlay, H.S. *et al.* (2010b). Ocean acidification. In: J.M. Baxter, P.J. Buckley, and M.T. Frost (eds), *MCCIP annual report card 2010–11, summary report*. MCCIP, Lowestoft.

Walther, K., Sartoris, F. J., Bock, C., and Pörtner, H.O. (2009). Impact of anthropogenic ocean acidification on thermal tolerance of the spider crab *Hyas araneus*. *Biogeosciences*, **6**, 2207–15.

Watson, S.A., Southgate, P.C., Tyler, P.A., and Peck, L.S. (2009). Early larval development of the Sydney Rock Oyster *Saccostrea glomerata* under near-future predictions of CO_2-driven ocean acidification. *Journal of Shellfish Research*, **28**, 431–7.

Wood, H.L., Spicer, J.I., and Widdicombe, S. (2008). Ocean acidification may increase calcification rates, but at a cost. *Proceedings of the Royal Society B: Biological Sciences*, **275**, 1767–73.

Worm, B., Barbier, E.B., Beaumont, N. *et al.* (2006). Impacts of biodiversity loss on ocean ecosystem services. *Science*, **314**, 787–90.

Zeebe, R.E., Zachos, J.C., Caldeira, K., and Tyrrell T. (2008). Carbon emissions and acidification. *Science*, **321**, 51–2.

CHAPTER 14

Impact of climate change mitigation on ocean acidification projections

Fortunat Joos, Thomas L. Frölicher, Marco Steinacher, and Gian-Kasper Plattner

14.1 Introduction

Ocean acidification caused by the uptake of carbon dioxide (CO_2) by the ocean is an important global change problem (Kleypas *et al.* 1999; Caldeira and Wickett 2003; Doney *et al.* 2009). Ongoing ocean acidification is closely linked to global warming, as acidification and warming are primarily caused by continued anthropogenic emissions of CO_2 from fossil fuel burning (Marland *et al.* 2008), land use, and land-use change (Strassmann *et al.* 2007). Future ocean acidification will be determined by past and future emissions of CO_2 and their redistribution within the earth system and the ocean. Calculation of the potential range of ocean acidification requires consideration of both a plausible range of emissions scenarios and uncertainties in earth system responses, preferably by using results from multiple scenarios and models.

The goal of this chapter is to map out the spatio-temporal evolution of ocean acidification for different metrics and for a wide range of multigas climate change emissions scenarios from the integrated assessment models (Nakićenović 2000; Van Vuuren *et al.* 2008b). By including emissions reduction scenarios that are among the most stringent in the current literature, this chapter explores the potential benefits of climate mitigation actions in terms of how much ocean acidification can be avoided and how much is likely to remain as a result of inertia within the energy and climate systems. The long-term impacts of carbon emissions are addressed using so-called zero-emissions commitment scenarios and pathways leading to stabilization of atmospheric CO_2. Discussion will primarily rely on results from the cost-efficient Bern2.5CC model (Plattner

et al. 2008) and the comprehensive carbon cycle–climate model of the National Centre for Atmospheric Research (NCAR), CSM1.4-carbon (Steinacher *et al.* 2009; Frölicher and Joos 2010).

The magnitude of the human perturbation of the climate system is well documented by observations (Solomon *et al.* 2007). Carbon emissions from human activities force the atmospheric composition, climate, and the geochemical state of the ocean towards conditions that are unique for at least the last million years (see Chapter 2). The current atmospheric CO_2 concentration of 390 ppmv is well above the natural range of 172 to 300 ppmv of the past 800 000 years (Lüthi *et al.* 2008). The rate of increase in CO_2 and in the radiative forcing from the combination of the well-mixed greenhouse gases CO_2, methane (CH_4), and nitrous oxide (N_2O) is larger during the Industrial Era than during any comparable period of at least the past 16 000 years (Joos and Spahni 2008). Ocean measurements over recent decades show that the increase in surface-ocean CO_2 is being paralleled by a decrease in pH (Doney *et al.* 2009). Ongoing global warming is unequivocal (Solomon *et al.* 2007): observational data show that global-mean sea level is rising, ocean heat content increasing, Arctic sea ice retreating, atmospheric water vapour content increasing, and precipitation patterns changing. The last decade (2000 to 2009) was, on a global average, the warmest in the instrumental record (http://data.giss.nasa.gov/gistemp/). Proxy reconstructions suggest that recent anthropogenic influences have widened the last-millennium multidecadal temperature range by 75% and that late 20th century warmth exceeded peak temperatures over the past millennium by 0.3°C (Frank *et al.* 2010).

The range of plausible 21st century emissions pathways leads to further global warming and ocean acidification (Van Vuuren *et al.* 2008b; Strassmann *et al.* 2009). Projections based on the scenarios of the Special Report on Emissions Scenarios (SRES) of the Intergovernmental Panel on Climate Change (IPCC) give reductions in average global surface pH of between 0.14 and 0.35 units over the 21st century, adding to the present decrease of 0.1 units since pre-industrial time (Orr *et al.* 2005; see also Chapters 1 and 3). Comprehensive earth system model simulations show that continued carbon emissions over the 21st century will cause irreversible climate change on centennial to millennial timescales in most regions, and impacts related to ocean acidification and sea level rise will continue to aggravate for centuries even if emissions are stopped by the year 2100 (Frölicher and Joos 2010). In contrast, in the absence of future anthropogenic emissions of CO_2 and other radiative agents, forced changes in surface temperature and precipitation will become smaller in the next centuries than internal variability for most land and ocean grid cells and ocean acidification will remain limited. This demonstrates that effective measures to reduce anthropogenic emissions can make a difference. However, continued carbon emissions will affect climate and the ocean over the next millennium and beyond (Archer *et al.* 1999; Plattner *et al.* 2008) and related climate and biogeochemical impacts pose a substantial threat to human society.

Thirty years ago, with their box-diffusion carbon cycle model, Siegenthaler and Oeschger (1978) demonstrated the long lifetime of an atmospheric CO_2 perturbation and pointed out that carbon emissions must be reduced 'if the atmospheric radiation balance is not to be disturbed in a dangerous way'. The United Nations Framework Convention on Climate Change (UNFCCC) that came into force in 1994 has the ultimate objective (article 2) 'to achieve . . . stabilization of greenhouse gas concentrations in the atmosphere at a level that would prevent dangerous anthropogenic interference with the climate system. Such a level should be achieved within a time frame sufficient to allow ecosystems to adapt naturally to climate change...' (UN 1992). Scenarios provide a useful framework for establishing policy-relevant information related to the UNFCCC. Here, the link between atmospheric CO_2 level and ocean acidification, and the timescales of change, are addressed.

The outline of this chapter is as follows. In the next section, we will discuss different classes of scenarios, their underlying assumptions, and how these scenarios are used. Metrics for assessing ocean acidification are also introduced. In Section 14.3, the evolution over this century of atmospheric CO_2, global-mean surface air temperature, and global-mean surface ocean acidification for the recent range of baseline and mitigation scenarios from integrated assessment models is presented. In Section 14.4, the minimum commitment, as a result of past and 21st century emissions, to long-term climate change and ocean acidification arising from inertia in the earth system alone is addressed. In Section 14.5, regional changes in surface ocean chemistry are discussed. In Section 14.6, delayed responses, irreversibility, and changes in the deep ocean are addressed using results from the comprehensive NCAR CSM1.4-carbon model. Finally, in Section 14.7, idealized profiles leading to CO_2 stabilization are discussed to further highlight the link between greenhouse gas stabilization, climate change, and ocean acidification. The overarching logic is to use the cost-efficient Bern2.5CC model to explore the scenario space and uncertainties for global-mean values and the NCAR CSM1.4-carbon model to investigate regional details for a limited set of scenarios. The set of baseline and emissions scenarios and associated changes in CO_2 and global-mean surface temperature have been previously discussed (Van Vuuren *et al.* 2008b; Strassmann *et al.* 2009). We also refer to the literature for a more detailed discussion of the NCAR CSM1.4-carbon ocean acidification results (Steinacher *et al.* 2009; Frölicher and Joos 2010).

14.2   Scenarios and metrics

Scenario-based projections are a scientific tool kit for investigating alternative evolutions of anthropogenic emissions and their influence on climate, the earth system, and the socio-economic system. Scenario-based projections are not to be misunderstood as predictions of the future, and as the time horizon increases the basis for the underlying assumptions becomes increasingly uncertain.

One class of scenarios includes idealized emissions or concentration pathways to investigate processes, feedbacks, timescales, and inertia in the climate system. These scenarios are usually developed from a natural science perspective and used for their illustrative power. Examples are a complete instantaneous reduction of emissions at a given year or idealized pathways leading to stabilization of greenhouse gas concentrations (Schimel *et al.* 1997). Another class of emissions scenarios has been developed using integrated assessment frameworks and integrated assessment models (IAMs) by considering plausible future demographic, social, economic, technological, and environmental developments. Examples include the scenarios of the IPCC SRES (Nakićenović *et al.* 2000) and the more recently developed representative concentration pathways (RCPs; Moss *et al.* 2008, 2010; Van Vuuren *et al.* 2008a).

The SRES emissions scenarios do not include explicit climate change mitigation actions. Such scenarios are usually called baseline or reference scenarios. In this chapter, results based on the two illustrative SRES scenarios B1 and A2, a low- and a high-emissions baseline scenario, are discussed (Fig. 14.1). The SRES scenarios have been widely

used in the literature and in the IPCC Fourth Assessment Report. They are internally consistent scenarios in the sense that each is based on a 'narrative storyline' that describes the relationships between the forces driving emissions. The 21st century emissions of the major anthropogenic greenhouse gases [CO_2, CH_4, N_2O, halocarbons, and sulphur hexafluoride (SF_6)], aerosols and tropospheric ozone precursors [sulphur dioxide (SO_2), carbon monoxide (CO), NO_x, and volatile organic compounds (VOCs)] are quantified with IAMs. The extent of the technological improvements contained in the SRES scenarios is not always appreciated. The SRES scenarios include already large and important improvements in energy intensity (energy used per unit of gross domestic product) and the deployment of non-carbon-emitting energy supply technologies compared with the present (Edmonds *et al.* 2004). By the year 2100, the primary energy demand in the SRES scenarios ranges from 55% to more than 90% lower than had no improvement in energy intensity occurred. In addition, in many of the SRES scenarios the deployment of non-carbon-emitting energy supply systems (solar, wind, nuclear, and biomass) exceeds the size of the global energy system in 1990.

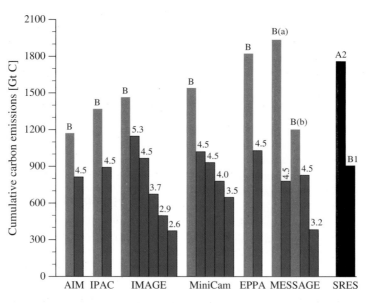

Figure 14.1 Cumulative CO_2 emissions in gigatonnes of carbon (Gt C) over the 21st century for a range of baseline (B; red), climate mitigation (blue), and the SRES A2 and B1 (black) scenarios. The numbers related to the mitigation scenarios indicate the radiative forcing targets in W m^{-2} imposed in the IAMs. The labels below the columns refer to the IAMs used to quantify the scenarios (Weyant *et al.* 2006) and to the SRES scenarios (Nakićenović 2000), respectively.

However, by design they do not include explicit policies to mitigate greenhouse gas emissions, which would lower the extent of climate change experienced over the 21st century.

Progress in developing multigas mitigation scenarios after the SRES report now allows for a comparison between consequences for the earth system of climate mitigation versus baseline scenarios. Figure 14.1 illustrates the relationship between scenarios and individual IAMs and how individual mitigation scenarios are linked to a specific baseline scenario for the post-SRES set of scenarios. Many mitigation scenarios were generated with several IAMs as part of the Energy Modeling Forum Project 21 (EMF-21; Weyant et al. 2006). The IAMs feature representations of the energy system and other parts of the economy, such as trade and agriculture, with varying levels of spatial and process detail. They also include formulations to translate emissions into concentrations and the associated radiative forcing (RF). The latter is a metric for the perturbation of the radiative balance of the lower atmosphere–surface system. Scenarios are generated by minimizing the total costs under the constraints set by societal drivers (e.g. population, welfare, and technological innovation) and most are related to SRES 'storylines'. Adding a constraint on RF in a baseline scenario leads to a scenario with policies specifically aimed at mitigation. The mitigation scenarios analysed here are constrained by stabilization of total RF in the period 2100 to 2150 with RF targets ranging from 2.6 to 5.3 W m^{-2}. From the wider set of baseline and mitigation scenarios described in the literature, four have been specifically selected and termed representative concentration pathways (RCPs). These include two mitigation scenarios with a RF target of 2.6 and 4.5 W m^{-2} and two baseline scenarios with a RF of around 6 and 8.5 W m^{-2} by the end of this century.

Two metrics appear particularly well suited for characterizing the outcome of a scenario in terms of ocean acidification. These are changes in pH and changes in the saturation state of water with respect to aragonite, a mineral form of calcium carbonate (CaCO$_3$) secreted by marine organisms. Ocean uptake of the weak acid CO$_2$ from the atmosphere causes a reduction in pH and in turn alters the CaCO$_3$ precipitation equilibrium (see Chapter 1).

Recent studies indicate that ocean acidification due to the uptake of CO$_2$ has adverse consequences for many marine organisms as a result of decreased CaCO$_3$ saturation, affecting calcification rates, and via disturbance to acid–base physiology (see Chapters 6–8). Vulnerable organisms that build shells and other structures of CaCO$_3$ in the relatively soluble form of aragonite or high-magnesian calcite, but also organisms that form CaCO$_3$ in the more stable form of calcite may be affected. Undersaturation as projected for the high-latitude ocean (Orr et al. 2005; Steinacher et al. 2009) has been found to affect pteropods for example, an abundant group of species forming aragonite shells (Orr et al. 2005; Comeau et al. 2009). Changes in CaCO$_3$ saturation are also thought to affect coral reefs (Kleypas et al. 1999; Langdon and Atkinson 2005; Hoegh-Guldberg et al. 2007; Cohen and Holcomb 2009). The impacts are probably not restricted to ecosystems at the ocean surface, but potentially also affect life in the deep ocean such as the extended deep-water coral systems and ecosystems at the ocean floor. The degree of sensitivity varies among species (Langer et al. 2006; Müller et al. 2010) and there is a debate about whether some taxa may show enhanced calcification at the levels of CO$_2$ projected to occur over the 21st century (Iglesias-Rodriguez et al. 2008). This wide range of different responses is expected to affect competition among species, ecosystem structure, and overall community production of organic material and CaCO$_3$. On the other hand, the impact of plausible changes in CaCO$_3$ production and export (Gangstø et al. 2008) on atmospheric CO$_2$ is estimated to be small (Heinze 2004; Gehlen et al. 2007). Other impacts of ocean acidification with potential influences on marine ecosystems include alteration in the speciation of trace metals as well as an increase in the transparency of the ocean to sound (Hester et al. 2008). The changes in the chemical composition of seawater such as higher concentrations of dissolved CO$_2$ are also likely to affect the coupled carbon and nitrogen cycle and the food web in profound ways (Hutchins et al. 2009), and the volume of water with a ratio of oxygen to CO$_2$ below the threshold for aerobic life is likely to expand (Brewer and Peltzer 2009).

The saturation state with respect to aragonite, Ω_a, is defined by:

$$\Omega_a = \frac{[Ca^{2+}][CO_3^{2-}]}{K^*_{sp}}$$

where brackets denote concentrations in seawater, here for calcium ions and carbonate ions, and K^*_{sp} is the apparent solubility product defined by the equilibrium relationship for the dissolution reaction of aragonite. Similarly, saturation can be defined with respect to calcite which is less soluble than aragonite. Uptake of CO_2 causes an increase in total dissolved inorganic carbon (C_T) and a decrease in the carbonate ion concentration and in saturation (see Chapter 1). Shells or other structures start to dissolve in the absence of protective mechanisms when saturation falls below 1 for the appropriate mineral phase. A value of Ω greater than 1 corresponds to supersaturation. Supersaturated conditions are possible, as the activation energy for forming aragonite or calcite is high.

The pH describes the concentration or, more precisely, the activity of the hydrogen ion in water, a_{H}, by a logarithmic function:

$$pH_T = -\log_{10} a_{H^+}.$$

The activity of hydrogen ion is important for all acid–base reactions. In this chapter, the total pH scale is used as indicated by the subscript T.

14.3 Baseline and mitigation emissions scenarios for the 21st century: how much acidification can be avoided?

Figure 14.1 shows the cumulative carbon emissions over this century for the recent set of baseline and mitigation scenarios (Van Vuuren et al. 2008b) and Fig. 14.2 their temporal evolution. Cumulative CO_2 emissions are in the range of 1170 to 1930 Gt C for the seven baseline scenarios and between 370 and 1140 Gt C for the mitigation scenarios, with the highest emissions associated with a high forcing and a weak mitigation target. In the baseline (no climate policy) scenarios, the range of increase in greenhouse gas emissions by 2100 is from 70 to almost 250% compared with the year 2000 (here, emissions are measured in CO_2-equivalent—CO_2-equivalent emissions of a forcing agent denote the

amount of CO_2 emissions that would cause the same radiative forcing over a time period of 100 years; IPCC 2007). Emissions growth slows down in the second half of the century in all baseline scenarios, because of a combination of stabilizing global population levels and continued technological change. The mitigation scenarios necessarily follow a different path, with a peak in global emissions between 2020 and 2040 at a maximum value of 50% above current emissions.

The projected CO_2 concentrations for the baseline cases calculated with the Bern2.5CC model (Plattner et al. 2008) range from 650 to 960 ppmv in 2100 using best-estimate model parameters (Fig. 14.2C). The CO_2 concentrations in the mitigation scenarios range from 400 to 620 ppmv in 2100. Uncertainties in the carbon cycle and climate sensitivity increase the overall range to 370 to 1310 ppmv (bars in Fig. 14.2C; Plattner et al. 2008). Uncertainties are particularly large for the high end. The two scenario sets, baseline and mitigation, are also distinct with respect to their trends. All baseline scenarios show an increasing trend in atmospheric CO_2, implying rising concentrations beyond 2100. In contrast, the mitigation scenarios show little growth or even a declining trend in CO_2 by 2100.

Projected global-mean surface air temperature changes by the year 2100 (relative to 2000) are 2.4 to 4.2°C (Fig. 14.2D) for the baseline scenarios and best-estimate Bern2.5CC model parameters. Uncertainties in the carbon cycle and climate sensitivity more than double the ranges associated with emissions. For the mitigation scenarios, the projected temperature changes by 2100 are 1.1 to 2.1°C using central model parameters. The mitigation scenarios bring down the overall range of CO_2 and temperature change substantially relative to the baseline range. As for CO_2, the greatest difference compared with the baseline is seen during the second part of the century, when the rate of temperature change slows considerably in all mitigation scenarios in contrast to the baseline scenarios. In several mitigation scenarios, surface air temperature has more or less stabilized by year 2100 (Van Vuuren et al. 2008b; Strassmann et al. 2009).

The evolution of the global-mean saturation state of aragonite (Fig. 14.2E) and pH_T (Fig. 14.2F) in the surface ocean mirrors the evolution of atmospheric

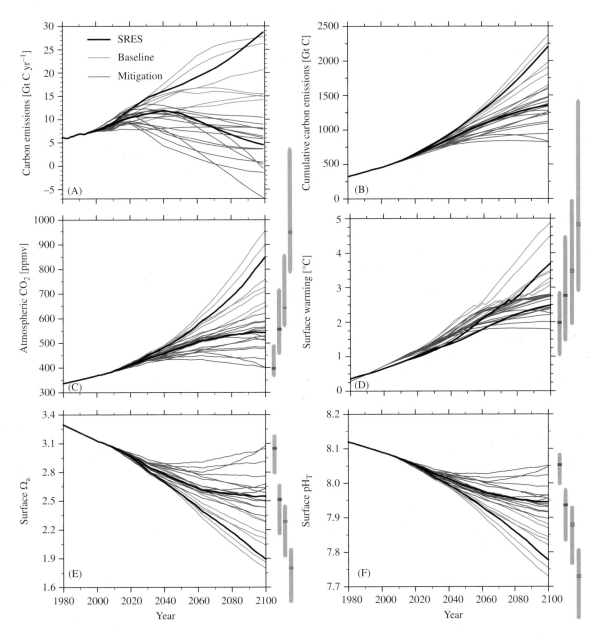

Figure 14.2   (A) Annual and (B) cumulative CO$_2$ emissions prescribed in the Bern2.5CC model and projected (C) atmospheric CO$_2$, (D) changes in global-mean surface air temperature, (E) global average surface saturation with respect to CaCO$_3$ in the form of aragonite, and (F) global average surface pH on the total pH scale (pH$_T$). Baseline scenarios are shown by red lines and mitigation scenarios by blue dotted lines. The SRES high-emissions A2 and low-emissions B1 marker scenarios are given by black lines. Bars indicate uncertainty ranges for the year 2100 and for the four representative concentration pathways (RCPs), marked for evaluation by climate modellers in preparation for the IPCC Fifth Assessment Report. The ranges were obtained by combining different assumptions about the behaviour of the CO$_2$ fertilization effect on land, the response of soil heterotrophic respiration to temperature, and the turnover time of the ocean, thus approaching an upper boundary of uncertainties in the carbon cycle, and additionally accounting for the effect of varying climate sensitivity from 1.5 to 4.5°C (Joos *et al.* 2001).

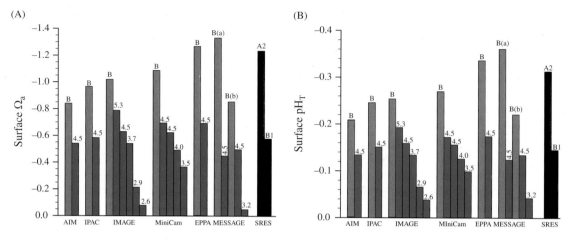

Figure 14.3 21st century change in global-mean surface saturation with respect to aragonite (Ω_a) and change in global-mean surface-ocean pH_T for the range of recent multigas mitigation (blue columns) and baseline (red, 'B') scenarios and the two illustrative SRES scenarios A2 and B1 (black). Numbers indicate the radiative forcing target in W m^{-2} associated with each mitigation scenario. The labels below the columns refer to individual IAMs or to the SRES scenarios. Results are from the standard set-up of the Bern2.5CC model.

CO_2. Ω_a decreases from a pre-industrial value of 3.7 to between 2.3 and 1.8 for the baseline scenarios and to between 3.1 and 2.4 for the mitigation scenarios using the standard model parameters. Again, uncertainties in the projections associated with the carbon cycle and climate sensitivity are largest for the high-emissions scenarios and lower-bound projected Ω_a becomes as low as a global average of 1.4 for the reference scenario with the highest emissions. Global average surface pH_T decreases from a pre-industrial value of 8.18 to 7.88–7.73 for the baseline scenarios and to 8.05–7.90 for the mitigation scenarios, with a lower bound value for the most extreme scenario of 7.6. The uncertainty ranges for Ω_a and pH_T stem almost entirely from uncertainties in the projection of atmospheric CO_2 as carbonate chemistry parameters are well defined and surface-water CO_2 follows the atmospheric rise relatively closely. Trends in surface saturation and pH_T are strongly declining in 2100 for the baseline scenarios, whereas the mitigation scenarios show small or even increasing trends.

The difference between baseline and mitigation scenarios is further highlighted by analysing the overall change in global-mean surface saturation state and pH_T over the 21st century (Fig. 14.3). Surface-ocean mean Ω_a changes over this century between –0.1 and –0.8 for the mitigation scenarios and between –0.9 and –1.4 for the baseline set.

Changes in pH_T are, with a change by –0.04 to –0.19 units, also much smaller for the mitigation than for the baseline set (–0.21 to –0.36).

The following conclusions emerge. Mitigation scenarios decisively lead to lower changes in atmospheric CO_2, to less climate change, and less ocean acidification. The difference in trends by 2100 implies that 21st century mitigation scenarios have a higher impact on the additional increase in CO_2, the additional warming, and the additional ocean acidification even beyond the 21st century than the differences between baseline and mitigation scenarios reported above for 2100. Assuming that these scenarios represent a lower bound on feasible emissions reductions, these results represent an estimate of the 'minimum warming' and of 'minimum ocean acidification' that considers inertia of both the climate system and socio-economic systems.

14.4 Inertia in the earth system: long-term commitment to ocean acidification by 21st century emissions

Simulations in which emissions of carbon and other forcing agents are hypothetically stopped in the year 2000 or 2100 allow us to investigate the legacy effects, i.e. the commitment, of historical and 21st century emissions. Three idealized 'emission-commitment' scenarios run with the NCAR

CSM1.4-carbon model have been selected to illustrate the long-term influence of anthropogenic carbon emissions on ocean acidification and climate (Fig. 14.4). In the first scenario ('Hist' case), emissions are hypothetically set to zero in 2000. In the other two scenarios, emissions follow the SRES B1 (low 'B1_c' case) and SRES A2 (high 'A2_c' case) path until 2100, when emissions are instantaneously set to zero. Setting emissions immediately to zero in 2000 or 2100 is not realistic, but it allows the quantification of the long-term impact of previous greenhouse gas emissions. The three scenarios roughly span the range of 21st century carbon emissions from baseline and mitigation scenarios (Fig. 14.2A). The 'Hist' case obviously features lower emissions (397 Gt C) than any of the mitigation scenarios. Cumulative emissions in the 'B1_c' case (1360 Gt C) are somewhat smaller than to those from the highest mitigation scenario and the lowest baseline scenario shown in Figs 14.1 and 14.2 and emissions in the 'A2_c' case (2210 Gt C) are close to those of the most extreme baseline scenario.

While projected atmospheric CO_2 and surface saturation in 2100 is similar for the Bern2.5CC

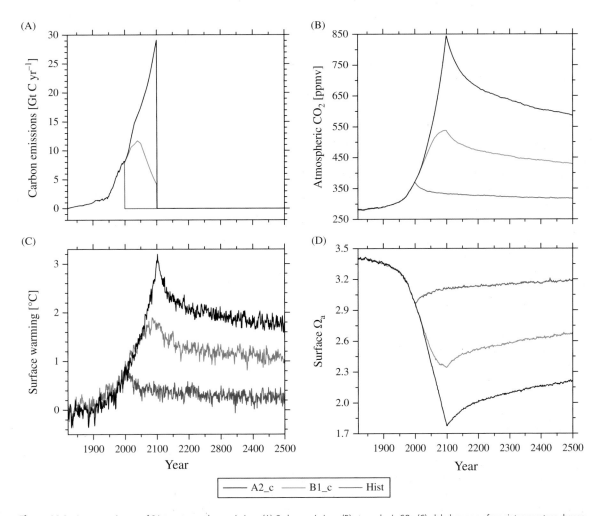

Figure 14.4   Long-term impact of 21st century carbon emissions. (A) Carbon emissions, (B) atmospheric CO_2, (C) global-mean surface air-temperature change, and (D) global average saturation state of surface waters with respect to aragonite (Ω_a) for three illustrative emissions commitment scenarios evaluated with the NCAR CSM1.4-carbon model (Frölicher and Joos 2010). In the high 'A2_c' case and the low 'B1_c' case, 21st century emissions follow the SRES A2 and SRES B1 business-as-usual scenario, respectively. Emissions are set to zero in both cases after 2100. In the 'Hist' case, emissions are stopped in the year 2000.

model (Fig. 14.2C) and the CSM1.4-carbon (Fig. 14.4B), projected 21st century warming is lower in CSM1.4-carbon than in the Bern2.5CC model. This difference is primarily related to the difference in climate sensitivity; 2°C for a nominal doubling of CO_2 in the CSM1.4 versus 3.2°C for the Bern2.5CC best estimate.

Atmospheric CO_2 concentration increases by 300% and by 190% over this century in the high 'A2_c' and low 'B1_c' case, respectively (Fig. 14.4B). Thereafter, atmospheric CO_2 decreases only very slowly, although carbon emissions are (unrealistically) reduced to zero in 2100. Atmospheric CO_2 concentration is still twice as high by 2500 than in pre-industrial times in the 'A2_c' case. On the other hand, CO_2 falls below 350 ppmv within a few decades in the 'Hist' case. The global-mean surface temperature anomaly peaks at 3°C in the 'A2_c' case and at 1.7°C in the 'B1_c' case and remains elevated for centuries (Fig. 14.4C). In the 'Hist' case, global-mean surface temperature remains only slightly perturbed (0.2°C warming) by 2500. The global average saturation state of aragonite in the surface ocean closely follows the evolution of atmospheric CO_2. Mean surface Ω_a is reduced by about half in 2100 for the 'A2_c' case and remains reduced over the next centuries.

The long perturbation lifetime of CO_2 is a consequence of the centennial to millennial timescales of overturning of various carbon reservoirs. Most of the excess carbon is taken up by the ocean and slowly (on a multicentury to millennial timescale) mixed down to the abyss. Ultimately, interaction with ocean sediments and the weathering cycle will remove the anthropogenic carbon perturbation from the atmosphere on timescales of millennia to hundreds of millennia (Archer *et al.* 1999; see Chapter 2).

In conclusion, the results from the commitment scenarios show that the magnitude of 21st century CO_2 emissions pre-determines the range of atmospheric CO_2 concentrations, temperature, and ocean acidification for the coming centuries, at least in the absence of the large-scale deployment of a technology to remove excess CO_2 from the atmosphere. In other words, the CO_2 emitted in the next decades will perturb the physical climate system, biogeochemical cycle, and ecosystems for centuries.

14.5 Regional changes in surface ocean acidification: undersaturation in the Arctic is imminent

The impacts of climate change and ocean acidification on natural and socio-economic systems depend on local and regional changes in climate and acidification rather than on global average metrics. It is important to recognize that the global-mean metrics discussed in the previous sections lead to different changes regionally (Fig. 14.5). Fortunately, the spatial patterns of change in pH_T and surface Ω_a scale closely with atmospheric CO_2 (Figs 14.6 and 14.7). This eases the discussion of local changes for the scenario range and enables us to make inferences for local and regional changes from projected atmospheric CO_2. This section presents regional changes in the saturation state of aragonite, Ω_a, and pH_T for the three commitment scenarios introduced in the previous section and as evaluated in the NCAR CSM1.4-carbon model.

There are large regional differences in the surface saturation state for pre-industrial conditions and in its change over time (Figs 14.5 and 14.6). The surface ocean was saturated with respect to aragonite in all regions under pre-industrial conditions (Kleypas *et al.* 1999; Key *et al.* 2004; Steinacher *et al.* 2009); the lowest saturation levels are simulated in the Arctic and in the Southern Ocean, whereas surface water with saturation values above 4 can be found in the tropics.

Surface-water saturation is projected to decrease rapidly in all regions until 2100 and remains reduced for centuries for all three zero-emission commitment scenarios (Fig. 14.6). The largest ocean-surface changes are found in the tropics and subtropics for Ω_a. In the high 'A2_c' case, Ω_a in the tropics and subtropics decreases from a saturation state of more than 4 in pre-industrial times to saturation below 2.5 at the end of the 21st century. The saturation of tropical and subtropical surface waters remains below 3 until 2500. Although experimental evidence remains scarce, these projected low saturation states in combination with other stress factors such as increased temperature pose the risk of the irreversible destruction of warm-water coral reefs (Kleypas *et al.* 1999; Hoegh-Guldberg *et al.* 2007).

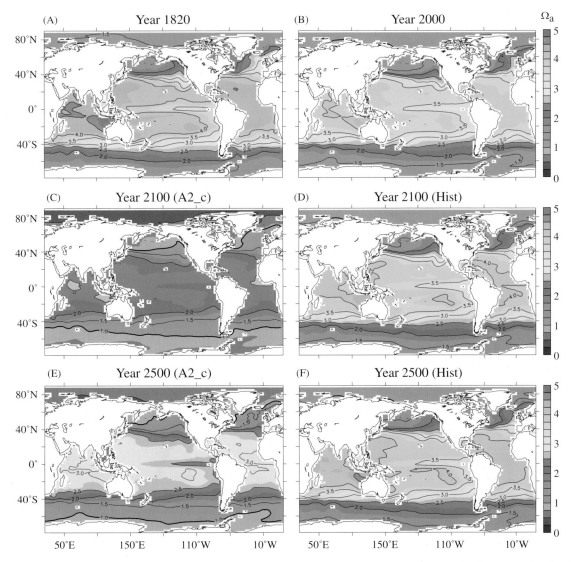

Figure 14.5 Regional distribution of the annual-mean saturation state with respect to aragonite (Ω_a) in the surface ocean for (A) pre-industrial conditions (here 1820), (B) by the year 2000, (C, D) by the end of the century, and (E, F) by 2500. The NCAR CSM1.4-carbon model was forced with reconstructed CO_2 emissions up to 2000. Emissions were set to zero after 2100 in the high 'A2_c' case (C, E) and after 2000 in the 'Hist' case (D, F). Blue colours indicate undersaturation and green to red colours supersaturation.

Undersaturation in the Arctic is imminent (Fig. 14.6C). By the time atmospheric CO_2 exceeds 490 ppmv (in 2040 in the 'A2_c' case), more than half of the Arctic Ocean will be undersaturated (annual mean; Steinacher *et al.* 2009). Undersaturation with respect to aragonite remains widespread in the Arctic Ocean for centuries even after cutting emissions in 2100 for both the 'A2_c' and the 'B1_c' cases (Frölicher and Joos 2010). The Southern Ocean becomes undersaturated on average when atmospheric CO_2 exceeds 580 ppmv (Orr *et al.* 2005) and remains undersaturated for centuries for the 'A2_c' commitment case. Large-scale undersaturation in the Southern Ocean is avoided in the 'B1_c' and 'Hist' cases.

The main reason for the vulnerability of the Arctic Ocean is its naturally low saturation state. In addition, climate change amplifies ocean acidification in

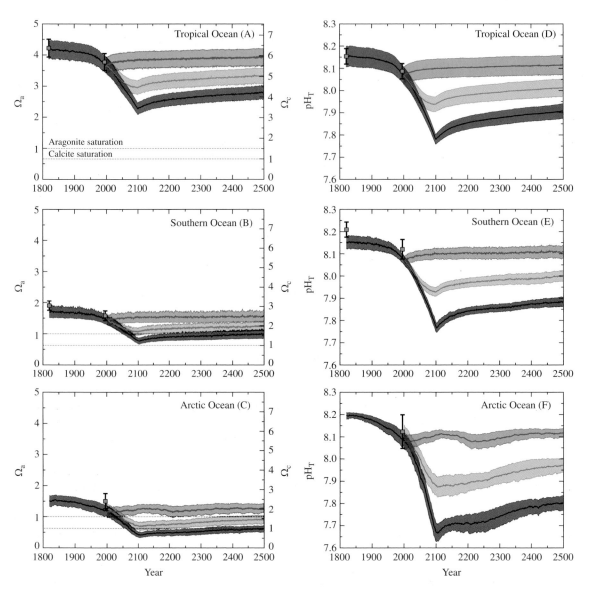

Figure 14.6 Projected evolution of CaCO$_3$ saturation states (left) and total pH (right) in the surface of the tropical ocean (30°N–30°S), Southern Ocean (60°S–90°S), and Arctic Ocean (65°N–90°N, except the Labrador and Greenland–Iceland–Norwegian seas) and for emissions commitment scenarios with no ('Hist' case, blue line, blue shading), low ('B1_c' case, red line, red shading) and high ('A2_c' case, black line, grey shading) emissions in the 21st century. Saturation with respect to aragonite, Ω_a, is indicated on the left y-axis and with respect to calcite, Ω_c, on the right y-axis. Shown are modelled annual means as well as the combined spatial and interannual variability of annual-mean values within each region (shading, ±1 SD). Observation-based estimates are shown by squares for the Southern Ocean and the tropics (GLODAP and World Ocean Atlas 2001, annual mean) and for summer conditions in the Arctic Ocean (CARINA database) with bars indicating the spatial variability. Model results are from the NCAR CSM1.4-carbon model. The level of $\Omega = 1$ separating supersaturated and undersaturated conditions for aragonite and calcite is shown by dashed, horizontal lines.

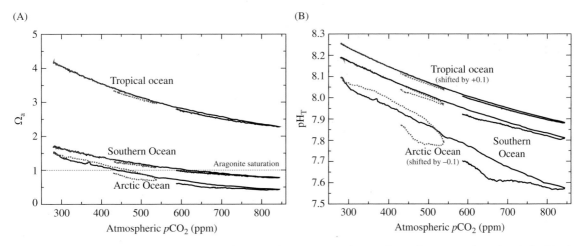

Figure 14.7 Saturation state (Ω_a) and total pH (pH$_T$) in surface water of three regions as a function of atmospheric CO_2. Results are from the low 'B1_c' (dashed) and high 'A2_c' (solid) commitment scenarios. The relation between atmospheric CO_2 and saturation state and pH$_T$ shows almost no path dependency in the tropical ocean and Southern Ocean. Some path dependency is found in the Arctic Ocean, with lower values in surface saturation and pH$_T$ for a given CO_2 concentration simulated after the peak in atmospheric CO_2. Note that the pH$_T$–CO_2 curves are shifted by +0.1 pH units for the tropical region and by –0.1 pH units for the Arctic region for clarity.

the Arctic, in contrast to other regions such as the Southern Ocean and the low-latitude oceans, where climate change has almost no effect on the saturation state in our simulations. Climate change amplifies the projected decrease in annual-mean Ω_a in the Arctic Ocean by 22% mainly due to surface freshening in response to the retreat of sea ice, causing local alkalinity to decrease and the uptake of anthropogenic carbon to increase (see also Chapter 3).

In summary, regional changes in the saturation state and pH$_T$ of surface waters are distinct. The largest decrease in pH$_T$ is simulated in the Arctic Ocean, where the lowest saturation is also found. Undersaturation is imminent in Arctic surface water (Figs 14.6 and 14.7) and remains widespread over centuries for 21st century carbon emissions of the order of 1000 Gt C or more.

14.6   Delayed responses in the deep ocean

Ocean acidification also affects the ocean interior as anthropogenic carbon continues to invade the ocean. Figure 14.8 displays how the saturation state and pH$_T$ changes along the transect from Antarctica, through the Atlantic Ocean to the North Pole for the 'A2_c' commitment scenario. The saturation

horizon separating supersaturated from undersaturated water rises from a depth between ~2000 and 3000 m all the way up to the surface at high latitudes. The volume of water that is supersaturated with respect to aragonite strongly decreases with time. In parallel, the volume of water with low pH$_T$ expands.

A general decrease in $CaCO_3$ saturation corresponds to a loss of volume providing habitat for many species that produce $CaCO_3$ structures. Following Steinacher et al. (2009), five classes of aragonite saturation levels are defined: (1) Ω_a above 4, considered optimal for the growth of warm-water corals, (2) Ω_a of 3 to 4, considered as adequate for coral growth, (3) Ω_a of 2 to 3, (4) Ω_a of 1 to 2, considered marginal to inadequate for coral growth, though experimental evidence is scarce, and finally (5) undersaturated water considered to be unsuitable for aragonite producers. Figure 14.9 shows the evolution of the ocean volume occupied by these five classes for the three commitment simulations. In the 'A2_c' case, water masses with saturation above 3 vanish by 2070 (CO_2 ~ 630 ppmv). Overall, the volume occupied by supersaturated water decreases from 40% in pre-industrial times to 25% in 2100 and 10% in 2300, and the volume of undersaturated water

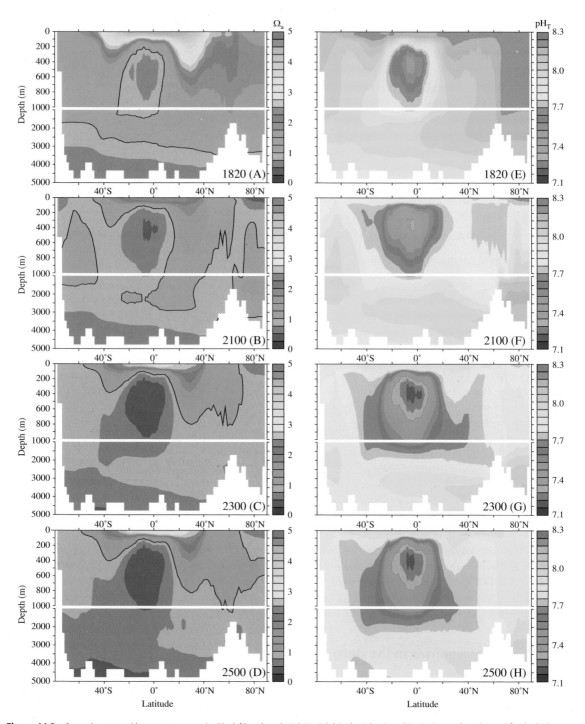

Figure 14.8 Saturation state with respect to aragonite (Ω_a, left) and total pH (pH$_T$, right) in the Atlantic and Arctic Oceans (zonal mean) for the high 'A2_c' commitment scenario by the year 1820 (A, E), 2100 (B, F), 2300 (C, G), and 2500 (D, H). Blue colours in the left panels indicate undersaturation. Note the different depth scales for the upper and the deep ocean, separated by the white horizontal line.

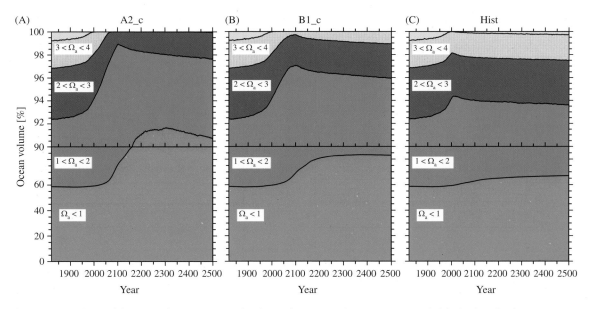

Figure 14.9 Evolution of the volume of water occupied by five classes of saturation with respect to aragonite (Ω_a) for the three illustrative emissions commitment scenarios. In the high 'A2_c' case and the low 'B1_c' case, 21st century emissions follow the SRES A2 and SRES B1 business-as-usual scenario, respectively. Emissions are set to zero in both cases after 2100. In the 'Hist' case, emissions are stopped in the year 2000. Note that the y-axis is stretched above 90%.

increases accordingly. The low 'B1_c' case also exhibits a large expansion of undersaturated water from 59 to 83% of ocean volume. In the 'Hist' case, the perturbations in volume fractions are much more modest and trends are largely reversed in the well-saturated upper ocean over the next few centuries.

The response of the saturation state is delayed in the ocean interior. This delay reflects the centennial timescales of the surface-to-deep transport of the anthropogenic carbon perturbation (Fig. 14.8). In the 'A2_c' case, the volume of undersaturated water reaches its maximum around 2300, 200 years after emissions have been stopped (Fig. 14.9). Even in the 'Hist' case, the volume fraction of supersaturated water continues to decrease. The fact that the volume fraction continues to change significantly after 2100 in the 'A2_c' and 'B1_c' cases demonstrates that some impacts of 21st century fossil fuel carbon emissions are strongly delayed and cause problems for centuries even for the extreme case of an immediate stop to emissions, i.e. the long-term commitment is substantial.

14.7 Pathways leading to stabilization of atmospheric CO$_2$

In this section, illustrative pathways leading to stabilization in atmospheric CO$_2$ are discussed in terms of their implications for projected ocean acidification and carbon emissions from the Bern2.5CC model (Fig. 14.10). The idea behind prescribing the CO$_2$ pathway is to illustrate how anthropogenic emissions have to develop if atmospheric CO$_2$ is to be stabilized as called for by the UNFCCC (UN 1992) and to illustrate the link between changes to the earth system and CO$_2$ stabilization levels.

Atmospheric CO$_2$ is prescribed to stabilize at levels ranging from 350 to 1000 ppmv. This causes surface-ocean saturation, Ω_a, to stabilize between 3.3 and 1.7 compared to a pre-industrial mean of 3.7. Surface-ocean pH$_T$ stabilizes at 8.1 to 7.7 (pre-industrial 8.2). This corresponds to an increase in the hydrogen ion concentration of 20 to 300% relative to pre-industrial times. Global-mean change of temperature of the surface ocean is between 1°C for the 350 ppmv stabilization level and 5°C for the 1000 ppmv stabilization level when the climate

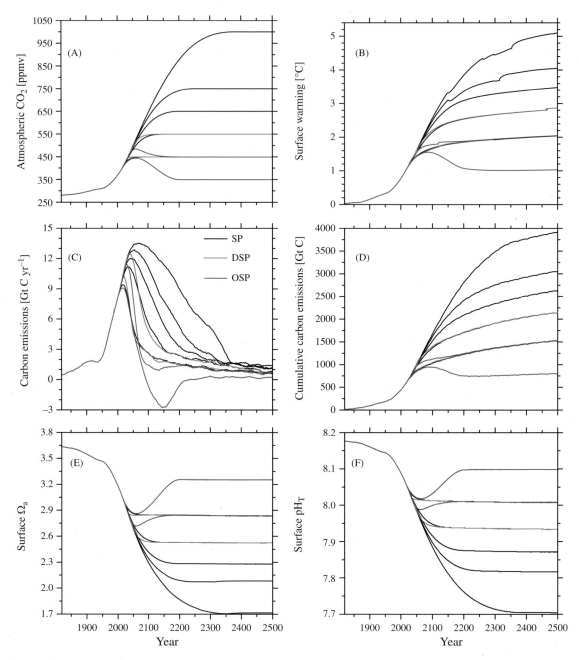

Figure 14.10   (A) Prescribed atmospheric CO_2 for pathways leading to stabilization and Bern2.5CC model projected (B) global-mean surface air temperature change, (C) annual and (D) cumulative carbon emissions, (E) global-mean surface saturation with respect to aragonite (Ω_a), and (F) global-mean surface total pH (pH$_T$). Pathways where atmospheric CO_2 overshoots the stabilization concentration are shown as blue lines and pathways with a delayed approach to stabilization as red lines; the different pathways to the same stabilization target illustrate how results depend on the specifics of the stabilization pathway. The label SP refers to stabilization profile, DSP to delayed stabilization profile, and OSP to overshoot stabilization profile.

sensitivity in the Bern2.5CC model is set to 3.2°C for a nominal doubling of CO_2.

Carbon emissions must drop if CO_2 is to be stabilized. Carbon emissions are allowed to increase for a few years to a few decades, depending on the pathway, but then have to drop in all cases and eventually become as low as the long-term geological carbon sink of a few tenths of a Gt C only. This is a consequence of the long lifetime of the anthropogenic perturbation. Atmospheric CO_2 thus reflects the sum of past emissions rather than current emissions. Cumulative emissions by 2500 for the Bern2.5CC model are in the range of 750 to 4000 Gt C for a stabilization between 350 and 1000 ppmv. Conventional fossil resources, mainly in the form of coal, are estimated to be about 5000 Gt C.

If emission reductions are delayed (DSP and OSP pathways in Fig. 14.10), more stringent reductions have to be implemented later in order to meet a certain stabilization target. This is illustrated by the two overshoot scenarios in which atmospheric CO_2 is prescribed to increase above the final stabilization levels and by the delayed scenarios where CO_2 is allowed to further increase initially.

The higher the emissions the larger the fraction of CO_2 emissions that stays airborne on timescales up to a few thousand years. This fraction is 20% for the 350 ppmv target and 40% for the 1000 ppmv target by the year 2500. The higher airborne fraction for high relative to low stabilization levels is primarily a consequence of the non-linearity of the seawater carbonate chemistry. The higher the partial pressure of CO_2, the smaller is the relative change in dissolved inorganic carbon for a given change in pCO_2. Thus, the partitioning of carbon between the atmosphere and the ocean shifts towards a higher fraction remaining airborne the greater the amount of carbon added to the ocean–atmosphere system.

14.8 Conclusions

We have examined a large set of projections for 21st century emissions for CO_2 and for a suite of non-CO_2 greenhouse and other air pollutant gases from the recent scenario literature (Van Vuuren *et al.* 2008b). Emissions scenarios provide an indication of the potential effects of mitigation policies. Most of the IAMs used to generate the set of baseline and

mitigation scenarios are idealized in many ways. New technologies and policies are assumed to be globally applicable and are often introduced over relatively short periods of time. Especially in the lowest mitigation scenarios, it is assumed that global climate policies can be implemented in the next few years to allow emissions to peak by 2020. These scenarios do not deal with the question of political feasibility and assume mitigation policies are implemented globally.

Physical impacts in terms of ocean acidification and climate change are lower in mitigation than baseline scenarios. Global average surface saturation with respect to aragonite is reduced to 3.1–2.4 by year 2100 in the mitigation compared to 2.3–1.8 in the baseline scenarios. The lowest scenarios result in a decrease in saturation state of 0.6 by 2100 compared with pre-industrial values and show only a small difference of 0.1 between current and end of century saturation conditions. These scenarios provide a guide to the range of global-mean surface acidification that may occur, assuming an ambitious climate policy. These low scenarios with forcing targets below 3 W m^{-2} depart from the corresponding no-climate policy baseline by 2015–2020 and incorporate the widespread development and deployment of existing carbon-neutral technologies. They require socio-political and technical conditions very different from those now existing.

Global emissions in the scenarios with a 4.5 W m^{-2} forcing target begin to diverge from baseline values by about 2020 to 2030, with emissions dropping to approximately present levels by 2100. CO_2, temperature, and ocean acidification start to diverge from the baseline projections later than emissions. This emphasizes the importance of early decisions to meet specific climate change mitigation targets.

Trends can be persistent and impacts of carbon emissions may continue for decades and centuries, long after carbon emissions have been reduced, due to the inertia in the climate–carbon system. This is exemplified by emissions commitment scenarios where carbon and other emissions are hypothetically set to zero and subsequent changes can be investigated. The projected global changes will affect different regions differently depending on their vulnerability to these changes. Widespread year-round undersaturation of surface waters in the

Arctic Ocean with respect to aragonite is likely to become reality in only a few years (Steinacher *et al.* 2009) and ocean acidification and Arctic undersaturation from baseline 21st century carbon emissions is irreversible on human timescales (Frölicher and Joos 2010). Globally, the volume of supersaturated water decreases for another two centuries after carbon emissions stop; the fraction of the ocean volume occupied by supersaturated water is as low as 8% in 2300 with the 'A2_c' case compared with 42% for pre-industrial conditions.

The focus of the analysis above is mainly on the magnitude of change. However, it should be stressed that rates of change are important. The rates of change of climate and ocean acidification co-determine the impacts on natural and socio-economic systems and their capabilities to adapt. Earlier analyses of the ice core and atmospheric records show that the 20th-century increase in CO_2 and its radiative forcing occurred more than an order of magnitude faster than any sustained change during at least the past 22 000 years (Joos and Spahni 2008). This implies that global climate change and ocean acidification, which are anthropogenic in origin, are progressing at high speed. It is evident from Fig. 14.2 that rates of change in surface-ocean pH_T and in Ω_a are much lower for the range of mitigation scenarios than for the range of baseline emissions scenarios.

A range of geoengineering options have been discussed to limit potential impacts of anthropogenic carbon emissions and climate change. Here, we summarize a few conclusions from a recent report (The Royal Society 2009). CO_2 removal techniques address the root cause of climate change by removing CO_2 from the atmosphere. Solar radiation management techniques attempt to offset the effects of increased greenhouse gas concentrations by causing the earth to absorb less solar radiation. Obviously, solar radiation management techniques do not contribute in a relevant way to mitigation of ocean acidification. Of the CO_2 removal methods assessed, none has yet been demonstrated to be effective at an affordable cost and with acceptable side-effects (The Royal Society 2009). If safe and low-cost methods can be deployed at an appropriate scale they could make an important contribution to reducing CO_2 concentrations and could provide a useful complement to conventional emissions reductions. Methods that remove CO_2 from the atmosphere without perturbing natural systems, and without requirements for large-scale land-use changes, such as CO_2 capture from air (IPCC 2005) and possibly also enhanced weathering, are likely to have fewer side-effects. Geoengineering techniques are currently not ready for application, in contrast to low-carbon technologies.

Experimental evidence has emerged in the past years that ocean acidification has negative impacts on many organisms and may severely affect cold- and warm-water corals or high-latitude species such as aragonite-producing pteropods. Considering the precautionary principle mentioned in the UNFCCC, our results may imply that atmospheric CO_2 should be stabilized somewhere around 450 ppmv or below in order to avoid the risk of large-scale disruptions in marine ecosystems. A stabilization of atmospheric CO_2 at or below 450 ppmv requires a stringent reduction in carbon emissions over the coming decades. The results from the IAMs suggest that such a low stabilization target is economically feasible.

14.9 Acknowledgements

This chapter is a contribution to the European Project on Ocean Acidification, EPOCA (FP7/2007–2013; no. 211384). We acknowledge support by the Swiss National Science Foundation and by the EU projects CARBOOCEAN (511176) and EUR-OCEANS (511106–2). Simulations with the NCAR CSM1.4-carbon model were carried out at the Swiss National Computing Center in Manno, Switzerland. We thank S. C. Doney, I. Fung, K. Lindsay, J. John, and colleagues for providing the CSM1.4-carbon code and J.-P. Gattuso, L. Hansson, and A. Oschlies for helpful comments.

References

Archer, D., Kheshgi, H., and Maier-Reimer, E. (1999). Dynamics of fossil fuel CO_2 neutralization by marine $CaCO_3$. *Global Biogeochemical Cycles*, **12**, 259–76.

Brewer, P.G. and Peltzer, E.T. (2009). Limits to marine life. *Science*, **324**, 347–8.

Caldeira, K. and Wickett, M.E. (2003). Anthropogenic carbon and ocean pH. *Nature*, **425**, 365.

Cohen, A.L. and Holcomb, M. (2009). Why corals care about ocean acidification: uncovering the mechanism. *Oceanography*, **2009**, 118–27.

Comeau, S., Gorsky, G., Jeffree, R., Teyssié, J.-L., and Gattuso, J.-P. (2009). Impact of ocean acidification on a key Arctic pelagic mollusc (*Limacina helicina*). *Biogeosciences*, **6**, 1877–82.

Doney, S.C., Fabry, V.J., Feely, R.A., and Kleypas, J.A. (2009). Ocean acidification: the other CO_2 problem. *Annual Review of Marine Science*, **1**, 169–92.

Edmonds, J., Joos, F., Nakićenović, N., Richels, R.G., and Sarmiento, J.L. (2004). Scenarios, targets, gaps and costs. In: C.B. Field and M.R. Raupach (eds), *The global carbon cycle: integrating humans, climate and the natural world*, pp. 77–102. Island Press, Washington, DC.

Frank, D.C., Esper, J., Raible, C.C. *et al.* (2010). Ensemble reconstruction constraints of the global carbon cycle sensitivity to climate. *Nature*, **463**, 527–30.

Frölicher, T.L. and Joos, F. (2010). Reversible and irreversible impacts of greenhouse gas emissions in multi-century projections with the NCAR global coupled carbon cycle-climate model. *Climate Dynamics*, **35**, 1439–59.

Gangstø, R., Gehlen, M., Schneider, B., Bopp, L., Aumont, O., and Joos, F. (2008). Modeling the marine aragonite cycle: changes under rising carbon dioxide and its role in shallow water $CaCO_3$ dissolution. *Biogeosciences*, **5**, 1057–72.

Gehlen, M., Gangstø, R., Schneider, B., Bopp, L., Aumont, O., and Ethe, C. (2007). The fate of pelagic $CaCO_3$ production in a high CO_2 ocean: a model study. *Biogesciences*, **4**, 505–19.

Heinze, C. (2004). Simulating oceanic $CaCO_3$ export production in the greenhouse. *Geophysical Research Letters*, **31**, L16308, doi:10.1029/2004GL020613.

Hester, K.C., Peltzer, E.T., Kirkwood, W.J., and Brewer, P.G. (2008). Unanticipated consequences of ocean acidification: a noisier ocean at lower pH. *Geophysical Research Letters*, **35**, L19601, doi:19610.11029/12008GL034913.

Hoegh-Guldberg, O., Mumby, P.J., Hooten, A.J. *et al.* (2007). Coral reefs under rapid climate change and ocean acidification. *Science*, **318**, 1737–42.

Hutchins, D.A., Mulholland, M.R., and Fu, F. (2009). Nutrient cycles and marine microbes in a CO_2-enriched ocean. *Oceanography*, **22**, 128–45.

Iglesias-Rodriguez, M.D., Halloran, P.R., Rickaby, R.E.M. *et al.* (2008). Phytoplankton calcification in a high-CO_2 world. *Science*, **320**, 336–40.

IPCC (2005). IPCC Special Report on Carbon Dioxide Capture and Storage. Prepared by Working Group III of the Intergovernmental Panel on Climate Change (eds Metz, B., O. Davidson, H. C. de Coninck, M. Loos, and L. A. Meyer). Cambridge University Press, Cambridge.

IPCC (2007). Climate Change 2007: Mitigation. Contribution of Working Group III to the Fourth Assessment Report of the Intergovernmental Panel on Climate Change (eds B. Metz, O.R. Davidson, P.R. Bosch, R. Dave, L.A. Meyer). Cambridge University Press, Cambridge.

Joos, F. and Spahni, R. (2008). Rates of change in natural and anthropogenic radiative forcing over the past 20,000 years. *Proceedings of the National Academy of Sciences USA*, **105**, 1425–30.

Joos, F., Prentice, I.C., Sitch, S. *et al.* (2001). Global warming feedbacks on terrestrial carbon uptake under the Intergovernmental Panel on Climate Change (IPCC) emission scenarios. *Global Biogeochemical Cycles*, **15**, 891–908.

Key, R.M., Kozyr, A., Sabine, C.L. *et al.* (2004). A global ocean carbon climatology: results from Global Data Analysis Project (GLODAP). *Global Biogeochemical Cycles*, **18**, GB4031, doi:10.1029/2004GB002247.

Kleypas, J.A., Buddemeier, R.W., Archer, D., Gattuso, J.P., Langdon, C., and Opdyke, B.N. (1999). Geochemical consequences of increased atmospheric carbon dioxide on coral reefs. *Science*, **284**, 118–20.

Langdon, C. and Atkinson, M.J. (2005). Effect of elevated pCO_2 on photosynthesis and calcification of corals and interactions with seasonal change in temperature/irradiance and nutrient enrichment. *Journal of Geophysical Research – Oceans*, **110**, C09S07, doi:10.1029/2004JC002576.

Langer, G., Geisen, M., Baumann, K. *et al.* (2006). Species-specific responses of calcifying algae to changing seawater carbonate chemistry. *Geochemistry, Geophysics, Geosystems*, **7**, Q09006, doi:09010.01029/02005GC001227.

Lüthi, D., Floch, M.L., Bereiter, B. *et al.* (2008). High-resolution carbon dioxide concentration record 650,000–800,000 years before present. *Science*, **453**, 379–82.

Marland, G., Boden, T.A., and Andres, R.J. (2008). *Global, regional, and national CO_2 emissions*. Carbon Dioxide Information Analysis Center, Oak Ridge National Laboratory, US Department of Energy, Oak Ridge, TN.

Moss, R.H., Babiker, M., Brinkman, S. *et al.* (2008). *Towards new scenarios for analysis of emissions, climate change, impacts, and response strategies*, 132 pp. Intergovernmental Panel on Climate Change, Geneva.

Moss, R.H., Edmonds, J.A., Hibbard, K.A. *et al.* (2010). The next generation of scenarios for climate change research and assessment. *Nature*, **463**, 747–56.

Müller, M.N., Schulz, K.G., and Riebesell, U. (2010). Effects of long-term high CO_2 exposure on two species of coccolithophores. *Biogesciences*, **7**, 1109–16.

Nakićenović, N. Alcamo, J., Davis, G. *et al.* (2000). *Special report on emissions scenarios*. Intergovernmental Panel on Climate Change, Cambridge University Press, New York.

Orr, J.C., Fabry, V.J., Aumont, O. *et al.* (2005). Anthropogenic ocean acidification over the twenty-first century and its impact on calcifying organisms. *Nature*, **437**, 681–6.

Plattner, G.-K., Knutti, R., Joos, F. *et al.* (2008). Long-term climate commitments projected with climate - carbon cycle models. *Journal of Climate*, **21**, 2721–51.

Schimel, D., Grubb, M., Joos, F. *et al.* (1997). *IPCC technical paper III. Stabilisation of atmospheric greenhouse gases: physical, biological, and socio-economic implications.* Intergovernmental Panel on Climate Change, Geneva.

Siegenthaler, U. and Oeschger, H. (1978). Predicting future atmospheric carbon dioxide levels. *Science*, **199**, 388–95.

Solomon, S., Qin, D., Manning, M. *et al.* (2007). Technical summary. In: S. Solomon, D. Qin, M. Manning, Z. Chen, M. Marquis, K.B. Averyt, M. Tignor, and H.L. Miller (eds), *Climate change 2007: the physical science basis. Contribution of Working Group I to the Fourth Assessment Report of the Intergovernmental Panel on Climate Change*, pp. 19–91. Cambridge University Press, Cambridge.

Steinacher, M., Joos, F., Frölicher, T.L., Plattner, G.-K., and Doney, S.C. (2009). Imminent ocean acidification in the Arctic projected with the NCAR global coupled carbon cycle-climate model. *Biogeosciences*, **6**, 515–33.

Strassmann, K.M., Joos, F., and Fischer, G. (2007). Simulating effects of land use changes on carbon fluxes: past contributions to atmospheric CO_2 increases and future commitments due to losses of terrestrial sink capacity. *Tellus B*, **60B**, 583–603.

Strassmann, K.M., Plattner, G.K., and Joos, F. (2009). CO_2 and non-CO_2 radiative forcing agents in twenty-first century climate change mitigation scenarios. *Climate Dynamics*, **33**, 737–49.

The Royal Society (2009). *Geoengineering the climate: science, governance and uncertainty.* The Royal Society, London.

UN (1992). *United Nations Framework Convention on Climate Change.* United Nations, New York.

Van Vuuren, D.P., Feddema, J., Lamarque, J.-F. *et al.* (2008a). *Work plan for data exchange between the integrated assessment and climate modeling community in support of phase-0 of scenario analysis for climate change assessment (representative community pathways).* http://cmip-pcmdi.llnl.gov/cmip5/docs/RCP_handshake.pdf

Van Vuuren, D.P., Meinshausen, M., Plattner, G.-K. *et al.* (2008b). Temperature increase of 21st century mitigation scenarios. *Proceedings of the National Academy of Sciences USA*, **105**, 15258–62.

Weyant, J.R., de la Chesnaye, F.C., and Blanford, G.J. (2006). Overview of EMF-21: multigas mitigation and climate policy. *Energy Journal*, **27**, 1–32.

CHAPTER 15

Ocean acidification: knowns, unknowns, and perspectives

Jean-Pierre Gattuso, Jelle Bijma, Marion Gehlen, Ulf Riebesell, and Carol Turley

15.1 Introduction

Although the changes in the chemistry of seawater driven by the uptake of CO_2 by the oceans have been known for decades, research addressing the effects of elevated CO_2 on marine organisms and ecosystems has only started recently (see Chapter 1). The first results of deliberate experiments on organisms were published in the mid 1980s (Agegian 1985) and those on communities in 2000 (Langdon et al. 2000; Leclercq et al. 2000). In contrast, studies focusing on the response of terrestrial plant communities began much earlier, with the first results of free-air CO_2 enrichment experiments (FACE) being published in the late 1960s (see Allen 1992). Not surprisingly, knowledge about the effects of elevated CO_2 on the marine realm lags behind that concerning the terrestrial realm. Yet ocean acidification might have significant biological, ecological, biogeochemical, and societal implications and decision-makers need to know the extent and severity of these implications in order to decide whether they should be considered, or not, when designing future policies.

The goals of this chapter are to summarize key information provided in the preceding chapters by highlighting what is known and what is unknown, identify and discuss the ecosystems that are most at risk, as well as discuss prospects and recommendation for future research.

15.2 Knowns and unknowns

The chemical, biological, ecological, biogeochemical, and societal implications of ocean acidification have been comprehensively reviewed in the previous chapters with one minor exception. Early work has shown that ocean acidification significantly affects the propagation of sound in seawater and suggested possible consequences for marine organisms sensitive to sound (Hester et al. 2008). However, subsequent studies have shown that the changes in the upper-ocean sound absorption coefficient at future pH levels will have no or a small impact on ocean acoustic noise (Joseph and Chiu 2010; Udovydchenkov et al. 2010).

The goal of this section is to condense the current knowledge about the consequences of ocean acidification in 15 key statements (Table 15.1). Each statement is given levels of evidence and, when possible, a level of confidence as recommended by the Intergovernmental Panel on Climate Change (IPCC) for use in its 5th Assessment Report (Mastrandrea et al. 2010). For the sake of brevity, the sections below do not provide the bibliographic citations which have already been given in the other chapters of this book. Readers are invited to refer to the relevant chapters indicated in Table 15.1 for a complete list of supporting references.

15.2.1 Chemical aspects

15.2.1.1 Ocean acidification occurred in the past
It is known with a very high level of confidence that ocean acidification occurred in the past. On geological timescales, the CO_2 concentration in the atmosphere and the carbonate chemistry of the oceans are constantly changing and adjusting to the forcings of tectonics, volcanism, weathering, biology, and, currently, human activity. Note, however, that atmospheric pCO_2 alone does not tell us much about the

Table 15.1 Summary of the knowns, unknowns, and challenges related to ocean acidification. The recommendations of Mastrandrea *et al.* (2010) were used for the levels of evidence ('limited', 'medium', or 'robust') and confidence ('very low', 'low', 'medium', 'high', and 'very high'). For statements related to projected impacts of anthropogenic ocean acidification, ranges of pCO_2, pH, $CaCO_3$ saturation state, etc. projected for 2100 under business-as-usual CO_2 emissions (e.g. generating atmospheric CO_2 concentrations of 793 ppmv in 2100) are considered. Question marks indicate that the effect is unknown. The chapters in which detailed information can be found are indicated.

Statement	Level of evidence	Level of confidence	Challenges	Chapter(s)
Chemical aspects				
Ocean acidification occurred in the past	Robust	Very high	Better constrain palaeoreconstructions of the carbonate system	2, 3
Ocean acidification is in progress	Robust	Very high	Better monitoring of key areas (e.g. coastal sites, coral reefs, polar regions, and the deep sea)	1, 3
Ocean acidification will continue at a rate never encountered in the past 55 Myr	Robust	Very high	Find two independent carbonate chemistry proxies to reconstruct the ocean carbonate chemistry with a high degree of confidence	2, 3
Future ocean acidification depends on emission pathways	Robust	Very high	Improve the representation of physical regimes at the regional scale to derive regional estimates	3, 14
The legacy of historical fossil fuel emissions on ocean acidification will be felt for centuries	Robust	Very high	Improve the representation of physical regimes at the regional scale to derive regional estimates	14
Biological and biogeochemical responses				
Ocean acidification will adversely affect calcification	Medium	High	Determine the mechanisms explaining that a few calcifiers are not affected or stimulated. Gain field evidence in addition to that available from CO_2 vents. Identify approaches to improve attribution on field observations	4, 6, 7
Ocean acidification will stimulate primary production	Medium	High	More work needed at the community level and under field conditions to better assess the global magnitude of the response	6, 7
Ocean acidification will stimulate nitrogen fixation	Medium	Medium	Investigate more species to test whether it is a widespread response. Determine the interaction with other variables in order to better assess the global magnitude and biogeochemical consequences	6
Some species or strains are tolerant	Robust	Very high	Gain a better understanding of the molecular and biochemical mechanisms underlying processes such as calcification	6, 7
Some taxonomic groups will be able to adapt	Limited	?	Identify approaches and tools to estimate the adaptation potential	4, 5, 6, 7, 8
Ocean acidification will change the composition of communities	Robust	High	Collect better information on non-calcifiers in the palaeorecord and determine the magnitude of the change in present key ecosystems	4, 5, 6, 7, 9, 10
Ocean acidification will impact food webs and higher trophic levels	Limited	?	Determine how species that may disappear will be replaced and whether the nutritional value of the replacement species may change	6, 7
Ocean acidification will have biogeochemical consequences at the global scale	Medium	Medium	A better understanding of key processes as a function of carbonate system variables is critically needed to improve model parameterization	11, 12

(continued)

Table 15.1   Continued

Statement	Level of evidence	Level of confidence	Challenges	Chapter(s)
Policy and socio-economic aspects				
There will be socio-economic consequences	Limited	?	Quantifying the monetary value of the goods and services that oceans provide and assessing how these may be impacted by ocean acidification	13
An ocean acidification threshold that must not be exceeded can be defined	Limited	?	Initiate and sustain an international effort to compile the increasing number of data being published in order to defined threshold(s). Investigate the need to consider thresholds based on geographical location, species, and ecosystems to advise decision-makers	13

saturation state of the ocean with respect to $CaCO_3$, as two parameters of the carbonate system are required to determine the seawater carbonate chemistry (see Box 1.1 in Chapter 1). There have been periods in earth's history during which the ocean had a lower pH than today (see Chapters 2 and 4), for instance at the end of the Permian 251 Myr ago, at the Palaeocene–Eocene Thermal Maximum (PETM) 55 Myr ago, and during the deglaciations that are characteristic of the Pleistocene epoch (which started ~1.8 Myr ago and ended 10 kyr ago, with the beginning of the Holocene). These events were a consequence of perturbations of the carbon cycle of different sizes, origins, and rates. The average saturation state with respect to $CaCO_3$ of surface waters was still favourable to calcifiers despite high CO_2 levels, due to higher concentrations of calcium and/or higher total alkalinity than today. In other words, pH and saturation state were decoupled during these events whereas both are declining together in the Anthropocene (the geological epoch that serves to mark the recent extent of human activities that have had a significant global impact on the earth's ecosystems).

All past ocean acidification events were accompanied by global warming, stronger stratification of the water column, and a decrease in the oxygenation of the deep sea. Attribution of the biological responses to one or several of these environmental factors is therefore difficult. It is important to realize that the climatic conditions prior to each of the events were very different from today (initial CO_2, ocean temperature, and chemistry). However, the most important difference between all previous geological events compared with that of the Anthropocene is the rate at which the human-induced carbon perturbation is proceeding (see Section 15.2.1.2).

Even though there is no perfect analogue to the present carbon perturbation, one should expect that the consequences of anthropogenic ocean acidification can only be worse than those recorded in the geological records, simply because the rate of change is unprecedented in the earth's history and marine ecosystems as we know them today have mainly evolved during a time of low atmospheric CO_2 and well-buffered seawater. The most pressing challenge remains to find two independent carbonate chemistry proxies that will allow us to reconstruct the ocean carbonate chemistry during earth's history.

15.2.1.2   Ocean acidification is in progress

It is known with a very high level of confidence that ocean acidification has been in progress since the beginning of the Industrial Revolution. The evidence is robust and comes from modelling but also from time-series stations and repeat measurements (see Chapter 3). Despite the short duration of the time series, the relatively short time interval between repeat measurements and a large seasonal variability in some sites, the decrease in pH and carbonate ion concentration and the increase in the concentration of dissolved inorganic carbon and pCO$_2$ are statistically significant. The decline in pH ranges from -0.0017 to -0.0019 units yr^{-1} in the surface waters of the open-ocean stations. The evidence is not as extensive in marginal seas, mostly because there are very few high-quality datasets covering a time span long enough to distinguish the declining trend against a seasonal and interannual variability which are much larger than in the open ocean. However, in Chapter 3, using an equilibrium approach and reasonable assumptions, Orr demonstrates that the changes of pH in marginal seas are generally within 10% of those of the open ocean. The situation is considerably more complex in the nearshore coastal ocean where pH is not only controlled by CO_2 uptake but also by changes in hydrodynamics and in the inputs of freshwater and organic matter. Nevertheless, it is certain that the uptake of anthropogenic CO_2 forces coastal systems towards a lower pH. Although ocean acidification is well documented in temperate oceanic waters, relatively little is known in the high latitudes, deep ocean, coastal areas, and marginal seas. A future challenge is to establish time-series measurements in these areas, not only to improve our understanding of present-day variability but also to help improve future projections.

15.2.1.3 Ocean acidification will continue at a rate never encountered in the past 55 Myr

The coupling between pH and the saturation state with respect to $CaCO_3$ (Ω) depends on both the magnitude and rate of CO_2 release (see Chapter 2). Ω is quite tightly regulated in a well-buffered ocean, even when pH changes. Two processes control Ω at different timescales: the dissolution of carbonate sediment ('seafloor carbonate neutralization' with a timescale of about 2 kyr) and rock weathering ('terrestrial carbonate neutralization' with a timescale of about 8 kyr; Ridgwell and Hargreaves 2007).

Releasing CO_2 over timescales shorter than ~1 kyr, as is happening today, is too fast for the two compensatory mechanisms to operate; consequently, Ω and pH decline in concert. The release of CO_2 on a timescale of ~2 kyr results in a partial buffering by seafloor carbonate neutralization and a partial decoupling of pH and Ω. On timescales longer than 10 kyr, Ω is more or less regulated, but the impacts of changes in climate and ocean circulation still need to be explored.

The carbon perturbation is estimated to have been of the order of 0.3 to 0.5 Gt C yr^{-1}(Kump et al. 2009) during the end Permian mass extinction. For the PETM, the closest analogue to the ongoing ocean acidification, estimates vary but a minimum value of 0.6 Pg C yr^{-1} during 5–10 kyr has been proposed (Zeebe et al. 2009). These past emissions are dwarfed by the current emissions of roughly 8.4 Pg C yr^{-1} (Friedlingstein et al. 2010) which are so large that the natural compensatory mechanisms cannot operate. Today's ocean is therefore slave to the atmosphere and its uptake of CO_2 is outstripping the buffering capacity of seawater, leading to a tight coupling between pH and Ω which decrease in concert. It is with a very high level of confidence that one can conclude that the current ocean acidification event is occurring on a timescale much faster than during the PETM and could therefore have more serious consequences. It is also worth mentioning that full chemical recovery after the PETM perturbation took about 100 000 yr. As already mentioned before, the most pressing challenge remains to find two independent carbonate chemistry proxies that will allow us to reconstruct the ocean carbonate chemistry during earth's history.

15.2.1.4 Future ocean acidification depends on emission pathways

Models project important changes in carbonate chemistry in response to historical and 21st century CO_2 emissions scenarios such as those selected for the 4th and 5th Assessment Reports of the IPCC (see Chapter 14). It is known with a very high level of confidence that future ocean acidification will depend on the emission pathways considered. The global surface-ocean mean Ω_a (global average saturation state of surface waters with respect to aragonite) is projected to decrease between 0.9 and 1.4 units from pre-industrial time to the end of the present century, while the corresponding change in pH ranges between 0.21 and 0.36 units. These changes in the global-mean properties such as Ω_a and pH hide large regional differences. Models project imminent undersaturation of surface waters in the Arctic ocean. Over 50% of the surface waters of the Arctic Ocean are projected to become undersaturated at atmospheric CO_2 levels exceeding 490 ppmv. Undersaturation of surface waters is also projected to be reached for atmospheric CO_2 levels exceeding 580 ppmv in the Southern Ocean. The overall largest changes in Ω_a are predicted for the tropics and subtropics, with a decrease from pre-industrial levels exceeding 4 to values below 2.5 at the end of the 21st century.

Mitigation leads to lower levels of atmospheric CO_2 and hence less climate change and ocean acidification. Global surface-ocean mean Ω_a decreases between 0.1 to 0.8 and pH decreases by 0.04 to 0.19 units. Mitigation options that lower atmospheric CO_2 will thus effectively alleviate the impacts of ocean acidification on marine systems. Due to the inertia of the earth system, the benefits from early mitigation of atmospheric CO_2 increase will become increasingly important after 2100. The main shortcoming of present coupled climate–carbon cycle models is their coarse resolution and over-simplistic representation of biological processes. This makes it difficult to evaluate different emission pathways and mitigation options at the regional scale and in particular in terms of biological impacts.

15.2.1.5 *The legacy of historical and 21st century fossil fuel emission will be felt for centuries*

It is known with a very high level of confidence that the legacy of historical and 21st century fossil fuel emission will be felt for centuries. Coupled climate–carbon cycle models are typically used to explore the long-term commitment to ocean acidification caused by the uptake of CO_2 since the onset of industrialization and over the 21st century following different emission scenarios. To this end, emissions are unrealistically set to zero in the year 2000 (historical scenario) or in 2100 for scenarios of the IPCC SRES B1 (low) and A2 (high) to explore the legacy of past emissions (see Chapter 14). For atmospheric CO_2 levels peaking around 850 ppmv (high scenario), Ω_a decreases by 50% in 2100. It increases subsequently as emissions are set to zero, but remains substantially lower than the pre-industrial value until the year 2500. Atmospheric CO_2 falls below 350 ppmv within a few decades in the historical scenario, and the perturbation of ocean chemistry remains relatively small. While the surface-ocean carbonate system responds quickly to changes in atmospheric CO_2 through air–sea gas exchange, the interior of the ocean shows a delayed response due to the centennial timescales involved in the surface-to-deep transport of anthropogenic

CO_2. In the high scenario, the volume occupied by undersaturated water increases from 60% in pre-industrial times to 75% in 2100 and reaches a maximum of 90% in 2300, 200 yr after the emissions are set to zero. The volume of undersaturated water increases up to 83% in the low scenario. Changes projected for the historical scenario are much more modest, yet the volume of supersaturated waters decreases too. These drastic and lasting changes in deep-water carbonate chemistry suggest the loss of suitable habitats for calcifying organisms.

Past and 21st century CO_2 emissions set the extent of ocean acidification over the coming centuries. Due to the inertia in the earth system, impacts of CO_2 emissions on seawater carbonate chemistry are delayed and will continue to perturb the biogeochemical cycles and marine ecosystems for centuries to come.

As mentioned in the previous section, the main shortcoming of present coupled climate–carbon cycle models for evaluating future changes in ocean carbonate chemistry is their coarse resolution. These models converge in their projections of the chemical consequences of ocean acidification at global as well as basin scale. However, studies mostly report changes in the mean state of properties (e.g. mean surface-ocean pH or saturation state averaged over a year or a decade). In order to integrate modelling studies with the growing understanding of the differential response of contrasting ocean regions, future research will need to focus on the assessment of impacts of ocean acidification at the regional scale. The interplay between ocean physics, chemistry, and biology will need to be investigated at seasonal and interannual timescales.

15.2.2 Biological and biogeochemical responses

15.2.2.1 *Ocean acidification will adversely affect calcification*

The precipitation, dissolution, and preservation of $CaCO_3$ are the processes which have been investigated most in the context of ocean acidification, both in the fossil record and in perturbation experiments. In Chapter 4 Knoll and Fischer provide an overview of the fate of calcifiers during earth's history. They show that several events in the geological past bearing the fingerprint of ocean acidification, global warming, and expanding anoxia, led to devastating changes in the abundance, diversity, and evolution of calcifying organisms. The bottom line from geological history is that the rate of the carbon perturbation is the key to the response of calcifying organisms (see Chapter 2 and above) as well as its amplification by global warming and declining levels of dissolved oxygen.

Despite the fact that some perturbation experiments reported no effect or a positive effect of ocean acidification on the rate of calcification of a few organisms (reviewed in Chapters 6 and 7), meta-analyses reveal an overall significant negative effect (Kroeker *et al.* 2010). The mean effect is negative and significant on corals, negative with a similar magnitude but non-significant on calcifying algae, coccolithophores, and molluscs, positive and significant on crustaceans, and positive but non-significant on echinoderms. Whether or not calcification decreases in response to elevated CO_2 and lower Ω, the deposition of $CaCO_3$ is thermodynamically less favourable under such conditions. Some organisms may have the capacity to up-regulate their metabolism and calcification to compensate for lower Ω. However, this would have energetic costs that would divert energy from other essential processes, and thus would not be sustainable in the long term. Full or partial compensation may be possible in certain organisms if the additional energy demand required to calcify under elevated CO_2 can be supplied as food, nutrients, and/or light (for those organisms dependent on photosynthesis).

Overall, calcifiers precipitating the less soluble aragonite and low-magnesian calcite are negatively affected, whereas those precipitating the more soluble high-magnesian calcite are not significantly affected. This may indicate that biological control of calcification is more important than the solubility of the mineral precipitated (Kroeker *et al.* 2010) but it seems that the analysis is flawed (Anderson, pers. comm.). Interestingly, Kroeker *et al.* (2010) found that calcifying organisms are more susceptible overall to ocean acidification than non-calcifying organisms across other processes.

Overall, the level of evidence that ocean acidification will adversely affect calcification is medium and the confidence level high. The remaining challenges are to estimate the energetic and physiological trade-offs in all life stages, and to improve our knowledge about the molecular and physiological mechanisms involved in calcification in order to better understand how the direction and magnitude of the response to ocean acidification are controlled.

15.2.2.2 Ocean acidification will stimulate primary production

In the oceans, photosynthesis is carried out primarily by microscopic phytoplankton, and to a lesser extent by macroalgae and seagrasses. Photosynthetic organisms must acquire inorganic carbon and a suite of major and trace nutrients from surface seawater. Dissolved CO_2, rather than the much more abundant bicarbonate ion, is the substrate used in the 'carbon fixation' step of photosynthesis. The enzyme responsible for carbon fixation, ribulose-1, 5-bisphosphate carboxylase oxygenase (RubisCO), has a low substrate affinity, achieving half-saturation of carbon fixation at concentrations well above those present in seawater. CO_2 must therefore be concentrated at the site of fixation against a concentration gradient and therefore with an energetic cost. As CO_2 diffuses readily through biological membranes, a large portion of the CO_2 transported into photosynthetic organisms is lost again via leakage. It is conceivable, therefore, that an increase in seawater CO_2 concentration reduces leakage and lowers the cost of concentrating CO_2, thereby stimulating primary production.

Stimulating effects of elevated CO_2 on photosynthesis and carbon fixation have indeed been observed in a variety of phytoplankton taxa, including diatoms, coccolithophores, cyanobacteria, and dinoflagellates. Modest increases of 10–30% in photosynthetic carbon fixation in response to elevated CO_2 were also observed in bioassay studies with oceanic plankton assemblages as well as in mesocosm experiments with coastal plankton communities. The extent to which phytoplankton respond to increased CO_2 depends to a large extent on the physiological mechanisms of inorganic carbon uptake and intracellular assimilation. Marine primary producers encompass phylogenetically very diverse groups, from prokaryotes to angiosperms, differing widely in their photosynthetic apparatus and carbon-concentrating mechanisms (CCMs). Species with effective CCMs are likely to be less sensitive to increased CO_2 levels than those lacking efficient CCMs. These differences are likely to alter competitive relationships among phytoplankton groups and result in shifts in plankton species composition as the oceanic CO_2 concentration continues to rise. The level of confidence of this stimulating effect is medium and its magnitude must be better constrained, especially at community level under *in situ* conditions.

15.2.2.3 Ocean acidification will stimulate nitrogen fixation

The fixation of atmospheric nitrogen gas (N_2) into ammonium (NH_4^+), a form readily available to the biota, is carried out by a small group of diazotrophic cyanobacteria. This process represents a major input of 'new' nitrogen to oligotrophic marine ecosystems. Because large parts of the global ocean are nitrogen-limited, nitrogen fixation plays a key role in determining primary production of the world's oceans. It is an energetically costly process which also requires the synthesis of a complex iron-rich enzyme. Additionally, cyanobacteria have to invest heavily in concentrating CO_2 at the site of carboxylation due to the low affinity of their RubisCO. Increasing surface-ocean CO_2 concentrations may therefore be beneficial for diazotrophic cyanobacteria. If the energy saved in the carbon acquisition process is reallocated to other processes, elevated CO_2 could stimulate nitrogen fixation (see Chapter 6). Enhanced nitrogen fixation in response to elevated CO_2 was indeed observed in the abundant filamentous diazotroph *Trichodesmium* sp. as well as the unicellular species *Crocosphaera watsonii* under iron-replete conditions. No stimulation was observed for *Nodularia spumigena*, a heterocystous diazotrophic cyanobacterium common in the Baltic Sea.

The level of confidence that ocean acidification will stimulate nitrogen fixation is medium because, considering the potential phylogenetic and metabolic diversity of marine nitrogen fixers, it is currently difficult to determine the representativeness for the global ocean of laboratory results on two of the three species investigated so far. Moreover, because of the high demand for iron and energy, the ecological niche of diazotrophs is restricted to iron-rich, nitrogen-poor waters in areas, or during periods of, high solar irradiance. With little information on the synergistic effects of CO_2, light, and iron, and

a small number of species tested so far, it is too early to assess the global significance of a stimulating effect of rising CO_2 on oceanic nitrogen fixation.

15.2.2.4 Some species or strains are tolerant

It is known with a very high level of confidence that some species or strains are tolerant to ocean acidification. Tolerance to ocean acidification in the range projected for the next century varies greatly among phyla and even species and is largely determined by the mechanisms and capacities of acid–base regulation. In metazoans, this is linked to the metabolic rate and, in turn, to the transport capacities for oxygen and CO_2. Several groups of animals (e.g. mammals, fishes, and some molluscs) with a high capacity for oxygen and CO_2 transport and exchange appear to be tolerant of lower pH, at least over short periods such as those of intense activity. In contrast, most marine invertebrate taxa have less developed gas exchange and acid–base regulatory capacities, and appear to have a lower tolerance to ocean acidification. Tolerance levels are also likely to be lower in early life stages, e.g. during egg and larval development.

In autotrophic organisms CO_2/pH sensitivity is linked to the carbon acquisition mechanisms supplying inorganic carbon to photosynthesis and, in the case of calcareous autotrophs, also to calcification. Differences in CO_2/pH sensitivity exist, for example, between diatoms, coccolithophores, and cyanobacteria, partly related to the efficiency of their CO_2 concentrating mechanisms. Different tolerances are also observed between species of the same group, or even strains of the same species, as seen for example in coccolithophores. To what extent these differences will affect overall competitive fitness and lead to the replacement of CO_2/pH-sensitive species by tolerant ones is currently unknown.

15.2.2.5 Some taxonomic groups will be able to adapt

Changes in environmental conditions generate two types of responses (Bradshaw and Holzapfel 2006): phenotypic plasticity, the ability of individuals to modify their behaviour, morphology, or physiology, and genetic (evolutionary) changes. Phenotypic plasticity is well known, but genetic changes have also been observed in populations of animals as diverse as birds, squirrels, and mosquitoes in response to the environmental changes that have occurred in the past decades. When the rate of environmental change in the geological record is fast, the probability of extinction is increased, suggesting that adaptation to slower changes is possible. Our understanding of the sensitivity of marine organisms to ocean acidification is almost entirely based on short-term perturbation experiments, lasting between a few hours and a few weeks. The results of such studies provide valuable insights into the phenotypic response of test organisms and communities to ocean acidification. However, such studies are generally too short to allow for evolutionary adaptation.

Ocean acidification occurs gradually over timescales of decades to centuries. With generation times of hours to days, unicellular algae and bacteria will go through tens of thousands of generations as pCO_2 increases to projected maximum levels, which may be sufficient for adaptive processes to become relevant. Moreover, most studies so far have focused only on selected phases of test organisms' life cycles. An individual may experience very different environmental conditions at different stages during its life cycle and its potential to adapt to stresses is likely to vary over different developmental phases.

The potential for adaptation can be investigated experimentally in studies using long-term exposure to elevated CO_2. This approach is most promising for organisms with short generation times, such as auto- and heterotrophic microorganisms. Adaptation potential can also be deduced from *in situ* observations at natural CO_2 venting sites, where benthic marine communities have experienced high pCO_2 and low pH conditions for several centuries or millennia. While this approach provides some valuable information on the potential range of adaptive responses at the organism, community, and ecosystem levels, the extrapolation of the observed responses to future ocean acidification is complicated for a number of reasons. The fact that pCO_2 and pH levels at these sites strongly vary over time and space (e.g. depending on water currents) and often exceed values projected for the next centuries makes it impossible to determine a dose–response relationship. As many benthic organisms are motile or have pelagic life stages, it is also difficult to distinguish between the local community, which has gone through multiple generations at the venting

site, and those individuals which have recently invaded the area. Apart from CO_2 venting sites, areas regularly experiencing upwelling of CO_2-enriched deep waters provide another natural laboratory for studying adaptation potential. While studies at these sites are facing some of the same difficulties as those at CO_2 venting sites (e.g. continuous invasion of non-exposed individuals), the range of pCO_2 is usually smaller, with less uncertainty regarding the dose–response relationship. Adaptation potential, or the lack thereof, can also be deduced from species extinction during high-CO_2 events during earth's history. A major difficulty lies in the uncertainty about whether these events provide a reasonable analogue for ocean acidification in the Anthropocene (see Chapters 2 and 4). In summary, the level of evidence that some species will adapt is limited. Due to the important role that adaptive processes may have on the response of the biota to ocean acidification, it is crucial to make the best use of all available approaches for addressing this critical issue.

15.2.2.6   Ocean acidification will change the composition of communities

Ecosystems are structured by the physico-chemical environment, including light, temperature, oxygen conditions, availability of nutrients and food, energy input to the mixed layer or, for benthic systems, substrate stability and heterogeneity. Ocean acidification directly modifies the chemical boundary conditions of marine communities and, in synergy with climate change, might result in the reorganization of certain marine ecosystems. For example, the poleward migration of species (Fields et al. 1993) could be impaired because ocean acidification will be particularly strong at high latitudes. The differential response of organisms to changes in carbonate chemistry is likely to modify the composition of communities through species loss and migration, changes in species succession, or, more generally, through altered competition between species.

Evidence for changes in community composition in response to ocean acidification is still scarce. However, comparative analysis of shallow-water benthic communities in the vicinity of volcanic CO_2 vents in the Mediterranean Sea demonstrates a 30% decrease in the overall diversity as well as a decrease in abundance or the loss of benthic calcifiers at lower pH. Similarly, changes in coral reef communities, in both cold and warm waters, appear likely as community calcification rates will decrease with decreasing saturation state of $CaCO_3$. In the case of tropical coral reefs, changes in the balance between calcification and erosion will ultimately result in the degradation of the reef structure and hence spatial heterogeneity, a factor likely to amplify shifts in community composition away from calcifiers towards algal-dominated ecosystems.

Potential shifts in community structure for pelagic open-ocean systems are more elusive. For example, while marine photosynthesis appears to benefit from increased levels of CO_2, the extent of this CO_2 fertilization effect depends on the physiological characteristics of individual phytoplankton groups. Groups with a comparable inefficient carbon acquisition pathway, such as certain coccolithophores, are favoured. Climate change is projected to enhance stratification and to reduce mixed layer depth, which in turn reduces the injection of macronutrients to the euphotic zone, thus creating conditions more favourable for nanophytoplankton, also including coccolithophores. On the other hand, several studies have reported reduced calcification in response to decreasing calcite saturation state for the coccolithophore *Emiliania huxleyi*, which ranks among the main calcifiers in the pelagic realm. From the preceding, it can be concluded that future physico-chemical conditions might favour nanophytoplankton at the expense of diatoms. The case of *E. huxleyi* exemplifies the interplay between climate change and ocean acidification in shaping future pelagic communities. Its outcome is highly uncertain.

The level of confidence that ocean acidification will change the composition of communities is high, mostly due to the robust evidence available for benthic communities. There is a need to collect better information on non-calcifiers in the palaeorecord and to determine the magnitude of the changes in present key ecosystems.

15.2.2.7 Ocean acidification will impact upon food webs and higher trophic levels
The evidence that ocean acidification will have an effect on food webs and higher trophic levels is limited. There is a lot more information on the effects of

ocean acidification on organisms at the base than at the top of the food web. The relatively low sensitivity of nektonic, active organisms to ocean acidification is related to their high capacity for acid–base regulation (see Chapter 8). Elevated CO_2 in body fluids could nevertheless reduce their fitness, for example by depressing foraging, growth, and reproduction. For example, ocean acidification was shown to alter the behaviour of larvae of two coral reef fish and dramatically decrease their survival during recruitment to adult populations, potentially decreasing the sustainability of fish populations (Munday *et al.* 2010).

Possible changes in the lower trophic levels (Section 15.2.2.6), either as shifts between dominant phytoplankton groups or altered food quality (e.g. altered elemental composition of phytoplankton), are likely to spread across the food chain and affect higher organisms that feed on them. For example, pteropods can represent as much as 93% of the total zooplankton biomass in high-latitude regions (Hunt *et al.* 2008) and are a key food resource for many predators such as herring, salmon, whales, or seabirds. The pteropod *Limacina helicina* can represent as much as 60% of the prey of the juvenile pink salmon in the northern Gulf of Alaska (Armstrong *et al.* 2005). Due to their key role as a prey, a decline in pteropod populations has the potential to generate significant changes at higher levels of the food web. However, if pteropods show a considerable decrease in abundance, it is certain that the ecological niche they occupy will not remain empty. Whether predators will be able to use the species that will replace pteropods and whether the new species will have a similar nutritional value is completely unknown.

Finally, it is important to note that the direct biological effects of ocean acidification operate at the cellular level but that it is the expression of these effects in populations and ecosystems that are of societal concern. Scaling from physiological data to the levels of populations and ecosystems remains to be done (Section 15.4.6).

15.2.2.8 Ocean acidification will have biogeochemical consequences at the global scale
Ocean acidification interacts with the major ocean biogeochemical cycles by modifying the rates of

key processes (see Chapter 12). The ongoing uptake of atmospheric CO_2 results in a decrease in the buffer capacity of ocean water. As a result, the strength of the ocean sink for CO_2 is going to decrease in the future, a direct positive feedback of ocean acidification to atmospheric CO_2 levels and hence the earth system. This positive feedback may be as large as 30% in the next 100 years for a business-as-usual scenario. The global rate of calcification is likely to decrease in response to increasing atmospheric CO_2. Model studies project a rather modest negative feedback associated with decreasing $CaCO_3$ production. If the export of particulate organic and inorganic carbon is tightly coupled through ballasting of the organic fraction by $CaCO_3$, then a decrease in the number of $CaCO_3$ particles will translate into a shallower remineralization depth for organic carbon, a positive feedback that could compensate the effects of reduced calcification. At timescales of several tens of thousands of years, the marine sedimentary reservoir of carbonates will provide the ultimate buffer against ocean acidification.

There is evidence that phytoplankton groups with an inefficient carbon acquisition pathway may benefit from increased levels of CO_2. However, in order to have an impact on the large-scale carbon cycle and feed back on atmospheric CO_2, enhanced photosynthesis needs to translate into a strengthening of the export of particulate organic carbon to depth. Changes in elemental stoichiometry, such as increasing the C:N ratio with increasing atmospheric CO_2 (see Chapter 6) might enhance future carbon export. The increased delivery of carbon to depth and its remineralization increases oxygen consumption and might result in a significant extension of oxygen minimum zones. Oxygen minimum zones are sites of intense denitrification, a suboxic metabolic pathway yielding N_2O, a potent greenhouse gas. An increase in the oceanic source of N_2O would correspond to a positive feedback on the earth's radiative balance.

Potential impacts of ocean acidification on the marine nitrogen cycle include enhanced nitrogen fixation and higher rates of water column denitrification. Both combine to reduce the mean residence time of fixed nitrogen in the ocean. However, given that both sources and sinks could be enhanced, it is not possible to make conclusions about the potential generation of imbalances which are required to cause net changes in the ocean's fixed nitrogen inventory and ultimately changes in the biological pump. Based on the present understanding, the magnitude of the feedbacks to the earth system involving nitrogen fixation and denitrification are unknown. The limited number of studies which have measured changes in the emissions of climate-active trace gases (e.g. dimethyl sulphide, halocarbon gases) to the atmosphere under elevated CO_2 (see Chapter 11) have yielded contrasting results. The effects thus remain poorly known and have not yet been included in coupled climate–marine biogeochemical models.

Overall, it is likely that ocean acidification will have biogeochemical consequences at the global scale but the magnitude is largely unknown and will remain so until the response of organisms and communities is better constrained.

15.2.3 Policy and socio-economic aspects

15.2.3.1 *There will be socio-economic consequences*

As this century progresses ocean acidification has the potential to affect a wide range of marine organisms, food webs, habitats, and ecosystems that supply important goods and services to humankind (see Chapter 14). Goods and services provided by oceans include the provision of food and food products and their significant contribution to global food security, the ocean's capacity for carbon storage and regulation of gas and climate, and the ability to regulate nutrients on a global scale. Marine habitats and ecosystems also provide leisure, recreation and well-being, and coral reefs, mangroves, and salt marshes protect our coastlines from inundation and erosion.

The provision of fish (including shellfish) is the most obvious service, currently supplying 15% of animal protein for 3 billion people worldwide, and a further 1 billion rely upon fisheries for their primary protein. The chemistry of CO_2 in seawater is such that polar regions, upwelling waters, deep oceans, and estuaries are likely to be affected first, and these are areas of importance for marine-based human food resources. Some organisms, many of which provide food or are key trophic links, will be affected by ocean acidification. With an increasing world population becoming more reliant upon marine resources for human consumption, and with other pressures, such as pollution, fisheries exploitation, and coastal development, any further stress to these resources through ocean acidification is of concern.

It is difficult to put a monetary value on the wealth of marine ecosystem goods and services because they are not all fully part of commercial markets or as easily quantifiable as economic services or manufacturing output, but a controversial attempt has been made with a total valuation of US$21 trillion yr^{-1}. For this reason marine goods and services do not carry as much weight in policy decisions as economic and manufacturing values. However, global fish production is valued at US$150 billion yr^{-1} and tropical coral reefs have an estimated value of US$30 billion yr^{-1} with the potential economic damage of ocean-acidification-induced coral reef loss estimated to be US$500–870 billion in this century. Despite the difficulty in obtaining an overall valuation, if marine systems become impacted the ramifications of loss of marine goods and services for society will be wide ranging. Overall, the socio-economic consequences of ocean acidification are still unknown and hampered by poor knowledge of the impacts at the ecosystem scale.

15.2.3.2 *An ocean acidification threshold that must not be exceeded can be defined*

One major question asked by policymakers is 'At what atmospheric CO_2 concentration will ocean acidification have significant, unacceptable, and irredeemable consequences?'. That is, is there a universal tipping point for ocean acidification which should be avoided? Observations performed around CO_2 vents show that a significant decline in biodiversity occurs below a mean pH_T of 7.8 (Hall-Spencer *et al.* 2008). It is difficult to use these data to define a critical threshold because the variance of pH around vents is large and the changes in biodiversity may be due to the lower end of the pH range rather than to the mean value.

Some attempts have been made to define global thresholds. The US Environmental Protection Agency quality criteria for water state: 'For open

ocean waters where the depth is substantially greater than the euphotic zone, the pH should not be changed more than 0.2 units outside the range of naturally occurring variation'. According to Schubert *et al.* (2006), the pH value of near-surface waters should not drop more than 0.2 units below the pre-industrial average value of 8.18 (pH scale not mentioned) in any larger ocean region (nor in the global mean). Rockström *et al.* (2009) have introduced the concept of planetary boundaries for estimating a safe operating space for humanity with respect to the functioning of the earth system. Ocean acidification is one of the nine boundaries proposed and the proposed control variable is the carbonate ion concentration. As a first estimate, Rockström *et al.* (2009) proposed a boundary where oceanic aragonite saturation state Ω_a is maintained at 80% or higher of the average global pre-industrial surface-seawater Ω_a of 3.44.

The spatial and temporal averaging to which the thresholds refer is often not mentioned, despite the fact that it is critical to do so because pH is subject to strong natural variability. All of these thresholds have been defined without any clear biological, ecological, or biogeochemical foundation. Furthermore, the nature of organismal responses is such that different species will react differently, each having different CO_2 and pH sensitivities which may depend on other factors such as the health of the ecosystem in which they dwell, food availability, and the presence of other stressors such as pollution and warming. Although ocean acidification is a global issue, different regions will be affected at different times (Fig. 15.1) making the definition of a threshold even more difficult.

One hundred members of the global network of science academies, the Interacademy Panel (IAP), signed up to a statement on ocean acidification which stated that 'even with stabilization of atmospheric CO_2 at 450 ppmv, ocean acidification will have profound impacts on many marine systems. Large and rapid reductions of global CO_2 emissions are needed globally by at least 50% by 2050'. Others have stated that even 450 ppmv is too high, and that atmospheric CO_2 should be stabilized at 350 ppmv to prevent catastrophic decline in coral reefs due to warming and acidification (e.g. Veron *et al.* 2009).

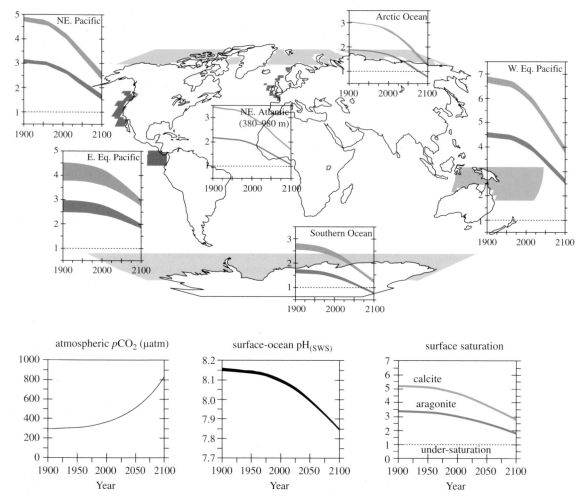

Figure 15.1 Projected regional changes in ocean chemistry likely to be experienced by particularly vulnerable ecosystems and compared with global-scale surface-ocean changes (Turley et al. 2010). The transient simulation of climate and carbonate chemistry was performed with the UVic Earth System Climate Model using observed historical boundary conditions to 2006 and the SRES A2 scenario to 2100 (Eby et al. 2009). For each of the six illustrative high-risk marine ecosystems [Arctic Ocean, Southern Ocean, north-east Pacific margin, intermediate-depth north-east Atlantic (500–1500 m), western equatorial Pacific, eastern equatorial Pacific] we have identified in this paper, the blue-shaded band indicates the annual range in ocean saturation state with respect to aragonite, while the green shaded band indicates the range for calcite saturation. Area-average surface-ocean conditions are calculated for all regions with the exception of the north-east Atlantic where area-average benthic conditions between 380 and 980 m have been used. The thickness of the line indicates the seasonal range, with the threshold of undersaturated environmental conditions marked as a horizontal dashed line. The varying evolution in the magnitude of the seasonal range between different regions is due to the complex interplay between changes in stratification, ocean circulation, and sea-ice extent, and distorted due to the non-linear nature of the saturation scale. The corresponding regions from which the annual ranges are calculated are shown shaded. Global ocean surface averages (bottom) are shown, from left to right: CO_2 partial pressure (pCO_2), pH_{SWS}, and calcite and aragonite saturation. Figure reproduced from Turley et al. (2010) with permission from Elsevier.

Although it is critical to provide one or a few acidity level(s) that should not be exceeded, it is a very challenging task which requires a large body of scientific evidence, including at the ecosystem level and on food webs, and the knowledge of the level of risk that society is prepared to accept.

15.3 Ecosystems at risk

Three areas of the global ocean are more susceptible to ocean acidification than others, either because ocean acidification will be more severe (polar regions and the deep sea) or because it acts synergistically with another major stressor (coral reefs).

15.3.1 Polar seas

High-latitude oceans are cold water bodies which have naturally low carbonate concentrations and are therefore most sensitive to ocean acidification (see Chapter 3). During the 21st century, their surface waters will become chemically corrosive to aragonite, first in the Arctic Ocean and a few years later the Southern Ocean. These severe conditions could prevail over much of the surface Arctic Ocean by the end of the century. Furthermore, the aragonite saturation horizon is shoaling (moving upwards) at a rate of 4 m yr^{-1} in the Iceland Sea, exposing each year 800 km^2 of seafloor to waters undersaturated with respect to aragonite. Despite the high vulnerability of polar areas to ocean acidification, the biological, ecological, and biogeochemical consequences are not well documented and very few perturbation experiments have been conducted. The first results were obtained on the pteropod (pelagic marine snail) *Limacina helicina* which has an important role in the food chain and the functioning of Arctic and sub-Arctic marine ecosystems. Its aragonitic shell serves as a ballast, enabling large vertical migrations, and as a protection against predators. The gross CaCO$_3$ precipitation of *L. helicina* decreases logarithmically as a function of decreasing aragonite saturation state, but still occurs, at a low rate, in undersaturated waters (Comeau *et al.* 2010). However, dissolution of CaCO$_3$ was not measured in this experiment and the saturation level up to which a positive balance between gross CaCO$_3$ precipitation and dissolution can be achieved is unknown. The recruitment of the benthic life stages of the spider crab *Hyas araneus* was shown to be affected by ocean warming and acidification (Walther *et al.* 2010). Knowledge about the response of polar organisms and ecosystems to ocean acidification is still in its infancy and it is critical to gather data on these particularly threatened ecosystems.

15.3.2 Deep-sea environments

The depths of the ocean are generally rich in CO$_2$ because much of the organic material exported by

the biological pump from the surface is mineralized, releasing CO$_2$ into waters close to the seabed. These deep CO$_2$-rich waters have low pH and a sufficiently low carbonate ion concentration that they are undersaturated with respect to CaCO$_3$ (see Chapter 3), making it difficult for calcifying organisms to live there. With additional atmospheric CO$_2$ entering the ocean surface, the horizon that separates saturated from undersaturated waters is shoaling. The global-mean depth of the aragonite saturation horizon is projected to shoal from its pre-industrial level of 1090 m to 280 m in 2100 under the IS92a scenario (see Chapter 3).

These changes in saturation may affect aragonitic cold-water corals. Nearly all of these corals now live in deep waters where $\Omega_a > 1$, but it is projected that by 2100 under the IS92a scenario 70% of them will be bathed in waters where $\Omega_a < 1$ (Guinotte *et al.* 2006). Shell- or skeleton-forming animals that live in deep water but above this horizon can currently calcify but may be amongst the earliest to be affected by ocean acidification (Turley *et al.* 2007). It is projected that as the saturation horizon moves past them they may no longer be able to calcify. Despite the sensitivity of deep-sea environments to ocean acidification, very few data are available on the biological responses. A recent short-term experiment has shown that the important cold-water coral *Lophelia pertusa* seems to be able to calcify in slightly undersaturated water but its rate of calcification decreases by 50% when kept in high-CO$_2$ seawater (Maier *et al.* 2009). While tropical coral reefs are built by a large number of species, cold-water coral communities are constructed by one or two species but provide shelter for many others. It is therefore likely that the combined effect of lower calcification and increased dissolution of pre-existing skeletons will have a negative impact on the biodiversity of cold-water coral communities. Deep-sea organisms other than corals may also be affected by ocean acidification as well as by the future reduction in the oxygen concentration (Brewer and Peltzer 2009).

15.3.3 Coral reefs

Coral reefs are CaCO$_3$ structures located at or near sea level constructed by scleractinian corals and coralline algae. The skeletons of both types of reef builders are particularly soluble because they are made of aragonite (corals) or high-magnesian calcite (coralline algae). Coral reefs are distributed in waters with a relatively high aragonite saturation state (Ω_a = 3.3 on average; Kleypas *et al.* 1999). The tropics and subtropics will see the biggest absolute changes in Ω_a with a projected drop from 4.2 in the year 1820 to 2.3 in 2100 under the A2 scenario (see Chapter 3). Steinacher *et al.* (2009) found that under the A2 scenario, waters with $\Omega_a > 3$ will disappear by 2070.

Coral reefs in the eastern tropical Pacific are good indicators of what the future of coral reefs could be. Manzello *et al.* (2008) observed that cements in intraskeletal pores were almost absent. Consequently, these reefs, which experience naturally high CO_2 and low Ω_a as a result of upwelling, are poorly developed, and are subject to high rates of bioerosion. The abundance of cement appears to be correlated to the seawater aragonite saturation state and inversely related to measured rates of bioerosion.

Corals and coralline algae are probably the organisms that have been investigated the most with respect to ocean acidification (see Chapter 7). Although the dependence of calcification rates on the carbonate chemistry is widespread in coral reef organisms, a few seem to be resistant to ocean acidification. Increased dissolution and lower calcification will lead, at some point in time, to a transition from net calcification and $CaCO_3$ accretion to net dissolution and net loss of $CaCO_3$. Additionally, zooxanthellate corals lose their endosymbiotic algae at elevated temperature, leading to coral bleaching and high rates of mortality (e.g. Hoegh-Guldberg 1999). The combination of ocean acidification and warming undoubtedly makes coral reefs the ecosystem most threatened by global environmental change (Hoegh-Guldberg *et al.* 2007). Several critical issues require better investigation: the mechanism(s) that enable some corals to resist ocean acidification, the interaction between coral bleaching and ocean acidification, and the response of natural communities through long-term perturbation experiments in the field.

15.4 Past limitations and future prospects

Synthesis of our present knowledge about ocean acidification and its consequences (Section 15.2),

demonstrates that many uncertainties remain. The present section briefly reviews the main reasons for this relatively poor knowledge and provides suggestions on what can be done to make faster progress in the near future.

15.4.1 Limited workforce and funding

In the 1990s, each year, between 7 and 42 individual authors published on the issue of changes to the carbonate system in seawater and their impacts on marine organisms and ecosystems (Fig. 15.2). A significant number of these papers were looking at the general effects of pH on physiological processes, sometimes at very high levels of acidity; fewer than 25 scientists were involved in investigating the effects of the changes in the carbonate system generated by the uptake of anthropogenic CO_2. Ocean acidification was not a topic of great interest to scientists, indeed the term had not even been coined. The number of individual authors increased considerably starting in 2005 and reached over 550 individual authors in 2010. This huge increase in the workforce

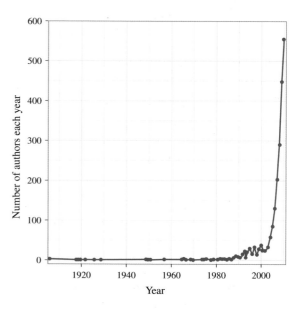

Figure 15.2 Number of individual authors having published at least one paper addressing ocean acidification per year. The bibliographic database compiled by Gattuso and Hansson (see Chapter 1) was used. The complete list of references is available as online Appendix 1.1 at http://ukcatalogue. oup.com/product/9780199591091.do. Only scientific papers were considered and duplicate authors were removed.

is the result of the increase in the number of national and international projects on ocean acidification and the consequent increase of the rate of publication (see Chapter 1). It now seems that the overall magnitude of the workforce is no longer a major impediment for ocean acidification research. Other difficulties, such as technological and conceptual limitations, hamper progress towards a better understanding of the consequences of ocean acidification. However, some key areas such as work at the community level as well as social and economic science are still limited by the number of scientists involved.

15.4.2   Inappropriate or inconsistent methods

As more and more results on the sensitivity of organisms and communities to ocean acidification become available, the breadth and depth of our understanding of causes and consequences grows. At the same time, the evolving picture becomes blurred by conflicting results. While part of this may simply reflect species-specific differences or the plasticity of organisms in responding to environmental stressors, part of the discrepancy may result from inappropriate methodology, poorly constrained experimental protocols, or misinterpretation of the data. Inconsistencies of this sort may originate from shortcomings such as:

- insufficient control and incomplete description of seawater carbonate chemistry,
- interference through confounding factors, such as temperature, nutrient and oxygen concentrations, irradiance and diurnal cycles,
- lack of acclimation of the test organisms to experimental conditions, causing stress- rather than treatment-related responses,
- pseudoreplication, i.e. the lack of independent, interspersed replicates at the treatment level,
- use of isolates kept in culture over years and decades which may have undergone phenotypic or genetic change unrelated to ocean acidification,
- reporting and interpreting observed responses in relation to mean values, e.g. of pH and pCO_2, when in fact large variations in carbonate chemistry occurred during the experiment,

- inappropriate normalization of measured data (e.g. normalizing per cell and per unit biomass may yield very different interpretations if cell size differs between treatments).

These and other possible pitfalls in ocean acidification research are addressed and recommendations for proper experimentation provided in the *Guide to best practices for ocean acidification research and data reporting* (Riebesell *et al.* 2010a). It is therefore hoped that future experiments will strictly follow these guidelines, hence helping to avoid confusion in the literature generated by inconsistent methods and data reporting.

15.4.3   Duration of experiments

The current anthropogenic ocean acidification is a very fast process compared with previous episodes in the geological past (Section 15.2.1.3; see also Chapter 2) but still long compared with the generation time of most marine organisms. Adaptation, i.e. adjustment to environmental change by genetic change, is much more likely in microbes than in multicellular marine organisms because they reproduce quickly relative to the timescale of global change and have large populations (Section 15.2.2.5). Their short generation time of a few days allows for thousands of generations by 2100, hence increasing the accumulation of mutations and, at least for prokaryotes, more efficient lateral gene transfer. Perturbation experiments have been the method of choice for investigating the effects of ocean acidification. They were conducted over periods of time (hours to weeks) that may have been long enough to reach a steady-state physiological response but too short to reach a 'steady-state' evolutionary response. There is a strong need to carry out longer experiments, encompassing hundreds or thousands of microbial generations. Such experiments have been performed in a green microalga found in soils and freshwater (Collins and Bell 2004, 2006), providing interesting insight for ocean acidification research. First, lines of this alga either increased photosynthesis without increased growth or both their growth and photosynthesis were insensitive to elevated CO_2. Specimens found in the vicinity of natural CO_2 springs exhibited a similar phenotype, suggesting that laboratory selection experiments can be used to predict the response of natural populations. Second, the evolutionary response differed in both direction and magnitude from the short-term (physiological) response to elevated CO_2 as it involved an increase in both photosynthesis and growth. The fact that short-term responses may not scale up predictably to longer timescales is worrisome, as current projections of the consequences of ocean acidification are almost exclusively based on short-term experiments. It highlights the need to perform experimental tests of evolutionary responses to elevated CO_2 in marine phytoplankton (Bell and Collins, 2008). Genomics, transcriptomics, proteomics, and assessment of the expression of specific marker genes for crucial functions are promising methods that are or soon will be available to tackle these issues.

15.4.4 Interactions with other stressors

Ocean acidification occurs simultaneously with changes in other environmental variables associated with anthropogenic climate change, including ocean warming, expanding hypoxia, changes in salinity, physical disturbance, mixing, and stratification. There are many examples in the geological record of pronounced extinction events when several environmental stressors were imposed at once (see Chapter 4). Future concurrent changes in some environmental variables may have synergistic effects with ocean acidification, amplifying or dampening the sensitivities of organisms to reduced pH and increased pCO_2. Several of these variables act together in constraining the window for performance of organisms (Pörtner 2010). For instance, as discussed by Pörtner and Farrell (2008), ocean acidification can narrow the thermal window of some animals, thereby further reducing their geographical distribution in a warming ocean. So far ocean acidification research has given little attention to multiple interacting environmental stressors. Future studies should therefore expand to test for synergetic effects of environmental stressors with ocean acidification.

In addition to interference from environmental stressors, the response of an organism to a perturbation such as ocean acidification may also be modi-

fied in the presence of competitive or trophic interactions. Hence, results obtained in isolation from other relevant influences need to be verified under more realistic conditions with multiple interacting variables. This calls for an extension of community-level investigations (see also Section 15.4.6). For an integrated assessment of the effects of ocean acidification on marine life it will therefore be crucial to cover the entire range of possible interacting influences and to employ a wide spectrum of approaches from subcellular to ecosystem-level experimentation and modelling.

15.4.5 Lack of field evidence other than around CO_2 vents

The abrupt changes and short duration of perturbation experiments are key limitations (Section 15.4.3) and it is critical that the predictions and projections made using such experiments can be corroborated using field data. Two approaches, observations across natural CO_2 gradients and retrospective studies, have been used with mixed success; just a few examples are provided in this section. The north–south shift from 'overcalcified' to weakly 'calcified' cells of the coccolithophore E. huxleyi in the Southern Ocean reflects a shift in dominance from one ecotype to another, rather than the environmental effect of decreased carbonate ion concentrations and calcite saturation state on a single 'apparently cosmopolitan' population (Cubillos et al. 2007). Observations at CO_2 vents (e.g. Hall-Spencer et al. 2008) have demonstrated that ocean acidification does have significant effects on benthic primary producers, calcifiers, and biodiversity. Retrospective studies have provided outcomes that are much less clear, because attribution to a single environmental factor has proven difficult. Grelaud et al. (2009) found an increasing carbonate mass in coccolithophore shells from 1917 to 2004 concomitant with rising pCO_2 and sea-surface temperature in the region of the Santa Barbara Basin, California. Although perturbation experiments suggest that coral calcification may have decreased by about 10% since the beginning of the Industrial Revolution, the evidence for this has not yet been found in field samples. Some (e.g. De'ath et al. 2009), but not all, retrospective studies show decreasing trends in calcification for the past several decades, but whether the decreases are due to ocean acidification, some other environmental factor (e.g. warming), or a combination of factors remains unclear.

The issue of attribution is a well-known problem in studies of the effects of climate change. Rosenzweig *et al.* (2008) compiled a huge database and used innovative statistical approaches to demonstrate that changes in physical and biological systems are pervasive and could probably be attributed to climate change caused by increasing concentrations of greenhouse gases. However, this approach is limited by the short time span of many datasets (Zwiers and Hegerl 2008), especially from marine systems, and does not provide the contribution from the major variables such as CO_2, temperature, or nutrients.

15.4.6   Limited work at the community level

Progress in our understanding of the possible impacts of ocean acidification on marine life is partly limited by the scarcity of information on responses at the community and ecosystem levels. Results obtained in single-species experiments are not easily extrapolated to natural systems due to the lack of information on the influence of competitive and trophic interactions on single-species responses. To close this gap two approaches are particularly promising: (1) community-level studies in natural high-CO_2 environments and (2) CO_2 perturbation experiments at the community and ecosystem level. Both approaches have provided important information on the effects of ocean acidification on natural marine communities. The best known example for a natural high-CO_2 environment is a CO_2 venting site in the Gulf of Naples, Italy, where a community shift was observed along a pCO_2 gradient, with calcifying organisms successively disappearing from the community towards the CO_2 venting site (Hall-Spencer *et al.* 2008). Community-level perturbation experiments have been conducted on both pelagic and benthic communities using mesocosm enclosures (Langdon *et al.* 2010; Riebesell *et al.* 2010b; Widdicombe *et al.* 2010). The so-called free ocean CO_2 enrichment (FOCE) system uses injection of acid or of CO_2-enriched air with controlled-loop pH feedback to maintain a set pH in a small volume (Kirkwood and Sano 2009).

Each of these approaches has strengths and weaknesses. Natural high-CO_2 environments capture the full scope of ecosystem interactions over long timescales (at least for the benthos), thereby providing crucial information on the effects on ocean acidification on trophic and competitive interactions and the potential for adaptation. The high spatial and temporal variability in pCO_2 and pH, however, makes it difficult to determine a reliable dose–response relationship, complicating the use of this information in projecting the response to future high-CO_2 scenarios. The interpretation is also complicated by the uncontrolled advection and recruitment of organisms from unperturbed adjacent areas. The latter problem also applies to FOCE-type experimental systems. Both approaches are restricted to benthic communities with non- or slowly migrating organisms. In contrast, mesocosm enclosure experiments, which range in size between one and a few hundred cubic metres and can be used for both benthic and pelagic communities, have the advantage of well-controlled carbonate chemistry and the absence of exchange with unperturbed surrounding waters. Their limitation lies in the fact that mesocosm experiments generally exclude migratory organisms and their duration is limited as a result of undesirable effects due to the growth of organisms on the walls, which limit their scope with respect to interaction at higher trophic levels and adaptational responses.

With none of the available approaches providing information on the full range of ecosystem responses, progress in our understanding of long-term, ecosystem-level impacts of ocean acidification requires the integration of: (1) community-level experimental studies, (2) field observations in high-CO_2 environments, and (3) ecosystem modelling with (4) single-species laboratory experiments addressing the mechanisms underlying the observed sensitivities and (5) long-term high-CO_2 exposure experiments examining the potential for adaptation.

15.4.7   Difficulties in performing meta-analysis

Meta-analysis, which statistically combines the results of several studies that address a shared research hypothesis, is a method of choice for assessing the overall effect of ocean acidification on marine organisms and ecosystems. The recent surge of experimental data has made possible the use of meta-analytical approaches (Hendriks *et al.* 2010; Kroeker *et al.* 2010; Liu *et al.* 2010). Despite their strengths, such approaches have several problems (Borenstein *et al.* 2009), the most serious of which is the inevitable use of a biased dataset. The 'file drawer problem' relates to the fact that studies reporting relatively high treatment effects can be easier to publish whereas those which are inconclusive never get published. This bias may not be extremely serious today as editors may find it exciting to publish data that are inconclusive as they go against earlier conclusions and could raise controversy. It is nevertheless critical that all datasets, inconclusive or not, are published in databases or in data journals (e.g. *Earth System Science Data*). The other reason for the use of a biased dataset is poor data reporting, which is a serious issue in ocean acidification research. The three recent meta-analyses mentioned above could not use all the data available because many are unavailable or unusable. In their data compilation, Nisumaa *et al.* (2010) identified 185 papers of interest but data from 85 of them could not be compiled for three reasons: only one parameter of the carbonate system was measured (49 papers), data could not be obtained from the authors (30 papers), or the data were lost (16 papers). The publication of the *Guide for best practices on ocean acidification research and data reporting* (Riebesell *et al.* 2010a) will hopefully lead to better data reporting in future publications.

15.4.8 Model development

Present-day state-of-the-art global biogeochemical models mostly rely on the representation of plankton functional types (PFTs). Following this approach, major biogeochemical functions are identified and assigned to specific groups, e.g. $CaCO_3$ production, biogenic silica formation, nitrogen fixation. The growth of each group is directly controlled by the availability of external nutrients. If environmental conditions (light, temperature, nutrients) cross a critical threshold, the plankton functional type disappears from the model world. This approach does not allow for shifts between species within a given

PFT, nor for adaptation. This simplistic representation contrasts with the emerging complexity of the biological responses to ocean acidification and challenges the modellers. The evaluation of the impacts of ocean acidification and climate change might well require the development of a new generation of ecosystem models. An alternative to the classical PFT approach consists in the random assignation of physiological traits to a large number of plankton types. This approach makes use of the variability of physiological data reported by experimental studies, rather than identifying *a priori* a limited number of PFTs which are described by representative mean parameter values. It allows for the emergence of marine communities.

While this approach appears promising when it comes to studying the environmental controls on microbial community structure and ecosystems, its capacity to project the reorganization of communities and ecosystems in response to global warming and ocean acidification remains to be demonstrated.

Future ocean acidification may alter the fitness of organisms, and their differential response is likely to modify the community composition, for example through altered competition between species (Section 15.2.2.6). Optimality-based adaptive models allow one to investigate the synergistic effects of temperature changes and ocean acidification on the basic life functions of organisms (e.g. growth, reproduction, maintenance, but also calcification), as well as intra- and interspecific competition (Irie *et al.* 2010). Conceptual frameworks such as this are widely used in evolutionary biology and it is anticipated that they will be used increasingly in the context of the effects of ocean acidification and climate change. However, extending similar conceptual approaches to the scale of the global ocean remains a challenge.

15.4.9 Need for a coordinated international effort

The ongoing and planned research projects at the national or international scale cover the field of ocean acidification well and have begun to generate critical data. Those research initiatives are at present uncoordinated, which has several consequences.

Although there is a risk of duplication of research which funding agencies might want to avoid during a dire economic situation, it has been minimized among some European projects (EPOCA, BIOACID, and UKOA) thanks to tight integration. A reasonable level of duplication is useful anyway because the replicability of results increases the level of confidence. There are also many activities that would be performed better and more efficiently at the international level. The SOLAS-IMBER Working Group on Ocean Acidification has outlined several priority areas for action: including biological variables in observing networks, launch of joint platforms and facilities, intercomparison exercises, the update of the *Guide to best practices for ocean acidification research and data reporting*, management of ocean acidification data, training, and outreach. However, there are currently no human resources and very limited funding to support such activities at the international level. The community may need to make a large effort in order to raise support for an international coordination office.

15.5   Conclusions

Some effects of the uptake of anthropogenic CO_2 by the oceans, such as the changes in the carbonate chemistry, are known with a high degree of certainty (Table 15.1). Most biological and ecological effects are much less certain. Nevertheless, there is no doubt that calcification, primary production and nitrogen fixation, and biodiversity will be altered but by a magnitude that is unknown. These changes will in turn generate changes in the biogeochemical cycles, society, and the economy. Whether these changes will be significant or not is also unknown. The levels of confidence for these changes, estimated for the first time in Table 15.1, can be evaluated either with a pessimistic or with an optimistic view. It is unfortunate that so much uncertainty remains, because human society needs to get information with a relatively high degree of confidence before it decides to regulate its activities. Even with a high degree of certainty on future climate change and its likely impacts on society and economy, the reduction of CO_2 emissions is proving extremely slow and difficult to implement. Ocean acidification and its impacts on marine ecosystems may well

provide an additional reason for reducing CO_2 emissions but the knowledge generated until now is patchy (many processes and functional groups have not been investigated) and sometimes uncertain and conflicting, making it difficult for policymakers to put ocean acidification high on the agenda.

There is a much more optimistic reading of the levels of certainty summarized in Table 15.1. Although it has been known for a relatively long time that ocean pH will decrease due to the uptake of CO_2, it is only very recently that the biological consequences have been identified. Considering the small number of scientists engaged in ocean acidification research during most of the last 20 years, the level of knowledge reached in less than 20 years is remarkable and compares very favourably with the development of research on the impact of climate change on terrestrial ecosystems. A considerable improvement in the levels of uncertainty can be expected in the coming years due to the recent launch of major national and international projects on ocean acidification, the build-up of a strong community of researchers, and ongoing synthesis efforts of the IPCC. There is no doubt that the near future will be exciting for ocean acidification research. Whether the outcome will confirm the pessimistic view of some scientists in the field or not, it is hoped that human society will consider ocean acidification together with climate change to decide on the best course for its future.

15.6   Acknowledgements

Lina Hansson and Anne-Marin Nisumaa are gratefully acknowledged for their assistance with Fig. 15.2. This work is a contribution to the 'European Project on Ocean Acidification' (EPOCA) which received funding from the European Community's Seventh Framework Programme (FP7/2007–2013) under grant agreement no 211384. C.T. received additional support from the UK Ocean Acidification Research Programme.

References

Agegian, C.R. (1985). *The biogeochemical ecology of Porolithon gardineri (Foslie)*, 178 pp. PhD Thesis, University of Hawaii, Honolulu , HI.

Allen, L.H. (1992). Free-air CO_2 enrichment field experiments: an historical overview. *Critical Reviews in Plant Sciences*, **11**, 121–34.

Armstrong, J.L., Boldt, J.L., Cross, A.D. et al. (2005). Distribution, size, and interannual, seasonal and diel food habits of northern Gulf of Alaska juvenile pink salmon, *Oncorhynchus gorbuscha. Deep-Sea Research (Part II, Topical Studies in Oceanography)*, **52**, 247–65.

Bell, G. and Collins, S. (2008). Adaptation, extinction and global change. *Evolutionary Applications*, **1**, 3–16.

Borenstein, M., Hedges, L.V., Higgins, J.P.T., and Rothstein, H.R. (2009). *Introduction to meta-analysis*, 421 pp. Wiley, New York.

Bradshaw, W.E. and Holzapfel, C.M. (2006). Evolutionary response to rapid climate change. *Science*, **312**, 1477–8.

Brewer, P.G. and Peltzer, E.T. (2009). Limits to marine life. *Science*, **324**, 347–8.

Collins, S. and Bell, G. (2004). Phenotypic consequences of 1,000 generations of selection at elevated CO_2 in a green alga. *Nature*, **431**, 566–9.

Collins, S. and Bell, G. (2006). Evolution of natural algal populations at elevated CO_2. *Ecology Letters*, **9**, 129–35.

Comeau, S., Jeffree, R., Teyssié, J.-L., and Gattuso, J.-P. (2010). Response of the Arctic pteropod *Limacina helicina* to projected future environmental conditions. *PLoS ONE*, **5**, e11362.

Cubillos, J.C., Wright, S.W., Nash, G. et al. (2007). Calcification morphotypes of the coccolithophorid *Emiliania huxleyi* in the Southern Ocean: changes in 2001 to 2006 compared to historical data. *Marine Ecology Progress Series*, **348**, 47–54.

De'ath, G., Lough, J.M., and Fabricius, K.E. (2009). Declining coral calcification on the Great Barrier Reef. *Science*, **323**, 116–19.

Eby, M., Zickfeld, K., Montenegro, A., Archer, D., Meissner, K.J., and Weaver, A.J. (2009). Lifetime of anthropogenic climate change: millennial time scales of potential CO_2 and surface temperature perturbations. *Journal of Climate*, **22**, 2501–11.

Fields, P.A., Graham, J.B., Rosenblatt, R.H., and Somero, G.N. (1993). Effects of expected global climate-change on marine faunas. *Trends in Ecology and Evolution*, **8**, 361–7.

Friedlingstein, P., Houghton, R.A., Marland, G. et al. (2010). Update on CO_2 emissions. *Nature Geoscience*, **3**, 811–12.

Grelaud, M., Schimmelmann, A., and Beaufort, L. (2009). Coccolithophore response to climate and surface hydrography in Santa Barbara Basin, California, AD 1917–2004. *Biogeosciences*, **6**, 2025–39.

Guinotte, J.M., Orr, J., Cairns, S., Freiwald, A., Morgan, L., and George, R. (2006). Will human-induced changes in seawater chemistry alter the distribution of deep-sea scleractinian corals? *Frontiers in Ecology and the Environment*, **4**, 141–6.

Hall-Spencer, J.M., Rodolfo-Metalpa, R., Martin, S. et al. (2008). Volcanic carbon dioxide vents show ecosystem effects of ocean acidification. *Nature*, **454**, 96–9.

Hendriks, I.E., Duarte, C.M., and Alvarez, M. (2010). Vulnerability of marine biodiversity to ocean acidification: a meta-analysis. *Estuarine, Coastal and Shelf Science*, **86**, 157–64.

Hester, K.C., Peltzer, E.T., Kirkwood, W.J., and Brewer, P.G. (2008). Unanticipated consequences of ocean acidification: a noisier ocean at lower pH. *Geophysical Research Letters*, **35**, L19601, doi:10.1029/2008GL034913.

Hoegh-Guldberg, O. (1999). Climate change, coral bleaching, and the future of the world's coral reefs. *Marine and Freshwater Research*, **50**, 839–66.

Hoegh-Guldberg, O., Mumby, P.J., Hooten, A.J. et al. (2007). Coral reefs under rapid climate change and ocean acidification. *Science*, **318**, 1737–42.

Hunt, B.P.V., Pakhomov, E.A., Hosie, G.W., Siegel, V., Ward, P., and Bernard, K. (2008). Pteropods in Southern Ocean ecosystems. *Progress in Oceanography*, **78**, 193–221.

Irie, T., Bessho, K., Findlay, H.S., and Calosi, P. (2010). Increasing costs due to ocean acidification drives phytoplankton to be more heavily calcified: optimal growth strategy of coccolithophores. *PLoS One*, **5**, e13436.

Joseph, J.E. and Chiu, C.-S. (2010). A computational assessment of the sensitivity of ambient noise level to ocean acidification. *Journal of the Acoustical Society of America*, **128**, EL144–EL149.

Kirkwood, W. and Sano, L. (2009). Developing new instrumentation for in situ experimentation related to ocean acidification - scaling up pH effects from the lab to the field. *Current, the Journal of Marine Education*, **25**, 13–14.

Kleypas, J.A., Buddemeier, R.W., Archer, D., Gattuso, J.-P., Langdon, C., and Opdyke, B.N. (1999). Geochemical consequences of increased atmospheric CO_2 on coral reefs. *Science*, **284**, 118–20.

Kroeker, K., Kordas, R.L., Crim, R.N., and Singh, G.G. (2010). Meta-analysis reveals negative yet variable effects of ocean acidification on marine organisms. *Ecology Letters*, **13**, 1419–34.

Kump, L.R., Bralower, T.J., and Ridgwell, A. (2009). Ocean acidification in deep time. *Oceanography*, **22**, 94–107.

Langdon, C., Takahashi, T., Marubini, F. et al. (2000). Effect of calcium carbonate saturation state on the rate of calcification of an experimental coral reef. *Global Biogeochemical Cycles*, **14**, 639–54.

Langdon, C., Gattuso, J.-P., and Andersson, A. (2010). Measurements of calcification and dissolution of benthic organisms and communities. In: U. Riebesell, V.J.

Fabry, L. Hansson, and J.-P. Gattuso (eds), *Guide to best practices for ocean acidification research and data reporting*, pp. 213–34. Publications Office of the European Union, Luxembourg.

Leclercq, N., Gattuso, J.-P., and Jaubert, J. (2000). CO_2 partial pressure controls the calcification rate of a coral community. *Global Change Biology*, **6**, 329–34.

Liu, J., Weinbauer, M.G., Maier, C., Dai, M., and Gattuso, J.-P. (2010). Effect of ocean acidification on microbial diversity, and on microbe-driven biogeochemistry and ecosystem functioning. *Aquatic Microbial Ecology*, **61**, 291–305.

Maier, C., Hegeman, J., Weinbauer, M.G., and Gattuso, J.-P. (2009). Calcification of the cold-water coral *Lophelia pertusa* under ambient and reduced pH. *Biogeosciences*, **6**, 1671–80.

Manzello, D.P., Kleypas, J.A., Budd, D.A., Eakin, C.M., Glynn, P.W., and Langdon, C. (2008). Poorly cemented coral reefs of the eastern tropical Pacific: possible insights into reef development in a high-CO_2 world. *Proceedings of the National Academy of Sciences USA*, **105**, 10450–5.

Mastrandrea, M.D., Field, C.B., Stocker, T.F. *et al.* (2010). *Guidance note for lead authors of the IPCC fifth assessment report on consistent treatment of uncertainties*. Intergovernmental Panel on Climate Change (IPCC). Available at: http://www.ipcc.ch

Munday, P.L., Dixson, D.L., McCormick, M.I., Meekan, M., Ferrari, M.C., and Chivers, D.P. (2010). Replenishment of fish populations is threatened by ocean acidification. *Proceedings of the National Academy of Sciences USA*, **107**, 12930–4.

Nisumaa, A.-M., Pesant, S., Bellerby, R.G.J. *et al.* (2010). EPOCA/EUR-OCEANS data compilation on the biological and biogeochemical responses to ocean acidification. *Earth System Science Data*, **2**, 167–75.

Pörtner, H.-O. (2010). Oxygen- and capacity-limitation of thermal tolerance: a matrix for integrating climate-related stressor effects in marine ecosystems. *Journal of Experimental Biology*, **213**, 881–93.

Pörtner, H.-O. and Farrell, A.P. (2008). Physiology and climate change. *Science*, **322**, 690–2.

Ridgwell, A. and Hargreaves, J.C. (2007). Regulation of atmospheric CO_2 by deep-sea sediments in an Earth system model. *Global Biogeochemical Cycles*, **21**, GB2008, doi:10.1029/2006GB002764.

Riebesell, U., Fabry, V.J., Hansson, L., and Gattuso, J.-P. (eds) (2010a). *Guide to best practices for ocean acidification research and data reporting*, 260 pp. Publications Office of the European Union, Luxembourg.

Riebesell, U., Lee, K., and Nejstgaad, J.C. (2010b). Pelagic mesocosms. In: U. Riebesell, V.J. Fabry, L. Hansson, and J.-P. Gattuso (eds), *Guide to best practices for ocean acidification research and data reporting*, pp. 95–112. Publications Office of the European Union, Luxembourg.

Rockström, J., Steffen, W., Noone, K. *et al.* (2009). Planetary boundaries: exploring the safe operating space for humanity. *Ecology and Society*, **14**, ART 32.

Rosenzweig, C., Karoly, D., Vicarelli, M. *et al.* (2008). Attributing physical and biological impacts to anthropogenic climate change. *Nature*, **453**, 353–7.

Schubert, R., Schellnhuber, H.-J., Buchmann, N. *et al.* (2006). *The future of oceans – warming up, rising high, turning sour*, 110 pp. German Advisory Council on Global Change, Berlin.

Steinacher, M., Joos, F., Frölicher, T.L., Plattner, G.-K., and Doney, S.C. (2009). Imminent ocean acidification projected with the NCAR global coupled carbon cycle-climate model. *Biogeosciences*, **6**, 515–33.

Turley, C.M., Roberts, J.M., and Guinotte, J.M. (2007). Corals in deep-water: will the unseen hand of ocean acidification destroy cold-water ecosystems? *Coral Reefs*, **26**, 445–8.

Turley, C., Eby, M., Ridgwell, A.J. *et al.* (2010). The societal challenge of ocean acidification. *Marine Pollution Bulletin*, **60**, 787–92.

Udovydchenkov, I.A., Duda, T.F., Doney, S.C., and Lima, I.D. (2010). Modeling deep ocean shipping noise in varying acidity conditions. *Journal of the Acoustical Society of America*, **128**, EL130–EL136.

Veron, J.E., Hoegh-Guldberg, O., Lenton, T.M. *et al.* (2009). The coral reef crisis: the critical importance < 350 ppm CO_2. *Marine Pollution Bulletin*, **58**, 1428–36.

Walther, K., Anger, K., and Pörtner, H.-O. (2010). Effects of ocean acidification and warming on the larval development of the spider crab *Hyas araneus* from different latitudes (54° vs. 79°N). *Marine Ecology Progress Series*, **417**, 159–70.

Widdicombe, S., Dupont, S., and Thorndyke, M. (2010). Laboratory experiments and benthic mesocosm studies. In: U. Riebesell, V.J. Fabry, L. Hansson, and J.-P. Gattuso (eds), *Guide to best practices for ocean acidification research and data reporting*, pp. 113–22. Publications Office of the European Union, Luxembourg.

Zeebe, R.E., Zachos, J.C., and Dickens, G.R. (2009). Carbon dioxide forcing alone insufficient to explain Palaeocene–Eocene Thermal Maximum warming. *Nature Geoscience*, **2**, 576–80.

Zwiers, F. and Hegerl, G. (2008). Attributing cause and effect. *Nature*, **453**, 296–7.

Index

Note: page numbers in *italics* refer to Figures and Tables

express

¡Mira!

Anneli McLachlan

www.heinemann.co.uk
✓ Free online support
✓ Useful weblinks
✓ 24 hour online ordering

01865 888058

Heinemann

Inspiring generations

Heinemann is an imprint of Pearson Education Limited, a company incorporated in England and Wales, having its registered office at Edinburgh Gate, Harlow, Essex, CM20 2JE. Registered company number: 872828

Heinemann is a registered trademark of Pearson Education Limited

© Harcourt Education Limited, 2006

First published 2006

12 11
10

British Library Cataloguing in Publication Data is available from the British Library on request.

ISBN: 978 0 435387 66 2

Editor: Kathryn Tate
Designer: Ken Vail Graphic Design, Cambridge
Managing Editor: Naomi Laredo
Publisher: Gillian Eades

Original illustrations © Harcourt Education Limited, 2006

Illustrated by Beehive Illustration (Mark Brierley, Tina McNaughton, Simon Rumble), Clive Goodyer, Sylvie Poggio Artists Agency (Tim Davies, Andy Elkerton, Mark Ruffle, Jo Taylor, Rory Walker), Ken Laidlaw, Young Digital Poland.

Printed in China (SWTC/10)

Cover photo © Getty Images

Picture research by Christine Martin and Liz Alexander

Acknowledgements

Anneli McLachlan and Harcourt Education Ltd would like to thank Alex Harvey, Christopher Lillington, Ana Machado, Esther Mallol, Gima León de Klein, Tracy Traynor and Liliana Vilalpando for their invaluable help in the development of this course. They would also like to thank IES La Albuera, CEIP Eresma, CEIP Martín Chico, Parador de Segovia, Álvaro Alonso, Pilar Centeno, Mariano Núñez, Irene Sanz, Víctor Vallejo, Iñaki Alegre Perez de Ciriza, José María Bazán of Nordqvist Productions in Alicante and all those involved with the recordings.

The author and publisher would like to thank the following individuals and organisations for permission to reproduce photographs:

Alamy Images pp**62** (Connor, Monique), **75** (Mexican hacienda, house in Nerja), **85** (Roger Federer), **93** (Camp Nou stadium), **98** (Ujué, Bilbao, Benidorm, Cazorla); Allstar Picture Library/Alamy p**51** (Antonio Banderas); Carlos Álvarez/Getty Images p**51** (Javier Bardem); Corbis pp**13** (Pedro), **62** (village); Getty Images pp**62** (Meryl, Shafiq, John, Joachim, Costas), **85** (Lance Armstrong, Paula Radcliffe, Magic Johnson, Raúl González); Getty Images/Stone p**99** (Cazorla); Instituto de Economía y Geografía p**98** (Gijón); Mario Anzuoni/Reuters/Corbis p**51** (Salma Hayek); Pat Behnke/Alamy p**99** (Madrid); Picture Contact/Alamy p**13** (Isabel); Rufus F. Folkks/Corbis p**51** (Penélope Cruz); Spanish Tourist Office pp**62** (countryside, coast), **106** (Alcázar Córdoba). All other photographs were provided by Jules Selmes and Harcourt Education Limited.

Every effort has been made to contact copyright holders of material reproduced in this book. Any omissions will be rectified in subsequent printings if notice is given to the publishers.

Tel: 01865 888058   www.heinemann.co.uk

Contenidos

En la clase

1 ¡Hola!

● Introducing yourself
● Getting used to Spanish pronunciation

 1 Escucha. ¿Quién habla? (1–5)
Listen. Who is speaking?

¿Cómo te llamas?

Ejemplo: 1 a

a
¡Hola! Me llamo Roberto.

b
¡Hola! Me llamo Natalia.

c
¡Hola! Me llamo María.

d
¡Hola! Me llamo Omar.

e
¡Hola! Me llamo Isabel.

 2 Escucha. ¿Dónde viven? (1–5)
Listen. Where do they live?

¿Dónde vives?

Vivo en …

Ejemplo: 1 Madrid

Málaga	Madrid	Buenos Aires	Sevilla	Barcelona

" Note how these letters are pronounced.

h   ¡**H**ola!   **h** is silent
ll   Me **ll**amo …   **ll** is a 'y' sound
v   **V**ivo en …   **v** is a 'b' sound

¡**H**ola! ¡**H**ola! ¡**H**ola!
Me **ll**amo **V**aleria.
Vivo en Se**v**illa.

Listen to the recording and say this five times as carefully as you can. **"**

Gramática

Spanish verb endings change to show who the verb refers to.

¿Cómo te llam**as**?   *What are **you** called?*
Me llam**o** …   *I am called …*
¿Dónde viv**es**?   *Where do **you** live?*
Viv**o** en …   *I live in …*

Para saber más   p.22 (ex 1); p.130

 3 Con tu compañero/a, haz cuatro diálogos.
With your partner, make four dialogues.

● ¡Hola! ¿Cómo te llamas?
■ ¡Hola! Me llamo <u>Justin</u>.
● ¿Dónde vives, Justin?
■ Vivo en <u>Los Ángeles</u>.

Justin Timberlake.............Los Ángeles
J-Lo...................................Nueva York
David Beckham ....Madrid
Avril Lavigne.....................Toronto

4 **Escribe los diálogos del ejercicio 3.**
Write out the dialogues from exercise 3.

 5 **Escucha y canta.**
Listen and sing along.

Look at how you write questions and exclamations in Spanish:

¡Hola!   ¿Cómo te llamas?   ¿Dónde vives?

¡Hola! Buenos días ...   ¡Hola! Buenos días ...   Buenas tardes ...   Buenas noches ...

Adiós. ¡Hasta luego!   Adiós. ¡Hasta luego!   Adiós, mi amigo ...   Adiós, mi amigo ...

6 **Escribe las frases correctamente.**
Write out the phrases correctly.

¡Hola!Buenosdías. ¿Cómotellamas?Mellamo... ¿Dóndevives?Vivoen... Buenastardes.Adiós.¡Hastaluego!Buenasnoches.

7 **Escucha y escribe:**
😊😊/😊/😐/☹. (1–6)
Listen and note down.

¿Qué tal?

¿Cómo estás?

😊😊 Fenomenal

😊 Bien

😐 Regular

☹ Fatal

8 **Con tu compañero/a, practica el diálogo. Luego, haz otro diálogo cambiando las palabras subrayadas.** *With your partner, practise the dialogue. Then make another dialogue, changing the underlined words.*

● ¡Hola! ¿Cómo te llamas?
■ Me llamo Patricia.
● ¿Qué tal, Patricia?
■ Bien, gracias. ¿Y tú?
● Regular.
● Hasta luego, Patricia.
■ Adiós.

Patricia	Isabel	Omar	Roberto
😐	😊😊	☹	😊

2 ¿Cuántos años tienes?

escuchar 1 Escucha y repite.
Listen and repeat.

0 cero	**4** cuatro	**8** ocho	**12** doce
1 uno	**5** cinco	**9** nueve	**13** trece
2 dos	**6** seis	**10** diez	**14** catorce
3 tres	**7** siete	**11** once	**15** quince

escuchar 2 Escucha y escribe los números que entiendes (a–j).
Listen and write down the numbers you hear.

Ejemplo: **a** 14

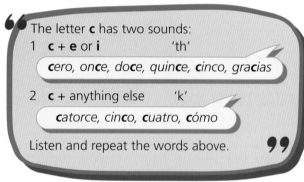

" The letter **c** has two sounds:
1 **c** + **e** or **i**   'th'
   cero, on**c**e, do**c**e, quin**c**e, **c**in**c**o, gra**c**ias

2 **c** + anything else   'k'
   cator**c**e, **c**in**c**o, **c**uatro, **c**ómo

Listen and repeat the words above. "

hablar 3 Con tus compañeros/as, juega al bingo.
With your partners, play bingo. Make a grid with six numbers.

Ejemplo:

12	8	✗
3	14	7

escribir 4 Escribe los números que vienen antes y después de estos números.
Write the numbers that come before and after these numbers.

Ejemplo: **a** dos, cuatro

a 3   **b** ▨   **c** 10   **d** 4   **e** 1   **f** 7   **g** 11   **h** 0

escuchar 5 Escucha. ¿Quién habla? (1–6)
Listen. Who is talking?

Ejemplo: **1** Débora

¿Cuántos años tienes?

Tengo … años.

Elena
5

Alejandro
11

Raúl
13

Débora
12

Laura
14

Miguel
4

Note how the **n** with a tilde is pronounced.

añ os	*anyos*
español	*espanyol*
Iñaki	*Inyaki*

Iñaki, Begoña y Toño viven en España.

Listen and repeat the sentence above.

When you give your age in English, you say 'I am 11'. In Spanish, you say:

Tengo once años. *I 'have' 11 years.*

tener	*to have*
tengo	*I have*
tienes	*you have*

hablar **6** **Con tu compañero/a, pregunta y contesta.**
With your partner, ask and answer questions.

- ¿Cómo te llamas?
- Me llamo <u>Alicia</u>.
- ¿Cuántos años tienes, <u>Alicia</u>?
- Tengo <u>doce</u> años.

Alicia (12)
Francisco (13)
Mercedes (11)
Iñaki (14)
Begoña (10)

leer **7** **Copia y rellena la tabla.**
Copy and fill in the grid.

	Name	Age	Lives in …
1	Ignacio		

¡Hola! ¿Qué tal?
Me llamo Ignacio.
Vivo en Caracas.
Tengo doce años.
¿Y tú? ¿Cómo te llamas?

¡Hola! Me llamo Belén. ¿Y tú?
¿Cómo te llamas?
Vivo en Cartagena. ¿Y tú?
¿Dónde vives?
Tengo trece años. ¿Y tú?
¿Cuántos años tienes?

2

¡Hola! ¿Cómo estás?
Vivo en Murcia y tengo once años.
Me llamo Alejandro.
¡Hasta luego!

escribir **8** **Escribe un email con tus datos.**
Write an email with your own details. Choose one of the emails above to use as a model.

1 Pon los meses en el orden correcto. Escucha y comprueba tus respuestas.
Put the months into the correct order. Listen and check your answers.

Ejemplo: enero, …

abril	enero	junio
julio	marzo	octubre
diciembre	mayo	noviembre
septiembre	febrero	agosto

In Spanish, the words for the months don't start with a capital letter.

enero, febrero, …

Note how these letters are pronounced.

| **j** | **j**unio, **j**ulio | as Scottish 'ch' in 'loch' |
| **z** | mar**z**o | as 'th' in 'thumb' |

Juanjo, Jaime y Javier juegan juntos en marzo.

Listen and repeat this sentence carefully.

2 Escucha y lee.
Listen and read.

16 dieciséis	**20** veinte	**24** veinticuatro	**28** veintiocho
17 diecisiete	**21** veintiuno	**25** veinticinco	**29** veintinueve
18 dieciocho	**22** veintidós	**26** veintiséis	**30** treinta
19 diecinueve	**23** veintitrés	**27** veintisiete	**31** treinta y uno

3 Escucha y escribe la fecha. (1–6)
Listen and write the date.

¿Cuál es la fecha de hoy?

Es el **dos** de **mayo**.

Ejemplo: **1** 2/5

| 1/4 | 2/5 | 20/12 | 14/9 | 15/1 |

| 6/3 | 30/7 | 9/8 | 26/10 | 11/2 |

3 ¡Feliz cumpleaños! 1

4 Escribe las cuatro fechas que quedan del ejercicio 3.
Write out the four dates that are left over from exercise 3.

5 Escucha y canta.
Listen and sing along.

¡Cumpleaños feliz,
cumpleaños feliz,
te deseamos todos,
cumpleaños feliz!

6 Escucha y escribe las fechas. (1–5)
Listen and write down the dates.

Ejemplo: **1** – 7/11

mi = *my*
tu = *your*
es = *is*

¿Cuándo es tu cumpleaños?				
Mi cumpleaños es	el	uno dos tres	de	enero febrero marzo

Pedro **Yolanda** **Enrique** **Silvia** **Raquel**

7 Haz un sondeo. Pregunta a diez compañeros/as de clase. Copia y rellena la tabla.
Do a survey. Ask ten classmates. Copy and fill in the grid.

Nombre	Edad	Cumpleaños
Michael	11	12 de mayo

● ¿Cuántos años tienes?
■ Tengo <u>once</u> años.
● ¿Cuándo es tu cumpleaños?
■ Mi cumpleaños es el <u>doce de mayo</u>.

Mini-test

I can
● say hello and goodbye
● say what I am called
● say how I am
● say where I live
● count up to 31
● say my age
● say when my birthday is
● ask four questions (name, where you live, age, birthday)

4 Hablamos español

escuchar 1 Escucha y lee el texto. Escribe estas frases en español.

Listen and read. Write the phrases below in Spanish.

Ejemplo: **1** idioma oficial

El español en el mundo

El español es el idioma oficial de veintiún países.
La mayoría de estos países están en el continente americano.
250 000 000 personas hablan español en el mundo.
17 339 000 personas hablan español como primer idioma en los Estados Unidos.

hablan = *speak*
como primer idioma
= *as a first language*

1 official language
2 21 countries
3 The majority
4 American continent
5 250 000 000 people speak Spanish in the world
6 United States

Honduras Cuba
México República Dominicana
Nicaragua Puerto Rico
Costa Panamá El Caribe
Rica
El Salvador Venezuela
Guatemala Colombia
Ecuador
Perú
El océano Bolivia
Pacífico Paraguay
Chile
Argentina
Uruguay
El océano
Atlántico

Spain (in Europe) and Equatorial Guinea (in Africa) also have Spanish as their official language.

How to deal with a longer text:
● look for all the words you **do** know, e.g. numbers;
● look for words which look like English words – cognates, e.g. official = **oficial**;
● use the glossary at the back of the book;
● use the questions to help you make sense of a text.

escribir 2 Copia y rellena la tabla con los países del mapa.
Copy the grid and fill in the countries from the map.

Norteamérica	Centroamérica	Sudamérica
México		

escuchar 3 Escucha y escribe los datos sobre cada persona. (1–6)
Listen and write down the information for each person.

Ejemplo:
Nombre: Pedro
País: Costa Rica
Edad: 14
Cumpleaños: 7/11

Nuria   Paulina
Paco   Elena
Pedro   Alejandro

¿Cómo **te** llam**as**?
¿Dónde viv**es**?
¿Cómo est**ás**?

¿Cómo **se** llama **usted**?
¿Dónde viv**e usted**?
¿Cómo est**á usted**?

4 Escucha y escribe si entiendes 'tú' o 'usted'. (1–5)
Listen and write down whether you hear 'tú' or 'usted'.

Ejemplo: 1 tú

 5 Con tu compañero/a, pregunta y contesta por Pedro e Isabel.
With your partner, ask and answer for Pedro and Isabel.

- ● ¿Cómo te llamas?
- ■ Me llamo …
- ● ¿Dónde vives?
- ■ Vivo en …
- ● ¿Cuántos años tienes?
- ■ Tengo … años.
- ● ¿Cuándo es tu cumpleaños, …?
- ■ Mi cumpleaños es …

**Pedro 14 7/11
Costa Rica**

**Isabel 12 4/8
Chile**

Gramática

The infinitive is the form of a verb you find in a dictionary or wordlist.
In Spanish, one group of verbs has infinitives ending in **-ir**, e.g.
viv**ir** *to live*

For the present tense, you take off the **-ir** and add these endings:

(yo)	viv**o**	*I live*
(tú)	viv**es**	*you live*
(él/ella)	viv**e**	*he/she/it lives*
(nosotros/nosotras)	viv**imos**	*we live*
(vosotros/vosotras)	viv**ís**	*you (plural) live*
(ellos/ellas)	viv**en**	*they live*

Para saber más   p.22 (exs 3–4); p.130

 6 Lee los textos y rellena el carné de identidad. ¿De quién es?
Read the texts and fill in the identity card. Whose is it?

Nombre: ?
País: ?
Edad: 14
Cumpleaños: 4/8

¡Hola! ¿Qué tal?
Me llamo Bea.
Vivo en Nicaragua y tengo catorce años.
Mi cumpleaños es el cuatro de agosto.
¿Y tú? ¿Dónde vives?
Escríbeme pronto.

¡Hola! ¿Cómo estás?
Me llamo Cristina.
Vivo en Ecuador y tengo catorce años.
Mi cumpleaños es el catorce de abril.
¿Y tú? ¿Dónde vives? ¿Cuántos años tienes?
¡Hasta luego!

 7 Escribe un diálogo como en el ejercicio 5 para Bea o Cristina.
Write a dialogue like the one in exercise 5 for Bea or Cristina.

1 **Escucha y canta.** *Listen and sing along.*

A	ah	**E**	eh	**J**	*hota	**N**	enneh	**R**	erre	**W**	uuveh dobleh
B	beh	**F**	efeh	**K**	kah	**Ñ**	enyeh	**S**	esseh	**X**	ekis
C	theh	**G**	*heh	**L**	eleh	**O**	oh	**T**	teh	**Y**	ee gri-ehgah
CH	cheh	**H**	acheh	**LL**	elyeh	**P**	peh	**U**	uuh	**Z**	theta
D	deh	**I**	ee	**M**	emmeh	**Q**	kuh	**V**	uuveh		

* **h** = an 'h' pronounced at the back of your throat

 2 **Escucha y pon los objetos en el orden correcto. (1–14)**
Listen and put the objects in the correct order.

¿Tienes …?   Tengo …

Ejemplo: **1** m

a
un bolígrafo

b
un cuaderno

c
un libro

d
un monedero

e
un diccionario

f
un lápiz

g
un estuche

h
un móvil

i
un sacapuntas

j
una goma

k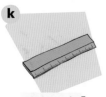
una regla

l
una agenda

m
una mochila

n
una calculadora

 3 **Con tu compañero/a, pregunta y contesta por los objetos del ejercicio 2.**
With your partner, ask and answer questions about the objects in exercise 2.

● ¿Cómo se escribe 'una mochila'?
■ Se escribe U-N-A M-O-C-H-I-L-A.

Gramática

In Spanish, all nouns are either masculine or feminine.

There are two words for 'a': **un** and una

Masculine	Feminine
un libro *a book*	una mochila *a bag*

Para saber más   p.23 (ex 6); p.129

4 **Empareja las descripciones con las mochilas.**
Match up the descriptions with the schoolbags.

1 En mi mochila tengo un libro, un cuaderno, un diccionario, una agenda, un móvil y también una calculadora, pero no tengo mi estuche.

2 En mi mochila, tengo un libro y mi estuche. En mi estuche, tengo un sacapuntas, un boli, una goma, un lápiz y también una regla.

y = *and*	también = *also*	pero = *but*

5 **Escucha y escribe el nombre correcto. (1–4)**
Listen and write down the correct name. (Below there is one person too many)

Ejemplo: **1** Eduardo

Luz Sergio Carolina María Eduardo

Tengo …	No tengo …	Necesito …

6 **Con tu compañero/a, pregunta y contesta por las personas en la tabla.**
With your partner, ask and answer for the people in the grid.

● ¿Tienes una regla, Carmelina?
■ Sí, tengo una regla.
● ¿Tienes una goma?
■ No. No tengo una goma. Necesito una goma.

Gramática

To make a sentence negative, put **no** before the verb.

No tengo un cuaderno.
I don't have an exercise book.
¿**No** tienes un lápiz?
Don't you have a pencil?

Para saber más   p.23 (ex 5)

	✏️	✏️	🧽	📕	✏️	📚	🖩	📏
Carmelina			✗					✔
Paco	✔	✗						
Liliana				✔			✗	
Alfonso			✗			✔		✗
Montse		✔		✗	✗	✔		

7 **Escribe una lista de lo que tienes/no tienes/necesitas para el instituto.**
Write a list of what you have/don't have/need for school.

En clase

escuchar

1 Escucha y escribe la letra correcta. (1–12)
Listen and write the correct letter.

Ejemplo: **1** e

Gramática

● The words for 'the' in Spanish:

	Singular		**Plural**
masculine	**el** libro	→	**los** libros
feminine	**la** mesa	→	**las** mesas

● The words for 'a/some' in Spanish are:

masculine	**un** libro	→	**unos** libros
feminine	**una** mesa	→	**unas** mesas

Para saber más   p.23 (exs 6–7); p.129

a	el alumno
b	el equipo de música
c	el profesor
d	el proyector
e	el ordenador
f	la ventana
g	la pizarra
h	la puerta
i	las mesas
j	las sillas
k	los libros
l	los rotuladores

hablar

2 Indica objetos en la clase.
Tu compañero/a los dice en español.
Point to objects in the classroom.
Your partner says them in Spanish.

● [points to chairs]
■ las sillas

escribir

3 Escribe la forma plural de las palabras.
Write the plural form of these words.

Ejemplo: **1** (el alumno) – los alumnos

1 el alumno		**5** la ventana
2 el profesor		**6** el libro
3 el proyector		**7** el rotulador
4 el ordenador		**8** el lápiz

Gramática

To make nouns plural in Spanish:
● If the noun ends in a vowel, add **-s**.

libro	→	libro**s**
mochila	→	mochila**s**
estuche	→	estuche**s**

● If the noun ends in a consonant, add **-es**.

móvil → móvil**es**

● If the noun ends in a **z**, change the **z** to **c** and add **-es**.

lápi**z** → lápi**ces**

Para saber más   p.23 (exs 6–7); p.129

4 Escucha y escribe en inglés qué hay (a) en la clase en el espacio y (b) en la clase acuática.

Listen and note down in English what there is in the space classroom and in the underwater classroom.

Ejemplo: **a** a board, …

¿Qué hay en la clase?

Hay una pizarra …

5 Con tu compañero/a, haz una frase muy larga.

With your partner, make a very long sentence.

- En mi clase hay <u>unos alumnos</u> …
- ■ En mi clase hay unos alumnos y <u>unas sillas</u> …
- En mi clase hay unos alumnos, unas sillas y <u>un profesor</u> …

en la clase hay …	
un profesor	una profesora
un equipo de música	una ventana
un proyector	una pizarra
un ordenador	una puerta
unos libros	unas sillas
unos rotuladores	unas mesas
unos alumnos	
unos lápices	

6 Copia y completa los textos con las palabras de los cuadros.

Copy the texts and fill in the gaps with words from the boxes.

Ejemplo: ¡Hola! Me llamo …

a

¡Hola! Me **(1)** _____ Miguel el marciano.

(2) _____ en el Planeta Marte con mi familia.

Mi **(3)** _____ es el trece de marzo.

Tengo quince **(4)** _____.

No tengo una mochila. No tengo un estuche pero tengo un ordenador.

En mi **(5)** _____ en la estación espacial, hay una pizarra, un profesor robot y un **(6)** _____.

¡Hasta luego!

b

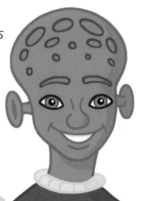

Me llamo **(1)** _____ y vivo en el planeta Venus.

(2) _____ catorce años. Mi cumpleaños es el

(3) _____ de **(4)** _____.

En mi clase en el planeta Venus, hay muchos

(5) _____. No tengo un profesor.

cumpleaños	clase	años	llamo
	ordenador	vivo	

Valeria	tengo	ordenadores
	veintidós	agosto

Resumen

Unidad 1

I can

- ask someone what they are called — ¿Cómo te llamas?
- say what I am called — Me llamo …
- ask someone where they live — ¿Dónde vives?
- say where I live — Vivo en …
- ask someone how they are — ¿Qué tal? ¿Cómo estás?
- say how I am — Fenomenal. Bien. Regular. Fatal.

Unidad 2

I can

- count up to 15 — uno, dos, tres, … quince
- ask someone their age — ¿Cuántos años tienes?
- say my age — Tengo … años.

Unidad 3

I can

- count up to 31 — veinte, veintiuno, … treinta, treinta y uno
- say what the date is — Es el veintidós de febrero.
- ask someone when their birthday is — ¿Cuándo es tu cumpleaños?
- say when my birthday is — Mi cumpleaños es el … de …

Unidad 4

I can

- name some Spanish-speaking countries — España, México, Argentina, …
- **G** understand when to use **tú** and **usted** — ¿Cómo te llamas? ¿Cómo se llama usted?
- **G** recognise all forms of **-ir** verbs — viv**o**, viv**es**, viv**e**, viv**imos**, viv**ís**, viv**en**

Unidad 5

I can

- say the Spanish alphabet — a, b, c, ch, …
- say what I have for school and what I need — Tengo un estuche. Necesito una mochila.
- **G** understand when to use **un** and **una** — un = *masculine*, una = *feminine*
- **G** make a sentence negative using **no** — No tengo una goma.

Unidad 6

I can

- say what there is in the classroom — Hay una pizarra …
- **G** use **el**, **la**, **los** and **las** correctly — el libro, los libros, la mesa, las mesas
- **G** understand how to form plural nouns — los libros, los rotuladores, los lápices
- **G** understand when to use **unos** and **unas** — unos alumnos, unas mesas

Prepárate

 1 **Escucha. Copia y rellena la tabla. (1–4)**
Listen. Copy and fill in the grid.

| Carlos | Eduardo | Ana | Julia |

Nombre	Edad	Ciudad
Carlos	13	Vigo

 2 **Con tu compañero/a, di los números de teléfono.**
With your partner, say the telephone numbers.

a 19 22 28 **b** 10 12 26 **c** 21 17 30 **d** 11 29 14 **e** 13 25 31

 3 **Con tu compañero/a, pregunta y contesta.**
With your partner, ask and answer.

¿Qué tal? ¿Dónde vives? ¿Cuántos años tienes? ¿Cuándo es tu cumpleaños?

4 **¿Cuál es la mochila de Omar?**
Which is Omar's schoolbag?

¡Hola! ¿Qué tal?
Me llamo Omar el organizado. Vivo en Sevilla. Tengo trece años.
En mi mochila, tengo unos cuadernos, una calculadora, un libro, una agenda, un monedero, un diccionario y mi estuche.
En mi estuche, tengo unos bolígrafos. También tengo unos rotuladores, un sacapuntas, una goma y una regla.
¿Y tú qué tienes?

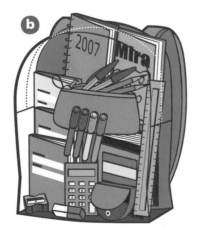
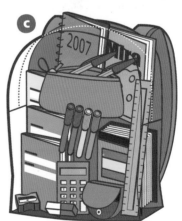

 5 **Describe otra mochila del ejercicio 4. Utiliza la descripción de la mochila de Omar como modelo.**
Describe another schoolbag from exercise 4. Use the description of Omar's schoolbag as a model.

¡Extra!

- ● Reading a story in Spanish
- ● Reading a poem in Spanish

1 Escucha y lee.
Listen and read.

1

T – ¡Hola! ¿Qué tal?
A – Bien, gracias. Y tú, ¿cómo estás?
T – Bien, gracias.

2

T – ¿Cómo te llamas?
A – Me llamo Angélica. Y tú, ¿cómo te llamas?
T – Me llamo Tomás.

3

T – ¿Cuántos años tienes, Angélica?
A – Tengo catorce años. Mi cumpleaños es el diez
de octubre.
T – ¡No me lo creo! Mi cumpleaños también es el
diez de octubre. Tengo catorce años también.

4

T – ¿Dónde vives, Angélica?
A – Vivo en Segovia. Y tú, ¿dónde vives?
T – Vivo en Madrid.

5

P – ¡Angélica!

A – ¡Hasta luego, Tomás!

6

T – Adiós, Angélica …

2 Con tu compañero/a, lee la historia de Tomás y Angélica.
With your partner, read the story of Tomás and Angélica out loud.

3 Contesta a las preguntas.
Answer the questions.

> vive = *he/she lives*
> tiene = *he/she has*

Ejemplo: 1 Tomás vive en Madrid.

1 ¿Dónde vive Tomás?
2 ¿Dónde vive Angélica?
3 ¿Cuántos años tiene Angélica?

4 ¿Cuántos años tiene Tomás?
5 ¿Cuándo es el cumpleaños de Angélica?
6 ¿Cuándo es el cumpleaños de Tomás?

4 Escucha y lee. Busca las palabras que no conoces en el vocabulario.
Listen and read. Look up the words you don't know in the glossary.

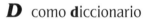

 ¿Qué tal? Regular.

A como **a**lfabeto
B como **b**olígrafo
C como **c**alculadora
CH como estu**ch**e

D como **d**iccionario
E como **e**scribir
F como ¡**f**eliz cumpleaños!

G como **g**oma en tu estuche
H como ¡**H**ola! ¡**H**ola!
I como **i**nteresante

¡Hola!

J como **j**unio o **j**ulio
K como **k**ilómetro
L como **l**eer un **l**ibro

LL como 'Me **ll**amo Pepe'
M como tu **m**ochila
N como **N**O **N**O **N**O **N**O

Ñ como espa**ñ**ol, por ejemplo
O como **o**ctubre – el mes
P como **p**rofesor o **p**royector

Q como '¿Y tú **q**ué tal?'
R como **r**egla o **r**egular
S como **s**eis **s**acapuntas

T como 'Tengo tres años'
U como **u**n y **u**na
V como **v**ivo o **v**einte

W como **W**ashington
X como '**X**avier vive en Cataluña'
Y como **Y**olanda o **y**
Z como un **z**umo que bebes

El alfabeto termina así.
¡Aprende! ¡Te toca a ti!

un una

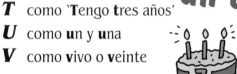

 20

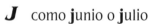

junio

julio

1 Kilómetro

 Me llamo Pepe.

 ¡NO!

 Me llamo Yolanda y tengo doce años.

1 The punctuation and accent keys on Isabel's computer aren't working. Can you help her out?

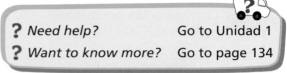

? Need help?   Go to Unidad 1

? Want to know more?   Go to page 134

- _Hola_ _C_mo te llamas_
- Me llamo Roberto.
- _Qu_ tal, Roberto_
- Bien, gracias. _Y t_ _
- Muy bien.
- Hasta luego, Roberto.
- Adi_s.

 2 Decode these questions and translate them into English. Then match up the questions with the answers below.

? Need help?   Go to Unidad 2 and 3

? Want to know more?   Go to page 134

Example: **1** ¿Cómo …

1 ^←=Υ❖ ☺♌ %$Υ$↘+
2 ^←#*○✖❖ ♌↘ ☺# ←#Υ!~♌$□❖↘+
3 ^✖=○✖♌ &@&♌↘+
4 ^←#*○☺❖↘ $□❖↘ ☺@♌○♌↘+
5 ^◇#■ ☺$~+

^	¿	@	i	=	ó
+	?	~	l	!	p
$	a	%	ll	◇	q
*	á	Υ	m	↘	s
←	c	○	n	☺	t
✖	d	□	ñ	#	u
♌	e	❖	o	&	v
■	é				

a Mi cumpleaños es el dieciocho de febrero.
b Me llamo Roberto.
c Bien, gracias.
d Vivo en Madrid.
e Tengo trece años.

 3 Unjumble these verbs, then write the English.

? Need help?   Go to Unidad 4

? Want to know more?   Go to page 131

Example: **1** vives – you live

1 vvise   **2** inevv   **3** vvio   **4** msoivvi   **5** vvie   **6** ísvvi

 4 Fill in the inkspots to apply the pattern from vivir to the verb escribir (to write).

1 escrib● *I write*   **4** escrib● *we write*

2 escrib● *you write*   **5** escrib● *you (plural) write*

3 escrib● *he/she/it writes* **6** escrib● *they write*

5 **Make these sentences positive to find out what Paco is saying about himself.**

> **?** *Need help?*   Go to Unidad 4 and 5
>
> **?** *Want to know more?*   Go to page 132

Example: **1** Me llamo Paco.

1 No me llamo Paco.
2 No vivo en Sevilla.
3 No tengo trece años.
4 Mi cumpleaños no es el nueve de enero.
5 No tengo una mochila.

 6 **Un, una, unos *or* unas?**
Copy and fill in the grid.

> **?** *Need help?*   Go to Unidad 5 and 6
>
> **?** *Want to know more?*   Go to page 129

Singular		Plural	
Masculine (un/el)	**Feminine (una/la)**	**Masculine (unos/los)**	**Feminine (unas/las)**
boli			

boli	sacapuntas	lápiz	proyector	goma	reglas
silla	agenda	móvil	monedero	diccionarios	mesas
equipo de música	cuaderno	libros	pizarra	estuche	
libro	alumno	ventana	puerta	ordenador	regla
rotuladores	sillas	cuadernos	diccionario		

leer 7 **Look at the picture of the space classroom and write out the sentences correctly.**

Example: **1** Hay una profesora y unos alumnos.

1 Hay **un profesor** / **una profesora** y unos alumnos.
2 Hay unas sillas y **unas mesas** / **unas mochilas**.
3 Hay **una pizarra** / **una pizza**.
4 Hay unos libros y **unos cuadernos** / **unas reglas**.
5 Hay un ordenador y **un proyector** / **unos proyectores**.
6 **No hay** / **Hay** rotuladores.
7 También hay **un equipo de música** / **unos equipos de música**.
8 **Hay** / **No hay** una ventana.

Palabras

Saludos	Greetings
¡Hola!	*Hello!*
¡Buenos días!	*Good morning!*
¡Buenas tardes!	*Good afternoon!*
¡Buenas noches!	*Good evening!*
¡Adiós!	*Goodbye!*
¡Hasta luego!	*See you later!*

Tú y yo	You and me
¿Cómo te llamas?	*What are you called?*
Me llamo Juan.	*I'm called Juan.*
¿Dónde vives?	*Where do you live?*
Vivo en Madrid.	*I live in Madrid.*
¿Qué tal?	*How are you?*
¿Cómo estás?	*How are you?*
Bien, gracias.	*Fine, thanks.*
fenomenal	*great*
regular	*not bad*
fatal	*awful*
¿Y tú?	*And you?*
¿Cuántos años tienes?	*How old are you?*
Tengo 13 años.	*I'm 13 years old.*
¿Cuándo es tu cumpleaños?	*When is your birthday?*
Mi cumpleaños es el uno de enero.	*My birthday is 1st January.*
¡Feliz cumpleaños!	*Happy birthday!*

Los números	Numbers
cero	*0*
uno	*1*
dos	*2*
tres	*3*
cuatro	*4*
cinco	*5*
seis	*6*
siete	*7*
ocho	*8*
nueve	*9*
diez	*10*
once	*11*
doce	*12*
trece	*13*
catorce	*14*
quince	*15*
dieciséis	*16*
diecisiete	*17*
dieciocho	*18*
diecinueve	*19*
veinte	*20*
veintiuno	*21*
veintidós	*22*
veintitrés	*23*
veinticuatro	*24*
veinticinco	*25*
veintiséis	*26*
veintisiete	*27*
veintiocho	*28*
veintinueve	*29*
treinta	*30*
treinta y uno	*31*

Los meses	The months
enero	January
febrero	February
marzo	March
abril	April
mayo	May
junio	June
julio	July
agosto	August
septiembre	September
octubre	October
noviembre	November
diciembre	December

¿Cuál es la fecha de hoy?	What date is it today?
Es el uno de agosto.	It's 1st August.
Es el dos de mayo.	It's 2nd May.

En mi mochila	In my schoolbag
un bolígrafo/boli	a pen
un cuaderno	an exercise book
un libro	a textbook
un monedero	a purse
un diccionario	a dictionary
un lápiz	a pencil
un estuche	a pencil case
un móvil	a mobile phone
un sacapuntas	a pencil sharpener
una agenda	a diary
una calculadora	a calculator
una goma	a rubber
una mochila	a schoolbag
una regla	a ruler

En clase	In the classroom
¿Cómo se escribe … ?	How do you spell … ?
Se escribe …	You spell it …
Tengo …	I have …
No tengo …	I don't have …
¿No tienes … ?	Don't you have … ?
Necesito …	I need …
¿Qué hay en la clase?	What is there in the classroom?
Hay …	There is/are …
No hay …	There isn't/There aren't …

el alumno	the pupil (male)
el profesor	the teacher (male)
el equipo de música	the stereo
el ordenador	the computer
el proyector	the overhead projector
la profesora	the teacher (female)
la pizarra	the board
la puerta	the door
la ventana	the window
los libros	the books
los rotuladores	the felt-tip pens
las mesas	the tables
las sillas	the chairs

Hay unos alumnos.	There are some pupils.
Hay unas sillas.	There are some chairs.
No hay rotuladores.	There are no felt-tip pens.

Palabras muy útiles	Very useful words
sí	yes
no	no
y	and
pero	but
también	also
tengo	I have
necesito	I need
hay	there is/are

Estrategia

Here are five simple steps to help you learn any word:

1 **LOOK** Look carefully at the word for 10 seconds or more.

2 **SAY** Practise saying the word to yourself – remember that some letters are pronounced differently in Spanish.

3 **COVER** Cover up the word, but only when you think you know it.

4 **WRITE** Write the word out from memory.

5 **CHECK** Did you write it correctly? If not, what did you get wrong? Repeat the five steps until you get it right – and try not to make the same mistake again.

2 · 1 ¿Qué estudias?

- Talking about your school subjects
- Using the -ar verb **estudiar** (to study)

escuchar 1 Empareja los dibujos con las palabras. Escucha y comprueba tus respuestas. (a–n)
Match up the pictures with the words. Listen and check your answers.

Ejemplo: a dibujo ✔

 a

 b

 c

 d

 e

 f

 g

 h

 i

 j

 k

 l

 m

 n

Estudio …

religión

educación física

tecnología

ciencias

geografía

inglés

francés

matemáticas

español

teatro

música

informática

historia

dibujo

> Note the two ways **g** is pronounced:
> 1 **g** + **e** or **i**   as 'ch' in 'loch'
> 2 **g** + anything else   as 'g' in 'great'
>
> reli**gi**ón   tecnolo**gí**a
> in**gl**és   **ge**o**gr**afía

escuchar 2 Escucha y escribe las letras correctas del ejercicio 1. (1–6)
Listen and write down the correct letters from exercise 1.

Ejemplo: 1 i, a ✔ c ✘

How did you work out the meanings of the subjects?
- Did you look for cognates – words which look like the English?
- Did you use a process of elimination?
- Did you pronounce the words out loud to see if you recognised them?

Make a list of strategies with your partner.

hablar 3 Con tu compañero/a, pregunta y contesta por estas personas.
With your partner, ask and answer for these people.

- ¿Qué estudias, <u>Carlos</u>?
- ■ Estudio <u>español, matemáticas y ciencias</u>.
- ¿Qué no estudias, <u>Carlos</u>?
- ■ No estudio <u>informática</u>.

 Carlos

 Dolores

 Conchi

 David

4 Escucha y canta.
Listen and sing along.

In Spanish, the words for the days don't start with a capital letter:
lunes, **m**artes …

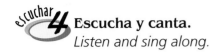

| lunes | martes | miércoles | jueves | viernes | sábado | domingo |

5 Copia la tabla. Escucha y escribe las letras correctas del ejercicio 1. (1–5)
Copy out the grid. Listen and fill in the correct letters from exercise 1.

1	lunes	a, …
2	martes	
3	miércoles	
4	jueves	
5	viernes	

If you do something 'every Monday …', use **los**:

Los lunes estudio francés y también …
Los sábados no estudio.

6 Lee el texto. ¿Qué estudia Sergio en la semana?
Read the text. What does Sergio study in the week?

Ejemplo: Monday: Spanish, maths and science

¡Hola! Me llamo Sergio. Tengo trece años.
Los lunes estudio español, matemáticas y ciencias.
Los martes estudio inglés, tecnología y también informática.
Los miércoles estudio educación física y religión.
Los jueves estudio teatro y música.
Los viernes estudio historia y dibujo.
No estudio francés en el instituto.
¿Y tú? ¿Qué estudias?

7 ¿Y tú? ¿Qué estudias? Utiliza el texto del ejercicio 6 como modelo.
What do you study? Use the text from exercise 6 as a model.

● say your name
● say how old you are
● say what you study on different days
● say what you don't study

Gramática

Another group of verbs has infinitives ending in **-ar**, e.g.
estudi**ar**   *to study*

For the present tense, you take off the **-ar** and add these endings:

(yo)	estudi**o**	*I study*
(tú)	estudi**as**	*you study*
(él/ella)	estudi**a**	*he/she/it studies*
(nosotros/nosotras)	estudi**amos**	*we study*
(vosotros/vosotras)	estudi**áis**	*you (plural) study*
(ellos/ellas)	estudi**an**	*they study*

Para saber más   p.40 (ex 1); p.130

2 ¿Qué haces en clase?

escuchar 1 Escucha y lee.
Listen and read.

¿Qué estudias?

Estudio inglés.

Estudio inglés también.

¿Qué haces en clase de inglés?

Hablo.

¡No como!

Escribo.

Escucho.

Leo.

Hablo con mis amigos o por teléfono.

Como chicle.

Escribo.

Escucho música.

¡No leo!

Spanish verbs are listed in a dictionary with these endings:

-ar e.g. habl**ar**   *to speak*
-er e.g. com**er**   *to eat*
-ir e.g. escrib**ir**   *to write*

Gramática

A third group of verbs has infinitives ending in **-er**, e.g. com**er**   *to eat*

Here are the **-ar**, **-er** and **-ir** verb endings:

	habl**ar**	com**er**	escrib**ir**
	(to speak)	*(to eat)*	*(to write)*
(I)	habl**o**	com**o**	escrib**o**
(you)	habl**as**	com**es**	escrib**es**
(he/she/it)	habl**a**	com**e**	escrib**e**
(we)	habl**amos**	com**emos**	escrib**imos**
(you – plural)	habl**áis**	com**éis**	escrib**ís**
(they)	habl**an**	com**en**	escrib**en**

Para saber más   p.40 (exs 1–3); p.130

2 Escucha y escribe las letras correctas. (1–5)
Listen and write down the correct letters.

Ejemplo: **1** e ✔, b ✔, a ✘

3 Con tu compañero/a, pregunta y contesta.
With your partner, ask and answer questions.

- ● ¿Qué estudias?
- ■ Estudio <u>francés</u>.
- ● ¿Qué haces en <u>francés</u>?
- ■ <u>Hablo, escucho, escribo y leo</u>.

4 Escribe los diálogos del ejercicio 3.
Write out the dialogues from exercise 3.

5 Lee el texto. ¿Verdadero o falso?
Escribe V o F.
Read the text. True or false? Write V or F.

Ejemplo: **1** F

1 Luis Miguel tiene tres años.
2 Luis Miguel vive en Cádiz.
3 Los viernes estudia español, inglés y francés.
4 No habla en clase.
5 Escucha y lee.
6 Escribe también.
7 Come chicle en clase.

> ¡Hola! Me llamo Luis Miguel. Tengo trece años y vivo en Cádiz. Los viernes estudio español, inglés y francés. Hablo mucho en clase. Escucho y leo. También escribo. No como chicle en clase. En dibujo pinto y en música canto – ¡qué bien!

mucho = *a lot*

> Look at the endings of the verbs in the questions in exercise 5. They are all in the 'he/she' form, e.g.
>
> Luis Miguel estudi**a**.   *Luis Miguel studies.*

6 Lee el texto otra vez. Copia y rellena la tabla.
Read the text again. Copy and fill in the grid.

estudiar	to study	I study	estudio
vivir	to live	I live	
hablar	to speak	I speak	
escuchar	to listen	I listen	
leer	to read	I read	
cantar	to sing	I sing	
pintar	to paint	I paint	

7 Describe tu semana. *Describe your week.*

Ejemplo: Los lunes estudio tecnología. En tecnología escucho, leo, hablo y también escribo.
Los martes estudio …

3 Los profesores

Talking about your teachers
Using adjectives that end in -o/-a

Escucha y elige el profesor correcto/la profesora correcta. (1–10)
Listen and choose the correct teacher.

Ejemplo: **1** d

a El profesor de inglés es **severo**.

b El profesor de tecnología es **aburrido**.

c El profesor de español es **divertido**.

d El profesor de teatro es **simpático**.

e El profesor de religión es **antipático**.

f La profesora de ciencias es **severa**.

g La profesora de francés es **aburrida**.

h La profesora de música es **divertida**.

i La profesora de geografía es **simpática**.

j La profesora de dibujo es **antipática**.

Gramática

Adjectives have masculine and feminine forms.

Many adjectives end in -o/-a in the singular.

Masculine	Feminine
divertido	divertida
severo	severa
simpático	simpática
aburrido	aburrida
antipático	antipática

Para saber más p.41 (exs 4, 6); p.129

Elige un profesor o una profesora. Tu compañero/a hace preguntas para saber quién es.
Choose a teacher. Your partner asks questions to find out who it is.

- ¿Profesor o profesora?
- Profesor.
- ¿Es simpático?
- No.
- ¿Es antipático?
- ¡Sí!
- Es el profesor de religión.
- ¡Sí!

> Remember to say words with the stress on an accented letter. Listen and practise these before you do speaking exercise 2.
>
> | simpático | inglés | religión |
> |-----------|--------|----------|
> | tecnología | música | francés |

3 Escucha y escribe la asignatura y la opinión en inglés. (1–6)

Listen and note the subject and opinion in English.

Ejemplo: 1 history ✔✔✔ strict

¿Cómo es el profesor de …?   ¿Cómo es la profesora de …?

> You can use **qualifiers** to make your sentences more detailed.
> ✔   **un poco** = *a bit*
> ✔✔   **bastante** = *quite*
> ✔✔✔ **muy** = *very*

4 Busca un profesor/una profesora …

Find a teacher who is …

Ejemplo: 1 Señor López

> señor = *Mr*
> señora = *Mrs*
> señorita = *Miss*

1 nice and quite funny

2 not very nice at all

3 quite boring

4 very nice and funny

5 a little strict

Español	**El señor López** es simpático y bastante divertido también. Habla mucho.
Música	**La señora Buitrago** es un poco severa pero canta muy bien.
Dibujo	**El señor Abad** es bastante aburrido. Pinta pero no habla mucho.
Inglés	**La señora Silgado** es muy, muy antipática. No escucha y habla rápidamente.
Matemáticas	**El señor Arranz** es divertido. También es muy simpático. Come chicle en clase.

rápidamente = *quickly*

5 Elige el verbo correcto para cada persona del ejercicio 4.

Choose the correct verb for each person from exercise 4.

Ejemplo: 1 El señor López <u>habla</u> mucho.

1 El señor López **habla** / **canta** mucho.
2 La señora Buitrago **escribe** / **canta** muy bien.
3 El señor Abad **pinta** / **habla** mucho.
4 La señora Silgado **no habla** / **no escucha**.
5 El señor Arranz **lee** / **come** en clase.

> You can extend your sentences using these words.
> y = *and*
> también = *also*
> pero = *but*

6 Describe a cinco profesores. Utiliza 'un poco', 'bastante', 'muy'.

Describe five teachers. Use the qualifiers un poco, bastante, muy.

Ejemplo: El señor Lamington es muy simpático y bastante divertido.

Mini-test

I can …
● say what I study
● name the days of the week
● say what I do in different lessons
● say what my teachers are like
G understand singular and plural forms of regular verbs
G use adjective agreements in the singular
G use qualifiers

1 **Escucha y escribe la letra correcta. (1–6)**
Listen and write down the correct letter.

¿Te gusta …? ¿Te gustan …?

Ejemplo: **1** b

a Me gusta el español.

b Me gusta la geografía.

c Me gustan las ciencias.

d Me gusta mucho la historia.

e No me gusta el inglés.

f No me gusta nada la educación física.

2 **Con tu compañero/a, pregunta y contesta.**
With your partner, ask and answer questions.

- ¿Te gusta la religión?
- No. No me gusta la religión.

1 ¿ ? ☹

2 ¿ ? ☺

3 ¿  ? ☺

4 ¿  ? ☺☺

5 ¿  ? ☹☹

6 ¿  ? ☹

Gramática

Singular subjects

¿Te gust**a** …? *Do you like …?*

☺ Me gust**a el** dibujo.

☺☺ Me gust**a mucho la** historia.

☹ **No** me gust**a el** dibujo.

☹☹ **No** me gust**a nada la** historia.

Plural subjects

¿Te gust**an** …? *Do you like …?*

☺ Me gust**an las** ciencia**s**.

☺☺ Me gust**an mucho las** ciencia**s**.

☹ **No** me gust**an las** matemática**s**.

☹☹ **No** me gust**an nada las** matemática**s**.

Para saber más p.132

When you give opinions on subjects using **(no) me gusta(n)**, make sure you use **el**, **la** or **las** before the subjects you mention.

el	**la**	**las**
dibujo	educación física	ciencias
español	geografía	matemáticas
francés	historia	
inglés	informática	
teatro	música	
	religión	
	tecnología	

leer 3 Escribe la asignatura y la opinión.
Note down the subject and the opinion.

Ejemplo: **1** drama – boring

1 El teatro es aburrido.
2 La tecnología es divertida.
3 La historia es difícil.
4 Las ciencias son interesantes.
5 El inglés es bueno.
6 La religión no es útil.
7 Me gusta el francés porque es fácil.
8 Me gustan las ciencias porque son importantes.

Gramática

Adjectives have masculine and feminine forms, and singular and plural forms.

● Many adjectives end in **-o** or **-a** in the singular.
● Some end in **-e**.
● Some end in a consonant.

	El/La ... es ...	Los/Las ... son ...
funny	divertid**o/a**	divertid**os/as**
good	buen**o/a**	buen**os/as**
boring	aburrid**o/a**	aburrid**os/as**
interesting	interesant**e**	interesant**es**
important	importante	important**es**
easy	fácil	fácil**es**
difficult	difícil	difícil**es**
useful	útil	útil**es**

Para saber más   p.41 (exs 4, 6); p.129

escuchar 4 Escucha y escribe la asignatura y la opinión. (1–8)

Ejemplo: **1** 🙂 religión – interesante

Me gusta el francés **porque** es fácil.
*I like French **because** it's easy.*

Me gustan las ciencias **porque** son útiles.
*I like science **because** it's useful.*

Use **es** for singular subjects.
Use **son** for plural subjects.

escribir 5 Escribe seis frases. *Write six sentences.*

Ejemplo: Me gusta la geografía porque es divertida.

escuchar 6 Escucha la canción y rellena los espacios en blanco.
Listen to the song and fill in the gaps.

divertido	buena	antipática
simpática	aburrido	interesante
	bueno	difícil

1 Me gusta la informática.
La profesora es muy **(1)** ____.
El francés es **(2)** ____.
Ah sí, me gusta bastante.

2 El teatro es **(3)** ____,
pero el dibujo es **(4)** ____.
La historia es muy **(5)** ____,
pero el profesor ¡qué pena!

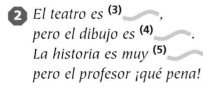

3 No me gustan las matemáticas.
La profesora es **(6)** ____.
El inglés es muy útil,
pero un poco **(7)** ____.

4 Aquí en el instituto,
escribo, escucho y hablo.
A mí me gusta mucho:
es un instituto muy
(8) ____. (×2)

1 **Escucha y escribe la letra correcta. (1–10)**

Ejemplo: **1** b

a *un bocadillo*	b *un plátano*
c *una hamburguesa*	d *una pizza*
e *una manzana*	
f *unas patatas fritas*	g *agua mineral*
h *un zumo de naranja*	i *una limonada*
j *una Coca-Cola*	

 2 **Escucha y escribe las letras correctas del ejercicio 1. (1–5)**

Ejemplo: **1** d, i

 3 **Haz un sondeo.**
Do a survey.

● ¿Qué comes en el recreo?
■ Como unas patatas fritas.
● ¿Qué bebes?
■ Bebo un zumo de naranja.

recreo = *lunch break*

4 **Empareja los números con las palabras. Escucha y comprueba tus respuestas. (a–h)**
Match up the numbers with the words. Listen and check your answers.

Ejemplo: **a** 30 – treinta

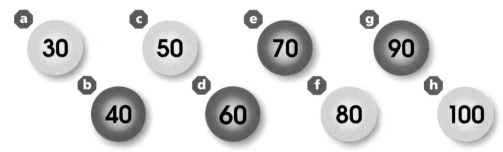

a **30**   c **50**   e **70**   g **90**
b **40**   d **60**   f **80**   h **100**

noventa
treinta
sesenta
cien
ochenta
cincuenta
setenta
cuarenta

5 Escucha y escribe los precios que entiendes. (a–h)
Listen and write down the prices you hear.

***Ejemplo:* a** 2€20

¿Cuánto es?

Son dos euros con …

Es un euro …

2€20 5€40 4€30 7€80

1€90 6€70 3€50 8€60

 6 Con tu compañero/a, pregunta y contesta por los precios del ejercicio 5.
With your partner, ask and answer questions about the prices in exercise 5.

● ¿Cuánto es?
■ Son dos euros con veinte.

 7 Escucha y lee el diálogo. Busca las frases en español en el texto.

***Ejemplo:* 1** Gracias

● Quiero una hamburguesa por favor.
■ Aquí tienes. ¿Algo más?
● Sí, un zumo de naranja por favor.
■ Aquí tienes. ¿Algo más?
● No, nada más. ¿Cuánto es?
■ Son tres euros con sesenta.
● Gracias.
■ De nada.

1 thank you
2 anything else?
3 here you are
4 please
5 how much is it?
6 no, nothing else
7 I would like
8 you're welcome

 8 Con tu compañero/a, haz diálogos que describen las bandejas.
With your partner, make dialogues to describe the trays.

● Quiero <u>dos hamburguesas</u> por favor.
■ Aquí tienes. ¿Algo más?
● Sí, <u>dos limonadas</u>, <u>una manzana</u> y <u>unas patatas fritas</u>, por favor. ¿Cuánto es?
■ Son <u>seis euros con cuarenta</u>.
● Gracias.
■ De nada.

1  6€40

2  2€60

3  5€30

4  2€40

 9 Escribe los diálogos del ejercicio 8.

Resumen

Unidad 1

I can

- list my school subjects — dibujo, ciencias, matemáticas, …
- name the days of the week — lunes, martes, …
- ask someone what they study — ¿Qué estudias?
- say what I study on different days — Los jueves estudio español, matemáticas, …
- G recognise all forms of **-ar** verbs — estudi**o**, estudi**as**, estudi**a**, estudi**amos**, estudi**áis**, estudi**an**

Unidad 2

I can

- ask someone what they do in a particular lesson — ¿Qué haces en clase de …?
- say what I do in different lessons — En español escucho, escribo y como.
- G recognise all forms of **-er** verbs — com**o**, com**es**, com**e**, com**emos**, com**éis**, com**en**

Unidad 3

I can

- ask what a teacher is like — ¿Cómo es el profesor de religión?
- say what my teachers are like — El profesor de religión es antipático.
- use accents to help me put the stress in the right place — simp**á**tico, ingl**é**s, religi**ó**n
- use connectives to extend sentences — La profesora de francés es severa **pero** canta muy bien.
- G use adjectives ending in **-o/-a** in the singular — La señora Silgado es antipática.
- G use qualifiers ('a bit', 'quite', 'very') — La profesora de inglés es **bastante** severa.

Unidad 4

I can

- ask someone if they like a subject — ¿Te gusta la historia?
- say what subjects I like or dislike — Me gusta mucho la historia. No me gustan nada las ciencias.
- give reasons for liking or disliking a subject — Es divertido. Es aburrido.
- make longer sentences using **porque** 'because' — Me gusta el francés **porque** es fácil.
- G use adjectives in the singular and plural — Las matemáticas son difíciles.

Unidad 5

I can

- order snacks and drinks — Quiero una hamburguesa y una limonada.
- ask how much something is — ¿Cuánto es?
- say and understand prices in Spanish — Son tres euros con treinta.
- count up to 100 — treinta, cuarenta, cincuenta, … cien

1 Copia la tabla. Escucha y escribe la asignatura, la opinión y la razón en inglés. (1–8)

Copy out the grid. Listen and note down the subject, the opinion and the reason, using English and symbols.

Ejemplo:

	Subject	Opinion	Reason
1	science	🙂	important

 2 Con tu compañero/a, pregunta y contesta.

With your partner, ask and answer questions.

- ● ¿Qué estudias los <u>lunes</u>?
- ■ Estudio <u>inglés</u>.
- ● ¿Qué haces en <u>inglés</u>?
- ■ <u>Hablo, escucho, escribo y leo</u>.

los lunes					
los martes					
los miércoles					
los jueves					
los viernes					

 3 Lee los textos y corrige el error en cada frase.

Read the texts and correct the mistake in each sentence.

1 Estudio inglés, matemáticas, español, informática, historia, teatro, música, religión y ciencias.

2 En el recreo escucho música o hablo con mis amigos. Como un bocadillo y bebo agua mineral.

3 En historia escucho con atención y leo mucho. También hablo y escribo. En música canto un poco. Me gusta mucho mi profesora de música porque es divertida y también muy simpática.

Nombre: Javier Reyes
Edad: 14 años
Fecha de nacimiento: el 2 de enero de 1992
Colegio: Lope de Vega, Almagro

Ejemplo: **1** Javier is fourteen.

1 Javier is thirteen.
2 His birthday is on 2nd February.
3 He studies English, maths, Spanish, CDT, history, drama, music, RE and science.
4 At lunch break he eats a sandwich and drinks a Coke.
5 In history he doesn't read much.
6 He doesn't like his music teacher.

 4 Describe tu instituto. Utiliza los textos del ejercicio 3 como modelo.

Describe your school. Use the texts from exercise 3 as a model.

escuchar

1 Escucha y lee.
Listen and read.

1

A – ¡Hola Tomás! ¿Qué tal?
T – ¡Angélica! … Bien, bien. Y tú ¿cómo estás?
A – Bien, gracias.

2

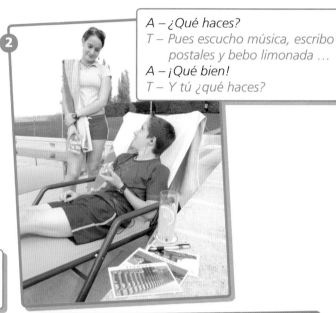

A – ¿Qué haces?
T – Pues escucho música, escribo postales y bebo limonada …
A – ¡Qué bien!
T – Y tú ¿qué haces?

3

A – Estudio inglés. Me gusta mucho porque es muy útil y bastante fácil.
T – ¿Cómo es tu profesor?
A – Tengo una profesora. Me gusta mucho porque es muy simpática y también muy divertida.

4

A – ¿Estudias inglés, Tomás?
T – Sí, pero no me gusta nada. Es muy difícil y no me gusta el profesor. Es muy antipático y también bastante aburrido.

5

P – ¡Angélica!
A – Ay, ¡qué horror! … ¡mi padre!
P – ¡Angélica!
A – ¡Hasta luego, Tomás!
T – Adiós, Angélica …

6

P – ¿Quién es, Angélica?
A – Se llama Tomás. Vive en Madrid. Es muy simpático …

¿Quién es? = *Who is that?*

2 Con tu compañero/a, lee la historia de Tomás y Angélica.
With your partner, read the story of Tomás and Angélica out loud.

 3 ¿Verdadero o falso? Escribe V o F. Corrige las frases falsas.
True or false? Write V or F. Correct the false sentences.

✔	un poco
✔✔	bastante
✔✔✔	muy

Ejemplo: **1** F – Tomás bebe una limonada.

1 Tomás bebe una Coca-Cola.
2 Angélica estudia francés.
3 Angélica tiene una profesora muy simpática y también muy divertida.
4 El profesor de inglés de Tomás es muy antipático y también bastante aburrido.
5 Tomás vive en Segovia.
6 Tomás es muy antipático.

 4 Lee el texto y busca las palabras que no conoces en el vocabulario.
Read the text and look up the words you don't know in the glossary.

a **La Nochebuena de Roberto el ratón**

b ¡Hola! ¿Qué tal? Me presento. Me llamo Roberto el ratón. Tengo dos años y vivo en Jerez en España. Vivo con la familia Núñez. Hay cuatro personas en la familia Núñez – Isabel, Teo, Manolito y Lola. Son muy, muy simpáticos.

c En España la Navidad es una fiesta importante y celebramos la Nochebuena – el veinticuatro de diciembre. Hay un árbol de Navidad. Hay guirnaldas y globos de todos los colores …

d La mesa es muy bonita. Isabel, Teo, Manolito y Lola comen una cena con pavo y beben cava – un tipo de champán. ¡Yo también como pavo!

la Nochebuena = *Christmas Eve*
la Navidad = *Christmas*
árbol de Navidad = *Christmas tree*
guirnaldas y globos = *streamers and balloons*
cena = *meal*
pavo = *turkey*
cava = *sparkling wine*

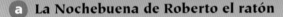 **5** Empareja los títulos con el párrafo correcto.
Match up the titles to the correct paragraph.

Decorations   **The Núñez family**   **The Christmas meal**

 6 Busca los verbos siguientes en el texto.
Look for the following verbs in the text.

Ejemplo: **1** me llamo

1 I am called   **2** I live   **3** we celebrate   **4** they eat   **5** they drink

leer 1 Find all the parts of **bailar** (to dance) and match them up with the English.

? Need help?   Go to Unidad 1 and 2
? Want to know more?   Go to page 130

1 I dance
2 you dance
3 he/she/it dances
4 we dance
5 you (plural) dance
6 they dance

escribir 2 Fill in the missing letters from the box to make the different parts of **leer** (to read).

? Need help?   Go to Unidad 2
? Want to know more?   Go to page 130

1 they read _ _ _ _ _
2 he reads _ _ _ _
3 you (singular) read _ _ _ _ _
4 I read _ _ _ _
5 you (plural) read _ _ _ _ _ _
6 we read _ _ _ _ _ _ _

s		e		e		l		e		s		l
	e		n		l		e		l		o	
e		é		e		o		e		e		l
	e		s		l		i		m		

leer 3 Which is the odd one out? Give a reason why.

? Need help?   Go to Unidad 2
? Want to know more?   Go to page 130

Example: **1** 'como' because the others are -ar verbs.

1	hablo	escucho	como
2	leo	como	escribo
3	estudio	escucho	vivo
4	escribes	escuchas	come
5	estudia	lees	habla
6	pinta	canta	bailas
7	hablamos	comemos	escriben
8	viven	escriben	hablan

4 **Follow the lines to find what the teachers are like, then unjumble the adjective and write the sentence in English.**

Example: **1** El profesor de geografía es un poco aburrido.
The geography teacher is a bit boring.

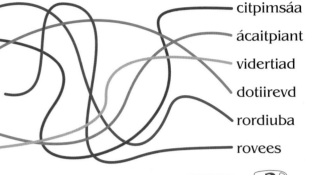
? *Need help?* Go to Unidad 3
? *Want to know more?* Go to page 129

1 El profesor de geografía es un poco — citpimsáa

2 La profesora de religión es muy — ácaitpiant

3 El profesor de inglés es bastante — vidertiad

4 El profesor de tecnología es un poco — dotiirevd

5 La profesora de español es muy — rordiuba

6 La profesora de francés es muy — rovees

 5 **Choose the right connective for each sentence.**

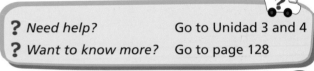
? *Need help?* Go to Unidad 3 and 4
? *Want to know more?* Go to page 128

Example: **1** porque

1 Me gusta el español **pero / porque** es muy interesante.
2 Me gustan mucho las matemáticas **pero / y** son muy importantes.
3 No me gusta el teatro **pero / porque** el profesor es muy severo.
4 No me gusta nada la historia porque es aburrida.
Leemos **pero / y** escribimos mucho en clase.
5 No, no me gusta nada la historia **y / pero** el profesor es simpático.

 6 **Choose the correct adjective for each sentence.**

? *Need help?* Go to Unidad 3 and 4
? *Want to know more?* Go to page 129

Example: **1** a

1 El profesor de francés es muy
　　a aburrido　**b** aburrida
　　c aburridos　**d** aburridas

2 La profesora de inglés es bastante
　　a simpáticos　**b** simpáticas
　　c simpática　**d** simpático

3 El profesor de historia es
　　a severos　**b** severas
　　c severa　**d** severo

4 No me gusta nada el dibujo porque es
　　a difícil　**b** difíciles

5 Me gustan las ciencias porque son
　　a divertidos　**b** divertidas
　　c divertida　**d** divertido

6 No me gustan las matemáticas. No son
　　a interesantes　**b** interesante

 7 **Find the following in the quiz above:**

● 6 nouns (school subjects);　● 2 qualifiers;　● 3 definite articles;
● 3 negative expressions;　● 1 connective;　● 6 adjectives.

Palabras

Las asignaturas	School subjects
¿Qué estudias?	What do you study?
Estudio …	I study …
No estudio …	I don't study …
Estudia …	He/She studies …
No estudia …	He/She doesn't study …
el dibujo	art
la educación física	PE
el español	Spanish
el francés	French
el inglés	English
el teatro	drama
la historia	history
la informática	ICT
la música	music
la religión	RE
la geografía	geography
la tecnología	technology/CDT
las ciencias	science
las matemáticas	maths

Los días de la semana	The days of the week
lunes	Monday
martes	Tuesday
miércoles	Wednesday
jueves	Thursday
viernes	Friday
sábado	Saturday
domingo	Sunday
los lunes	every Monday

Los números 30–100	Numbers 30–100
treinta	30
cuarenta	40
cincuenta	50
sesenta	60
setenta	70
ochenta	80
noventa	90
cien	100

¿Qué haces en inglés?	What do you do in English?
En inglés escucho, hablo, leo y escribo.	In English, I listen, speak, read and write.
Escucho música.	I listen to music.
Hablo con mis amigos.	I speak with my friends.
Hablo por teléfono.	I speak on the phone.
No leo.	I don't read.
Escribo mucho.	I write a lot.
No como.	I don't eat.
Como chicle.	I chew gum. (I eat chewing gum.)
En dibujo pinto.	In art, I paint.
En música canto.	In music, I sing.
cantar	to sing
escuchar	to listen
hablar	to speak
pintar	to paint
comer	to eat
leer	to read
escribir	to write
vivir	to live

¿Cómo es tu profesor?	What's your teacher like?
El profesor de … es …	The … teacher (male) is …
aburrido	boring
antipático	unpleasant
divertido	amusing
severo	strict
simpático	nice, kind
La profesora de … es …	The … teacher (female) is …
aburrida	boring
antipática	unpleasant
divertida	amusing
severa	strict
simpática	nice, kind

Opiniones	Opinions
¿Te gusta el español?	Do you like Spanish?
Me gusta el español.	I like Spanish.
Me gusta la geografía.	I like geography.
Me gusta mucho la historia.	I really like history.
No me gusta el inglés.	I don't like English.
No me gusta nada la educación física.	I don't like PE at all.
¿Te gustan las ciencias?	Do you like science?
Me gustan las ciencias.	I like science.
bueno/buena	good
difícil	difficult
fácil	easy
importante	important
interesante	interesting
útil	useful
¿Qué te gusta?	What do you like?
¿Por qué?	Why?
Me gusta la informática porque es fácil.	I like ICT because it's easy.
Me gustan las ciencias porque son útiles.	I like science because it's useful.

¿Qué comes?	Snacks
¿Qué comes en el recreo?	What do you eat at lunch break?
Como …	I eat …
Come …	He/She eats …
un bocadillo	a sandwich
un plátano	a banana
una hamburguesa	a hamburger
una manzana	an apple
una pizza	a pizza
unas patatas fritas	some crisps
¿Qué bebes?	What do you drink?
Bebo …	I drink …
Bebe …	He/She drinks …
agua mineral	a mineral water
un zumo de naranja	an orange juice
una limonada	a lemonade
una Coca-Cola	a Coca-Cola

¿Cuánto es?	How much is it?
Quiero dos hamburguesas por favor.	I'd like two hamburgers please.
¿Cuánto es?	How much is it?
Son dos euros con veinte.	It's two euros twenty.
Aquí tienes.	Here you are.
¿Algo más?	Anything else?
No, nada más.	No, nothing else.
Gracias.	Thank you.
De nada.	You're welcome.

Palabras muy útiles	Very useful words
un poco	a bit
bastante	quite
muy	very
me gusta	I like
no me gusta	I don't like

Estrategia

Working with cognates

A **cognate** is a word that is spelt the same way in English and Spanish. A **near-cognate** is spelt almost the same.

- In Chapter 2 there are a lot of near-cognates. Can you find five on this page? Do they all mean exactly the same as the English?

Words like these make learning easier. Just remember that their spelling and pronunciation are slightly different from the English words.

- Study the five words you spotted for 10 seconds each. Then shut the book.
- Try to write the words correctly, remembering any spelling differences.
- Now try to say the words correctly, pronouncing the letters in the Spanish way.

Mi familia
3

1 ¿Tienes hermanos?

● Talking about your family
● Using **tener** (to have)

escuchar **1** Escucha y escribe la letra correcta. (1–6)

¿Tienes hermanos?

Ejemplo: **1** a

a Tengo un hermano.

b Tengo una hermana.

c Tengo un hermano y una hermana.

d Tengo dos hermanos y dos hermanas.

e No tengo hermanos. Soy hija única.

f No tengo hermanos. Soy hijo único.

hablar **2** Haz un sondeo.
Do a survey.

● ¿Tienes hermanos, David?
■ Sí, tengo dos hermanos.

Nombre	¿Hermanos?	¿Hermanas?
David	2	–

escuchar **3** Escucha y escribe los nombres y las edades de los hermanos. (1–6)
Listen and write down the names and ages of the brothers and sisters.

Ejemplo: **1** una hermana – Silvia – 13

Marta	Carlos	Alfredo
Sergio	Miriam	Josefa
Lola	Silvia	Paco

Gramática

tener	to have
tengo	I have
tienes	you have
tiene	he/she/it has
tenemos	we have
tenéis	you (plural) have
tienen	they have

In English, you say 'I am 11' or 'He/She is 12'. Remember, in Spanish, you use **tener**:

Tengo once años.
   I 'have' eleven years.
¿Cuántos años **tienes**?
   How many years do you 'have'?

Para saber más   p.58 (ex 1); p.131

44 cuarenta y cuatro

escuchar 4 Escucha y lee.

En la foto hay cuatro personas:
**mi madre, mi padre,
mi hermano** y **mi hermana.**

En la foto hay ...
mi abuelo y **mi abuela.**

... **mi tío, mi tía, mi primo**
y **mi prima.**

> hay = *there is/there are*

leer 5 Busca las frases en los textos del ejercicio 4.
Find the phrases in the texts in exercise 4.

Ejemplo: 1 my father – mi padre

1	my father	**6**	my brother
2	my mother	**7**	my uncle
3	my grandmother	**8**	my aunt
4	my sister	**9**	my cousin (m)
5	my grandfather	**10**	my cousin (f)

Gramática

The words for 'my', 'your', 'his' and 'her' change according to whether the noun they describe is singular or plural.

	Singular	Plural
my	**mi** hermana	**mis** hermana**s**
your	**tu** hermano	**tus** hermano**s**
his/her	**su** hermano	**sus** hermano**s**

Para saber más   p.58 (ex 2); p.130

hablar 6 Con tu compañero/a, pregunta y contesta sobre esta familia.

With your partner, ask and answer about this family.

● ¿Cómo se llama tu padre?
■ Mi padre se llama Paco.
● ¿Cuántos años tiene Paco?
■ Paco tiene cuarenta y dos años.

> ¿Cómo se llama tu ...?
> Mi ... se llama ...
> ¿Cuántos años tiene ...?
>
> ¿Cómo se llama**n** tus ...?
> Mis ... se llama**n** ...
> ¿Cuántos años tiene**n** ...?

Paco 42
Maribel 39
Timoteo 18
Carmen 15
Luisa 71
Virginia 13
Mateo 70

escribir 7 Escribe una descripción de tu familia.

Write a description of your family.

Ejemplo: En mi familia hay cinco personas.
Mi madre se llama Sarah. Tiene cuarenta años.
Mi padre se llama ... Tiene ...

2 ¿Tienes animales?

1 Escucha y escribe la letra correcta. (1–12)

Ejemplo: **1** k

¿Tienes animales?

a

un gato

b

un pájaro

c

un perro

d

una cobaya

e

un caballo

f

un conejo

g

un pez

h

un ratón

i

un hámster

j

una tortuga

k

una serpiente

l

No tengo animales.

2 Con tu compañero/a, pregunta y contesta.

● ¿Tienes animales?
■ Tengo <u>dos serpientes</u>.

	Singular	Plural
● Ending in vowel:	conejo	conejo**s**
● Ending in consonant:	animal	animal**es**
	pe**z**	pe**ces**

3 Escribe las respuestas del ejercicio 2.

Ejemplo: Tengo dos serpientes, …

4 Escucha y escribe la letra correcta. (1–12)

Ejemplo: **1** c

¿Cómo es?

a blanco
b negro
c gris
d azul rojo
e verde y amarillo
f naranja
g marrón
h rosa
i pequeño
j grande
k bonito
l feo

5 Empareja los animales con las descripciones.

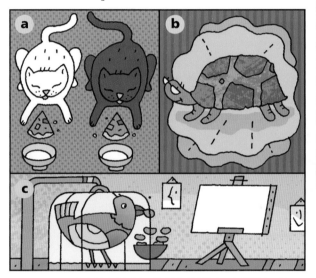

a

b

c

Gramática

Most colour words behave just like other adjectives:

Singular masculine/feminine	Plural masculine/feminine
amarill**o**/amarill**a**	amarill**os**/amarill**as**
verde	verde**s**
azul	azul**es**

These colour words only change in the plural.

rosa	rosas
naranja	naranjas

Para saber más p.130

2
Ana
Mis gatos se llaman Ricardo y Daniela.
Ricardo es negro y Daniela es blanca.
Ricardo y Daniela son grandes y son muy inteligentes.
Tienen ocho años.
Beben leche y comen pizza.

1
Pepe
Mi tortuga se llama 'Speedy Gonzalez'.
Es verde y marrón.
Speedy es muy fea y es bastante aburrida, pero es inteligente.
Tiene veinte años.
Come mucha ensalada.

es = *he/she is*
son = *they are*

3
Paco
Mi pájaro se llama Picasso.
Es verde y amarillo, rosa y azul.
Es muy pequeño y muy simpático.
Picasso tiene cuatro años.
Pinta y habla mucho. También come patatas fritas.

6 Copia y rellena el formulario sobre los animales del ejercicio 5.

Owner:	Pepe
Animal:	tortoise
Name:	Speedy Gonzalez
Colour:	green and brown
Description:	...
Age:	...
Other details:	...

7 Dibuja y describe un animal imaginario.

escuchar 1 Escucha y escribe el nombre correcto. (1–12)

Ejemplo: **1** Antonio

¿Cómo eres?

Alfredo

Soy alto.

Soy alta.

Silvia

Antonio

Soy bajo.

Soy baja.

Ramona

Carlos

Soy delgado.

Soy delgada.

Marga

David

Soy gordo.

Soy gorda.

Lola

Alonso

Soy guapo.

Soy guapa.

Miriam

Leonardo

Soy feo.

Soy fea.

Berta

Gramática

ser 'to be' is an important irregular verb. Learn it by heart!

ser	to be
soy	I am
eres	you are
es	he/she/it is
somos	we are
sois	you (plural) are
son	they are

Para saber más   p.59 (ex 4); p.132

hablar 2 ¡Juega! Elige una persona del ejercicio 1.
Play a game! Choose a person from exercise 1.

- ¿Eres un chico o una chica?
- Un chico.
- ¿Eres bajo?
- No, no soy bajo.
- ¿Eres alto?
- Sí.
- ¿Eres Alfredo?
- Sí, soy Alfredo.

un chico = *a boy*
una chica = *a girl*

" Try to sound as Spanish as possible when you are pronouncing Spanish names!

Listen and repeat these names:

Alfredo	Silvia
Antonio	Ramona
Carlos	Marga
David	Lola
Alonso	Miriam
Leonardo	Berta

"

escribir 3 Escribe frases que describen a tres miembros de tu familia.
Write sentences describing three members of your family.

Ejemplo: Mi herman**a** es guap**a**.
Mi herman**o** es gord**o**, pero guap**o**.

4 Escucha y escribe la edad y los dos adjetivos para cada persona. (1–8)

Listen and note down the age and the two adjectives for each person.

Ejemplo: 1 Roberto – 12 – alto, simpático

| Roberto | Sadiq | Alonso | Pepe |
| Natalia | Elena | Montserrat | Silvia |

antipático	antipática
aburrido	aburrida
divertido	divertida
simpático	simpática
tímido	tímida
perezoso	perezosa
inteligente	

 5 Lee el texto y corrige las frases.

1 Javier is 19.
2 He has three sisters.
3 The band is called The Mad Boys.
4 Rafael is short and slim.
5 Pepe plays the guitar.
6 Miguel is very fat.
7 Kiki the snake is very lazy.

loco = *crazy*
tocar = *to play (an instrument)*
una canción = *song*

http://www.Losgatoslocos.es/detalles

LOS GATOS LOCOS

JAVIER
KIKI
PEPE
RAFAEL
MIGUEL

¡Hola! Me llamo Javier. Tengo diecisiete años. Soy bastante inteligente y hablo mucho. Soy alto y delgado y muy guapo. Canto en un grupo con mis tres hermanos. El grupo se llama 'Los gatos locos'.

Mi hermano Rafael tiene quince años. Es alto, delgado y simpático. Toca la guitarra muy bien. Pepe tiene trece años. Es bajo y bastante divertido. Escribe canciones.

Miguel tiene doce años. Es bajo y es un poco gordo también. Toca la guitarra y canta. Tenemos una serpiente. Se llama Kiki. Es bastante grande y muy divertida. Kiki es un miembro muy importante de 'Los gatos locos'.

Mini-test

I can ...

● talk about the members of my family
● say what they are called and how old they are
● say what pets I have and what they are like
G use the possessive adjectives **mi(s)**, **tu(s)**, **su(s)**
G use the correct endings on colour adjectives
G use the irregular verb **ser** (to be)
G use the irregular verb **tener** (to have)

 6 Describe un grupo de música imaginario con seis miembros: 'Los chicos y las chicas'.

Describe an imaginary pop group with six members: 'The boyz and girlz'.

4 Tengo los ojos azules

● Talking about eyes and hair
● Using adjectives after nouns

1 Escucha y escribe la letra correcta. (1–5)

¿De qué color son tus ojos?

Ejemplo: **1** d

Tengo los ojos azules.
a

Tengo los ojos verdes.
b

Tengo los ojos marrones.
c

Tengo los ojos grises.
d

Tengo los ojos rojos.
e

2 Escucha y escribe la letra correcta. (1–9)

Ejemplo: **1** c

¿Cómo es tu pelo?

Tengo el pelo liso y castaño.
a

Tengo el pelo largo y rubio.
b

Tengo el pelo corto y negro.
c

Tengo el pelo rizado y pelirrojo.
d

Tengo el pelo corto y gris.
e

Tengo el pelo ondulado y blanco.
f

Tengo barba.
g

Tengo bigote.
h

Tengo gafas.
i

3 Juega a las tres en raya con tu compañero/a.
Play noughts and crosses with your partner.

● 'a' Tengo el pelo corto y rubio.

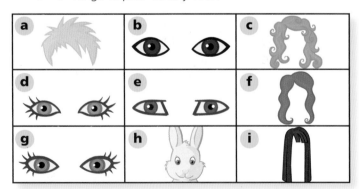

Gramática

● In Spanish, most adjectives come after the noun they are describing.

Tengo el pelo **corto**.
*I have **short** hair.*

¿Tienes una serpiente **verde**?
*Do you have a **green** snake?*

Tiene los ojos **azules**.
*He/She has **blue** eyes.*

Para saber más p.59 (ex 5); p.130

4 ¿Cómo eres? Escribe una descripción.
What do you look like? Write a description.

Ejemplo: Tengo los ojos …. Tengo el pelo … y ….

5 Escucha y escribe la letra del dibujo y el nombre correcto. (1–5)

Ejemplo: **1** b Paco

Paco   Sergio   Angelina   Arturo   Mariela

6 Lee los textos y rellena una ficha sobre cada persona.
Read the texts and fill in a form for each person.

Hollywood habla español ...

Penélope Cruz tiene los ojos marrones.
Tiene el pelo liso y castaño y muy largo.
Es una persona simpática. También es guapa y delgada.
Lee mucho. Escucha música y baila.
Y Penélope habla español ...

Salma Hayek vive en Los Ángeles en los Estados Unidos. Es mexicana.
Es muy guapa. Tiene los ojos marrones y el pelo liso y castaño.
Es bastante alta. Es inteligente y divertida.
Y Salma habla español ...

Javier Bardem tiene los ojos marrones y el pelo corto y castaño y
un poco ondulado. Tiene bigote. Es muy guapo y muy fuerte. También
es divertido e inteligente. Le gusta el dibujo y el deporte. Pinta mucho.
Y Javier habla español ...

Antonio Banderas no es muy alto pero es muy, muy guapo.
Tiene los ojos marrones y el pelo corto y castaño.
Es simpático e inteligente. Vive en Los Ángeles.
Y Antonio habla español ...

> fuerte = *strong*
> le gusta = *he/she likes*
> el deporte = *sport*

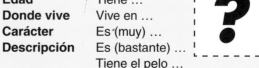

Nombre	Se llama ...
Edad	Tiene ...
Donde vive	Vive en ...
Carácter	Es (muy) ...
Descripción	Es (bastante) ...
	Tiene el pelo ...
	Tiene los ojos ...
Otros detalles	Le gusta ...

7 Describe a una persona famosa.
Utiliza 'qualifiers'. Incluye los
datos siguientes:
*Describe a famous person. Use qualifiers
and include the following details:*

5 ¿Cómo es?

- Using **tener** and **ser** in the he/she form
- Using texts as a model for creative writing

escuchar 1 **Escucha y escribe la descripción. Copia y rellena la tabla. (1–4)**

Listen and note down the description of the wanted people. Copy the grid and fill it in.

Nombre	Descripción	Ojos	Pelo
1 Lola Jiménez	*baja, gorda*	*azules*	*rubio, largo y liso*
2 Alonso Núñez			
3 Berta Mallol			
4 Rafael Márquez			

hablar 2 **Con tu compañero/a, describe a las personas.**

- ¿Cómo es?
- Es …
- ¿Cómo es su pelo?
- Tiene el pelo …
- ¿Cómo son sus ojos?
- Tiene los ojos …
- Es dibujo (a).
- ¡Correcto!

✔ ¡Correcto!
✗ ¡Incorrecto!

es	alt**o**/alt**a**
	baj**o**/baj**a**
	delgad**o**/delgad**a**
	gord**o**/gord**a**
	guap**o**/guap**a**
	fe**o**/fe**a**
tiene los ojos	azules
	verdes
	marrones
	grises
tiene el pelo	liso
	largo
	corto
	rizado
	ondulado
	castaño
	rubio
	negro
	pelirrojo
	gris
	blanco
tiene	barba
	bigote
	gafas

3 Empareja las descripciones con los dibujos del ejercicio 2.

SE BUSCA ...

1 Fernando Palacios de Navarra
- Hombre de cuarenta años
- Es grande y gordo
- Tiene los ojos marrones
- Tiene el pelo corto, negro y liso
- Tiene bigote
- Tiene una disposición muy violenta
- Armado
- Le gustan las patatas fritas
- Bebe mucho café

2 Carolina Alazraki
- Mujer de veinticinco años
- Es alta y delgada
- Tiene los ojos azules
- Tiene el pelo largo, rubio y ondulado
- Tiene gafas
- Armada
- Tiene un ratón blanco
- Fuma cigarros
- Habla francés
- Escucha música todo el tiempo

3 Diego García Igoa
- Hombre de treinta años
- Es alto y delgado
- Tiene los ojos verdes
- No tiene pelo
- Tiene barba
- Homicidio
- Escribe poesías
- No le gustan los animales
- Come chicle todo el tiempo

4 Marta Sanz Ruiz
- Mujer de treinta años
- Es baja y gorda
- Tiene los ojos grises
- Tiene el pelo largo, negro y ondulado
- Tiene una disposición muy violenta
- Estudia literatura
- Tiene un perro negro
- No habla mucho

le gusta(n) = he/she likes
fuma = he/she smokes
todo el tiempo = all the time
literatura = literature

4 Busca estas frases en español en el texto.

1 He drinks a lot of coffee.
2 She smokes cigars.
3 She speaks French.
4 He writes poems.
5 He chews gum all the time.
6 She doesn't talk much.

5 Lee los textos y decide quién habla del ejercicio 3.
Read the texts and decide who is speaking from exercise 3.

a A ver, no fumo, y no bebo café. No me gusta nada. Nunca como chicle. Me gustan los animales y la música. Y me gustan mucho los libros. Leo mucho y me gusta la historia. Es muy interesante.

b Como mucho. Como patatas fritas, hamburguesas, bocadillos … No fumo – no me gusta nada. Hablo español e inglés pero no hablo francés. No estudio mucho. No tengo animales.

nunca = never

 6 Dibuja y describe un póster de 'se busca'.
Draw and describe a 'wanted' poster.

Resumen

Unidad 1

I can

- ask someone if they have any brothers and sisters
- say whether I have any brothers and sisters
- say how many people there are in my family
- say what they are called

- say how old they are
- G use the possessive adjectives **mi(s)**, **tu(s)**, **su(s)**

- G recognise all forms of the irregular verb **tener**

¿Tienes hermanos?
Tengo dos hermanos. Soy hija única.
En mi familia hay cinco personas.
Mi hermano se llama Pepe.
Mi madre se llama Clara.
Pepe tiene trece años.
¿Cómo se llama **tu** hermano?
Mi hermano se llama …

tengo, tienes, tiene, tenemos, tenéis, tienen

Unidad 2

I can

- ask someone if they have any pets
- say what pets I have
- say what my pets are called
- say what my pet looks like
- say what colour it is
- G use the correct endings on colour adjectives

¿Tienes animales?
Tengo dos serpientes. No tengo animales.
Mi perro se llama Paco.
Es grande. Es bonito.
Es rojo y blanco.
Mi perro es negro. Mi serpiente es negra.

Unidad 3

I can

- talk about my appearance
- talk about someone else's appearance
- talk about my character
- talk about someone else's character
- G recognise all forms of the irregular verb **ser**

Soy alto y delgado.
Es gorda y baja.
Soy divertido. Soy tímida.
Es aburrido. Es simpática.
Soy inteligente. ¿**Eres** alto? **Es** bajo.

Unidad 4

I can

- talk about eyes
- talk about hair
- G use adjectives correctly (after nouns)

Tengo los ojos azules.
Tengo el pelo largo, rubio y ondulado.
Tengo el pelo **corto**.

Unidad 5

I can

- describe someone else's appearance

- G use the irregular verbs **tener** and **ser** in the he/she form

Es gordo. Tiene los ojos verdes y el pelo negro y liso.
Es feo. **Tiene** bigote.

Prepárate

 1 Escucha y escribe la letra correcta. (1–3)

a   **b**   **c**   **d**

 2 Escucha otra vez. Copia y rellena la tabla. (1–3)

Name	Age	Appearance	Personality
1 Mariana	18	slim, …	very intelligent
2 Ana María			
3 Julia			

3 Con tu compañero/a, pregunta y contesta.

1 ¿Tienes hermanos?   Tengo …   4 ¿Cómo eres?   Soy …

2 ¿Cuántas personas hay en tu familia?   Hay …   5 ¿Cómo es tu pelo?   Tengo el pelo …

6 ¿De qué color son tus ojos?   Tengo los ojos …

3 ¿Tienes animales?   Tengo …

 4 Mira los dibujos. ¿Quién es? ¿Paolo o Claudia? Escribe P o C.

Ejemplo: **1** P

Me llamo **Paolo**. Tengo trece años. Vivo en Orense.
Soy bajo y gordo. Tengo el pelo castaño y los ojos marrones.
Soy bastante inteligente y hablo mucho.
Tengo una hermana. Se llama Laurita. Es muy simpática. Vivo con mi madre y mi padre.
Me gustan mucho los animales. Tengo un gato, un perro y dos cobayas.

Me llamo **Claudia**. Tengo catorce años. Vivo en Albacete.
Soy alta y delgada. Tengo el pelo negro y los ojos verdes.
Soy un poco tímida.
No tengo animales, pero tengo tres hermanos. Se llaman Tito, Noelia y Lorenzo.
Tito y Noelia son gemelos. Tienen ocho años. Lorenzo tiene quince años.

 5 ¿Qué dice Javier?

Ejemplo: Me llamo Javier.
Soy …
Tengo …

guapo, inteligente
ojos marrones
pelo corto, negro y rizado
dos hermanos – Luis y Diego
un perro – Dalí
un gato – Buñuel

¡Extra!

 escuchar 1 Escucha y lee.

1

T – ¡Angélica! ¡Hola! ¿Qué tal?
A – ¡Hola, Tomás! Estoy bien, muy bien. ¿Y tú?
T – Bien, bien, gracias.

2

T – ¿Tienes un perro?
A – Sí … Vive aquí en el hotel.
T – ¿Cómo se llama?
A – Se llama Hielito y es muy divertido.

3

T – ¡Qué bonito! Tiene los ojos azules.
A – De carácter es muy simpático. ¿Tienes animales, Tomás?
T – No, no tengo animales pero tengo un hermano.

4

T – Se llama Víctor. Es pequeño y no es tan divertido como Hielito … Tiene nueve años y me irrita mucho.

5

T – ¡Ay, no! ¡Aquí está! ¡Hasta luego, Angélica!

6

V – Tomás tiene una amiga. Se llama Angélica. Es muy guapa. Es alta y delgada. Tiene el pelo castaño y los ojos marrones …
M – Hmm, ¡qué interesante!

 leer 2 Copia las frases. Escribe positivo (P) o negativo (N).

Ejemplo: **1** P

1 Hielito es muy divertido.
2 Hielito es bonito.
3 De carácter es muy simpático.
4 Víctor no es tan divertido como Hielito.
5 Víctor irrita mucho a Tomás.

 escribir 3 Escribe descripciones de estas personas.

1 Tomás   **2** Angélica   **3** Hielito   **4** Víctor

 hablar 4 Con tu compañero/a, lee la historia de Tomás y Angélica.

5 Elige las respuestas correctas.

La geografía de España

1 *Elige tres países que tienen una frontera con España.*
a *Francia*
b *Italia*
c *Portugal*
d *Andorra*

2 *Elige las islas españolas.*
a *las islas Eolias*
b *las islas Baleares*
c *las islas Canarias*
d *las islas Caimán*

3 *Elige las montañas que no están en España.*
a *los Pirineos*
b *los Alpes*
c *la Sierra Morena*
d *la Sierra Nevada*
e *los Picos de Europa*

4 *Los ríos principales de España se llaman:*
a *el Ebro*
b *el Sena*
c *el Guadalquivir*
d *el Tajo*

5 *Las tres costas principales de España se llaman:*
a *la costa Mediterránea*
b *la costa Brava*
c *la costa Atlántica*
d *la costa Cantábrica*

6 *La capital de España se llama:*
a *Barcelona*
b *Madrid*
c *Sevilla*
d *Bilbao*

6 Escucha y comprueba tus respuestas.

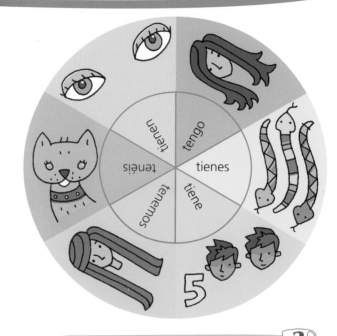

? Need help?   Go to Unidad 1
? Want to know more? Go to page 131

1 Use each segment of the wheel to write a sentence, then translate them into English.

Example: Tengo los ojos azules.
I have blue eyes.

2 Choose the correct word. Write out the English.

? Need help?   Go to Unidad 1
? Want to know more? Go to page 130

Example: **1** mi abuelo – my grandfather

1 mi / mis abuelo
2 tus / tu madre
3 mis / mi abuela
4 su / sus hermanos
5 tus / tu padre
6 mi / mis primos

3 Describe this crowded classroom in Spanish.

? Need help?   Go to Unidad 2
? Want to know more? Go to page 135

Example: En la clase hay diez peces …

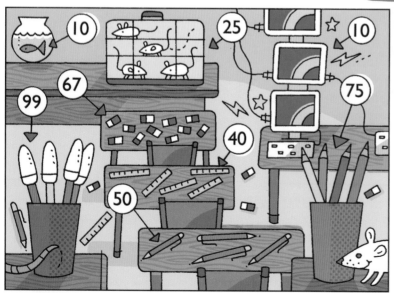

rotulador
pez
lápiz
bolígrafo
regla
goma
ratón
ordenador

escribir 4 *Translate these sentences into Spanish.*

1 I am tall.
2 You (singular) are short.
3 He is fat.
4 She is thin.
5 We are good-looking.
6 You (plural) are tall.
7 They are funny.
8 They are intelligent.

> **?** *Need help?*   Go to Unidad 3
> **?** *Want to know more?*   Go to page 129

leer 5 *Look at the picture and fill in the gaps in the text.*

Example: **1** naranjas

> **?** *Need help?*   Go to Unidad 4
> **?** *Want to know more?*   Go to page 129

Jordi

Emilia

Jorge

César

Valeria

Luciana

Mi familia es multicolor y es muy interesante. Yo me llamo Emilia. Tengo trece años. Tengo los ojos **(1)** _____ y el pelo **(2)** _____ muy, muy **(3)** _____.
En mi familia, hay seis personas. Mi madre, mi padre, mi hermana, mi hermano, el perro y yo.
Mi hermana se llama Luciana. Tiene catorce años. Tiene el pelo **(4)** _____, liso y largo. Tiene los ojos **(5)** _____ muy preciosos. ¡Mi hermana no es tímida!
Mi hermano se llama Jordi. Tiene dieciocho años. No tiene pelo pero tiene gafas. Tiene los ojos **(6)** _____. Es muy divertido.
Mi padre se llama Jorge. Tiene los ojos **(7)** _____ y el pelo **(8)** _____, largo y liso. Tiene gafas.
Mi madre se llama Valeria. Tiene cuarenta años. Tiene los ojos **(9)** _____ y el pelo amarillo, largo y **(10)** _____. Mi madre es una señora muy inteligente.
El perro se llama César. Es **(11)** _____ y tiene gafas. Tiene tres años. Es muy inteligente, pero es un poco perezoso. Le gustan las hamburguesas.

Palabras

Mis hermanos

My brothers and sisters

¿Tienes hermanos?

Do you have any brothers or sisters?

tener — *to have*

Tengo … — *I have …*

Tiene … — *He/She has …*

un hermano — *one brother*

una hermana — *one sister*

dos hermanos — *two brothers*

tres hermanas — *three sisters*

No tengo hermanos. — *I don't have any brothers or sisters.*

Soy hijo único. — *I'm an only child. (male)*

Soy hija única. — *I'm an only child. (female)*

¿Cómo se llama tu hermano? — *What's your brother called?*

¿Cómo se llama tu hermana? — *What's your sister called?*

Mi hermano se llama … — *My brother is called …*

Mi hermana se llama … — *My sister is called …*

¿Cómo se llaman tus hermanos? — *What are your brothers (and sisters) called?*

¿Cómo se llaman tus hermanas? — *What are your sisters called?*

Mis hermanos se llaman … — *My brothers (and sisters) are called …*

Mis hermanas se llaman … — *My sisters are called …*

su hermano — *his/her brother*

sus hermanos — *his/her brothers (and sisters)*

su hermana — *his/her sister*

sus hermanas — *his/her sisters*

¿Cuántos años tiene tu hermano? — *How old is your brother?*

Tiene nueve años. — *He's nine years old.*

En mi familia

In my family

¿Cuántas personas hay en tu familia? — *How many people are there in your family?*

En mi familia hay tres personas. — *In my family there are three people.*

mi madre — *my mother*

mi padre — *my father*

mi abuelo — *my grandfather*

mi abuela — *my grandmother*

mi tío — *my uncle*

mi tía — *my aunt*

mi primo — *my cousin (male)*

mi prima — *my cousin (female)*

Los números 30–100 — *Numbers 30–100*

treinta — *30*

cuarenta — *40*

cincuenta — *50*

sesenta — *60*

setenta — *70*

ochenta — *80*

noventa — *90*

cien — *100*

treinta y uno — *31*

cuarenta y dos — *42*

cincuenta y tres — *53*

sesenta y cuatro — *64*

setenta y cinco — *75*

ochenta y siete — *87*

noventa y nueve — *99*

Los animales — *Pets*

¿Tienes animales? — *Do you have any pets?*

Tengo … — *I have …*

un caballo — *a horse*

una cobaya — *a guinea pig*

un conejo — *a rabbit*

un gato — *a cat*

un hámster — *a hamster*

un pájaro — *a bird*

un perro — *a dog*

un pez — *a fish*

un ratón — *a mouse*

una serpiente — *a snake*

una tortuga — *a tortoise*

dos conejos	two rabbits
tres peces	three fish
No tengo animales.	I don't have any pets.

Los colores — *Colours*

amarillo/amarilla	yellow
blanco/blanca	white
negro/negra	black
rojo/roja	red
azul	blue
gris	grey
marrón	brown
naranja	orange
rosa	pink
verde	green
El perro es blanco.	The dog is white.
La serpiente es amarilla.	The snake is yellow.

¿Cómo es? — *What's he/she/it like?*

bonito/bonita	cute, pretty
feo/fea	ugly
pequeño/pequeña	small
grande	big

¿Cómo eres? — *What are you like?*

ser	to be
Soy …	I'm …
Eres …	You're …
Es …	He's/She's …
un chico	a boy
una chica	a girl
alto/alta	tall
bajo/baja	short
delgado/delgada	thin
gordo/gorda	fat
guapo/guapa	good-looking
feo/fea	ugly
aburrido/aburrida	boring
antipático/antipática	unpleasant
divertido/divertida	amusing
simpático/simpática	nice, kind
perezoso/perezosa	lazy
tímido/tímida	shy
inteligente	intelligent

Mis ojos y mi pelo — *My eyes and my hair*

¿De qué color son tus ojos?	What colour are your eyes?
Tengo los ojos …	I have … eyes.
Tiene los ojos …	He/She has … eyes.
azules	blue
grises	grey
marrones	brown
verdes	green

¿Cómo es tu pelo?	What's your hair like?
Tengo el pelo …	I have … hair.
Tiene el pelo …	He/She has … hair.
blanco	white
castaño	brown
gris	grey
negro	black
pelirrojo	red/ginger
rubio	blond
liso	straight
largo	long
corto	short
rizado	curly
ondulado	wavy

Tengo barba.	I have a beard.
Tengo bigote.	I have a moustache.
Tengo gafas.	I wear (have) glasses.

Estrategia

Words that you see everywhere!

In every language, there are some words that you will see and hear again and again in different situations. Because of this, they are called **high-frequency** words. The good news is that you can learn them once and use them again and again, too!

Have another look at Chapter 3. Can you find two or three sentences containing each of the words below?

- tengo
- y
- no
- muy

escuchar 1 Escucha y escribe el país correcto. (1–10)

Ejemplo: **1** Francia

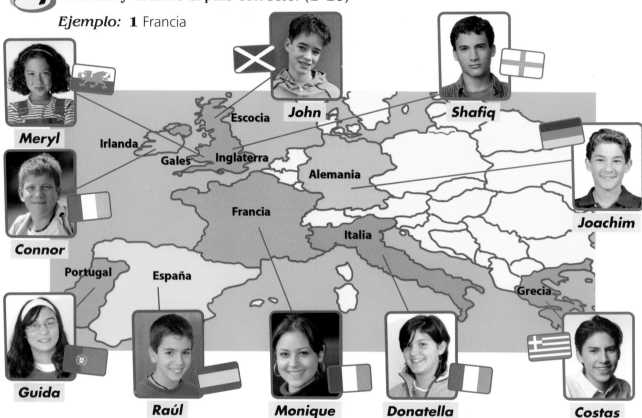

Meryl

John

Shafiq

Escocia

Irlanda

Gales

Inglaterra

Alemania

Joachim

Connor

Francia

Italia

Portugal

España

Grecia

Guida

Raúl

Monique

Donatella

Costas

hablar 2 Elige una persona del ejercicio 1. Con tu compañero/a, pregunta y contesta.

● ¿Dónde vives?
■ Vivo en <u>Gales</u>.
● ¿Eres <u>Meryl</u>?
■ Sí, soy <u>Meryl</u>.

escuchar 3 Escucha y escribe a o b. (1–5)

Ejemplo: **1** a

a *Vivo en una casa.*

b *Vivo en un piso.*

escuchar 4 Escucha y escribe la letra correcta. (1–5)

Ejemplo: **1** a

a *en el campo*   b *en la montaña*   c *en la costa*   d *en una ciudad*   e *en un pueblo*

5 **Con tu compañero/a, haz cinco diálogos.**

● ¿Dónde vives?
■ Vivo en Francia, en una casa en el campo.

	País						
1			✔	✔			
2		✔				✔	
3		✔					✔
4			✔		✔		
5			✔				✔

escribir 6 **Escribe las frases del ejercicio 5.**

Ejemplo: Vivo en Francia, en una casa en el campo.

> To make your work better, give as much detail as you can in your answers:
>
> ¿Dónde vives?
> Vivo en Inglaterra … en una casa … en una ciudad.
>
> Show off what you know!

leer 7 **Lee los textos. ¿Conchita o Roberto? Escribe C o R.**

Ejemplo: **1** R

1 Vivo en una casa.
2 No vivo en una casa.
3 Vivo en la costa.
4 Vivimos en una ciudad.
5 Mis hermanas viven en Francia en la costa.
6 Mi hermana vive en Irlanda en un piso.
7 Mi hermana estudia inglés.
8 Mi hermana estudia francés.

¡Hola! Me llamo Roberto.
Vivo en una casa en una ciudad
– Madrid.
Madrid es la capital de España.
¡Me gusta mucho!
Tengo doce años y estudio inglés
en el colegio. Me gusta mucho
porque es muy fácil.
Mi hermana estudia inglés en la
universidad en Irlanda. Vive en
un piso.

¡Hola! Me llamo Conchita. Vivo
en un piso en la costa.
Vivo en España, en Santander.
Mi padre vive en Francia en una
casa en la costa.
Tengo dos hermanas. Se llaman
Cintia y Arancha. Mis hermanas
viven con mi padre en Francia.
Cintia tiene veintidós años.
Arancha tiene diecinueve años.
Estudia francés.

2 ¿Cómo es tu casa?

1 Escucha y escribe la letra correcta. (1–6)

Ejemplo: **1** c

a

Vivo en un piso moderno.

b

Vivo en un piso antiguo.

¿Cómo es tu piso?		¿Cómo es tu casa?	
(No) Es	antiguo	(No) Es	antigua
	moderno		moderna
	bonito		bonita
	nuevo		nueva
	feo		fea
	cómodo		cómoda
	pequeño		pequeña
	viejo		vieja
	grande		grande

c

Vivo en una casa pequeña y vieja.

d

Vivo en una casa grande y cómoda.

e

Vivo en una casa bonita.

f

Vivo en un piso nuevo pero muy feo.

2 Escucha y escribe. (1–3)

Ejemplo: **1** piso – en una ciudad – grande – ☺

está = it is

3 Escucha y lee la poesía. Pon los dibujos en el orden correcto.
Listen and read the poem out loud. Then put the drawings into the correct order.

Ejemplo: d, b, …

a

b
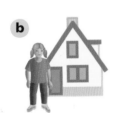

Paco vive en Nicaragua. Su casa es antigua.
Beate vive en Alemania, en una casa muy pequeña.
Ricardo vive en Oviedo, en un piso un poco feo.
Luz vive en Zaragoza – no está en la costa.
Vivimos en la montaña. Vivimos en España.
Y tú, ¿vives en una casa? ¿Es bonita y vieja?
Escríbeme y cuéntame: ¿Cómo es y dónde está?

c

Zaragoza

d

Nicaragua

e

f

Oviedo

g
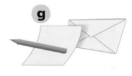

4 Con tu compañero/a, lee la poesía.
Read the poem out loud with your partner.

5 Escucha y escribe las letras en el orden correcto. (1–3)

Ejemplo: **1** c, a, …

¡Hola! Me llamo Almudena Moderna. Vivo en una casa muy moderna.

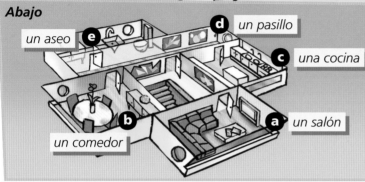

Arriba

h mi dormitorio

g el dormitorio de mi hermano

i un cuarto de baño

f el dormitorio de mis padres

Fuera

k un jardín

l un garaje

j una terraza

Abajo

d un pasillo

e un aseo

c una cocina

b un comedor

a un salón

 6 Con tu compañero/a, haz diálogos.

● ¿Qué hay abajo?
■ Hay <u>un salón</u>, <u>una cocina</u> y <u>un comedor</u>.
● ¿Qué hay arriba?
■ Hay <u>un cuarto de baño</u>, <u>mi dormitorio</u> y <u>el dormitorio de mis padres</u>.
● ¿Qué hay fuera?
■ Hay <u>un jardín</u>.
● ¿Es la casa número <u>11</u>?
■ Sí, es la casa número <u>11</u>.

¿Qué hay	abajo? arriba? fuera?
Hay …	

 7 Describe dónde vives.

● *say which country you live in* — Vivo en Gales/Inglaterra.
● *say where your house/flat is exactly* — Vivo en una ciudad.
● *say what your house/flat is like* — Vivo en una casa vieja y grande.
● *say what rooms there are* — Abajo hay un comedor, …
Arriba hay un cuarto de baño, …
Fuera hay …

3 ¿Qué haces?

 1 Escucha y escribe la letra correcta. (1–10)

Ejemplo: **1** e

¿Qué haces en tu dormitorio?

Mando mensajes.

Escucho música.

Bebo.

Duermo.

Veo la televisión.

Juego con el ordenador.

Estudio.

Hablo por teléfono.

Leo libros.

Como bocadillos.

Gramática

In stem-changing verbs, the middle letters change in some forms of the present tense.

jug**ar**	*to play*	dorm**ir**	*to sleep*
j**ue**go	*I play*	d**ue**rmo	*I sleep*
j**ue**gas	*you play*	d**ue**rmes	*you sleep*
j**ue**ga	*he/she/it plays*	d**ue**rme	*he/she/it sleeps*
jugamos	*we play*	dormimos	*we sleep*
jugáis	*you (plural) play*	dormís	*you (plural) sleep*
j**ue**gan	*they play*	d**ue**rmen	*they sleep*

Juego con el ordenador. *I play on the computer.*
Duermo en mi dormitorio. *I sleep in my bedroom.*

Para saber más p.77 (ex 4); p.131

 2 Con tu compañero/a, haz diálogos.

● ¿Qué haces en tu dormitorio?
■ Mando mensajes, bebo Coca-Cola y escucho música.

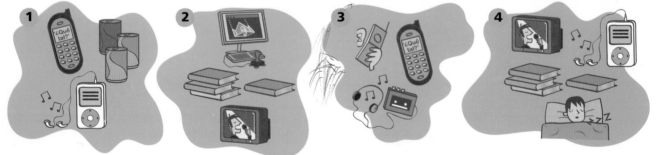

 3 Escucha y escribe las letras correctas del ejercicio 1. (1–5)

Ejemplo: **1** e, d

 4 ¿Qué haces en tu dormitorio?

Ejemplo: En mi dormitorio …

66 sesenta y seis

5 Lee el texto y corrige las frases.

Ejemplo: **1** Diego Desorden vive en <u>un piso</u>.

> ¡Hola! Me llamo Diego Desorden. Vivo en un piso en la ciudad. En el piso hay dos dormitorios, un salón, una cocina y un cuarto de baño. Mi dormitorio es muy grande – me gusta mucho. Normalmente en mi dormitorio, juego con mi ordenador y navego por internet. También hablo mucho por teléfono. Duermo mucho en mi dormitorio, por supuesto.
>
> En el salón, veo la televisión. Comemos en la cocina.
>
> Tengo un ratón blanco. Se llama Raúl. Duerme en el cuarto de baño. Tiene un año. Es muy bonito y simpático.

normalmente = *normally*
navego por internet = *I surf the net*
por supuesto = *of course*

1 Diego Desorden vive en una casa.
2 El dormitorio de Diego Desorden es muy pequeño.
3 En su dormitorio Diego juega con su perro.
4 Diego no habla mucho por teléfono.
5 Diego no duerme mucho.
6 Tiene un ratón negro.
7 El ratón duerme en la cocina.
8 Es muy antipático.

6 Escucha y canta.

En mi dormitorio hay un ordenador,
es muy, muy importante.
Juego mucho con mi ordenador.
Mi ordenador me gusta bastante.

Escucho música con mi ordenador,
Navego por internet por supuesto.
Hablo mucho con mi ordenador,
y leo con él también.

Estudio con mi ordenador,
con mi ordenador como pizza.
Bebo agua con mi ordenador,
y duermo con él por la noche.

Mi ordenador se llama Tim,
mi ordenador es mi amigo.
Ordenador, te quiero mucho.
Estás siempre en mis sueños.

te quiero = *I love you*
siempre = *always*
mis sueños = *my dreams*

Mini-test

I can ...
- ask someone where they live
- say where I live
- name some European countries
- name the rooms in a house
- describe my house or flat
- **G** use stem-changing verbs

7 List the verbs in the 'I' form in the song. Write the English.

Ejemplo: juego – I play

4 En mi dormitorio

● Describing your bedroom
● Using prepositions

1 **Escucha y escribe las letras correctas. (1–12)**

Ejemplo: **1** d

a un armario

b un equipo de música

c una lámpara

d una cama

e una ventana

f pósters

g una estantería

h un ordenador

i una mesa

j una silla

k una televisión

l una puerta

m una alfombra

en las paredes = *on the walls*

2 **Escucha y completa las palabras. (1–7)**
Listen and complete the words that are beeped out.

Ejemplo: **1** silla, cama

3 **Con tu compañero/a, describe los dormitorios.**

● ¿Qué hay en tu dormitorio, <u>Ana</u>?
■ En mi dormitorio hay <u>un armario</u>, …

Ana	Ignacio	Yolanda	Gonzalo

escuchar 4 Escucha y lee. ¿Qué significan las palabras en verde? (1–8)
Listen and read. What do the words in green mean?

Ejemplo: **1** on (top)

1 El pez está **encima** de la estantería.

2 La serpiente está **a la derecha** del equipo de música.

3 La tortuga está **a la izquierda** del ordenador.

4 El gato está **debajo** de la cama.

5  El conejo está **delante** de la puerta.

6 El perro está **al lado** de la silla.

7 Los ratones están **detrás** del armario.

8  La cobaya está **entre** la televisión y la lámpara.

Gramática

Many prepositions end with **de** ('of') in Spanish.
If **de** and **el** come together, they join up to make **del**.

La mesa está **a la derecha del** armario. — The table is **on the right of the** wardrobe.
El ordenador está **al lado de la** cama. — The computer is **beside the** bed.

Describing position:
está = *is*
están = *are*

Para saber más — p.77 (ex 5); p.133

leer 5 Lee el texto y escribe los objetos que faltan en el dibujo.
Read the text and write down the items that are missing in the picture.

Ejemplo: **a** un armario

Te voy a describir mi dormitorio. Hay **una estantería** al lado de la puerta. Hay **una alfombra** azul en el suelo. Hay **un armario** a la derecha de la ventana.

A la izquierda de la ventana hay **una mesa**. Encima de la mesa hay **un ordenador** y **una lámpara**. Hay **una silla** detrás de la mesa. **Mi gato** duerme debajo de la silla.

Tengo muchos pósters en las paredes. Tengo **un póster** de Tobey Maguire entre **dos pósters** de Cameron Diaz. **Mi cama** está al lado del armario. Hay **un equipo de música** delante de la cama.

Mi dormitorio es muy bonito. Me gusta mucho porque es bastante grande.

escribir 6 Describe tu dormitorio.

Ejemplo: En mi dormitorio hay …. No tengo ….
La cama está al lado de …. Me gusta mucho porque es ….

5 Mi rutina diaria

- Talking about your daily routine
- Using reflexive verbs

1 Escucha y lee el texto.

¿Qué haces por la mañana?

Por la mañana

a Me despierto.

b Me levanto.

c Me ducho.

d Me peino.

e Me visto.

f Desayuno.

g Voy al instituto.

Gramática

Reflexive verbs describe an action you do to yourself,
e.g. to get yourself up, to shower yourself.

ducharse	*to take a shower*	**despertarse**	*to wake up*
(yo) **me** ducho	*I take a shower*	(yo) **me** despierto	*I wake up*
(tú) **te** duchas	*you take a shower*	(tú) **te** despiertas	*you wake up*
(él/ella) **se** ducha	*he/she/it takes a shower*	(él/ella) **se** despierta	*he/she/it wakes up*

Para saber más   p.77 (ex 6); p.132

2 Haz un sondeo.
Do a survey.

- ¿Qué haces por la mañana, Isabel?
- Me peino, desayuno y voy al instituto.

Nombre	
Isabel	d, f, g

3 Escucha y lee.

¿Qué haces por la tarde?

Por la tarde

h Hago mis deberes.

i Ceno.

j Veo la televisión.

k Me lavo los dientes.

l Me acuesto.

4 Escucha y escribe las letras correctas de los ejercicios 1 y 3. (1–5)

Ejemplo:

1	a, f	h, j

5 Lee los textos y escribe la letra o las letras correctas.

Read the texts and write the correct letter or letters for each one.

Ejemplo: 1 a

> mi vida = *my life*
> agradable = *pleasant*
> luego = *then*
> después = *afterwards*
> atún = *tuna*

Abajo

Arriba

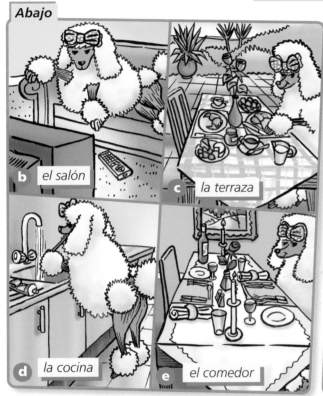

el salón — **b**
la terraza — **c**
la cocina — **d**
el comedor — **e**

el dormitorio
el cuarto de baño
f **g** **h** **i** **j**

1 ¡Hola! Me llamo Kiki. Vivo en una casa pequeña en un pueblo con mi madre. Mi vida es muy agradable.

2 A ver … por la mañana, me despierto y me levanto.

3 Luego desayuno en la terraza. Después me lavo los dientes en la cocina.

4 Veo la televisión abajo en el salón.

5 Por la tarde, me ducho en el cuarto de baño. Me peino también.

6 Me visto para cenar. Ceno con mi madre en el comedor. Me gustan los bocadillos de atún.

7 Después de cenar, me acuesto.

6 Describe tu rutina diaria.

> ● Use these words and phrases in your answer.
>
> | | |
> |---|---|
> | A ver … | *Let's see …* |
> | Por la mañana … | *In the morning …* |
> | Luego … | *Then …* |
> | Después … | *Afterwards …* |
> | Por la tarde … | *In the evening …* |
> | También … | *Also …* |

Resumen

Unidad 1

I can

- ask someone where they live — ¿Dónde vives?
- say where I live — Vivo en Gales.
- name some European countries — Francia, España, Alemania, Italia, Grecia, Portugal
- ask someone whether they live in a house or a flat — ¿Vives en una casa o en un piso?
- say whether I live in a house or a flat — Vivo en una casa. Vivo en un piso.
- say where my house or flat is exactly — Vivo en Francia, en una casa en la montaña.

Unidad 2

I can

- name rooms in a house — un dormitorio, un comedor, una cocina, …
- say what is upstairs — Arriba hay un cuarto de baño, mi dormitorio y el dormitorio de mis padres.
- say what is downstairs — Abajo hay un salón, una cocina y un comedor.
- G use adjectives to describe my house or flat — Vivo en un piso antiguo. Vivo en una casa pequeña.

Unidad 3

I can

- ask someone what they do in their bedroom — ¿Qué haces en tu dormitorio?
- say what I do in my bedroom — Mando mensajes, bebo Coca-Cola y escucho música.
- G use stem-changing verbs — **Jue**go con el ordenador. D**ue**rmo en mi dormitorio.

Unidad 4

I can

- ask someone to describe their bedroom — ¿Qué hay en tu dormitorio, Ana?
- describe my bedroom — En mi dormitorio hay un armario, una cama, un ordenador, …
- say where things are in my room — La lámpara está encima de la mesa. El ordenador está al lado del armario.
- G shorten **de** + **el** to **del** — El equipo de música está al lado **del** ordenador.

Unidad 5

I can

- ask someone about their daily routine — ¿Qué haces por la mañana? ¿Qué haces por la tarde?
- talk about my daily routine — Me peino, desayuno y voy al instituto.
- say where in the house I do things — Me ducho en el cuarto de baño.
- G use reflexive verbs — **Me** despierto, **me** ducho y **me** visto.

escuchar **1** Escucha y escribe las letras correctas. (1–8)

Ejemplo: **1** b

leer **2** Empareja las preguntas y las respuestas.

Ejemplo: **1** e

1 ¿Dónde vives?
2 ¿Vives en una casa o en un piso?
3 ¿Cómo es?
4 ¿Qué haces por la mañana?
5 ¿Qué hay abajo, en tu casa?
6 ¿Qué hay arriba, en tu casa?
7 Describe tu dormitorio.
8 ¿Qué haces en tu dormitorio?

a En mi dormitorio escucho música, veo la televisión y mando mensajes a mis amigos.
b Es pequeña.
c Arriba hay tres dormitorios y un cuarto de baño.
d Vivo en una casa.
e Vivo en Burgos.
f Me despierto y me levanto. Desayuno y voy al instituto.
g En mi dormitorio hay una cama, un armario, una televisión y un equipo de música.
h Abajo hay una cocina, un comedor, un salón y un aseo.

hablar **3** Con tu compañero/a, haz las preguntas del ejercicio 2 y contesta.

● ¿Dónde vives?
■ Vivo en <u>Oxford</u>.

leer **4** Lee el texto y contesta a las preguntas.

Ejemplo: **1** In Madrid.

1 Where does Rodrigo live?
2 Who does he live with?
3 What's his house like?
4 Where is his house?
5 What is there downstairs?
6 What is there upstairs?
7 What is there outside?
8 How does he feel about his house?

Vivo en Madrid con mi padre y mis tres hermanos. Vivimos en una casa antigua. Es muy cómoda.

Está en la ciudad. Abajo hay un pasillo, un salón, un comedor, un cuarto de baño y la cocina. Arriba hay cuatro dormitorios y dos cuartos de baño. Fuera hay una terraza y un jardín muy grande. Me gusta mucho mi casa porque es muy grande.

Y tú, ¿vives en una casa o en un piso?

Rodrigo

escribir **5** Escribe un texto como el texto de Rodrigo.

¡Extra!

 **1 Escucha y lee.**

T – ¡Hola Angélica!
A – ¡Hola Tomás! ¿Qué tal?
T – Bien, gracias. ¿Y tú?
A – Muy bien, gracias.

A – ¿Qué haces?
T – Leo un libro. Es muy interesante.
A – Vives en Madrid, Tomás, ¿verdad?
T – Sí, vivo en Madrid.

T – Vivimos en una casa muy antigua y muy pequeña en el centro. Es muy vieja, pero me gusta mucho. Es bastante cómoda.

T – ¿Y cómo es tu casa, Angélica?
A – A ver, es muy grande. Hay noventa dormitorios y noventa aseos. Hay muchos pasillos también.

A – ¡Es muy moderna y también muy cómoda! Hay también un jardín y una piscina muy grande …

T – ¿Qué? ¿Noventa dormitorios? ¡Ah, ya lo entiendo! ¿Vives aquí en el hotel?
A – Sí. Mi padre es el director. Y tú, ¿qué haces aquí?

A – Ahora lo entiendo perfectamente.

T – Mi madre trabaja aquí también. Es profesora de yoga. Vengo con mi madre y con mi hermano Víctor.

P – ¡Angélica, Angélica!

 2 Con tu compañero/a, lee la historia de Tomás y Angélica.

Lee los textos y contesta a las preguntas en inglés.

A Me llamo María. Vivo en una hacienda en Yucatán en México. Se llama 'La Tigra' y está en el campo. Es una hacienda típica de la región. Hay plantaciones de café, de banano y de caña.

La casa es muy grande. Abajo hay un salón grande, una cocina y un comedor. La cocina es muy importante y en el comedor comemos con los visitantes.

Arriba hay diez dormitorios y cinco cuartos de baño. Fuera hay un jardín con plantas, flores y palmares. En el jardín leo o escucho los pájaros. También hay un patio interior con una fuente.

B Me llamo Sergio y vivo en una casa en las afueras de Nerja en España. Es una casa antigua y muy pequeña. Abajo hay una cocina, un comedor y un salón. Por la tarde hago mis deberes en la cocina, luego ceno en el comedor y después veo la televisión en el salón. Arriba hay dos dormitorios y un cuarto de baño.

Me gusta mucho mi dormitorio porque es muy cómodo, pero no es muy grande. Tengo una cama grande y una alfombra gris en el suelo. No tengo ordenador.

En mi dormitorio leo libros y duermo. Paso mi tiempo fuera con mis tres gatos. Mi gato preferido se llama Salman. Es muy inteligente pero a veces es impaciente.

afueras = *suburbs*

Cognates are Spanish words that are the same as or similar to English words. Finding cognates in a text can help you work out what it's about.

1 What crops are grown on María's hacienda?
2 Describe María's garden.
3 Find words in Maria's text that are cognates of English words.
4 Does Sergio like his bedrooom?
5 What does he do in his bedroom?

Contesta a las preguntas en español.

A 1 ¿Dónde vive María?
2 ¿Qué cultivan en la hacienda La Tigra?
3 ¿Qué hay abajo, en la casa?
4 ¿Qué hay arriba, en la casa?
5 ¿Qué hace María en el jardín?

B 1 ¿Dónde vive Sergio?
2 ¿Cómo es su casa?
3 Describe su dormitorio.
4 ¿Cuántos gatos tiene Sergio?
5 Describe su gato preferido.

1 Match up the dialogues to the people.

> **?** *Need help?*   Go to Unidad 1
> **?** *Want to know more?*   Go to page 130

1 ¡Hola! Estamos en el Camp Nou y hay mucha gente de Europa aquí … ¿Dónde vives, Rocco?

> Vivo en Italia.

2 ¿Y vosotros, dónde vivís?

> Vivimos en Portugal.

3 ¿Y vosotras, dónde vivís?

> Nosotras vivimos en Francia.

4 Yo vivo en Grecia. Me gusta mucho el fútbol español – sobre todo el Barça.

5 Y él, vive en Alemania. Se llama Kurt.

6 ¿Y ellos, dónde viven?

> Viven en Escocia …

2 Find the Spanish for the following verbs (1–6) and pronouns (7–12) in exercise 1.

1 I live
2 you (singular) live
3 he lives
4 we live
5 you (plural) live
6 they live
7 I
8 he
9 you (plural, masculine)
10 we (feminine)
11 them

3 Antonio has made six mistakes in his Spanish. Rewrite the text and correct his mistakes.

> **?** *Need help?*   Go to Unidad 2
> **?** *Want to know more?*   Go to page 129

Vivo en un piso, en Barcelona, en la ciudad. El piso es bastante grande y cómoda. A mí me gusta mucho. Es muy antigua también.

Hay tres dormitorios. El dormitorio de mis padres es grandes, pero mi dormitorio es muy pequeña. Hay una cocina pequeño pero el salón es grande y cómoda.

4 **Choose the correct verb forms.**

1 **Juego** / **Jugo** con el ordenador.
2 ¿**Dormes** / **Duermes** en tu dormitorio?
3 ¿**Jugas** / **Juegas** mucho con el ordenador?
4 **Duerme** / **Dorme** en la cocina.
5 **Juegamos** / **Jugamos** juntos.
6 **Duermen** / **Dormen** juntos.

? *Need help?* Go to Unidad 3
? *Want to know more?* Go to page 131

 5 **Choose the correct preposition to fill in the gaps.**

Example: **1** El perro está <u>debajo</u> de la cama.

1 El perro está ⁓ de la cama.
2 El pez está ⁓ de la silla.
3 El conejo está ⁓ de la estantería.
4 Las serpientes están ⁓ del armario.
5 La tortuga está ⁓ del equipo de música.
6 El gato está ⁓ de la puerta.
7 La cobaya está ⁓ del ordenador.
8 El pájaro está ⁓ la televisión y la lámpara.

? *Need help?* Go to Unidad 4
? *Want to know more?* Go to page 133

encima	detrás	debajo
delante	entre	
a la izquierda		a la derecha
al lado		

 6 **Copy out the text, changing the infinitives to the 'I' form.**

? *Need help?* Go to Unidad 5
? *Want to know more?* Go to page 132

Example: **1** Por la mañana <u>me despierto</u> …

Por la mañana **(1)** (despertarse) y **(2)** (levantarse). Después desayuno en la cocina y luego **(3)** (ducharse), **(4)** (peinarse) y **(5)** (lavarse) los dientes en el cuarto de baño. **(6)** (vestirse) en mi dormitorio. Por la tarde, hago mis deberes y ceno en el comedor. Después de cenar, **(7)** (acostarse).

despertarse	*to wake up*	me despierto …
vestirse	*to get dressed*	me visto …
acostarse	*to go to bed*	me acuesto …

Palabras

Los países

¿Dónde vives?
vivir
Vivo en …
Vive en …
Vivimos en …
Viven en …

Alemania
Escocia
España
Francia
Gales
Grecia
Inglaterra
Irlanda
Italia
Portugal

Countries

Where do you live?
to live
I live in …
He/She lives in …
We live in …
They live in …

Germany
Scotland
Spain
France
Wales
Greece
England
Ireland
Italy
Portugal

¿Dónde vives?

¿Vives en una casa o
  en un piso?
Vivo en una casa.
Vivo en un piso.

en el campo
en la montaña
en la costa
en una ciudad
en un pueblo

Where do you live?

Do you live in a house
  or a flat?
I live in a house.
I live in a flat.

in the countryside
in the mountains
on the coast
in a city/town
in a village

Mi casa

¿Cómo es tu piso?
Es …
  antiguo
  moderno
  bonito
  feo
  nuevo
  viejo
  pequeño
  cómodo
  grande

My house

What's your flat like?
It's …
  old(-fashioned)
  modern
  pretty
  ugly
  new
  old
  small
  comfortable
  big

¿Cómo es tu casa?
Es …
  antigua
  moderna
  bonita
  fea
  nueva
  vieja
  pequeña
  cómoda
  grande

What's your house like?
It's …
  old(-fashioned)
  modern
  pretty
  ugly
  new
  old
  small
  comfortable
  big

Las habitaciones

¿Qué hay en tu
  casa/piso?
¿Qué hay abajo?

¿Qué hay arriba?
¿Qué hay fuera?
Hay …
  un comedor
  un cuarto de baño
  un aseo
  un pasillo
  un salón
  una cocina
  un dormitorio
  un garaje
  un jardín
  una terraza
el dormitorio de
  mis padres
el dormitorio de
  mi hermano

Rooms

What is there in your
  house/flat?
What is there
  downstairs?
What is there upstairs?
What is there outside?
There's …
  a dining room
  a bathroom
  a toilet
  a corridor
  a living room
  a kitchen
  a bedroom
  a garage
  a garden
  a terrace
my parents' bedroom

my brother's bedroom

En mi dormitorio

¿Qué haces en tu
  dormitorio?
Mando mensajes.
Escucho música.
Bebo Coca-Cola.
Duermo mucho.
Veo la televisión.
Juego con el
  ordenador.
Estudio.

In my bedroom

What do you do in
  your bedroom?
I send text messages.
I listen to music.
I drink Coca-Cola.
I sleep a lot.
I watch television.
I play on the
  computer.
I study.

Hablo por teléfono.	I talk on the phone.
Leo libros.	I read books.
Como bocadillos.	I eat sandwiches.
Navego por internet.	I surf the net.

Mi dormitorio — *My bedroom*

¿Qué hay en tu dormitorio?	What is there in your bedroom?
En mi dormitorio hay …	In my bedroom there's/there are …
un armario	a wardrobe
un equipo de música	a hi-fi
un ordenador	a computer
una alfombra	a rug
una cama	a bed
una estantería	a shelf/shelves
una lámpara	a lamp
una mesa	a table
una puerta	a door
una silla	a chair
una televisión	a television
una ventana	a window
pósters	posters

Las preposiciones — *Prepositions*

encima de	on
a la derecha de	to the right of
a la izquierda de	to the left of
debajo de	under
delante de	in front of
al lado de	beside
detrás de	behind
entre	between
a la derecha del armario	to the right of the wardrobe
al lado de la cama	beside the bed
en las paredes	on the walls

Mi rutina diaria — *My daily routine*

¿Qué haces por la mañana?	What do you do in the morning?
Por la mañana …	In the morning …
me despierto	I wake up
me levanto	I get up
me ducho	I shower
me peino	I comb my hair
me visto	I get dressed
desayuno	I have breakfast
voy al instituto	I go to school
¿Qué haces por la tarde?	What do you do in the evening?
Por la tarde …	In the evening …
hago mis deberes	I do my homework
ceno	I have supper
veo la televisión	I watch TV
me lavo los dientes	I brush my teeth
me acuesto	I go to bed
Desayuno en la terraza.	I have breakfast on the terrace.
Me ducho en el cuarto de baño.	I shower in the bathroom.

Palabras muy útiles — *Very useful words*

por la mañana	in the morning
por la tarde	in the evening
luego	then
después	afterwards

Estrategia

Spot the stems!

Spanish verbs can seem very complicated, because they have a lot of different endings. You'll find them easier to learn if you can recognise the first part of the verb, which usually stays the same. For example, **vivo**, **vives**, **vive**, **vivimos** all start with **viv-**. This is called the **stem** of the verb.

Here are some other stems from Chapter 4. Which verbs do they belong to?

- est-
- habl-
- com-

1 Mi tiempo libre

● Saying what you do in your free time
● Using **salir** (to go out) and **hacer** (to do)

1 Escucha y escribe la letra correcta. (1–10)

¿Qué haces en tu tiempo libre?

Ejemplo: **1** d

a
Voy al cine.

b
Voy a la piscina.

c
Voy de compras.

d
Salgo con mis amigos.

e
Hago mis deberes.

f
Monto en bicicleta.

g
Escucho música.

h
Veo la televisión.

i
Navego por internet.

j
Juego con mi ordenador.

2 Con tu compañero/a, haz diálogos.
Utiliza los dibujos del ejercicio 1.

● ¿Qué haces en tu tiempo libre?
■ Voy a la piscina y también monto en bicicleta.

1 bf

2 ac

3 ijd

4 ecgh

5 gcbe

Gramática

If **a** (*to*) and **el** come together, they join up to make **al**.

a + el → Voy **al** cine.
a + la → Voy **a la** piscina.

Para saber más   p.133

3 ¿Qué haces en tu tiempo libre?
Escribe cinco frases.

Ejemplo:
Salgo con mis amigos.
No navego por internet. …

Gramática

Salir and **hacer** are two important irregular verbs.

salir	*to go out*	hacer	*to do*
sal**g**o	*I go out*	ha**g**o	*I do*
sales	*you go out*	haces	*you do*
sale	*he/she/it goes out*	hace	*he/she/it does*
salimos	*we go out*	hacemos	*we do*
salís	*you (plural) go out*	hacéis	*you (plural) do*
salen	*they go out*	hacen	*they do*

Para saber más   p.94 (exs 1–2); p.131

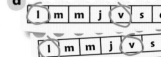

 leer 4 Empareja las expresiones españolas con los dibujos.

Ejemplo: **1** b

1 Todos los días …
2 Los lunes …
3 Una vez por semana …
4 Los fines de semana …
5 Dos veces a la semana …

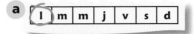
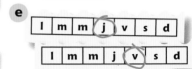

a | l | m | m | j | v | s | d |

d | l | m | m | j | v | s | d |
| l | m | m | j | v | s | d |

b | l | m | m | j | v | s | d |

e | l | m | m | j | v | s | d |
| l | m | m | j | v | s | d |

c | l | m | m | j | v | s | d |

 escuchar 5 Escucha y escribe. (1–6)

Ejemplo: **1** Julio – los lunes – piscina

| Mauricio | Nora | Iván | Claudia | Julio | Juana |

 leer 6 Lee el texto. ¿Verdadero o falso? Escribe V o F.

Ejemplo: **1** V

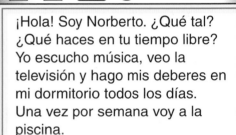

1 Norberto escucha música todos los días.
2 Norberto ve la televisión una vez por semana.
3 Hace sus deberes en el comedor.
4 Una vez por semana va a la piscina.
5 Los fines de semana va al cine con sus amigos.
6 Va de compras una vez por semana.

¡Hola! Soy Norberto. ¿Qué tal? ¿Qué haces en tu tiempo libre? Yo escucho música, veo la televisión y hago mis deberes en mi dormitorio todos los días.
Una vez por semana voy a la piscina.
Los fines de semana monto en bicicleta y voy al cine con mis amigos – es muy divertido.
No voy de compras. No me gusta nada. Es muy aburrido.

 escribir 7 Escribe una respuesta a Norberto. Incluye los detalles siguientes:

● *Say when you do an activity:* Todos los días … Los lunes … Una vez por semana …
● *Say where you do the activity:* en mi dormitorio … en el instituto …
● *Say who you do an activity with:* con mis amigos … con mi hermana …
● *Say which activities you don't do:* No …
● *Give an opinion:* (No) me gusta. Es …
● *Use connectives:* y, también …

To say you don't do something, put **no** in front of the verb.
No hago mis deberes. *I don't do my homework.*

2 ¿Qué hora es?

escuchar 1 Escucha y repite.

¿Qué hora es?

Es la una.
Son las dos. (etc.)

menos cinco · y cinco
menos diez · y diez
menos cuarto · y cuarto
menos veinte · y veinte
menos veinticinco · y veinticinco
y media

escuchar 2 Escucha y escribe la letra correcta. (1–8)

Ejemplo: **1** c

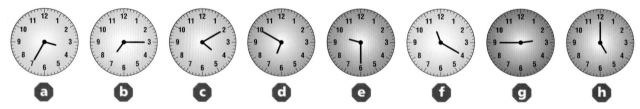

a b c d e f g h

escribir 3 Escribe ocho diálogos sobre las horas del ejercicio 2.

Ejemplo: **a** – ¿Qué hora es?
– Son las cuatro menos veinticinco.

escuchar 4 Escucha y escribe la letra y la hora. (1–6)

Ejemplo: **1** f – 8 a.m.

a b c d
e f g

de la mañana

de la tarde

de la noche

5 Lee las preguntas y escribe la hora. Luego, pregunta y contesta con una frase entera.
Read the questions and note down a time for each. Then ask and answer the questions using full sentences.

¿**A** qué hora comes?

Como **a** las dos.

● ¿A qué hora vas al cine?
■ Voy al cine a las cuatro de la tarde.

1	¿A qué hora vas al cine? 4 p.m.
2	¿A qué hora escuchas música?
3	¿A qué hora sales con tus amigos?
4	¿A qué hora vas de compras?
5	¿A qué hora navegas por internet?
6	¿A qué hora vas a la piscina?
7	¿A qué hora ves la televisión?

Gramática

Ir is an important irregular verb.

ir	to go
voy	I go
vas	you go
va	he/she/it goes
vamos	we go
vais	you (plural) go
van	they go

Para saber más p.94 (ex 3); p.131

Los fines de semana locos

Me llamo Juana.

Los fines de **(1)** ～ a las nueve de la **(2)** ～ me voy a la montaña con mi tortuga pequeña, Juana.

 6 Adivina las palabras que faltan. Luego escucha la canción.
Guess the missing words. Then listen to the song and check.

Me llamo Cristina.

Los fines de semana a la una de la **(3)** ～ con mi serpiente Cristina voy a la **(4)** ～.

 7 Copia y rellena la tabla con palabras de la canción.

Horas	
Expresiones de frecuencia	Los fines de semana
Verbos	
Adjetivos	
Nombres	

Me llamo Ramón.

Los fines de semana a las cinco de la **(5)** ～ veo la televisión con mi cobaya Ramón.

Los fines de semana escucho música en mi **(6)** ～ a las once de la noche. Luego duermo, duermo, duermo.

 8 Escribe otra estrofa para la canción.

loco = *mad*

3 ¿Qué deportes haces?

 1 Escucha y escribe el nombre correcto. (1–11)

¿Qué deportes haces?

Ejemplo: **1** Juan

Hago atletismo.
Sergio

Hago ciclismo.
Juan

Hago equitación.
Yasmina

Hago esquí.
Eduardo

Hago natación.
Irene

Hago patinaje.
Yasmina

Juego al baloncesto.
Alejandro

Juego al fútbol.
Eduardo

Juego al tenis.
Amaya

Juego al voleibol.
Alejandro

Juego al hockey.
Irene

> **h**   **h**ago   **h** is silent
> **qu**   ¿**Qu**é?, es**qu**í   **qu** is a 'k' sound
>
> ¿**Qu**é **h**aces?   Hago es**qu**í y e**qu**itación.
>
> Listen and repeat carefully.

 2 Con tu compañero/a, elige personas del ejercicio 1. Pregunta y contesta.

● Eres <u>Irene</u>. ¿Qué deportes haces?
■ <u>Hago natación</u> y <u>juego al hockey</u>.

 3 Escucha y escribe las palabras que faltan.

Me llamo Raúl. **(1)** Hago muchos deportes.
Hago ciclismo y **(2)** ____ también.
Los fines de **(3)** ____ juego al fútbol y al voleibol con mis amigos.
Todos los días, a las siete de la **(4)** ____, voy a la piscina y hago **(5)** ____.
Los jueves **(6)** ____ al baloncesto. Todos los días, después de la cena,
(7) ____ música, veo la **(8)** ____ y juego con la Playstation en mi dormitorio.

después de = *after*
cena = *supper*

4 Escribe las frases en un orden lógico.

Ejemplo: **1** Todos los días juego al fútbol.

1 días al juego los fútbol. Todos
2 viernes Los patinaje. hago
3 fines de juego Los al semana tenis.
4 voleibol. Los al martes juego

5 hago veces esquí. A
6 hockey. Una al vez por juego semana
7 veces A juego baloncesto. al
8 de Los semana hago equitación. fines

leer 5 Read the questions with your partner and answer with a, b or c. Work out your score and read the solution.

> o más = *or more*
> nunca = *never*

① ¿Haces mucho deporte?
  a Sí, hago deporte dos veces o más a la semana.
  b Sí, hago deporte una vez por semana.
  c No, no hago deporte.

② ¿Qué haces los fines de semana?
  a Juego al baloncesto y al tenis. También hago ciclismo.
  b Juego al fútbol en el parque con mis amigos.
  c Voy de compras con mis amigos.

③ ¿En el colegio te gusta la educación física?
  a Me gusta mucho.
  b Me gusta bastante.
  c No me gusta nada.

④ ¿Haces natación?
  a Sí, voy a la piscina todos los días.
  b Sí, hago natación una vez por semana.
  c Nunca hago natación.

⑤ Todos los días …
  a … juego al fútbol.
  b … veo el fútbol en la televisión.
  c … juego con mi ordenador.

⑥ ¿Qué deporte hacen estas personas famosas?

a Lance Armstrong **b** Paula Radcliffe **c** Roger Federer

d Magic Johnson **e** Raúl González

escribir 6 Escribe tus respuestas del ejercicio 5 en un texto.

Ejemplo: Hago deporte una vez por semana. Juego al fútbol en el parque con mis amigos. …

Soluciones

10–15 puntos: Eres muy deportista. Eres activo/a y te gusta mucho el deporte.

5–9 puntos: Eres un poco deportista. Te interesa el deporte y haces un poco de vez en cuando.

0–4 puntos: No eres deportista. No te gusta nada el deporte. Prefieres jugar con tu ordenador o hacer compras con tus amigos.

Notas

Preguntas 1–5: a = 2 puntos, b = 1 punto, c = 0

Pregunta 6: 1 punto para cada deporte correcto:
a = ciclismo, b = atletismo, c = tenis,
d = baloncesto, e = fútbol

Mini-test

I can
  ● say what I do in my free time
  ● say what sports I play/do
  ● tell the time
  ● use expressions of frequency
  G use **al** and **a la** correctly
  G use the irregular verbs **ir**, **salir** and **hacer**

4 Me gusta ir al cine

 1 Escucha y lee. Escribe la letra correcta. (1–10)

Ejemplo: **1** b

1 **jugar** al fútbol
2 **hacer** atletismo
3 **navegar** por internet
4 **ir** al cine
5 **salir** con mis amigos
6 **ver** la televisión
7 **hacer** mis deberes
8 **escuchar** música
9 **ir** de compras
10 **hacer** natación

a · b · c · d · e · f · g · h · i · j

The verbs above are in the infinitive. You can spot infinitives because they end in **-ar**, **-er** or **-ir**. This is the form you find in a dictionary or word list.

2 Pon los dibujos del ejercicio 1 en el orden del texto.

Ejemplo: **1** 😊 c

¿Qué te gusta hacer en tu tiempo libre?
1 *A ver, me gusta ir de compras.*
2 *Me gusta salir con mis amigos.*
3 *Me encanta ir al cine.*
4 *Me gusta mucho escuchar música.*
5 *Odio ver la televisión.*
6 *No me gusta hacer mis deberes.*
7 *No me gusta nada navegar por internet.*
8 *Prefiero hacer atletismo.*
9 *Me gusta mucho hacer natación.*
10 *Pero sobre todo me encanta jugar al fútbol.*

sobre todo = *above all*

😊	Me gusta
😊😊	Me gusta mucho
😊😊😊	Me encanta
😊 +	Prefiero
😞	No me gusta
😞😞	No me gusta nada
😞😞😞	Odio

 3 Busca los infinitivos en el texto del ejercicio 2. Escribe una lista.

Ejemplo: hacer = to do
              ir = to go

Gramática

Me gusta and similar expressions are followed by the **infinitive** of the verb.

Me gusta **jugar** al fútbol.   *I like playing football.*
No me gusta **hacer** natación.   *I don't like swimming.*
Prefiero **ir** de compras.   *I prefer going shopping.*

Para saber más   p.95 (ex 6); p.132

4 Escucha y escribe. (1–8)

Ejemplo: **1** 😊 – ver la televisión

¿Qué te gusta hacer en tu tiempo libre? 😊

¿Qué **no** te gusta hacer? 😞

 5 Escucha otra vez. Escribe la letra correcta. (1–8)

Ejemplo: **1** b

¿Por qué?　Porque es …　Porque **no** es …

a — sano

b — barato

c — interesante

d — fácil

e — divertido

f — bueno

g — aburrido

h — caro

 6 Con tu compañero/a, haz diálogos.

- ● ¿Qué te gusta hacer en tu tiempo libre?
- ■ Me gusta mucho <u>hacer natación</u>.
- ● ¿Por qué?
- ■ Porque es <u>sano</u>.
- ● ¿Qué no te gusta hacer?
- ■ Odio <u>ir de compras</u>.
- ● ¿Por qué?
- ■ Porque es <u>muy aburrido</u>.

 **7 Copia y completa el texto.**

dinero = *money*

 8 Escribe un texto sobre ti.
Write a text about yourself.

En mi tiempo libre, me encanta **(1)** escuchar *música* porque es interesante y también es **(2)** [monedas]. No tengo mucho dinero.

(3) 😊😊😊 navegar por internet – es **(4)** ✔✔✔ interesante.

No me gusta ir al cine porque es bastante **(5)** [monedas]. No me gusta nada ir de compras porque es muy, muy **(6)** .

(7) 😊😊 salir con mis amigos porque es **(8)** , pero prefiero ver la televisión.

(9) 😞😞😞 hacer mis deberes – no es **(10)** .

5 ¿Qué vas a hacer?

 1 Escucha y escribe las letras correctas. (1–8)

Ejemplo: 1 **i** , **a**

¿Qué vas a hacer?

Voy a …

i

Mañana

ii

La semana que viene

iii

Este fin de semana

iv

En las vacaciones

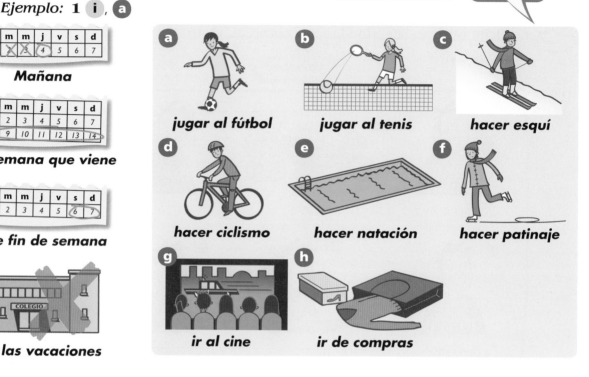

a jugar al fútbol **b** jugar al tenis **c** hacer esquí

d hacer ciclismo **e** hacer natación **f** hacer patinaje

g ir al cine **h** ir de compras

2 Con tu compañero/a, haz cinco diálogos.

● ¿Qué vas a hacer <u>en las vacaciones</u>?
■ Voy a <u>hacer esquí</u>.

Gramática

To say what you are 'going to' do, use a form of **ir** followed by **a** plus the **infinitive**. This is called the near future tense.

voy a	**jugar** al fútbol	*I am going to play football*
vas a	**hacer** esquí	*you are going to go skiing*
va a	**ir** de compras	*he/she is going to go shopping*
vamos a	**ir** al cine	*we are going to go to the cinema*
vais a	**salir**	*you (plural) are going to go out*
van a	**escuchar** música	*they are going to listen to music*

Para saber más   p.95 (ex 7); p.132

3 Describe tus planes para las vacaciones.

Ejemplo:
En las vacaciones, voy a …

 Escribe las letras correctas de las actividades del ejercicio 1 mencionadas en cada texto.

Ejemplo: **1** d, …

1

La semana que viene voy a hacer ciclismo. Me encanta hacer ciclismo – es muy muy sano. Por la tarde voy a jugar al fútbol y al tenis. Voy a ir al cine con mis amigos también.

2

En las vacaciones voy a ir a Marbella con mi familia. Voy a jugar al fútbol en la playa con mi hermano. Mi madre va a hacer natación y mi hermana va a ir de compras.

3

Este fin de semana voy a jugar al fútbol con mis amigos. El domingo, voy a hacer natación y patinaje también. No voy a hacer ciclismo porque no me gusta nada. Es muy aburrido. ¿Y tú? ¿Qué vas a hacer este fin de semana?

 Escucha y escribe 'present' o 'near future'. (1–8)

Ejemplo: **1** present

> Sentences talking about the holidays (i.e. the near future) will have the infinitive.

6 **Lee la carta de Teo y contesta a las preguntas en inglés.**

¡Hola! ¿Qué tal?

Te voy a hablar de mis pasatiempos. Normalmente hago ciclismo todos los días y hago patinaje una vez por semana. No voy de compras. No me gusta nada. A veces juego al tenis, pero no mucho.

En las vacaciones voy a ir a Calella con mi familia. Voy a jugar al fútbol y jugar al voleibol en la playa con mi hermano.

Mi madre va a hacer natación y mi hermana va a escuchar música.

Los fines de semana mi madre y mi hermana van a ir de compras a la ciudad. Yo voy a hacer ciclismo o escuchar música.

No vamos a ver la televisión, porque no hay televisión en el piso.

Todos los días a las ocho de la mañana mi padre va a jugar al tenis con su amigo Pepe – ¡qué horror! ¡Yo voy a dormir!

¿Y tú? ¿Qué vas a hacer en las vacaciones? ¡Escríbeme pronto!

Un abrazo

Teo

> a veces = *sometimes*
> la playa = *beach*

1 What does Teo generally do in his free time? (3 things)
2 What doesn't he normally do?
3 What is each of these people going to do during the holidays?
- Teo
- his brother
- his mother
- his sister
- his father

Resumen

Unidad 1

I can

- ask someone what they do in their free time — ¿Qué haces en tu tiempo libre?
- say what I do in my free time — Monto en bicicleta y navego por internet.
- say what I don't do — No voy de compras.
- G use **al** and **a la** correctly — Voy **al** cine. Voy **a la** piscina.
- G use expressions of frequency — todos los días, los lunes, una vez por semana
- G understand all forms of **salir** (to go out) — salgo, sales, sale, salimos, salís, salen
- G understand all forms of **hacer** (to do) — hago, haces, hace, hacemos, hacéis, hacen

Unidad 2

I can

- ask and tell the time — ¿Qué hora es? Son las dos y media.
- talk about different times of day — de la mañana, de la tarde, de la noche
- ask someone at what time they do an activity — ¿A qué hora comes?
- say at what time I do something — Escucho música a las nueve.
- G use all forms of the irregular verb **ir** — voy, vas, va, vamos, vais, van

Unidad 3

I can

- ask someone what sports they do — ¿Qué deportes haces?
- say what sports I do — Hago atletismo. Hago ciclismo.
- say what games I play — Juego al baloncesto. Juego al tenis.

Unidad 4

I can

- say what I like or don't like doing — Me gusta ir de compras. No me gusta hacer mis deberes.
- ask someone what they like doing — ¿Qué te gusta hacer en tu tiempo libre?
- ask them why — ¿Por qué?
- give reasons why — porque es sano, porque no es caro
- G use **me gusta**, **me encanta**, **odio** and **prefiero** with an infinitive — Me encanta ir al cine. Odio ver la televisión. Prefiero jugar al fútbol.

Unidad 5

I can

- say what I am going to do — Voy a ir de compras. Voy a hacer natación.
- use expressions about the future — mañana, la semana que viene, en las vacaciones
- ask someone what they are going to do — ¿Qué vas a hacer?
- G understand all forms of the near future tense — Voy a jugar …, Vas a jugar …, Va a jugar …
- G spot sentences in the present and near future tenses — Normalmente juego al tenis. En las vacaciones voy a ir a Calella.

1 Escucha. ¿Verdadero o falso? (1–6)

Ejemplo: **1** V

Alba

Ernesto

Elisa

Ramón

Marcos

Irene

 2 Escucha otra vez y escribe los adjetivos que entiendes.

Ejemplo: divertido, …

 3 Con tu compañero/a, pregunta y contesta.

- ¿Qué hora es?
- Son las <u>dos y media</u>.
- ¿Qué vas a hacer?
- Voy a <u>jugar al tenis</u>.

 4 Empareja los dibujos con las frases apropiadas. Luego escribe las frases en inglés.

a Este fin de semana voy a ir al cine.
b Voy a la piscina todos los días – me gusta mucho.
c Normalmente voy de compras todos los fines de semana.

d Monto en bicicleta dos veces a la semana, porque es muy sano.
e ¿A qué hora ves la televisión?
f Me encanta navegar por internet porque no es caro.

 5 Escribe un texto sobre tus planes para la semana que viene.

Ejemplo: La semana que viene voy a hacer natación. Me gusta mucho porque es sano.

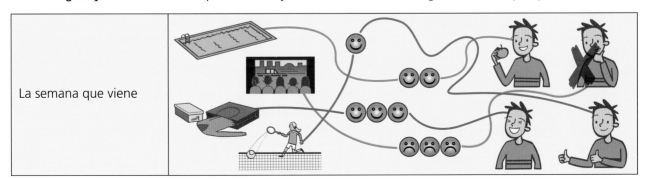

La semana que viene

¡Extra!

escuchar 1 Escucha y lee.

T – ¡Hola Angélica!
A – ¡Hola Tomás! ¿Qué tal?
T – Hmm, regular. ¿Y tú?
A – Fenomenal, gracias.

A – ¿Qué haces en tu tiempo libre?
T – Juego al voleibol y al fútbol.
¿Y tú? ¿Qué deportes haces?

T – Me gusta hacer natación también.

hablar 2 Con tu compañero/a, lee la historia de Tomás y Angélica.

T – ¿Te gusta jugar al tenis?
A – Ah sí, me encanta. Es muy divertido.
Y tú, ¿juegas todos los días?
T – Todos los días no, pero a veces sí.

A – Pues juego al tenis y me gusta jugar al baloncesto, pero prefiero hacer natación.

T – ¿Qué vas a hacer este fin de semana, Angélica?
A – A ver, voy a ir de compras con mi madre …

P – ¡Angélica, Angélica!
T – ¡Hasta luego, Angélica!

leer 3 Empareja el inglés con el español del texto.

Ejemplo: 1 ¿Qué tal?

1 How are you?
2 Great.
3 Do you like playing tennis?
4 Do you play every day?

5 What do you do in your free time?
6 What sports do you do?
7 I prefer swimming.
8 I'm going to go shopping.

4 Pon los dibujos en el orden correcto del texto.

Ejemplo: **1** c

El estadio Camp Nou

1 Me **llamo** Evangelista Futbolista. **Soy** muy guapo. **Tengo** los ojos marrones y el pelo castaño. **Soy** inteligente y hablador, pero a veces **soy** un poco impaciente. No **soy** perezoso. **Vivo** para el fútbol – **es** mi vida. **Soy** Evangelista Futbolista …

2 **Vivo** en Barcelona y **juego** para el equipo famoso de Barcelona, FC Barcelona – el Barça.

3 **Hago** mucho deporte. Normalmente a las ocho de la mañana **monto** en bicicleta y **hago** ciclismo durante dos horas.

4 Luego, a las diez de la mañana **voy** al estadio y **juego** al fútbol con los otros miembros del equipo Barça.

5 **Como** a las doce y después **hago** atletismo hasta las cuatro. Me **gusta** mucho.

6 Normalmente **voy** a la piscina a las cuatro de la tarde y **hago** natación durante dos horas.

7 Ayer **fue** un día diferente. ¡**Fue** mi cumpleaños!

8 Por la tarde **escuché** música y **jugué** con mi Playstation. También **navegué** un poco por internet.

9 A las ocho y media **hablé** con mi madre por teléfono durante veinte minutos. Mi madre **es** muy habladora.

10 A las nueve y media **comí** una pizza y **bebí** cerveza en un restaurante con mis amigos.

hablador = *talkative*   fue = *was*
equipo = *team*   cerveza = *beer*
ayer = *yesterday*

5 Copy out the grid and put the verbs in red in the text into the correct column. Fill in the English.

PRESENT TENSE			
Regular verbs	Irregular verbs	Stem-changing verbs	English

6 Write down the infinitives of the verbs in the grid (in green in the text) and match them up with the English.

Past tense	English	Infinitive
escuché	I listened	escuchar
bebí		
hablé		
jugué		
navegué		
comí		

I surfed
I spoke
I drank
I played
I ate
I listened

El tiempo libre
5 Gramática

escribir 1 Susi has spilt Coke all over her verb tables. Can you write them out again for her?

? **Need help?** Go to Unidad 1
? **Want to know more?** Go to page 131

sa●ir	to go out
sal●o	I go out
sale●	you go out
sa●e	he/she/it goes out
sal●mos	we go out
●alís	you (plural) go out
sal●n	they go out

hace●	to do
ha●o	I do
ha●es	you do
hac●	he/she/it does
hace●os	we do
ha●éis	you (plural) do
hace●	they do

escribir 2 Translate these sentences into Spanish.

1 I go out with my friends.
2 I do my homework.
3 She does her homework.
4 He is going out with his father.
5 They are going out in the evening.

su = 'his/her' + singular noun
sus = 'his/her' + plural noun

escribir 3 Write six sentences, each containing one element from each circle. Translate your sentences into English.

? **Need help?** Go to Unidad 2
? **Want to know more?** Go to page 131

Example: Los fines de semana va al cine. At the weekend she goes to the cinema.

Todos los días	voy	
Los lunes	vas	de compras
Los jueves	va	al cine
Una vez por semana	vamos	a la piscina
Los fines de semana	vais	
Dos veces a la semana	van	

escribir 4 Write sentences using the grid.

? **Need help?** Go to Unidad 3
? **Want to know more?** Go to page 131

Example: Hago equitación.   I go riding.

I	you	she	we	you (plural)	they
equitación	esquí	natación	atletismo	ciclismo	patinaje

5 Find the infinitives and the 'I' verb forms in the word snakes, then fill in the grid.

Infinitive	English	'I' form
jugar al fútbol	to play football	juego al fútbol

jugaralfútbolhaceratletismonavegarpori nt
ernetalcinesalirconmisamigosverlatelevis
iónhacermisdeberesescucharmúsicairdecomp
rashacernatación

hagomisdeberesvoyalcinevoydecomprashago
naci nataciónjuegoalfútbolnavegoporinternetsalgo
conmisamigosescuchomúsicahagoatletismoveolatele
visión

6 Choose the correct verb to complete each sentence. Then translate the sentences into English.

? Need help? Go to Unidad 4
? Want to know more? Go to page 132

1 No me gusta **hacer** / **hago** / **hace** mis deberes.
2 Me encanta **voy** / **ir** / **vamos** al cine.
3 Prefiero **hacemos** / **hacer** / **hago** natación.
4 Me gusta mucho **juego** / **jugamos** / **jugar** con mi conejo.
5 No me gusta nada **estudiar** / **estudio** / **estudian** todos los días.
6 Los sábados me gusta **hablo** / **hablas** / **hablar** por teléfono.
7 ¿Te gusta **beber** / **bebes** / **bebo** agua mineral?
8 Me encanta **leo** / **leer** / **leemos** los libros de J. K. Rowling.
9 No me gusta **dormir** / **duermo** / **duermes** en el salón.
10 Me gusta mucho **como** / **comer** / **comes** chicle en clase.

7 Complete the sentences with words from the box.

? Need help? Go to Unidad 5
? Want to know more? Go to page 132

1 Voy a jugar al tenis.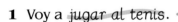
2 ¿ _____ ir a la piscina?
3 Enrique va a _____ .
4 _____ ir al cine.
5 ¿Vais a _____ ?
6 Van a _____ .

jugar al fútbol	Vamos a
hacer ciclismo	jugar al tenis
Vas a	hacer esquí

Palabras

En mi tiempo libre
¿Qué haces en tu tiempo libre?
Voy al cine.
Voy a la piscina.

Voy de compras.
Salgo con mis amigos.

Hago mis deberes.
Monto en bicicleta.
Escucho música.
Veo la televisión.
Navego por internet.
Juego con mi ordenador.
No voy al cine.

In my free time
What do you do in your free time?
I go to the cinema.
I go to the swimming pool.
I go shopping.
I go out with my friends.
I do my homework.
I ride my bike.
I listen to music.
I watch television.
I surf the net.
I play on my computer.
I don't go to the cinema.

¿Con qué frecuencia?
todos los días
los lunes
una vez por semana
dos veces a la semana
los fines de semana
nunca

How often?
every day
on Mondays
once a week
twice a week
at weekends
never

Los deportes
¿Qué deportes haces?

Hago atletismo.
Hago ciclismo.
Hago equitación.
Hago esquí.
Hago natación.
Hago patinaje.
Juego al baloncesto.
Juego al fútbol.
Juego al hockey.
Juego al tenis.
Juego al voleibol.
No hago deporte.

Sports
What sports do you do?
I do athletics.
I do/go cycling.
I do/go riding.
I do/go skiing.
I do/go swimming.
I do/go skating.
I play basketball.
I play football.
I play hockey.
I play tennis.
I play volleyball.
I don't do any sports.

¿A qué hora … ?
¿Qué hora es?
Es la una.
Son las dos.
Es la una y cinco.
Son las dos y diez.
Son las tres y cuarto.
Son las cuatro y veinte.
Son las cinco y veinticinco.
Son las seis y media.
Son las siete menos veinticinco.
Son las ocho menos veinte.
Son las nueve menos cuarto.
Son las diez menos diez.
Son las once menos cinco.
Son las doce.
de la mañana
de la tarde
de la noche

At what time … ?
What time is it?
It's one o'clock.
It's two o'clock.
It's five past one.
It's ten past two.
It's quarter past three.
It's twenty past four.
It's twenty-five past five.
It's half past six.
It's twenty-five to seven.
It's twenty to eight.
It's quarter to nine.
It's ten to ten.
It's five to eleven.
It's midday/midnight.
in the morning
in the afternoon
in the evening

¿A qué hora comes?

¿A qué hora vas al cine?

¿A qué hora escuchas música?
¿A qué hora sales con tus amigos?

¿A qué hora vas de compras?
¿A qué hora navegas por internet?
¿A qué hora ves la televisión?
A las dos de la tarde.

At what time do you eat?
At what time do you go to the cinema?
At what time do you listen to music?
At what time do you go out with your friends?
At what time do you go shopping?
At what time do you surf the net?
At what time do you watch TV?
At two p.m.

¿Qué te gusta hacer?	What do you like doing?
¿Qué te gusta hacer en tu tiempo libre?	What do you like doing in your free time?
¿Qué no te gusta hacer?	What don't you like doing?
Me gusta …	I like …
Me gusta mucho …	I really like …
No me gusta …	I don't like …
No me gusta nada …	I don't like … at all.
Me encanta …	I love …
Odio …	I hate …
Prefiero …	I prefer …
jugar al fútbol	playing football
hacer atletismo	doing athletics
navegar por internet	surfing the internet
ir al cine	going to the cinema
salir con mis amigos	going out with my friends
ver la televisión	watching television
hacer mis deberes	doing my homework
escuchar música	listening to music
ir de compras	going shopping
hacer natación	going swimming
¿Por qué?	Why?
Porque es …	Because it's …
Porque no es …	Because it isn't …
aburrido	boring
barato	cheap
bueno	good
caro	expensive
divertido	amusing
fácil	easy
interesante	interesting
sano	healthy

¿Qué vas a hacer mañana?	What are you going to do tomorrow?
¿Qué vas a hacer?	What are you going to do?
Voy a jugar al tenis.	I'm going to play tennis.
Va a escuchar música.	He/She's going to listen to music.
Vamos a ir de compras.	We're going to go shopping.
Vais a hacer natación.	You're going to go swimming. (pl)
Van a ver la televisión.	They're going to watch television.
mañana	tomorrow
la semana que viene	next week
este fin de semana	this weekend
en las vacaciones	in the holidays

Palabras muy útiles — *Very useful words*
sobre todo — *above all*

Estrategia
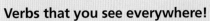

Verbs that you see everywhere!

You can use the verb **tener** in lots of situations:

Tengo una serpiente.
Tengo dos hermanas.
Tengo doce años.

Tener is what we call a **high-frequency** verb. Learning verbs like this will help you to say a lot more in Spanish!

There are some other very useful verbs in Chapter 5. Try to find four different ways of finishing these sentences:

- Voy …
- Hago …
- Juego …
- Es …

En la ciudad

1 ¿Cómo es tu ciudad?

- Saying what your town is like
- Using comparatives

Escucha y escribe la letra correcta. (1–5)

Ejemplo: **1** d

a
*Vivo en una ciudad.
Es **importante** e **industrial**.*

b
*Vivo en un pueblo.
Es **tranquilo** y **bonito**.*

c
*Vivo en una ciudad.
Es **grande** y **moderna**.*

d
*Vivo en un pueblo.
Es **pequeño** e **histórico**.*

e
*Vivo en una ciudad.
Es **turística** y un poco **fea**.*

¿Cómo es tu pueblo? 	¿Cómo es tu ciudad? 
Es histórico, moderno, pequeño, tranquilo, turístico, bonito, feo, industrial, importante, grande	Es histórica, moderna, pequeña, tranquila, turística, bonita, fea, industrial, importante, grande

Con tu compañero/a, pregunta y contesta.

- ¿Vives en una ciudad o en un pueblo?
- ■ Vivo en <u>un pueblo</u>.
- ¿Cómo es tu <u>pueblo</u>?
- ■ Es <u>pequeño</u> e <u>histórico</u>.

When the word **y** (*and*) is followed by a word beginning with **i** or **hi**, it changes to **e** to make it easier to pronounce.

escuchar **3** Escucha y repite. (1–8)

1 Madrid es más importante que Cazorla.
2 Madrid es más grande que Cazorla.
3 Madrid es más moderna que Cazorla.
4 Madrid es menos tranquila que Cazorla.

5 Cazorla es más pequeño que Madrid.
6 Cazorla es más tranquilo que Madrid.
7 Cazorla es más bonito que Madrid.
8 Cazorla es menos importante que Madrid.

Gramática

| **más** | + adjective | + **que** | *more … than* |
| **menos** | + adjective | + **que** | *less … than* |

Madrid es **más turística que** Cazorla.
Cazorla es **menos importante que** Madrid.

*Madrid is **more touristy than** Cazorla.*
*Cazorla is **less important than** Madrid.*

Para saber más   p.112 (ex 2); p.130

leer **4** Haz una lista de las ciudades en España, desde la más grande hasta la más pequeña.

Make a list of Spanish cities from the largest to the smallest.

> Sevilla es más grande que Benidorm.
> Córdoba es menos grande que Sevilla.
> Gijón es más grande que Benidorm.
> Barcelona es más grande que Sevilla.
> Madrid es más grande que Barcelona.
> Gijón es menos grande que Córdoba.

escribir **5** Escribe cuatro frases.

Ejemplo: Me gusta Sevilla porque es histórica, pero prefiero Granada porque es más bonita.

	☺	¿Por qué?	☺ +	¿Por qué?
1	Sevilla		Granada	**más**
2	Madrid		Cazorla	**más**
3	Bilbao		Benidorm	**menos**
4	Barcelona		Córdoba	**menos**

2 ¿Qué hay?

 1 **Escucha y escribe las letras correctas. (1–8)**

Ejemplo: **1** n ✔, h ✔

¿Qué hay en la ciudad?

✔ Hay …

✘ No hay …

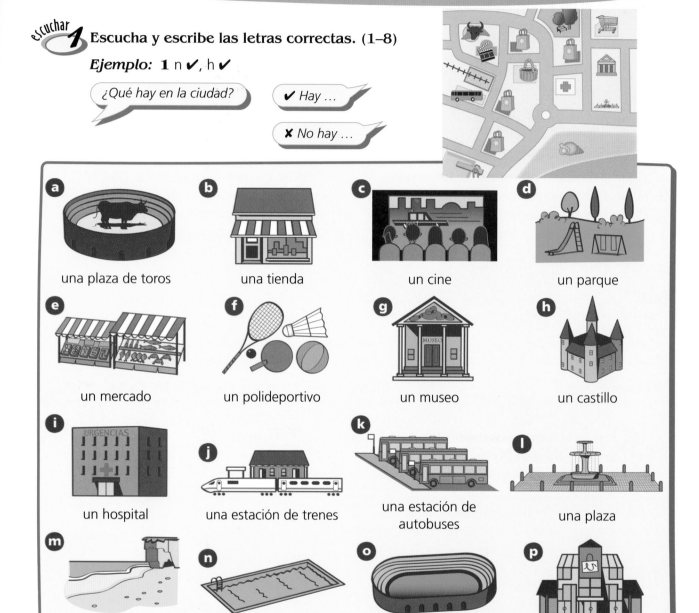

a una plaza de toros

b una tienda

c un cine

d un parque

e un mercado

f un polideportivo

g un museo

h un castillo

i un hospital

j una estación de trenes

k una estación de autobuses

l una plaza

m una playa

n una piscina

o un estadio

p un centro comercial

2 **Juego de memoria.** *Memory game.*

● ¿Qué hay en la ciudad?
■ Hay <u>un museo</u>, pero no hay <u>un estadio</u>.
● Hay un museo y <u>un parque</u>, pero no hay un estadio.

When you are listening, you sometimes have to listen out for more than one item at a time. Try not to write while listening, but note your answer as soon as the recording is paused.

 3 **Describe tu ciudad o tu pueblo.**

Ejemplo: Mi ciudad se llama Canterbury.
En Canterbury hay un centro comercial … pero no hay …

4 Escucha y repite las direcciones. (a–h)

a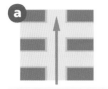

Sigue todo recto.

b

Dobla a la derecha.

c

Dobla a la izquierda.

d

Cruza la plaza.

e

Toma la segunda calle a la derecha.

f

Toma la segunda calle a la izquierda.

g

Está a la derecha.

h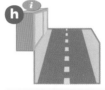

Está a la izquierda.

 **5** Escucha y escribe el lugar y las letras correctas del ejercicio 4. (1–6)

Listen and write down the place and the correct letters from exercise 4.

Ejemplo: 1 un parque d, a

¿Hay un/una …?
¿Dónde está?

¿Hay unos/unas …?
¿Dónde están?

¿Dónde está el/la …?
¿Dónde están los/las …?

Gramática

Like **ser**, **estar** also means 'to be'. **Estar** describes positions and temporary states.

estoy	*I am*
estás	*you are*
está	*he/she/it is*
estamos	*we are*
estáis	*you (plural) are*
están	*they are*

Para saber más p.132

 6 Con tu compañero/a, mira el mapa. Pregunta y contesta.

● ¿Dónde está el polideportivo?
■ Cruza la plaza. Dobla a la derecha. Está a la izquierda.

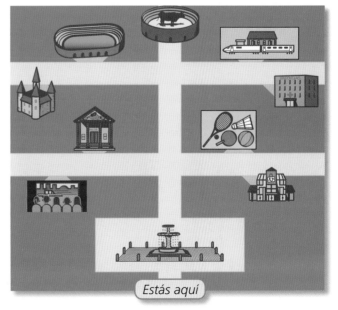

Estás aquí

1 Escucha y lee. Empareja los diálogos con los dibujos correctos. (1–3)

1 ● ¿Quieres ir al cine?
　■ ¿Cuándo?
　● El viernes.
　■ ¿A qué hora?
　● A las cuatro de la tarde.
　■ De acuerdo.

2 ● ¿Quieres ir a la playa?
　■ ¿Cuándo?
　● El lunes.
　■ ¿A qué hora?
　● A las cinco de la tarde.
　■ Vale.

3 ● ¿Quieres ir al polideportivo?
　■ ¿Cuándo?
　● El domingo.
　■ ¿A qué hora?
　● A las diez de la mañana.
　■ Lo siento, no puedo.

2 Juega con tu compañero/a y haz diálogos.

Ejemplo:

　● ¿Quieres ir al polideportivo?
　● ¿Cuándo?
　⋮ El miércoles.
　● ¿A qué hora?
　∷ A las cuatro.

Remember that
　a + el = al
but 　**a + la = a la**

al cine 　　**to the** cinema
a la piscina 　**to the** swimming pool

● lunes

∴ martes

⋰ miércoles

∷ jueves

⁙ viernes

⁙ sábado

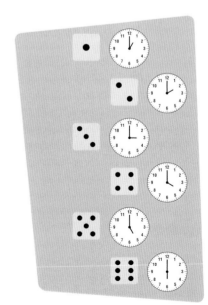

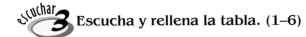

escuchar 3 Escucha y rellena la tabla. (1–6)

	Quiere ir …	Día	Hora	✔/✗
1	al estadio	viernes	3:00	✔

Gramática

	querer	to want
(yo)	qu**ie**ro	I want
(tú)	qu**ie**res	you want
(él/ella)	qu**ie**re	he/she/it wants
(nosotros/nosotras)	queremos	we want
(vosotros/vosotras)	queréis	you (plural) want
(ellos/ellas)	qu**ie**ren	they want

Querer is a stem-changing verb. It can be used with an **infinitive**:

Quiero **ir** a la piscina. — I want to go to the swimming pool.
¿Quieres **ir** al cine? — Do you want to go to the cinema?
Quiere **ir** al polideportivo. — He wants to go to the sports centre.

Para saber más — p.113 (ex 5); p.131, 133

You can improve the sound of your Spanish by reusing things you hear. Listen to the conversations again and try to pick out these expressions:

pues
a ver
oye
ay

hablar 4 Con tu compañero/a, haz tres diálogos.

● ¡Hola! ¿Qué tal? Oye … ¿Quieres ir <u>al parque</u>?
■ ¿Cuándo?
● A ver … el <u>miércoles</u>.
■ ¿A qué hora?
● Pues … a <u>las siete</u> de la <u>noche</u>.
■ Ah sí, bueno, sí. ¿Dónde está <u>el parque</u>?
● Mira, estás aquí. <u>Cruza la plaza y toma la segunda calle a la izquierda</u>.

escribir 5 Escribe uno de los diálogos del ejercicio 4.

Mini-test

I can
● say what my town or village is like
● compare two places
G use **hay** to say what there is/isn't in a town
● give directions to places
● make and respond to invitations
G use **querer** (to want) with an infinitive

4 ¿Qué tiempo hace?

- Talking about the weather
- Using **cuando** (when) to join bits of information

Escucha y escribe la letra correcta. (1–10)

Ejemplo: **1** d

> ¿Qué tiempo hace?

 Hace buen tiempo.

 Hace mal tiempo.

 Hace calor.

> **Hay niebla** means 'there is fog'.
> **Hace sol** means 'it does sun'.
>
> These are examples of phrases that you can't translate word-for-word from Spanish to English. Make a list of phrases like this to help you learn them.

 Hace frío.

 Hace sol.

 Hace viento.

 Hay niebla.

Hay tormenta.

Llueve.

Nieva.

2 Escucha y escribe la ciudad y las letras correctas del ejercicio 1.

Ejemplo: Madrid – b, h

| Bilbao | Madrid | Sevilla |
| Sierra Nevada | Toledo | Zaragoza |

3 Mira el mapa. Con tu compañero/a, pregunta y contesta.

- ¿Qué tiempo hace en Madrid?
- Hace buen tiempo y también hace calor.

4 ¿Qué tiempo hace en cada estación?
What is the weather like in each season?

Ejemplo:
En primavera hace buen tiempo …

en primavera

en verano

en invierno

en otoño

Bilbao
ESPAÑA Zaragoza
Madrid
Toledo
Sevilla Sierra Nevada

104 ciento cuatro

5 Escucha y completa el texto con las palabras del cuadro. (1–12)
Aprende la poesía de memoria.

Cuando hace **(1)** ~~~~, voy a la piscina con Carolina,

y cuando **(2)** ~~ frío, voy siempre al castillo.

Cuando hace mucho **(3)** ~~, voy a la playa con mi amiga Flor,

pero cuando hace **(4)** ~~ tiempo … ¡Ay, qué aburrido!

Cuando hace **(5)** ~~, ¡nunca voy al centro comercial!

Cuando hay **(6)** ~~, escucho música con Manuela.

Cuando a veces **(7)** ~ tormenta, no me gusta ir de compras.

Cuando llueve, **(8)** ~~ mucho: juego en mi dormitorio.

¿Quieres ir al centro? No, lo siento – ¡hace **(9)** ~~!

Cuando nieva en **(10)** ~~, hago esquí con mi hermano.

¿Qué tiempo hace? Dime … ¿**(11)** ~~ tiempo hace?

Hace calor y hace sol – **(12)** ~~ – ¡fenomenal!

hace	hay	niebla	qué	invierno	sol
viento		buen tiempo	mal	pues	
	calor	llueve			

cuando = *when*

6 Haz frases con 'cuando'.

Ejemplo: 1 Hace buen tiempo. Voy a la piscina.
→ Cuando hace buen tiempo, voy a la piscina.

7 Prepara una presentación sobre tu pueblo/ciudad, el tiempo y tus pasatiempos.
Prepare a presentation about your village/town, the weather and your hobbies.

Me llamo …	No hay …
Vivo en …	Cuando …
Es …	Me gusta …
Hay …	No me gusta …

5 Este fin de semana

● Using two tenses together
● Saying what you do in town

escuchar 1 Escucha y escribe las letras correctas. (1–5)

Ejemplo: **1** d, i

Normalmente …
Los fines de semana …

a **voy** al centro comercial

b **juego** al fútbol

c **hago** ciclismo

d **voy** al cine

e **hago** mis deberes

Mañana …
Este fin de semana …
En las vacaciones …

f **voy a ir** de compras

g **voy a jugar** con mi ordenador

h **voy a hacer** natación

i **voy a ir** al estadio

j **voy a hacer** equitación

The sentences in exercise 1 are in two different tenses: the **present** (on the left) and the **near future** (on the right).

Expressions like these can also help you to work out whether a sentence is about the present or the future:

Present
normalmente = *normally*
los fines de semana = *every weekend*
ahora = *now*

Future
en las vacaciones = *in the holidays*
este fin de semana = *this weekend*
mañana = *tomorrow*

hablar 2 Con tu compañero/a, haz frases utilizando el presente y el futuro.

Ejemplo: **a** Ahora hago mis deberes. Mañana voy a hacer equitación.

a
b
c
d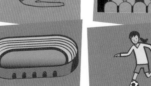
e
f

escuchar 3 Escucha y escribe las respuestas de Mireya. (1–6)

1 ¿Cómo es tu ciudad?
2 ¿Qué hay?
3 ¿Te gusta?
4 ¿Qué haces cuando llueve?
5 ¿Qué haces cuando hace buen tiempo?
6 ¿Qué vas a hacer este fin de semana?

Read Paz's web page. Match each paragraph to one of the headings below.

A this weekend
B what Paz does in town
C what there is in town

gimnasio = *gymnasium, gym*
torneo = *tournament*
después = *afterwards*

http://www.laciudaddepaz.es

¡Hola! Me llamo Paz.
Bienvenidos a mi página web.

1 Vivo en una ciudad grande y moderna. Me gusta mucho porque es muy bonita y bastante tranquila. Hay muchos parques y una piscina. Hay un centro comercial muy grande e importante. ¡En el centro hay muchos restaurantes y tres cines! También hay un polideportivo con un gimnasio muy grande.

2 En la ciudad hago mucho deporte. Normalmente voy a la piscina y hago natación. Hago ciclismo en los parques. Cuando hace mal tiempo, voy al

polideportivo y juego al voleibol o al baloncesto con mis amigos. A veces voy de compras al centro comercial. Me gusta mucho vivir aquí.

3 Este fin de semana, el sábado, voy a ir al polideportivo, pero no voy a jugar al voleibol, voy a jugar al tenis. El domingo voy a ir de compras con mi madre. Vamos a ir al centro comercial. Después voy a ir al restaurante y al cine con mi hermana Luz.

 Lee el texto otra vez. Haz una lista de los verbos en el presente y en el futuro.
Read the text again and make a list of verbs in the present and in the future.

Verbos en el presente	Verbos en el futuro
Me llamo	voy a ir

 Empareja las mitades de las frases.

Ejemplo: **1** d

1 Paz vive
2 Hay muchos
3 Paz hace
4 Va
5 Este fin de semana
6 El domingo, va a ir

a a la piscina.
b de compras, al restaurante y al cine.
c va a ir al polideportivo.
d en una ciudad grande y moderna.
e parques y una piscina.
f mucho deporte.

 Habla de tu ciudad y las actividades posibles a tu compañero/a. Contesta a las preguntas del ejercicio 3.

 Escribe una página web sobre tu ciudad. Utiliza la página web de Paz como modelo.

Resumen

Unidad 1

I can

- ■ *say where I live* — Vivo en un pueblo.
- ■ *ask someone what their town or village is like* — ¿Cómo es tu ciudad?
- G *use adjectives to describe places* — Es **pequeño** e **histórico**.
- G *use comparatives to compare places* — Cazorla es **más pequeño que** Madrid. Madrid es **menos tranquila que** Cazorla.

Unidad 2

I can

- ■ *ask about places in town* — ¿Qué hay en la ciudad? ¿Hay un centro comercial?
- G *use **hay** to say what there is* — Hay un polideportivo.
- G *use **no hay** to say what there isn't* — No hay una piscina.
- ■ *ask for directions* — ¿Dónde está el polideportivo?
- ■ *give directions* — Sigue todo recto. Está a la derecha.
- G *understand all forms of the irregular verb **estar*** — estoy, estás, está, estamos, estáis, están

Unidad 3

I can

- ■ *invite someone to do something* — ¿Quieres ir al polideportivo?
- ■ *ask when (day and time)* — ¿Cuándo? ¿A qué hora?
- ■ *give a day and time* — El domingo. A las dos.
- ■ *accept or turn down an invitation* — De acuerdo. Lo siento, no puedo.
- G *understand all forms of **querer*** — quiero, quieres, quiere, queremos, queréis, quieren

Unidad 4

I can

- ■ *ask what the weather is like* — ¿Qué tiempo hace?
- ■ *talk about the weather* — Hace buen tiempo. Hace calor.
- ■ *name the seasons* — primavera, verano, otoño, invierno
- ■ *say what I do in different weathers* — Cuando hace sol, voy a la playa.

Unidad 5

I can

- ■ *understand expressions referring to the present or what normally happens* — normalmente, los fines de semana, …
- ■ *understand expressions referring to the future* — en las vacaciones, este fin de semana, …
- G *use the present tense to talk about things I normally do* — Normalmente voy de compras y juego al fútbol.
- G *use the near future tense to talk about what I'm going to do* — Este fin de semana voy a ir al cine y voy a hacer natación.

 1 Copia y rellena la tabla. (1–5)

	Town/Village?	Description	Opinion
1	village	quiet, pretty	–

 2 Lee el texto y contesta a las preguntas en inglés.

Vivo en una ciudad bastante grande. Es industrial y muy importante.
Es menos grande que Madrid, pero es más moderna.
Hay un estadio y un centro comercial y un mercado importante.
También, hay muchas tiendas y una plaza de toros.
Hay un hospital, un polideportivo, un cine y una estación de trenes.
Hay muchos museos bonitos también.
Me gusta mucho vivir aquí.
Cuando hace buen tiempo, juego al fútbol en el parque con mis amigos.
Este fin de semana voy a hacer natación en la piscina.   Lola

1 What is the city like? (3)
2 What is the city like compared to Madrid? (2)
3 Name five places you can find in the city. (5)
4 What does Lola do when the weather is fine? (2)
5 What does she say about this weekend? (3)

 3 Con tu compañero/a, haz cinco diálogos.

● ¿Dónde está el polideportivo?
■ Cruza la plaza y toma la segunda calle a la derecha.

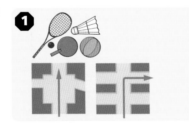

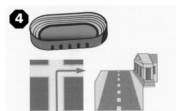

 4 Escribe las frases.

Ejemplo: **1** Cuando hace buen tiempo, voy al parque.

¡Extra!

● Using a text as a model for writing
● Speaking at length

escuchar **1** Escucha y lee.

hablar **2** Con tu compañero/a, lee la historia de Tomás y Angélica.

1

T – ¡Hola Angélica!
A – ¡Hola Tomás! ¿Cómo estás?
T – ¡Bien! Angélica, ¿quieres dar un paseo conmigo en el centro?

A – De acuerdo. ¿Cuándo?
T – Mañana.
A – ¿A qué hora?
T – A las once.
A – Vale.

2

T – Me gusta mucho Segovia. Es muy histórica y muy bonita.
A – Sí, me gusta mucho también. En Segovia hay castillos, una catedral e iglesias, pero también hay restaurantes y bares muy buenos.

3

T – También me encantan los sitios modernos.

4

5

6

T – Angélica, ¿quieres ir al restaurante mañana?

A – Lo siento, no puedo. Voy a ir de compras con una amiga.

A – ¡Pero el viernes sí!

dar un paseo = *to go for a walk*

3 Escribe una versión alternativa de la historia de Tomás y Angélica.

- Angélica doesn't feel well …
- She hates Segovia …
- She is going to play tennis with a friend tomorrow.

4 Escucha a Esperanza y escribe lo que dice.

Think about all the topics you have covered in *¡Mira! 1*. Esperanza will talk about a lot of them. As you listen, you could make notes in a spider diagram to get as much detail down as possible.

asignaturas:
😊 Le gusta(n)/ 😞 No le gusta(n)

edad/cumpleaños

familia

vacaciones

animales

Esperanza

tiempo libre

descripción de E.

casa

ciudad

5 Prepara una presentación de tus datos personales.

Alba
Sevilla
12 años,
4/1

- give your name (p. 6)
- say how old you are (p. 8)
- say when your birthday is (p. 10)
- say what you study (p. 26)
- say which subjects you like and dislike and why (p. 32)
- say how many people there are in your family (p. 45)
- say who they are (p. 45)
- say what they are like (p. 48)
- say whether you have any pets (p. 46)
- describe them (p. 47)
- say where you live (p. 62)
- describe your town/village (pp. 98, 100)
- describe your house/flat (pp. 64, 65)
- describe your room (p. 68)
- say what you do in your free time (p. 80)
- say what you are going to do in the holidays (pp. 88, 106)

ciento once **111**

En la ciudad
6 Gramática

leer 1 Which adjective in each group is the odd one out? Give a reason for your answer (there may be more than one answer!). Then write a sentence in Spanish using the adjective.

? *Want to know more?*   Go to page 129

Example: **1** bonita – It's feminine. – Vivo en una ciudad bonita.

1 a moderno	**b** tranquilo	**c** bonita
2 a histórico	**b** histórica	**c** históricas
3 a moderno	**b** grande	**c** moderna
4 a industrial	**b** importante	**c** grandes
5 a pequeña	**b** bonito	**c** tranquila
6 a feo	**b** turístico	**c** fea

leer 2 Match up the phrases which mean the same.

? *Want to know more?*   Go to page 130

Example: **a** más bonito = **i** menos feo

a más bonito	**f** más pequeño
b más histórico	**g** menos moderno
c menos grande	**h** más interesante
d más feo	**i** menos feo
e menos aburrido	**j** menos bonito

escribir 3 Translate these sentences into Spanish.

? *Want to know more?*   Go to page 129

Example: **1** En el centro comercial hay unas tiendas.

1 In the shopping centre there are some shops.
2 In the square there is a cinema.
3 There is a train station and a bus station.
4 There isn't a castle, but there is a museum.
5 There are some parks.
6 There is a stadium and a sports centre.
7 There is a swimming pool in the park.
8 There aren't any cinemas.

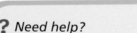

Do you need:
el or **la**?
un or **una**?
unos or **unas**?

4 **Al *or* a la?** *Choose the correct one to complete each sentence.*

Example: **1** Voy al cine.

? *Need help?*   Go to Unidad 3
? *Want to know more?*   Go to page 133

1 Voy **al** / **a la** cine.
2 Voy **al** / **a la** piscina.
3 Voy **al** / **a la** estación de trenes.
4 Voy **al** / **a la** plaza de toros.
5 Voy **al** / **a la** parque.

6 Voy **al** / **a la** museo.
7 Voy **al** / **a la** polideportivo.
8 Voy **al** / **a la** mercado.
9 Voy **al** / **a la** estadio.
10 Voy **al** / **a la** centro comercial.

 5 *Write out the sentences correctly, then translate them into English.*

Example: Quieren ir al mercado.
They want to go to the market.

? *Need help?*   Go to Unidad 3
? *Want to know more?*   Go to page 131, 133

quieren ir

queremos ir

queréis ir

quiere ir

quiero ir

quieres ir

al

a la

al

al

al

al

 6 *Choose the right answer to each question.*

Example: **1** d

? *Need help?*   Go to Unidad 5
? *Want to know more?*   Go to page 132

1 ¿Qué vas a hacer mañana?
  a Juega al fútbol.
  b Juego con mi ordenador.
  c Va a ir al estadio.
  d Voy a ir de compras.

2 ¿Qué haces por la tarde?
  a Hago ciclismo.
  b Van al cine.
  c Voy a hacer equitación.
  d Va a hacer natación.

3 ¿Qué vas a hacer en las vacaciones?
  a Vas en la ciudad.
  b Hago mis deberes.
  c Voy a ir a la playa.
  d Hago natación.

4 ¿Qué vas a hacer este fin de semana?
  a Hace buen tiempo.
  b Voy a ir al estadio.
  c Juego con mi ordenador.
  d Hago mis deberes.

5 ¿Qué haces por la mañana?
  a Voy a ir al cine.
  b Normalmente, me levanto y voy al instituto.
  c Hace sus deberes.
  d Voy a jugar al fútbol.

The tense of the answer should match the tense of the question.

Palabras

Mi ciudad	**My town**
Vivo en …	I live in …
un pueblo	a village
una ciudad	a town/city
¿Cómo es tu pueblo?	What's your village like?
Es …	It's …
bonito	pretty
feo	ugly
histórico	historic
moderno	modern
pequeño	small
tranquilo	peaceful
turístico	appealing to tourists
industrial	industrial
importante	important
grande	big
¿Cómo es tu ciudad?	What's your town like?
Es …	It's …
bonita	pretty
fea	ugly
histórica	historic
moderna	modern
pequeña	small
tranquila	peaceful
turística	appealing to tourists
industrial	industrial
importante	important
grande	big
Es más importante que …	It's more important than …
Es menos industrial que …	It's less industrial than …
Prefiero … porque es más/menos …	I prefer … because it's more/less …

En la ciudad	**In town**
¿Qué hay en la ciudad?	What is there in town?
¿Hay … ?	Is/Are there … ?
Hay …	There is/are …
No hay …	There isn't/aren't …

un castillo	a castle
un centro comercial	a shopping centre
un cine	a cinema
un estadio	a stadium
un hospital	a hospital
un mercado	a market
un museo	a museum
un parque	a park
un polideportivo	a sports centre
una estación de autobuses	a bus station
una estación de trenes	a train station
una piscina	a swimming pool
una playa	a beach
una plaza	a square
una plaza de toros	a bullring
una tienda	a shop
unos/muchos museos	some/many museums
unas/muchas tiendas	some/many shops

Direcciones	**Directions**
¿Dónde está el/la … ?	Where is the … ?
¿Dónde están los/las … ?	Where are the … ?
Estás aquí.	You are here.
Sigue todo recto.	Go straight on.
Dobla a la derecha.	Turn right.
Dobla a la izquierda.	Turn left.
Cruza la plaza.	Cross the square.
Toma la segunda calle a la derecha.	Take the second road on the right.
Toma la segunda calle a la izquierda.	Take the second road on the left.
Está a la derecha.	It's on the right.
Está a la izquierda.	It's on the left.

Invitaciones	**Invitations**
¿Quieres ir … ?	Do you want to go … ?
Quiero ir …	I want to go …
al castillo	to the castle
al centro comercial	to the shopping centre
al cine	to the cinema
al estadio	to the stadium
al mercado	to the market
al museo	to the museum
al parque	to the park

al polideportivo	*to the sports centre*
a la piscina	*to the swimming pool*
a la playa	*to the beach*
a la plaza de toros	*to the bullring*

¿Cuándo?	*When?*
El lunes.	*On Monday.*
El martes.	*On Tuesday.*
El miércoles.	*On Wednesday.*
El jueves.	*On Thursday.*
El viernes.	*On Friday.*
El sábado.	*On Saturday.*
El domingo.	*On Sunday.*

¿A qué hora?	*At what time?*
A las diez de la mañana.	*At ten in the morning.*

De acuerdo.	*OK.*
Está bien.	*Fine.*
Bueno.	*Good.*
Vale.	*OK.*
Lo siento, no puedo.	*I'm sorry, I can't.*

El tiempo	*Weather*
¿Qué tiempo hace (en Madrid)?	*What's the weather like (in Madrid)?*
Hace buen tiempo.	*It's nice.*
Hace mal tiempo.	*It's bad.*
Hace calor.	*It's hot.*
Hace frío.	*It's cold.*
Hace sol.	*It's sunny.*
Hace viento.	*It's windy.*
Hay niebla.	*It's foggy.*
Hay tormenta.	*It's stormy.*
Llueve.	*It's raining.*
Nieva.	*It's snowing.*

Cuando llueve, voy al cine.	*When it rains, I go to the cinema.*
Cuando hace sol, voy a la playa.	*When it's sunny, I go to the beach.*

Las estaciones	*The seasons*
en primavera	*in spring*
en verano	*in summer*
en otoño	*in autumn*
en invierno	*in winter*

¿Cuándo?	*When?*
normalmente	*normally*
ahora	*now*
los fines de semana	*every weekend*
mañana	*tomorrow*
este fin de semana	*this weekend*
en las vacaciones	*in the holidays*

Palabras muy útiles	*Very useful words*
aquí	*here*
cuando	*when*
está	*it is*
pero	*but*
a ver	*let's see*
ay	*oh*
oye	*listen*
pues	*well*

Estrategia

Unfamiliar words

How can you find out the meaning of a word you don't know?

1 Look at the *Palabras* at the end of each chapter;
2 Look in the *Vocabulario* at the back of the book;
3 Use a dictionary.

You can also work out meanings

4 by using picture clues;
5 by looking at the context;
6 by thinking of a cognate or near-cognate;
7 by working out what type of word it is: adjective, verb, noun …

● Find out the meaning of the words in bold:

En mi pueblo hay un **mercado**.

No hay un **polideportivo**.

Hay muchas tiendas en la **plaza**.
Cuando hace calor, voy a la **playa**.

● Which of the seven strategies above did you use?

escribir 1 *Fill in the gaps.*

Example: 1 ¡Hola!

1 ¡H_l_!
2 B_en_s dí_s.
3 _ue_as t_rd_s.
4 Bu_na_ n_c_es.
5 Adi_s.
6 ¡H_st_ l__go!

leer 2 *Match up the birthdays to the cards.*

Example: 1 el doce de abril

el veintiocho de septiembre
el diecisiete de junio
el ocho de diciembre
el cuatro de enero
el doce de abril

escribir 3 *What's in the classroom? Complete the words to write the labels.*

Example: a el proyector

a el pro____
b el al____
c la piz____
d el equ____
e el ord____
f las me____
g las sil____
h los li____

escribir 4 *Put these sentences into the right order.*

Example: 1 Me llamo Luis. Vivo en Madrid.

1 llamo Me Luis. Madrid en Vivo.
2 Me llamo Sonia. en Vivo Sevilla.
3 llamo Alberto Me. Málaga Vivo en.
4 Ignacio Me llamo. Buenos Vivo en Aires.
5 llamo Me Rosario. Barcelona Vivo en.

1 Copy the email and fill in the gaps with words from the box.

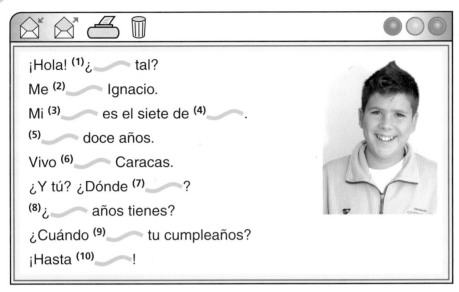

¡Hola! **(1)** ¿___ tal?

Me **(2)** ___ Ignacio.

Mi **(3)** ___ es el siete de **(4)** ___.

(5) ___ doce años.

Vivo **(6)** ___ Caracas.

¿Y tú? ¿Dónde **(7)** ___?

(8) ¿___ años tienes?

¿Cuándo **(9)** ___ tu cumpleaños?

¡Hasta **(10)** ___!

Tengo	es	Cuántos	en	cumpleaños
vives	llamo	mayo	luego	Qué

2 What does each person say?

Example: Fabián: Tengo un sacapuntas, un diccionario, un libro y una agenda, pero no tengo una goma. Necesito una goma.

	✏️	✏️	▭	📓	✂️	diccionario	📗	📖
Fabián			✗		✔	✔	✔	✔
Mar	✔	✗	✔	✔				✔
Paz	✔	✔		✔	✗		✔	
Julio			✔		✔	✔	✔	✗

✔ = Tengo   ✗ = No tengo … Necesito …

3 Write out and complete these sums.

más	+
menos	−
son	=

Example: **a** Uno más trece son catorce.

a 1 + 13 =

b 10 + 11 =

c 12 – 3 =

d 15 – 9 =

e 31 – 3 =

f 22 + 5 =

g 4 + 5 =

h 6 + 17 =

i 3 + 10 =

j 20 – 6 =

leer 1 Match up the symbols with the correct subjects.

Example: **1** f

a dibujo
b español
c inglés
d francés
e teatro
f historia
g música
h tecnología
i informática
j geografía
k educación física
l religión
m ciencias
n matemáticas

leer 2 Who is speaking? What day is it?

Example: **1** Pedro, Tuesday

> Los lunes estudio inglés.
> Los martes estudio historia.
> Los miércoles estudio informática.
> Los jueves estudio teatro.
> Los viernes estudio español.
>
> **Pedro**

> Los lunes estudio matemáticas.
> Los martes estudio dibujo.
> Los miércoles estudio geografía.
> Los jueves estudio tecnología.
> Los viernes estudio educación física.
>
> **Fátima**

escribir 3 Match up the halves of the words. Then write a sentence using each word.

Example: hamburguesa – Me gustan las hamburguesas.

hambur   **tano**   **pi**   **Cola**

man   **zza**   **limo**   **zana**

boca   **nada**      **Coca-**   **guesa**

plá   **tas**   **patatas fri**   **dillo**

1 Write out the verbs correctly and match them up with the English.

Example: **1** escribo = I write

1 bricoes	**6** eol	I study	I drink
2 ioduste	**7** ebob	I live	I speak
3 olabh	**8** nitop	I listen	I write
4 ueccosh	**9** tonca	I paint	I sing
5 mooc	**10** ovvi	I read	I eat

2 Which sentence in each group is the odd one out? Give a reason.

Example: **1** b is the only negative sentence.

1   **a** Belén: Me gusta mucho el español porque es fácil.

     b Marcelo: No me gusta la historia porque es muy aburrida.

     c Ramón: Me gusta mucho el dibujo porque la profesora es buena.

2   **a** Aurora: Me gustan mucho las ciencias porque son interesantes.

     b Mateo: No me gustan las matemáticas porque son difíciles.

     c Claudia: Me gusta la historia porque el profesor es divertido.

3   **a** Ana: No me gusta el teatro porque es aburrido.

     b Javier: Me gusta la religión porque es interesante.

     c Isabel: No me gusta la informática porque es muy difícil.

4   **a** Rafael: Me gustan las ciencias porque son interesantes.

     b Antonio: No me gustan las matemáticas.

     c Pilar: No me gusta la historia.

3 Copy out and complete the email using the words from the box.

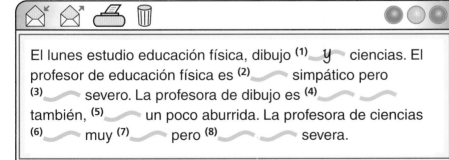

El lunes estudio educación física, dibujo **(1)** _y_ ciencias. El profesor de educación física es **(2)** ‿‿‿ simpático pero **(3)** ‿‿‿ severo. La profesora de dibujo es **(4)** ‿‿‿ también, **(5)** ‿‿‿ un poco aburrida. La profesora de ciencias **(6)** ‿‿‿ muy **(7)** ‿‿‿ pero **(8)** ‿‿‿ severa.

muy	es	bastante	divertida	un poco
	pero	y	muy simpática	

Te toca a ti A

leer 1 *True or false?*

Example: 1 F

1 Pablo tiene tres caballos.
2 Conchi tiene dos pájaros.
3 Juan tiene un perro y dos gatos.
4 Belén tiene dos serpientes y una tortuga.
5 Carlos tiene tres cobayas y un hámster.

escribir 2 *Write out the sentence correctly.*

Example: Soy alto y …

Soyaltoydelgadoymuyguapoymuyinteligenteperotambiénsoyantipáticoymuyperezoso.

escribir 3 *Copy out the texts and fill in the gaps with words from the boxes.*
(There is one word too many each time.)

hermano	Polita	hermana
	gatos	años

Me llamo **(1)** _____.
Tengo dos **(2)** _____. Se llaman Piki y Riki.
Tengo una **(3)** _____. Se llama Pilar. Tiene quince **(4)** _____.

hámsters	se llaman	diez
hermano	Ricardo	perro

Me llamo **(5)** _____.
Tengo un pájaro y dos **(6)** _____. Mi pájaro se llama Pablo y mis
hámsters **(7)** _____ Ping y Pong.
Tengo un **(8)** _____. Se llama Alejandro. Tiene **(9)** _____ años.

hija	hermana	caballo

Me llamo Sara.
Tengo un **(10)** _____. Se llama Nono.
No tengo hermanos. Soy **(11)** _____ única.

1 Copy and complete.

Mi familia

Diego – 3
Margarita – 1
Enrique – 6
Adrián – 4
Rafael – 1
Conchita
Claudia – 4
Pilar – 2
Penélope – 3
José – 5
Bartolomé – 2

¡Hola! Me ⁽¹⁾___ Conchita. Tengo una familia enorme. Tengo seis ⁽²⁾___ y ⁽³⁾___ hermanas.

Mis ⁽⁴⁾___ se llaman Diego, ⁽⁵⁾___, ⁽⁶⁾___, ⁽⁷⁾___, ⁽⁸⁾___ y ⁽⁹⁾___.

Diego tiene ⁽¹⁰⁾___ años. Bartolomé ⁽¹¹⁾___ dos años. ⁽¹²⁾___ tiene cinco años.

Adrián tiene cuatro ⁽¹³⁾___. Rafael ⁽¹⁴⁾___ un año. Enrique ⁽¹⁵⁾___.

⁽¹⁶⁾___ hermanas se llaman Claudia, ⁽¹⁷⁾___, ⁽¹⁸⁾___ y ⁽¹⁹⁾___.

Claudia tiene ⁽²⁰⁾___ años. ⁽²¹⁾___ tiene dos años. Penélope ⁽²²⁾___ tres ⁽²³⁾___.

Margarita ⁽²⁴⁾___.

 2 Write the opposites of these adjectives.

Example: **1** alto – bajo

1 alto **3** pequeño **5** bonito **7** divertido **9** liso
2 delgado **4** guapo **6** simpático **8** largo **10** blanco

 3 Read Raúl's email and answer the questions in Spanish.

Example: **1** Raúl tiene doce años.

1 ¿Cuántos años tiene Raúl?
2 ¿Dónde vive?
3 ¿De qué color son sus ojos?
4 ¿Cómo es su pelo?
5 ¿Cómo es Raúl?
6 ¿Cómo se llama su hermana?
7 ¿Cómo es su hermana?
8 ¿Tiene animales?

¡Hola! ¿Qué tal?

Me llamo Raúl. Tengo doce años. Vivo en Barcelona. Tengo el pelo rubio y los ojos azules.

Soy bastante inteligente y hablo mucho. Soy bastante impaciente. ¿Eres interesante e inteligente? ¿Te gusta hablar por internet? Tengo una hermana. Se llama Mónica. Es muy simpática y también muy guapa.

Vivo con mi padre. Es muy divertido.
Me gustan los animales. Tengo tres gatos y dos peces.
¿Y tú? ¿Cómo eres? ¿Tienes hermanos?

Raúl

Te toca a ti A

escribir

1 *Unjumble these countries.*

Example: 1 Portugal

1 oPgtlrua
2 daIrnal
3 tIenrgrlaa
4 osEciac
5 aEpñas
6 mAaliena
7 Ialita
8 sGlae
9 aiFracn
10 cGarie

leer

2 *Match up the houses to the descriptions.*
(There is one description too many.)

Example: 1 d

a Vivo en una casa antigua en el campo.
b Vivo en una casa grande en la montaña.
c Vivo en un piso moderno en la costa.
d Vivo en un piso viejo en la ciudad.
e Vivo en una casa pequeña en un pueblo.
f Vivo en un piso nuevo en la ciudad.
g Vivo en un piso viejo en la costa.
h Vivo en un piso cómodo en la montaña.
i Vivo en una casa bonita en la costa.

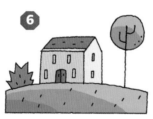

leer

3 *Choose the right preposition.*

1 El perro está **debajo** / **encima** de la cama.
2 El gato está **al lado** / **delante** de la silla.
3 El pez está **debajo** / **encima** de la mesa.
4 Las cobayas están **detrás** / **delante** del armario.
5 La tortuga está **al lado** / **delante** de la puerta.
6 La serpiente está **entre** / **detrás de** la televisión y la lámpara.

1 Put these sentences into the correct order.

Example: **1** Vivo en Francia.

1 Francia. en Vivo

2 ¿dónde tú, vives Y?

3 su en con Vive Madrid hermano.

4 un Vivimos en pequeño. piso

5 ¿una Vivís casa en?

6 grande Viven casa en en una costa. la

 2 Who is talking? Salma or Salim, or both?

Example: **1** Salma

Salma

En mi dormitorio tengo una mesa con mi ordenador. Juego con el ordenador y navego un poco por internet. Tengo un equipo de música también. Escucho música y hablo mucho por teléfono o mando mensajes a mis amigos.

Hay una alfombra verde en el suelo. Hay un armario y una estantería. ¿Qué más? Una cama pequeña. Hay muchos pósters en las paredes. No me gusta mi dormitorio porque es muy pequeño.

A ver, tengo una cama muy grande en mi dormitorio porque duermo mucho.
Hay una lámpara encima de la estantería. Hay una alfombra amarilla muy bonita en el suelo. También hay una mesa y una silla. Tengo un armario muy grande.

En mi dormitorio leo libros y estudio. No tengo ordenador y no hay pósters en las paredes. Por la tarde veo la televisión en mi dormitorio.
Mi dormitorio me gusta bastante porque es grande y cómodo.

Salim

 3 Copy out the text and fill in the gaps with words from the box. (There is one word too many.)

Vivo	me ducho
escucho	es	Como
juego	come	hablo
Me llamo	me levanto
Me gusta	me lavo	veo

Mi día ideal

(1)_____ Antonio. (2)_____ en Almagro, en una casa muy grande. Por la mañana no (3)_____ de la cama. (4)_____ una hamburguesa y (5)_____ la televisión en mi dormitorio. (6)_____ mucho mi dormitorio porque (7)_____ muy cómodo. Luego, (8)_____ en el baño pero no (9)_____ los dientes. Por la tarde (10)_____ por teléfono y (11)_____ con mi ordenador en la terraza. Después, como una pizza con mis padres en el comedor pero mi perro (12)_____ un bocadillo de atún en el salón.

1 *Fill in the gaps in these time expressions.*

1 T•d•s l•s d•as …
2 L•s lu•e• …
3 U•a v•z p•r s•m•n• …
4 L•s f•n•s d• s•m•n• …
5 A v•c•s …
6 L•s •u•v•s …

2 *Complete the crossword.*

Horizontalmente

1 😊 😊 😊
2 😊 +
3 😊 😊
4 😊

Verticalmente

1 ☹
2 ☹ ☹

3 *Write eight sentences saying whether or not you like these activities.*

Example: **1** Me gusta escuchar música.

ir a la piscina
escuchar música
ir de compras
hacer mis deberes
ir al cine
navegar por internet
ver la televisión
salir con mis amigos

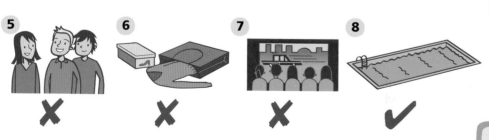

✔ = Me gusta …
✘ = No me gusta …

leer 1 *Who is speaking? Passive Pablo or Active Alana? Write P or A.*

Example: **1** A

1 Hago atletismo y hago ciclismo dos veces por semana.

2 Los miércoles hago natación y juego al baloncesto.

3 Me gusta mucho dormir.

4 Los sábados salgo con mis amigos.

5 Monto en bicicleta tres veces por semana.

6 Me gusta escuchar música y ver la televisión.

7 Navego por internet y juego con mi ordenador todos los días.

8 Los fines de semana juego al tenis y juego al hockey.

 leer 2 *Answer the questions on Julio's sports timetable in Spanish.*

Example: **1** Va a hacer natación.

	Mañana			Tarde		
	8:30	*10:00*	*11:05*	*12:15*	*2:30*	*3:15*
lunes	natación	hockey	voleibol	fútbol	baloncesto	patinaje
martes	baloncesto	ciclismo	natación	hockey	tenis	voleibol
miércoles	natación	fútbol	tenis	ciclismo	baloncesto	patinaje
jueves	tenis	patinaje	hockey	natación	fútbol	voleibol
viernes	natación	voleibol	fútbol	hockey	ciclismo	tenis

1 ¿Qué va a hacer el lunes a las ocho y media?

2 ¿Qué va a hacer el martes a las diez?

3 ¿Qué va a hacer el miércoles a las doce y cuarto?

4 ¿Qué va a hacer el jueves a las dos y media de la tarde?

5 ¿Qué va a hacer el viernes a las tres y cuarto de la tarde?

6 ¿A qué hora va a jugar al fútbol el lunes?

7 ¿A qué hora va a jugar al hockey el martes?

8 ¿A qué hora va a hacer natación el viernes?

 escribir 3 *Put these sentences into a logical order.*

Example: Mañana voy a dormir hasta las once. …

Luego, a las diez de la noche, vamos a comer una pizza o hamburguesas.

Mañana voy a dormir hasta las once.

A las ocho voy a salir con mis amigos otra vez.

A las cinco voy a ver la televisión: hay un partido de fútbol muy importante en la televisión.

Vamos a ir al cine.

A las doce voy a hacer natación con mi hermano.

A las tres de la tarde voy a ir de compras con mis amigos.

Te toca a ti A

leer **1** ***Match up the symbols with the descriptions.***

Example: **1** g

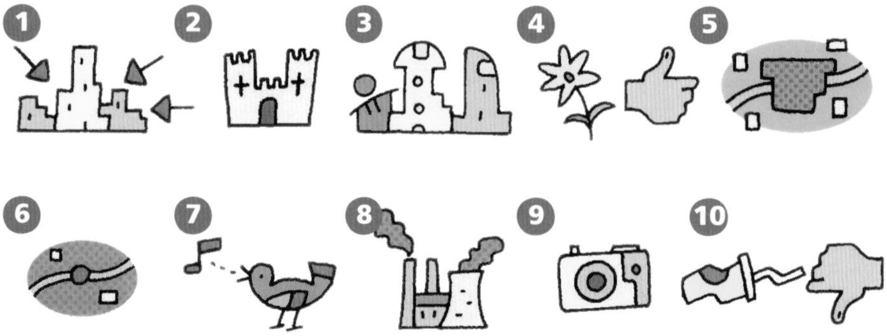

a Es histórico.
b Es moderno.
c Es industrial.
d Es pequeño.
e Es grande.
f Es tranquilo.
g Es importante.
h Es turístico.
i Es bonito.
j Es feo.

escribir **2** ***Use the code to work out the weather expressions.***

Example: **1** Nieva

1 ➔♊□★❖
2 ♦❖■□ ⊙❖⅄ #♊□⊙❍♉
3 ♦❖■□ ★♊□➔#♉
4 ♦❖♈ ➔♊□🡇⅄❖
5 ♦❖■□ 🡇⭘□➔ #♊□⊙❍♉
6 ♦❖♈ #♉🡅⊙□➔#❖
7 🡄⭘□★□
8 ♦❖■□ ■❖⅄♉🡅
9 ♦❖■□ ✱🡅✪♉
10 ♦❖■□ %♉⅄

❖	**a**	⊙	**m**
🡇	**b**	➔	**n**
■	**c**	♉	**o**
□	**e**	❍	**p**
✱	**f**	🡅	**r**
♦	**h**	%	**s**
♊	**i**	#	**t**
✪	**í**	⭘	**u**
⅄	**l**	★	**v**
🡄	**ll**	♈	**y**

escribir **3** ***Look at the diary and write short dialogues.***

Example:
– ¿Quieres ir al cine?
– ¿Cuándo?
– El lunes.
– ¿A qué hora?
– A las tres de la tarde.

a + el = al

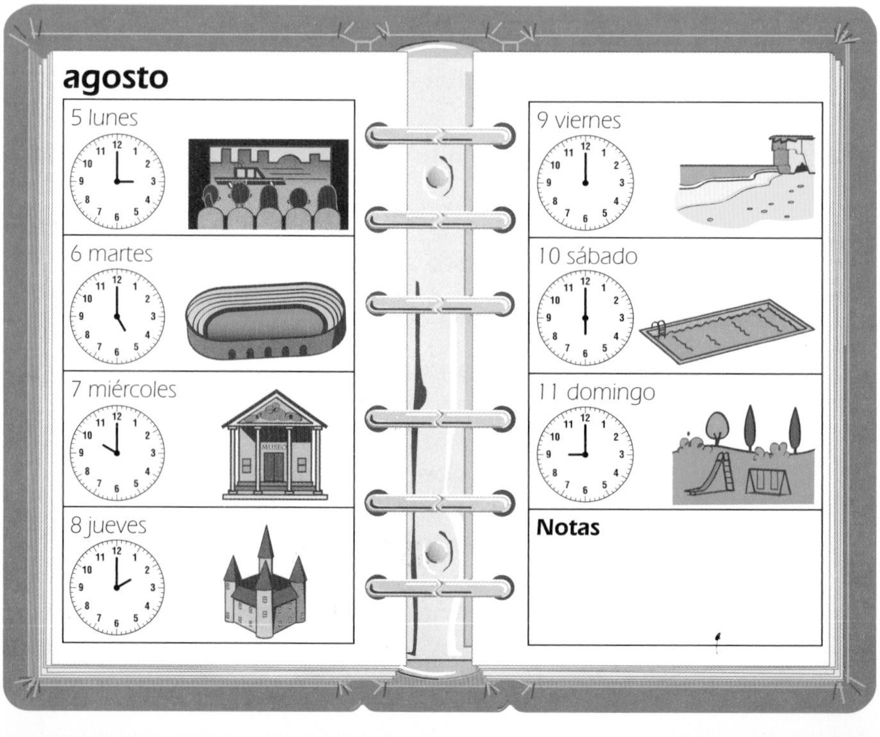

1 Read what Antonio says and correct the mistakes in the sentences.

Example: **1** El pueblo ideal de Antonio es pequeño.

Mi pueblo ideal es pequeño y antiguo. Es más tranquilo que la ciudad y eso me gusta mucho. No es turístico. Hay un parque y una piscina pequeña. Hay una tienda y un restaurante.

Está en la costa. Hay una playa muy bonita pero no hay un polideportivo. Salgo mucho y hago mucho deporte. Voy a la piscina y hago natación. Hago ciclismo en el parque. Juego al voleibol con mis amigos.

No voy de compras porque no me gusta ir de compras, entonces no lo hago. Siempre hace buen tiempo en mi pueblo ideal. A mí me gusta mucho.

1 El pueblo ideal de Antonio es grande.
2 El pueblo ideal de Antonio no es antiguo.
3 Es menos tranquilo que la ciudad.
4 El pueblo ideal de Antonio es turístico.
5 Hay parques.
6 Hay muchas tiendas.
7 Está en la montaña.
8 Hay un polideportivo.
9 Antonio no hace mucho deporte.
10 Le gusta ir de compras.

 2 Write out the questions and answers.

Example:
1 ¿Dónde está el museo?
Sigue todo recto y cruza la plaza.

1

2

3

4

 3 Write sentences.

Example: En enero, cuando hace frío, hago patinaje.

Gramática

Glossary of grammar terms

adjective
a word describing a noun (*divertido* – entertaining, *interesante* – interesting)

connective
a joining word (*y* – and, *pero* – but)

definite article
the word for 'the' *(el/la/los/las)*

infinitive
the dictionary form of a verb, ending in **-ar**, **-er** or **-ir** in Spanish (*escuchar* – to listen, *leer* – to read, *vivir* – to live, *ser* – to be)

indefinite article
the word for 'a' *(un/una)* or 'some' *(unos/unas)*

gender
whether a word is masculine or feminine (*un bocadillo* – m, *una pizza* – f)

irregular verb
a verb that doesn't follow the regular patterns of any of the three verb groups (**-ar**, **-er** or **-ir**), e.g. *ser* – to be

noun
a word naming a person or thing (*bolígrafo* – pen, *mochila* – schoolbag)

plural
the form of a word used to refer to more than one of something (*libros* – books, *gomas* – rubbers)

preposition
a word or phrase used to show where something is in relation to something else (*debajo de* – under, *entre* – between) or to show possession (*de* – of, 's)

pronoun
a word that stands for a noun; a personal pronoun shows who does an action (*yo* – I, *tú* – you, *él* – he, *ella* – she)

possessive adjective
an adjective showing who something belongs to (*mi* – my, *tu* – your, *su* – his/her)

qualifier
a word that makes a description more specific (*un poco* – a bit, *bastante* – quite, *muy* – very)

regular verb
a verb that follows the patterns of one of the three verb groups (**-ar**, **-er** or **-ir**)

singular
the form of a word used to refer to only one of something (*libro* – book, *goma* – rubber)

verb
a word that refers to what someone is doing or what is happening (*como* – I eat, *vivo* – I live); it can also refer to an ongoing state (**es** *guapo* – he **is** good-looking, **está** *aquí* – it **is** here)

1   Nouns

1.1   Gender

A noun is a word that names a person or thing. In Spanish all nouns have a gender: they are either masculine or feminine.

bolígrafo (*pen*) – masculine
mochila (*schoolbag*) – feminine

The words for people are the gender you would expect:

padre (*father*) – masculine
madre (*mother*) – feminine

With other nouns you need to learn their gender. Here are some tips to help you:
● Nouns ending in **-o** are usually masculine, e.g. **bolígrafo**
● Nouns ending in **-a** are usually feminine, e.g. **mochila**

1.2   Singular/Plural

The plural is used when referring to more than one of something. The form of the plural depends on the noun's ending.

ends in a vowel	add **-s**	libro**s** *books*
ends in a consonant	add **-es**	móvil**es** *mobile phones*
ends in **-z**	change **z** to **c** and add **-es**	lápi**z**c**es** *pencils*

1.3   The indefinite article ('a', 'some')

In Spanish, the words for 'a' and 'some' change according to the gender of the noun and whether it is singular or plural.

Singular		Plural	
masculine	feminine	masculine	feminine
un libro *a book*	**una** mochila *a schoolbag*	**unos** libro**s** *some books*	**unas** mochilas *some schoolbags*

1.4   The definite article ('the')

The Spanish for 'the' also changes according to the gender of the noun and whether it is singular or plural.

Singular		Plural	
masculine	feminine	masculine	feminine
el alumn**o** *the pupil*	**la** ventana *the window*	**los** alumno**s** *the pupils*	**las** ventanas *the windows*

2   Pronouns

A pronoun takes the place of a noun.

Jenny went home. → *She* went home.
The game is fun. → *It* is fun.

Spanish has words for 'I', 'you', 'he', 'she', etc., but generally they are not used with verbs: the verb on its own is enough. However, you do need to be able to recognise them.

yo	*I*
tú	*you (singular)*
él	*he*
ella	*she*
nosotros	*we (male)*
nosotras	*we (female)*
vosotros	*you (plural, male)*
vosotras	*you (plural, female)*
ellos	*they (male)*
ellas	*they (female)*

3   Adjectives

3.1   Agreement of adjectives

Adjectives describe nouns. Their endings change to agree with the noun they describe, and they fall into three groups:
● ending in **-o** or **-a**
● ending in **-e**
● ending in a consonant.

	Singular	
	masculine	feminine
-o or **-a**	divertid**o**	divertid**a**
-e	interesant**e**	interesant**e**
consonant	fácil	fácil

	Plural	
	masculine	feminine
-o or **-a**	divertid**os**	divertid**as**
-e	interesant**es**	interesant**es**
consonant	fácil**es**	fácil**es**

Colours

Colour words are adjectives, and most of them change according to the rules above.

El caballo es negr**o** y la serpiente es negr**a**.
La tortuga es verd**e** y las serpientes son verd**es**.
El pájaro es azul y los peces son azul**es**.

A few colour words change only in the plural, e.g. **rosa** – **rosas** (pink) and **naranja** – **naranjas** (orange).

3.2   Position of adjectives

In Spanish, most adjectives follow the noun they are describing.

Tengo dos peces **amarillos**.
   *I have two **yellow** fish.*
Me gustan los profesores **interesantes**.
   *I like **interesting** teachers.*

3.3   Comparatives

When comparing two or more things, to say 'more … than' or 'less … than', use **más/menos** + adjective + **que**:

Madrid es **más moderna que** Cazorla.
   *Madrid is **more modern than** Cazorla.*
Cazorla es **menos importante que** Madrid.
   *Cazorla is **less important than** Madrid.*

As usual, the adjective agrees with the noun it is describing.

3.4   Possessive adjectives

Possessive adjectives are the words for 'my', 'your', 'his', 'her', etc. They show who something belongs to. These words change to agree with the noun they are describing: you add **-s** when it is plural.

	Singular	Plural
my	mi hermano mi hermana	mi**s** hermano**s** mi**s** hermana**s**
your	tu hermano tu hermana	tu**s** hermano**s** tu**s** hermana**s**
his/her	su hermano su hermana	su**s** hermano**s** su**s** hermana**s**

4   Verbs

4.1   The infinitive

The infinitive is the form of a verb used in a dictionary or wordlist. In Spanish, verbs fall into three groups according to the ending of the infinitive: **-ar**, **-er** or **-ir**.

4.2   The present tense

The present tense is used to talk about what usually happens (e.g. I **go** to school every day) or about how things are (e.g. French lessons **are** very boring). In Spanish, it can also be used to talk about what is happening now (e.g. I **am doing** my homework).

(a) regular verbs

To use a verb in the present tense, you remove the infinitive ending and add a new ending for each person. Each of the three verb groups (**-ar**, **-er**, **-ir**) has a different set of endings that you need to learn.

-ar verbs

hablar – *to speak*		
(yo)	habl**o**	*I speak*
(tú)	habl**as**	*you speak (singular)*
(él/ella)	habl**a**	*he/she/it speaks*
(nosotros/nosotras)	habl**amos**	*we speak*
(vosotros/vosotras)	habl**áis**	*you speak (plural)*
(ellos/ellas)	habl**an**	*they speak*

-er verbs

comer – to eat

(yo)	com**o**	I eat
(tú)	com**es**	you eat (singular)
(él/ella)	com**e**	he/she/it eats
(nosotros/nosotras)	com**emos**	we eat
(vosotros/vosotras)	com**éis**	you eat (plural)
(ellos/ellas)	com**en**	they eat

-ir verbs

escribir – to write

(yo)	escrib**o**	I write
(tú)	escrib**es**	you write (singular)
(él/ella)	escrib**e**	he/she/it writes
(nosotros/nosotras)	escrib**imos**	we write
(vosotros/vosotras)	escrib**ís**	you write (plural)
(ellos/ellas)	escrib**en**	they write

(b) stem-changing verbs

Some Spanish verbs are called stem-changing verbs. These are usually regular in their endings, but some forms of the verb in the present tense have a vowel change in the 'stem' (the part to which the endings are added). Some useful stem-changing verbs are given below.

jugar – to play

(yo)	j**ue**go	I play
(tú)	j**ue**gas	you play (singular)
(él/ella)	j**ue**ga	he/she/it plays
(nosotros/nosotras)	jugamos	we play
(vosotros/vosotras)	jugáis	you play (plural)
(ellos/ellas)	j**ue**gan	they play

dormir – to sleep

(yo)	d**ue**rmo	I sleep
(tú)	d**ue**rmes	you sleep (singular)
(él/ella)	d**ue**rme	he/she/it sleeps
(nosotros/nosotras)	dormimos	we sleep
(vosotros/vosotras)	dormís	you sleep (plural)
(ellos/ellas)	d**ue**rmen	they sleep

querer – to want

(yo)	qu**ie**ro	I want
(tú)	qu**ie**res	you want (singular)
(él/ella)	qu**ie**re	he/she/it wants
(nosotros/nosotras)	queremos	we want
(vosotros/vosotras)	queréis	you want (plural)
(ellos/ellas)	qu**ie**ren	they want

(Note that **querer** is often followed by another verb in the infinitive.)

(c) irregular verbs

Some verbs are not regular in the present tense: they don't follow the usual patterns for **-ar**, **-er** or **-ir** verbs. Some useful irregular verbs are given below.

hacer – to do

(yo)	ha**go**	I do
(tú)	haces	you do (singular)
(él/ella)	hace	he/she/it does
(nosotros/nosotras)	hacemos	we do
(vosotros/vosotras)	hacéis	you do (plural)
(ellos/ellas)	hacen	they do

tener – to have (also stem-changing)

(yo)	t**engo**	I have
(tú)	t**ie**nes	you have (singular)
(él/ella)	t**ie**ne	he/she/it has
(nosotros/nosotras)	tenemos	we have
(vosotros/vosotras)	tenéis	you have (plural)
(ellos/ellas)	t**ie**nen	they have

ir – to go

(yo)	**voy**	I go
(tú)	**vas**	you go (singular)
(él/ella)	**va**	he/she/it goes
(nosotros/nosotras)	**vamos**	we go
(vosotros/vosotras)	**vais**	you go (plural)
(ellos/ellas)	**van**	they go

Gramática

salir – *to go out*

(yo)	sal**g**o	*I go out*
(tú)	sales	*you go out (singular)*
(él/ella)	sale	*he/she/it goes out*
(nosotros/nosotras)	salimos	*we go out*
(vosotros/vosotras)	salís	*you go out (plural)*
(ellos/ellas)	salen	*they go out*

ser – *to be* (permanent states)

(yo)	**soy**	*I am*
(tú)	**eres**	*you are (singular)*
(él/ella)	**es**	*he/she/it is*
(nosotros/nosotras)	**somos**	*we are*
(vosotros/vosotras)	**sois**	*you are (plural)*
(ellos/ellas)	**son**	*they are*

estar – *to be* (position and temporary conditions)

(yo)	est**oy**	*I am*
(tú)	est**ás**	*you are (singular)*
(él/ella)	est**á**	*he/she/it is*
(nosotros/nosotras)	estamos	*we are*
(vosotros/vosotras)	estáis	*you are (plural)*
(ellos/ellas)	est**án**	*they are*

ser and *estar*

In Spanish, there are two verbs meaning 'to be': **ser** and **estar**.

ser is used to refer to ongoing or permanent states:

soy tímido – *I'm shy*, **eres** alto – *you're tall*, **es** negro – *it's black*, mi cumpleaños **es** – *my birthday is*

It is also used for telling the time:

¿Qué hora **es**?	*What time is it?*
Son las cuatro.	*It's 4 o'clock.*

estar is used to refer to position and temporary conditions:

¿Dónde **está**?	*Where is it?*
¿Cómo **estás**?	*How are you?*

(d) reflexive verbs

Reflexive verbs often describe an action done to yourself. They are like ordinary verbs but with an extra part in front of them which is called a reflexive pronoun.

duchar**se**	*to shower*
Me ducho.	*I shower.*
Te duchas.	*You shower.*
Se ducha.	*He/She showers.*
Nos duchamos.	*We shower.*
Os ducháis.	*You (plural) shower.*
Se duchan.	*They shower.*

The reflexive verbs that you have seen in this book are these:

ducharse (*to shower*), despertarse (*to wake up*), levantarse (*to get up*), acostarse (*to go to bed*), vestirse (*to get dressed*), peinarse (*to brush your hair*).

Some reflexive verbs are also stem-changing:

despertarse – Me desp**ie**rto.	*I wake up.*
acostarse – Me ac**ue**sto.	*I go to bed.*
vestirse – Me v**i**sto.	*I get dressed.*

4.3 The near future tense

The near future tense is used to talk about what you are going to do in the near future. It is formed with the appropriate present tense form of **ir + a** + a verb in the infinitive.

En las vacaciones, **voy a jugar** al fútbol.
   In the holidays I'm going to play football.
Mañana **voy a ir** al cine.
   Tomorrow I'm going to go to the cinema.

4.4 Making verbs negative

To make a sentence or a question negative, put **no** before the verb.

No tengo diez años.
   I'm not 10 years old.
¿**No** vives en Barcelona?
   Don't you live in Barcelona?

4.5 *me gusta/me gustan*

me gusta (*I like*) literally means *it is pleasing to me.*

***me gusta/me gustan* + noun**
Use **me gust<u>a</u>** with singular nouns:

Me gusta el profesor.   *I like the teacher.*
No me gusta el francés.   *I don't like French.*

Use **me gustan** with plural nouns:

Me gustan los rotuladores.   *I like the felt-tips.*
No me gustan las ciencias.   *I don't like science.*

To ask someone else what they like, use
¿Te gusta …? or **¿Te gustan …?**

When giving an opinion with **(no) me gusta(n)**, you need to include the definite article **el**, **la**, **los**, **las**:

Me gusta **el** dibujo.   *I like art.*
No me gustan **los** perros.   *I don't like dogs.*

me gusta + *verb*
me gusta can also be followed by another verb in the infinitive.

Me gusta **navegar** por internet.
   I like surfing the net.
No me gusta nada **hacer** natación.
   I don't like swimming at all.

4.6   Verbs with the infinitive

Some verbs can be followed by another verb in the infinitive, e.g. **me gusta** *(I like)*, **me encanta** *(I love)*, **odio** *(I hate)*, **prefiero** *(I prefer)* and **quiero** *(I want)*.

Me gusta **ver** la televisión.
   I like watching television.
No me gusta nada **ir** de compras.
   I really don't like going shopping.
Me encanta **escuchar** música.
   I love listening to music.
Odio **hacer** mis deberes.
   I hate doing my homework.
Prefiere **salir** con sus amigos.
   He prefers going out/to go out with his friends.
¿Quieres **ir** al cine?
   Do you want to go to the cinema?

5   Prepositions

A preposition is a word or phrase showing the relationship of one thing to another. Many prepositions refer to position, e.g. 'on top of', 'behind'.

a	*to*
a la derecha de	*to the right of*
a la izquierda de	*to the left of*
al lado de	*beside*
de	*from, of*
debajo de	*under*
delante de	*in front of*
detrás de	*behind*
encima de	*on*
entre … y	*between … and*

The preposition **de** can also refer to who/what something belongs to:

la habitación **de** mi hermano   *my brother's room*

de + el
In Spanish, many prepositions include **de** *(of)*. When this is followed by **el**, **de + el** merge to make **del**. But when it is followed by **la**, the two words remain separate.

a la izquierda **del** ordenador
   on the left of the computer
delante de la casa *in front of the house*

a + el
When the preposition **a** *(to)* is followed by **el**, they merge to make **al**. But when it is followed by **la**, the two words remain separate.

Quiero ir **al** polideportivo.
   I want to go to the sports centre.
Voy a la playa. *I am going to the beach.*

6   Extras

6.1   Question forms
Questions in Spanish are easy to recognise, because they have two question marks – one (upside down) at the beginning and another at the end.

¿Dónde vives**?**   *Where do you live?*

Gramática

Question words

As in English, you can ask a question by using a question word, e.g. **¿dónde?** (where?), **¿cómo?** (how?), **¿cuántos?** (how many?). Note that Spanish question words always have an accent.

¿Cómo estás?	*How are you?*
¿Cuántos años tienes?	*How old are you?*

'Yes/no' questions

In Spanish, you can turn a sentence into a 'yes/no' question simply by using punctuation **¿ … ?** when writing and a rising intonation when speaking.

¿Los libros son interesantes?
  Are the books interesting?
¿Vives en una ciudad? *Do you live in a town?*

6.2   Punctuation

In Spanish, questions have two question marks – one (upside down) at the beginning and another at the end. The beginning of a question may come in the middle of a sentence.

¿Dónde vives**?** *Where do you live?*
Y tú, **¿**tienes hermanos**?**
  And you, have you any brothers or sisters?

Similarly, exclamations have two exclamation marks – one (upside down) at the beginning and another at the end.

¡Hasta luego**!**	*See you later!*
Hmm, ¡qué interesante**!**	*Hmm, how interesting!*

6.3   Pronunciation and stress

Pronunciation

In Spanish, the spelling of a word shows you how to pronounce it. These rules will help you work out how to say a Spanish word.

c + a, o, u or a consonant	*k*	**ca**sa, **co**mo, **cu**mpleaños, **cl**ase
c + e, i	*th*	**ce**ro, **ci**nco
g + a, o, u or a consonant	*g* as in *great*	**ga**to, **go**rdo, re**gu**lar, **gr**ande
g + e, i	*h* at the back of the throat (or *ch* as in *loch*)	a**ge**nda, reli**gi**ón
h	always silent	**h**istoria, **h**ay
j	*h* at the back of the throat (or *ch* as in *loch*)	**j**ulio
ll	*y*	**ll**amo
ñ	*ny*	espa**ñ**ol
qu	*k*	**qu**é
v	*b*	**v**i**v**es
z	*th*	mar**z**o

Stress

If you know three basic rules, you will be able to pronounce any Spanish word with the stress (emphasis) on the correct syllable.

- If the word ends in **a vowel**, **n** or **s**, the stress falls on the penultimate (next to last) syllable: **<u>ca</u>sa**, **<u>gus</u>tan**.
- If the word ends in **any other letter**, the stress falls on the last syllable: **re<u>gu</u>lar**, **a<u>zul</u>**.
- For any word not following these rules, the stressed syllable is shown by an accent: **es<u>tá</u>**, **ma<u>má</u>**, **in<u>glés</u>**, **edu<u>ca</u>ción**, **<u>hám</u>ster**, **<u>fá</u>cil**, **<u>mú</u>sica**, **bo<u>lí</u>grafo**.

Vocabulario español–inglés

A

abajo   *downstairs*
el abrazo   *hug*
abril   *April*
la abuela   *grandmother*
el abuelo   *grandfather*
aburrido/a   *bored/boring*
acostarse   *to go to bed*
la actividad   *activity*
activo/a   *active*
de acuerdo   *OK*
me acuesto   *I go to bed*
adiós   *goodbye*
adivina   *guess (command)*
el adjetivo   *adjective*
las afueras   *outskirts*
la agenda   *diary*
agosto   *August*
agradable   *pleasant*
el agua mineral   *mineral water*
¡ah!   *ah!*
ahora   *now*
al/a la   *to the*
Alemania   *Germany*
el alfabeto   *alphabet*
la alfombra   *carpet/rug*
¿algo más?   *anything else?*
allí   *there*
la alternativa   *alternative*
alto/a   *tall*
la alumna   *pupil (f)*
el alumno   *pupil (m)*
amarillo/a   *yellow*
la amiga   *friend (f)*
el amigo   *friend (m)*
el animal   *animal/pet*
el año   *year*
antiguo/a   *old/ancient*
antipático/a   *unpleasant*
apago   *I switch off*
aprende   *learn (command),*
    he/she learns
apropiado/a   *appropriate*
apunta   *note down (command)*
aquí   *here*
aquí tienes   *here you are*
el árbol   *tree*
armado/a   *armed*
el armario   *wardrobe*
arqueológico/a   *archaeological*
la arquitectura   *architecture*
arriba   *upstairs*
el aseo   *toilet*
la asesina   *murderess*
el asesino   *murderer*
así   *like this*
la asignatura   *school subject*
la atención   *attention*
el atletismo   *athletics*
el atún   *tuna*
el autobús   *bus*
¡ay!   *oh!*
a ver   *let's see*

ayer   *yesterday*
azul   *blue*

B

baila   *dance (command),*
    he/she dances
bailar   *to dance*
bajo/a   *short*
el baloncesto   *basketball*
el banano   *banana tree*
la bandeja   *tray*
el baño   *bathroom*
el bar   *pub*
barato/a   *cheap*
la barba   *beard*
bastante   *quite*
bebe   *drink (command),*
    he/she drinks
beber   *to drink*
bebes   *you drink*
bebí   *I drank*
bebo   *I drink*
la bicicleta   *bicycle*
bien   *well*
bienvenido/a   *welcome*
el bigote   *moustache*
la biología   *biology*
blanco/a   *white*
el bocadillo   *sandwich*
el boli/bolígrafo   *pen*
bonito/a   *pretty*
buen   *good*
Buenas noches   *Good night*
Buenas tardes   *Good*
    afternoon
bueno/a   *good*
Buenos días   *Good morning*
busca   *look for/look up*
    (command)
se busca   *wanted*
buscan   *they look for/look up*
buscar   *to look for/look up*
busco   *I look for/look up*

C

el caballo   *horse*
cada   *each*
el café   *coffee*
la calculadora   *calculator*
el calor   *heat*
la cama   *bed*
el campo   *countryside*
la caña   *sugar cane*
la canción   *song*
canta   *sing (command),*
    he/she sings
la capital   *capital*
caro/a   *expensive*
la carta   *letter*
la casa   *house*
castaño   *brown (hair)*

el castillo   *castle*
la catedral   *cathedral*
la cava   *sparkling wine*
celebrar   *to celebrate*
la cena   *supper; meal*
el centro   *town centre*
el centro comercial   *shopping*
    centre
la cerveza   *beer*
chatear   *to chat (on-line)*
la chica   *girl*
el chicle   *chewing gum*
el chico   *boy*
el chocolate   *chocolate*
el ciclismo   *cycling*
las ciencias   *science*
las cifras   *figures*
el cigarro   *cigar*
el cine   *cinema*
la ciudad   *city*
la clase   *lesson*
el cobaya   *guinea pig (m)*
la cobaya   *guinea pig (f)*
la Coca-Cola   *Coca-Cola*
la cocina   *kitchen*
el colegio   *school*
el color   *colour*
¿de qué color son tus ojos?
    what colour are your eyes?
come   *eat (command)*
come   *he/she eats*
el comedor   *dining room*
comemos   *we eat*
comer   *to eat*
comercial   *commercial*
comes   *you eat*
comí   *I ate*
como   *I eat; as*
¿cómo?   *how?*
¿cómo eres?   *what are you*
    like?
¿cómo es?   *what's ... like?*
¿cómo se escribe ...?   *how*
    do you spell ... ?
cómodo/a   *comfortable*
la compañera   *partner (f)*
el compañero   *partner (m)*
completa   *complete*
    (command)
compra   *he/she buys*
ir de compras   *to go shopping*
compro   *I buy*
comprueba   *check*
    (command)
con   *with*
el conejo   *rabbit*
conmigo   *with me*
conoce   *he/she knows*
conocer   *to know*
conoces   *you know*
conocido/a   *well-known*
conozco   *I know*
construye   *build/make*
    (command)
contesta   *answer (command)*

contéstame *answer me*
contestar (a) *to answer*
el continente *continent*
copia *copy (command)*
correctamente *correctly*
correcto/a *correct*
corrige *correct (command)*
corto/a *short (hair)*
la cosa *thing*
la costa *coast*
creo *I believe*
el criminal *criminal*
cruza *cross (command)*
el cuaderno *exercise book*
el cuadro *picture*
¿cuál? *which?*
cuando *when*
¿cuándo? *when?*
¿cuánto es? *how much is it?*
¿cuantos/as? *how many?*
¿cuántos años tienes? *how old are you?*
el cuarto de baño *bathroom*
cultivan *they cultivate*
cultural *cultural*
el cumpleaños *birthday*

D

da *give (command)*
da *he/she gives*
el dado *die*
dar *to give*
los datos *details*
de *of*
debajo (de) *below*
los deberes *homework*
decir *to say/tell*
decorativo/a *decorative*
del/de la *of the*
delante (de) *in front of*
delgado/a *thin*
depende (de) *it depends (on)*
el deporte *sport*
deportista *sporty*
a la derecha (de) *on the right (of)*
desayunar *to have breakfast*
desayuno *I have breakfast*
describe *describe (command), he/she describes*
describen *they describe*
describir *to describe*
la descripción *description*
desde *from*
deseamos *we wish*
desear *to wish*
el desorden *mess*
despertarse *to wake up*
me despierto *I wake up*
después *afterwards*
los detalles *details*
detrás (de) *behind*
di *say/tell (command)*
el día *day*

el diálogo *dialogue*
dibuja *draw (command)*
el dibujo *drawing, art*
el diccionario *dictionary*
dice *he/she says*
diciembre *December*
diferente *different*
difícil *difficult*
dime *tell me (command)*
el dinero *money*
el director *manager (m)*
la directora *manager (f)*
la disposición *disposition*
divertido/a *amusing*
dobla *turn (command)*
el domingo *Sunday*
los domingos *on Sundays*
donde *where*
¿dónde? *where?*
¿dónde está? *where is it?*
¿dónde vives? *where do you live?*
dormimos *we sleep*
dormir *to sleep*
dormís *you sleep (pl)*
el dormitorio *bedroom*
ducharse *to shower*
me ducho *I shower*
duerme *he/she sleeps*
duermen *they sleep*
duermo *I sleep*
durante *while*

E

e *and (before i/hi)*
la edad *age*
el edificio *building*
la educación física *physical education (PE)*
el ejemplo *example*
el ejercicio *exercise*
el *the (m) (sg)*
elige *choose (command)*
el email *email*
empareja *pair up (command)*
en *in/on/at*
le encanta(n) *he/she loves …*
les encanta(n) *they love …*
encima (de) *on top (of)*
enero *January*
enorme *enormous*
la ensalada *salad*
entiendes *you understand*
entiendo *I understand*
entonces *then*
entre *between*
el equipo *team*
el equipo de música *stereo*
la equitación *horse riding*
eres *you are*
el error *error*
es *he/she is*
es la una *it's one o'clock*

Escocia *Scotland*
escribe *write (command)*
escribe *he/she writes*
escríbelo *write it*
escríbeme *write to me*
escribes *you write*
escribir *to write*
escribo *I write*
escucha *listen (command), he/she listens*
escuchamos *we listen*
escuchar *to listen (to)*
escuché *I listened to*
escucho *I listen*
la escuela *school*
eso *that*
el espacio *space/gap*
España *Spain*
el español *Spanish (lang)*
espléndido/a *splendid*
el esquí *skiing*
esta *this (f)*
está (en) *he/she is (in)*
está bien *it's OK*
la estación de autobuses *bus station*
la estación de trenes *train station*
la estación espacial *space station*
el estadio *stadium*
los Estados Unidos *United States*
estáis *you are (pl)*
estamos *we are*
están *they are*
la estantería *shelf*
estar *to be*
estas *these (f)*
estás *you are*
¿estás? *are you?*
estás aquí *you are here*
este *this (m)*
estos *these (m)*
estoy *I am*
la estrofa *verse*
el estuche *pencil case*
estudia *he/she studies*
estudiamos *we study*
estudiar *to study*
estudias *you study*
estudio *I study*
estupendo/a *marvellous*
la expresión *expression*

F

fácil *easy*
falso/a *false*
falta *… is missing*
faltan *… are missing*
la familia *family*
famoso/a *famous*
fatal *awful*

Vocabulario español—inglés

febrero *February*
la fecha *date*
feliz cumpleaños *happy birthday*
fenomenal *great*
feo/a *ugly*
la ficha *form*
el fin de semana *weekend*
la flor *flower*
la foto *photo*
el francés *French (lang)*
Francia *France*
la frase *sentence*
con frecuencia *frequently*
el frío *cold*
frito/a *fried*
la frontera *frontier/border*
fue *he/she/it was*
la fuente *fountain*
fuera *outside*
fuerte *strong*
fuma *he/she smokes*
fumar *to smoke*
el fútbol *football*
el futbolista *footballer*
el futuro *future*

G

las gafas *spectacles*
Gales *Wales*
el garaje *garage*
el gato *cat*
los gemelos *twins (m)*
la geografía *geography*
el gimnasio *gymnasium*
el globo *balloon*
la goma *rubber*
gordo/a *fat*
gracias *thank you*
la gramática *grammar*
grande *big*
Grecia *Greece*
gris *grey*
el grupo de música *music band*
guapo/a *good-looking*
la guirnalda *streamer*
le gusta(n) *he/she likes*
me gusta(n) (mucho) *I like … (a lot)*
no me gusta(n) (nada) *I don't like … (at all)*
¿te gusta(n)…? *do you like …?*

H

la habitación *room*
habla *he/she speaks*
hablador(a) *talkative*
habláis *you speak (pl)*
hablamos *we speak*
hablan *they speak*

hablar *to speak/to talk*
hablas *you speak*
hablé *I spoke*
hablo *I speak*
hace *he/she/it makes/does*
hace buen tiempo *it's good weather*
hace calor *it's hot*
hace frío *it's cold*
hace mal tiempo *it's bad weather*
hace sol *it's sunny*
hace viento *it's windy*
hacéis *you do/make (pl)*
hacemos *we do/make*
hacen *they do/make*
hacer *to do/make*
hacer los deberes *to do homework*
haces *you do/make*
la hacienda *Mexican farm*
hago *I do/make*
hago atletismo *I do athletics*
la hamburguesa *hamburger*
el hámster *hamster*
hasta *until*
¡Hasta luego! *See you later!*
hay *there is/are*
hay de todo *it's got everything*
hay niebla *it's foggy*
hay tormenta *it's stormy*
haz *do/make (command)*
la hermana *sister*
el hermano *brother*
la hija *daughter*
la hija única *only child (f)*
el hijo *son*
el hijo único *only child (m)*
la historia *history*
histórico/a *historic*
el hockey *hockey*
¡hola! *hello!*
el hombre *man*
el homicidio *homicide*
la hora *hour/the time*
horizontalmente *horizontally*
¡qué horror! *how awful!*
el hospital *hospital*
el hotel *hotel*
hoy *today*
las humanidades *humanities*

I

ideal *ideal*
el idioma *language*
el idioma oficial *official language*
el primer idioma *first language*
el segundo idioma *second language*
la iglesia *church*

imaginario/a *imaginary*
impaciente *impatient*
importante *important*
increíble *incredible*
industrial *industrial*
infantil *infant (school)*
la información *information*
la informática *ICT*
Inglaterra *England*
el inglés *English (language)*
el instituto *school*
inteligente *intelligent*
me interesa *it interests me*
interesante *interesting*
el interior *interior*
el internet *internet*
por internet *on the net*
el invierno *winter*
invita *he/she invites*
invitar *to invite*
ir a *to go to*
ir al cine *to go to the cinema*
Irlanda *Ireland*
irregular *irregular*
me irrita *he/she/it irritates me*
la isla *island*
Italia *Italy*
a la izquierda (de) *on the left (of)*

J

el jardín *garden*
juega *play (command)*
juega *he/she plays*
juegan *they play*
juegas *you play*
juego *I play*
juego con mi ordenador *I play on my computer*
el jueves *Thursday*
los jueves *on Thursdays*
jugáis *you play (pl)*
jugamos *we play*
jugar *to play*
jugar al fútbol *to play football*
jugué *I played*
julio *July*
junio *June*
juntos *together*

L

la *the (f) (sg)*
al lado (de) *beside*
la lámpara *lamp*
los lápices *pencils*
el lápiz *pencil*
largo/a *long*
las *the (f) (pl)*
lavarse los dientes *to brush your teeth*

Vocabulario español–inglés

me lavo los dientes *I brush my teeth*
la leche *milk*
lee *read (command), he/she reads*
leemos *we read*
leer *to read*
leo *I read*
la letra *letter*
levantarse *to get up*
me levanto *I get up*
libre *free*
el libro *(text) book*
la limonada *lemonade*
liso/a *straight*
listo/a *ready*
llama *he/she calls*
se llama *he/she/it is called*
se llaman *they are called*
me llamo *I am called*
llueve *it rains*
loco/a *mad, crazy*
la lógica *logic*
los *the (m) (pl)*
luego *then*
el lugar *place*
el lunes *Monday*
los lunes *on Mondays*

M

la madre *mother*
mal *bad*
mañana *tomorrow*
la mañana *morning*
de la mañana *in the morning*
mandar *to send*
mando mensajes *I send text messages*
la manzana *apple*
el mapa *map*
el marciano *Martian*
marrón *brown*
Marte *Mars*
el martes *Tuesday*
los martes *on Tuesdays*
marzo *March*
más *more*
más … que *more … than*
las matemáticas *maths*
mayo *May*
mayor *older*
la mayoría *majority*
y media *half past*
mediterráneo/a *Mediterranean*
la memoria *memory*
menciona *mention (command)*
mencionar *to mention*
menos *less*
menos … que *less … than*
menos cuarto *a quarter to*

menos veinticinco *twenty-five to*
el mensaje *message*
el mercado *market*
el mes *month*
la mesa *table*
mi *my (+ singular noun)*
a mí me gusta *I like*
el miembro *member*
el miércoles *Wednesday*
los miércoles *on Wednesdays*
el minuto *minute*
mira *look (command), he/she looks*
mirar *to look*
mis *my (+ plural noun)*
la mitad *half*
la mochila *schoolbag/rucksack*
el modelo *model*
moderno/a *modern*
el módulo *module*
el monedero *purse*
monta *he/she rides*
la montaña *mountain*
montar (a caballo) *to ride (a horse)*
montar (en bicicleta) *to ride (a bicycle)*
monto *I ride …*
monumental *monumental*
el móvil *mobile*
mucho *a lot*
muchos/as *many*
el mueble *piece of furniture*
los muebles *furniture*
la mujer *woman*
las mujeres *women*
multicolor *multicoloured*
el mundo *world*
el museo *museum*
la música *music*
muy *very*

N

el nacimiento *birth*
nada *nothing*
de nada *you're welcome*
nada más *nothing else*
naranja *orange*
la natación *swimming*
hacer natación *to do/go swimming*
naval *naval*
navegar por internet *to surf the net*
navegas *you surf*
navego *I surf*
navegué *I surfed*
la Navidad *Christmas*
necesitas *you need*
necesito *I need*
negativo/a *negative*
negro/a *black*

la niebla *fog*
nieva *it snows*
no *no/not*
no hay *there isn't/aren't*
no me gusta (nada) *I don't like (at all)*
no puedo *I can't*
no tengo *I haven't got*
la noche *night*
de noche *by night*
la Nochebuena *Christmas Eve*
el nombre *name*
normalmente *normally*
nosotros/as *we*
la nota *note*
la novela *novel*
noviembre *November*
nuevo/a *new*
el número *number*
nunca *never*

O

o *or*
octubre *October*
odio … *I hate …*
el Oeste *west*
el ojo *eye*
ondulado/a *wavy*
la ópera *opera*
la opinión *opinion*
el orden *order*
el ordenador *computer*
el otoño *autumn*
otro/a *other (sg)*
otros/as *other (pl)*
¡oye! *listen! (command)*

P

el padre *father*
la página (web) *(web) page*
el país *country*
los países *countries*
el pájaro *bird*
la palabra *word*
el palacio *palace*
los palmares *palm groves*
para *for*
parecido/a *alike*
la pared *wall*
el parque *park*
el pasatiempo *hobby*
pasear *to walk*
dar un paseo *to go for a walk*
el pasillo *corridor/hall*
las patatas fritas *crisps*
el patinaje *ice skating, roller skating*
el patio *patio*
el pavo *turkey*
los peces *fishes*
peinarse *to comb your hair*

Vocabulario español–inglés

me peino *I comb my hair*
pelirrojo/a *red-haired*
el pelo *hair*
pequeño/a *small*
perezoso/a *lazy*
perfectamente *perfectly*
pero *but*
el perro *dog*
la persona *person*
el personaje *character*
el pez *fish*
pinta *he/she paints*
pintar *to paint*
pinto *I paint*
la piscina *swimming pool*
el piso *flat*
la pizarra *(white)board*
la pizza *pizza*
el planeta *planet*
el plano *plan*
la planta *plant*
la plantación *plantation*
el plátano *banana*
la playa *beach*
la plaza *square*
la plaza de toros *bullring*
el plural *plural*
un poco *a little*
poder *to be able to*
el poema *poem*
la poesía *poetry*
el polideportivo *sports centre*
pon *put (command)*
por *for*
por eso *because of this*
por favor *please*
por la mañana *in the morning*
por la tarde *in the evening*
¿por qué? *why?*
por supuesto *of course*
porque *because*
Portugal *Portugal*
posible *possible*
positivo/a *positive*
la postal *postcard*
el póster *poster*
preferido/a *favourite*
preferir *to prefer*
prefiere *he/she prefers*
prefieres *you prefer*
prefiero *I prefer*
la pregunta *question*
preguntar *to ask*
pregunto *I ask*
prepara *prepare (command), he/she prepares*
prepárate *prepare yourself (command)*
la presentación *presentation*
presentarse *to introduce oneself*
el presente *the present (tense)*
la prima *cousin (f)*
primaria *primary (school)*

la primavera *spring*
primer/o/a *first*
el primo *cousin (m)*
principal *principal (adj)*
el profesor *teacher (m)*
la profesora *teacher (f)*
pronto *soon*
el proyector *projector*
el pueblo *village/small town*
puedo *I can*
la puerta *door*
pues *well*
el punto *point*

Q

que *that*
¿qué? *what?*
¿qué bebes? *what are you drinking?*
¿qué comes? *what are you eating?*
¿qué hay en la clase? *what is there in the classroom?*
¿qué hora es? *what time is it?*
¿a qué hora vas? *what time are you going?*
¡qué pena! *what a pain!*
¿qué tal? *how are you?*
¿qué te gusta? *what do you like?*
¿qué te gusta hacer? *what do you like to do?*
¿qué vas a hacer? *what are you going to do?*
queréis *you want (pl)*
queremos *we want*
querer *to want*
¿quién es? *who is it?*
quiere *he/she wants*
quieren *they want*
quieres *you want*
quiero *I want*
te quiero *I love you*

R

rápidamente *quickly*
el ratón *mouse*
los ratones *mice*
la raya *line*
las tres en raya *noughts and crosses*
la razón *reason*
el recreo *lunch break*
la región *region*
la regla *rule/ruler*
regular *regular*
la religión *religion (RE)*
rellena *fill in (command)*
repite *repeat (command)*
la respuesta *answer*

el restaurante *restaurant*
el resumen *summary*
los resúmenes *summaries*
el río *river*
rizado/a *curly*
el robot *robot*
rojo/a *red*
rosa *pink*
el rotulador *felt-tip pen*
rubio/a *blond*

S

el sábado *Saturday*
los sábados *on Saturdays*
saber *to know*
sabes *you know*
el sacapuntas *sharpener*
sale *he/she goes out*
salen *they go out*
sales *you go out*
salgo *I go out*
salimos *we go out*
salir *to go out*
salir con (mis amigos) *to go out with (my friends)*
salís *you go out (pl)*
el salón *living room*
el salón-comedor *lounge-diner*
la salud natural *human biology*
sano/a *healthy*
seguir *to continue, carry on*
la semana *week*
la semana que viene *next week*
señor *Mr*
señora *Mrs*
señorita *Miss*
septiembre *September*
ser *to be*
la serpiente *snake*
severo/a *severe*
sí *yes*
siempre *always*
lo siento *I'm sorry*
la sierra *mountain range*
significan *they mean*
sigue todo recto *go straight on (command)*
la silla *chair*
similar *similar*
simpático/a *kind/nice*
el singular *singular*
sobre *on*
sobre todo *above all*
el sol *sun*
la solución *solution*
somos *we are*
son *they are*
son las siete *it's 7 o'clock*
el sondeo *survey*
soy *I am*

su *his/her/their (+ singular noun)*
subraya *underline (command)*
subrayado/a *underlined*
el suelo *floor*
el sueño *dream*
la suerte *luck*
¡qué suerte! *how lucky!*
por supuesto *of course*
el Sur *south*
sus *his/her/their (+ plural noun)*
suyo/a *his/hers/theirs*

T

la tabla *grid*
también *also*
tan *so*
la tarde *afternoon*
de la tarde *in the afternoon*
taurino *bullfighting related*
el teatro *drama*
la tecnología *technology/CDT*
el teléfono *telephone*
por teléfono *on the 'phone*
la televisión *television*
tenemos *we have*
tener *to have*
tengo *I've got …*
tengo … años *I'm … years old*
el tenis *tennis*
termina *finish (command)*
termina *he/she finishes*
la terraza *terrace*
el texto *text*
la tía *aunt*
el tiempo libre *free time*
la tienda *shop*
tiene *he/she has*
tienen *they have*
tienes *you have*
¿tienes …? *have you …?*
tímido/a *shy*
el tío *uncle*
típico/a *typical*
tira *throw (command)*
tirar *to throw*
toca *play/touch (command)*
te toca a ti *it's your turn*
toca (la trompeta) *he/she plays (the trumpet)*
todo el tiempo *all the time*
todos los días *every day*
toma *take (command)*
tomar *to take*
la tormenta *storm*
el torneo *tournament*
los toros *bulls/bullfight*
la torre *tower*
la tortuga *tortoise*

trabaja *he/she works*
trabajar *to work*
traduce *translate (command)*
traducir *to translate*
tranquilo/a *quiet/peaceful*
el tren *train*
la trompeta *trumpet*
tú *you*
tu *your (+ singular noun)*
el turista *tourist*
turístico/a *touristy*
tus *your (+ plural noun)*
tuyo/a *yours*

U

un *a/an (m)*
una *a/an (f)*
unas *some (f) (pl)*
unos *some (m) (pl)*
la universidad *university*
útil *useful*
utiliza *use (command)*
utilizando *using*

V

va *he/she goes*
las vacaciones *holidays*
vais *you go (pl)*
¡vale! *OK!*
van *they go*
van a *they're going to*
vas *you go*
ve *he/she sees*
las veces *times*
a veces *sometimes*
dos veces *twice*
vemos *we see*
vengo *I come*
venir *to come*
la ventana *window*
veo *I see*
ver *to see*
ver la tele *to watch TV*
el verano *summer*
el verbo *verb*
¿verdad? *is it true?*
verdadero/a *true*
verde *green*
la versión *version*
verticalmente *vertically*
ves *you see*
vestirse *to get dressed*
la vez *time*
una vez *once*
una vez por semana *once a week*
la vida *life*
viejo/a *old*
viene *he/she comes*
el viento *wind*
el viernes *Friday*

los viernes *on Fridays*
violento/a *violent*
el visitante *visitor*
visitar *to visit*
me visto *I get dressed*
vive *he she lives*
viven *they live*
vives *you live*
vivimos *we live*
vivir *to live*
vivís *you live (pl)*
vivo *I live*
vivo en *I live in*
el voleibol *volleyball*
vosotros/as *you (pl)*
voy *I go*
voy al instituto *I go to school*
voy a … *I'm going to …*

Y

y *and*
y cuarto *a quarter past*
y media *half past*
ya *already*
yo *I*
yoga *yoga*

Z

el zumo de naranja *orange juice*

Vocabulario inglés—español

A

a/an   *un/una*
a lot   *mucho*
above all   *sobre todo*
adjective   *el adjetivo*
afternoon   *la tarde*
afterwards   *después*
age   *la edad*
I agree/OK   *de acuerdo*
already   *ya*
also   *también*
always   *siempre*
I am   *soy/estoy*
amusing   *divertido/a*
ancient   *antiguo/a*
and   *y/e (before i/hi)*
animal   *el animal*
to answer   *contestar (a)*
answer   *la respuesta*
anything else?   *¿algo más?*
apple   *la manzana*
April   *abril*
they are   *son/están*
we are   *somos/estamos*
you are   *eres/estás*
you (pl) are   *sois/estáis*
are you?   *¿eres?/¿estás?*
as   *como*
to ask   *preguntar*
athletics   *el atletismo*
August   *agosto*
aunt   *la tía*
autumn   *el otoño*
awful   *fatal*

B

bad   *mal*
banana   *el plátano*
basketball   *el baloncesto*
bath   *el baño*
bathroom   *el cuarto de baño*
to be   *ser/estar*
beach   *la playa*
beard   *la barba*
because   *porque*
because of this   *por eso*
bed   *la cama*
bedroom   *el dormitorio*
beer   *la cerveza*
behind   *detrás (de)*
below   *debajo (de)*
beside   *al lado (de)*
between   *entre*
bicycle   *la bicicleta*
big   *grande*
biology   *la biología*
bird   *el pájaro*
birthday   *el cumpleaños*
happy birthday   *feliz cumpleaños*

black   *negro/a*
blond   *rubio/a*
blue   *azul*
board   *la pizarra*
book   *el libro*
bored/boring   *aburrido/a*
boy   *el chico*
brother   *el hermano*
brown (hair)   *castaño*
to brush your teeth   *lavarse los dientes*
building   *el edificio*
bus station   *la estación de autobuses*
but   *pero*
to buy   *comprar*

C

calculator   *la calculadora*
I can   *puedo*
capital   *la capital*
carpet   *la alfombra*
cat   *el gato*
(town) centre   *el centro*
chair   *la silla*
to chat (on-line)   *chatear*
cheap   *barato/a*
to check   *comprobar*
chewing gum   *el chicle*
chocolate   *el chocolate*
Christmas   *la Navidad*
Christmas Eve   *la Nochebuena*
cinema   *el cine*
city   *la ciudad*
coast   *la costa*
Coca-Cola   *la Coca-Cola*
coffee   *el café*
cold   *frío/a*
colour   *el color*
to comb your hair   *peinarse*
to come   *venir*
I come   *vengo*
he/she comes   *viene*
computer   *el ordenador*
corridor   *el pasillo*
country   *el país*
countryside   *el campo*
cousin   *el primo/la prima*
crisps   *las patatas fritas*
curly   *rizado/a*
to cross   *cruzar*
cycling   *el ciclismo*

D

to dance   *bailar*
date   *la fecha*
daughter   *la hija*
day   *el día*
December   *diciembre*
it depends (on)   *depende (de)*

to describe   *describir*
diary   *la agenda*
dictionary   *el diccionario*
die   *el dado*
different   *diferente*
difficult   *difícil*
dining room   *el comedor*
to do   *hacer*
dog   *el perro*
door   *la puerta*
downstairs   *abajo*
drama   *el teatro*
to draw   *dibujar*
drawing   *el dibujo*
dream   *el sueño*
to drink   *beber*

E

each   *cada*
easy   *fácil*
to eat   *comer*
email   *el email*
English   *el inglés (lang)*
enough   *bastante*
every day   *todos los días*
example   *el ejemplo*
exercise book   *el cuaderno*
expensive   *caro/a*

F

family   *la familia*
famous   *famoso/a*
fat   *gordo/a*
father   *el padre*
favourite   *preferido/a*
February   *febrero*
felt-tip pen   *el rotulador*
to fill in   *rellenar*
to finish   *terminar*
fish   *el pez/los peces*
flat   *el piso*
floor   *el suelo*
flower   *la flor*
fog   *la niebla*
football   *el fútbol*
for   *por/para*
form   *la ficha*
free time   *el tiempo libre*
French   *el francés (lang)*
Friday   *el viernes*
friend   *el amigo/la amiga*
from   *de*
furniture   *los muebles*
future   *el futuro*

G

garage   *el garaje*
garden   *el jardín*
geography   *la geografía*

to get dressed *vestirse*
to get up *levantarse*
girl *la chica*
to give *dar*
glasses *las gafas*
to go *ir*
I go *voy*
to go out *salir*
to go shopping *ir de compras*
to go straight on *seguir todo recto*
to go to bed *acostarse*
good *buen/bueno/a*
Good afternoon *Buenas tardes*
good-looking *guapo/a*
Good morning *Buenos días*
Good night *Buenas noches*
Goodbye *Adiós*
grandfather *el abuelo*
green *verde*
grey *gris*
grid *la tabla*
group *el grupo*
guinea-pig *el cobaya/la cobaya*
guitar *la guitarra*
gymnasium *el gimnasio*

H

hair *el pelo*
half *la mitad*
half past *y media*
hamburger *la hamburguesa*
happy *feliz*
I hate *odio*
to have *tener*
I have *tengo*
to have breakfast *desayunar*
hello *hola*
her *su*
here *aquí*
here you are *aquí tienes*
hers *suyo/a*
his *su, suyo/a*
historic *histórico/a*
history *la historia*
hobbies *los pasatiempos*
holidays *las vacaciones*
homework *los deberes*
horse *el caballo*
horse riding *la equitación*
hour *la hora*
house *la casa*
how? *¿cómo?*
how are you? *¿qué tal?*
how do you spell …? *¿cómo se escribe …?*
how many? *¿cuántos/as?*
how much is it? *¿cuánto es?*

I

I *yo*
ice skating *el patinaje*
ICT *la tecnología*
important *importante*
in *en*
in front (of) *delante (de)*
incredible *increíble*
intelligent *inteligente*
interesting *interesante*
to invite *invitar*
is *es*

J

January *enero*
July *julio*
June *junio*

K

kind *simpático/a*
kitchen *la cocina*
to know *saber/conocer*

L

lamp *la lámpara*
language *el idioma*
lazy *perezoso/a*
to learn *aprender*
left *a la izquierda*
lemonade *la limonada*
less *menos*
lesson *la clase*
letter *la carta*
life *la vida*
I like *me gusta(n)*
to listen (to) *escuchar*
a little *un poco*
flat *el piso*
to live (in) *vivir (en)*
living room *el salón*
long *largo/a*
to look *mirar*
to look for/look up *buscar*
a lot *mucho*
I love *me encanta(n)*
lunch break *el recreo*

M

mad *loco/a*
majority *la mayoría*
to make *hacer*
I make *hago*
you make *haces*
man *el hombre*
many *muchos/muchas*

map *el mapa*
March *marzo*
maths *las matemáticas*
May *mayo*
they mean *significan*
member *el miembro*
message *el mensaje*
I send messages *mando mensajes*
milk *la leche*
minute *el minuto*
is/are missing *falta(n)*
mobile *el móvil*
modern *moderno/a*
module *el módulo*
Monday *el lunes*
money *el dinero*
month *el mes*
more *más*
morning *la mañana*
mother *la madre*
mouse *el ratón*
much *mucho/a*
music *la música*
my *mi*

N

name *el nombre*
my name is *me llamo*
to need *necesitar*
never *nunca*
new *nuevo/a*
no *no*
normally *normalmente*
not *no*
to note down *escribir*
nothing *nada*
nothing else *nada más*
noughts and crosses *las tres en raya*
November *noviembre*
now *ahora*
number *el número*

O

October *octubre*
it's 1 o'clock *es la una*
of *de*
of course *por supuesto*
it's OK *vale*
old *viejo/a, antiguo/a*
how old are you? *¿cuántos años tienes?*
on *sobre*
on top (of) *encima (de)*
once a week *una vez por semana*
only child *hijo único/hija única*
opinion *la opinión*
orange *naranja*

orange juice   *el zumo de naranja*
other   *otro/otra*
outside   *fuera*
outskirts   *las afueras*

P

page   *la página*
to paint   *pintar*
park   *el parque*
partner   *el compañero/la compañera*
P.E.   *la educación física*
pen   *el bolígrafo/boli*
pencil   *el lápiz*
pencil case   *el estuche*
people   *las personas*
pet   *el animal*
photo   *la foto*
pink   *rosa*
place   *el lugar*
to play   *jugar*
I play   *juego*
please   *por favor*
to prefer   *preferir*
presentation   *la presentación*
pretty   *bonito/a*
projector   *el proyector*
pupil   *el alumno/la alumna*

Q

a quarter past   *... y cuarto*
a quarter to   *... menos cuarto*
question   *la pregunta*
quickly   *rápidamente*
quiet   *tranquilo/a*
quite   *bastante*

R

rabbit   *el conejo*
to rain   *llover*
to read   *leer*
reason   *la razón*
red   *rojo/a*
red-haired   *pelirrojo/a*
restaurant   *el restaurante*
right   *a la derecha*
to ride   *montar (a, en)*
river   *el río*
roller skating   *el patinaje*
room   *la habitación*
rubber   *la goma*
rucksack   *la mochila*
ruler   *la regla*

S

sandwich   *el bocadillo*
Saturday   *el sábado*
to say/tell   *decir*
he/she says   *dice*
school   *el colegio/la escuela/el instituto*
schoolbag   *la mochila*
science   *las ciencias*
to search for   *buscar*
season   *la estación*
to see   *ver*
to send   *mandar*
sentence   *la frase*
September   *septiembre*
sharpener   *el sacapuntas*
shelf(ves)   *la estantería*
shop   *la tienda*
short   *corto/a*
to shower   *ducharse*
shy   *tímido/a*
to sing   *cantar*
singular   *singular*
sister   *la hermana*
skating   *el patinaje*
skiing   *el esquí*
to sleep   *dormir*
small   *pequeño/a*
to smoke   *fumar*
snake   *la serpiente*
it snows   *nieva*
so   *tan*
solution   *la solución*
some   *unos/unas*
sometimes   *a veces*
son   *el hijo*
song   *la canción*
soon   *pronto*
I'm sorry   *lo siento*
Spanish (lang)   *el español*
to speak   *hablar*
sport   *el deporte*
spring   *la primavera*
square   *la plaza*
stadium   *el estadio*
stereo   *el equipo de música*
storm   *la tormenta*
straight on   *todo recto*
to study   *estudiar*
summary   *el resumen*
sun   *el sol*
Sunday   *el domingo*
stereo   *el equipo de música*
to surf the net   *navegar*
swimming   *la natación*
swimming pool   *la piscina*

T

to take   *tomar*
to talk   *hablar*
tall   *alto/a*

teacher   *el profesor/la profesora*
team   *el equipo*
telephone   *el teléfono*
to tell   *decir*
tennis   *el tenis*
terrace   *la terraza*
text   *el texto*
thank you   *gracias*
that   *que*
the   *el/la/los/las*
their   *su*
theirs   *suyo/a*
then   *entonces/luego*
there   *allí*
there is /are ...   *hay ...*
these   *estos/estas*
they   *ellos/ellas*
things   *las cosas*
this   *este/esta*
Thursday   *el jueves*
today   *hoy*
together   *juntos*
toilet   *el aseo*
tomorrow   *mañana*
tortoise   *la tortuga*
town   *el pueblo/la ciudad*
train station   *la estación de trenes*
translate   *traducir*
Tuesday   *el martes*
to turn   *doblar*
twins   *gemelos/as*

U

ugly   *feo/a*
uncle   *el tío*
to understand   *entender*
until   *hasta*
upstairs   *arriba*
to use   *utilizar*
useful   *útil*

V

verb   *el verbo*
very   *muy*
village   *el pueblo*
vocabulary   *el vocabulario*
volleyball   *el voleibol*

W

to wake up   *despertarse*
to walk   *pasear*
wall   *la pared*
to want   *querer*
wardrobe   *el armario*
to watch TV   *ver la tele*
water   *el agua*
wavy   *ondulado/a*

Vocabulario inglés—español

we   *nosotros/nosotras*
weather   *el tiempo*
Wednesday   *el miércoles*
week   *la semana*
weekend   *el fin de semana*
welcome   *bienvenido/a*
well   *bien*
what?   *¿qué?*
what time is it?   *¿qué hora es?*
when   *cuando*
when?   *¿cuándo?*
where is ...?   *¿dónde está ...?*
which?   *¿cuál?*
white   *blanco/a*
whiteboard   *la pizarra*
who is ...?   *¿quién es ...?*
why?   *¿por qué?*
wind   *el viento*
window   *la ventana*
winter   *el invierno*
with   *con*
with me   *conmigo*
woman   *la mujer*
word   *la palabra*
to   work   *trabajar*
world   *el mundo*
I   would like ...   *quiero ...*
to   write   *escribir*

Y

year   *el año*
yellow   *amarillo/a*
yes   *sí*
yesterday   *ayer*
you   *tú*
you (pl)   *vosotros/as*
you're welcome   *de nada*
your   *tu*
yours   *tuyo/a*